An understanding of how stars evolve is central to astrophysics. The basic theory is well established. However, the subject has undergone a renaissance in recent years as powerful computers have become widely available and allowed complex evolutionary models to be developed and compared in great detail with observations from the latest instruments. This timely volume presents the review articles from an international meeting in Elba, Italy, where experts gathered to review how our understanding of stellar evolution has advanced.

Topics covered include fundamentals of stellar evolution, star clusters, variable stars, asymptotic giant branch stars, degenerate stars, the evolution of binary stars, and chemical and galactic evolution. Throughout, theory and observation are closely compared. The book also emphasises the critical role stars have on our understanding of how galaxies evolve.

In this book we are provided with both the fundamentals and the latest research. In this way, it will provide an invaluable supplement for graduate students, and a timely review for researchers.

CAMBRIDGE CONTEMPORARY ASTROPHYSICS

Advances in Stellar Evolution

CAMBRIDGE CONTEMPORARY ASTROPHYSICS

Series editors
José Franco, Steven M. Kahn, Andrew R. King and Barry F. Madore

Titles available in this series

Gravitational Dynamics, *edited by O. Lahav. E. Terlevich and R. J. Terlevich* (ISBN 0 521 56327 5)

High-sensitivity radio astronomy, *edited by N. Jackson and R. J. Davis* (ISBN 0 521 57350 5)

Relativistic Astrophysics, *edited by B. J. T. Jones and D. Marković* (ISBN 0 521 62113 5)

Advances in Stellar Evolution, *edited by R. T. Rood and A. Renzini* (ISBN 0 521 59184 8)

Advances in Stellar Evolution

Proceedings of the workshop *Stellar Ecology*,
held in Marciana Marina, Elba, Italy
23–29 June, 1996

Edited by

R. T. ROOD
University of Virginia

A. RENZINI
Università di Bologna and
European Southern Observatory

CAMBRIDGE
UNIVERSITY PRESS

PUBLISHED BY THE PRESS SYNDICATE OF THE UNIVERSITY OF CAMBRIDGE
The Pitt Building, Trumpington Street, Cambridge CB2 1RP United Kingdom

CAMBRIDGE UNIVERSITY PRESS
The Edinburgh Building, Cambridge CB2 2RU, United Kingdom
40 West 20th Street, New York, NY 1001–4211, USA
10 Stamford Road, Oakleigh, Melbourne 3166, Australia

First published 1997

Printed in the United Kingdom at the University Press, Cambridge

A catalogue record for this book is available from the British Library

ISBN 0 521 59184 8 hardback

Contents

Variable Stars

AGB Stars and Planetary Nebulae

White Dwarfs

Binary Star Evolution

Stars in Galaxies

Participants

Organizing Committee:

Robert T. Rood	rtr@virginia.edu
Alvio Renzini	arenzini@eso.org
Federico Ferrini	federico@astrpi.difi.unipi.it
John Faulkner	johnf@lick.ucsc.edu
Mario Livio	mlivio@stsci.edu
James W. Truran	truran@nova.uchicago.edu
Alexander V. Tutukov	atutukov@anubis.inasan.rssi.ru

Attendees:

Taft Armandroff	armand@noao.edu
Dave Arnett	dave@bohr.physics.arizona.edu
Stephen Becker	sab@lanl.gov
Betty Blanco	70313.1655@compuserve.com
Victor Blanco	70313.1655@compuserve.com
Enzo Brocato	brocato@astrte.te.astro.it
Carla Cacciari	cacciari@astbo3.bo.astro.it
Victoria Caloi	caloi@victoria.ias.fra.cnr.it
Santi Cassisi	cassisi@astrte.te.astro.it
Marcio Catelan	catelan@stars.gsfc.nasa.gov
Adrienne Cool	cool@spark.berkeley.edu
Roberto D.D. Costa	costa@obs-nice.fr
Robert Cumming	robert@astro.su.se
Noella D'Cruz	nld2n@virginia.edu
Ben Dorman	ben.dorman@gsfc.nasa.gov
John Faulkner	johnf@lick.ucsc.edu
Donald H. Ferguson	73251.3043@compuserve.com
Federico Ferrini	federico@astrpi.difi.unipi.it
Enrique García–Berro	garcia@benard.fa.upc.es
Laura Greggio	greggio@hal1.usm.uni-muenchen.de
Masayuki Y. Fujimoto	fujimoto@phys.hokudai.ac.jp
Izumi Hachisu	hachisu@chianti.c.u-tokyo.ac.jp
Richard C. Henry	rch@pha.jhu.edu
Icko Iben, Jr.	icko@sirius.astro.uiuc.edu
Thais Idiart	thais@purcell.obs-nice.fr
Takashi Iijima	iijima@astras.pd.astro.it
Naoki Itoh	n_itoh@hoffman.cc.sophia.ac.jp
Mariko Kato	mariko@sunrise.hc.keio.ac.jp
Ivan King	king@glob.berkeley.edu
T. D. Kinman	kinman@noao.edu
Attay Kovetz	attay@etoile.tau.ac.il
Francesca Lanteri Cravet	lanteri@astbo3.bo.astro.it
John Lattanzio	johnl@flash.maths.monash.edu.au
Gregory Laughlin	gpl@boris.physics.lsa.umich.edu
Jim Liebert	liebert@as.arizona.edu
K. Y. Lo	kyl@astro.uiuc.edu
Claudia Maraston	claudiam@astbo3.bo.astro.it
Marcella Marconi	marcella@astr1pi.difi.unipi.it

Romas Mitalas	mitalas@uwovax.uwo.ca
Sergio Ortolani	ortolani@astrpd.pd.astro.it
Anna Piersimoni	apiersimoni@astrte.te.astro.it
Giampaolo Piotto	piotto@astrpd.pd.astro.it
Philipp Podsiadlowski	podsi@astro.ox.ac.uk
Onno Pols	onno@ast.cam.ac.uk
Dina Prialnik	dina@comet.tau.ac.il
Narayan Rana	ncrana@iucaa.ernet.in
Saul Rappaport	sar@mit.edu
Alvio Renzini	arenzini@eso.org
R. Michael Rich	rmr@carmen.phys.columbia.edu
Vincenzo Ripepi	ripepi@fis.unical.it
Robert T. Rood	rtr@virginia.edu
Alfred Rosenberg	rosenberg@astrpd.pd.astro.it
Maria-Teresa Ruiz	mtruiz@das.uchile.cl
Stephen C. Russell	sr@cp.dias.ie
Stefano Sandrelli	sandrelli@astbo4.bo.astro.it
Ivo Saviane	saviane@astrpd.pd.astro.it
Detlef Schönberner	DeSchoenberner@aip.de
Ed Sion	emsion@ucis.vill.edu
Norman Simon	nsimon@unlinfo.unl.edu
Iglika Spassovska	iglika@hanul.issa.re.kr
Ted Stecher	stecher@uit.gsfc.nasa.gov
Daiichiro Sugimoto	sugimoto@chianti.c.u-tokyo.ac.jp
Allen Sweigart	sweigart@bach.gsfc.nasa.gov
Alexander V. Tutukov	atutukov@anubis.inasan.rssi.ru
David Tytler	tytler@ucsd.edu
Don A. VandenBerg	vandenberg@vstars.phys.uvic.ca
Kim Venn	venn@hal2.usm.uni-muenchen.de
Volker Weidemann	supas058@astrophysik.uni-kiel.d400.de
Klaus Werner	werner@astrophysik.uni-kiel.d400.de
Peter Wood	wood@mso.anu.edu.au
Lev R. Yungelson	lry@inasan.rssi.ru

Preface

Robert T. Rood & Alvio Renzini
Charlottesville, Virginia, USA & Garching bei München, Germany
February, 1997

The workshop *Stellar Ecology* was held 23–29 June, 1996 at the Elba International Physics Center in Marciana Marina, Elba, Italy. The meeting celebrated the 65th birthday of Icko Iben, Jr., and we organized a scientific program which stressed both stellar evolution and the role of stars in galactic evolution. For the workshop name we had chosen the word ecology—"the totality or pattern of relations between organisms and their environment"—to emphasize our goal of looking at the entire picture.†

The study of stellar structure is one of the cornerstones of modern astrophysics. In no other branch of astrophysics is there such an intimate contact between theory and observations. In some cases, one might argue that the first order problem is solved. For example, we know the basic interior structure of the stars in the various parts of globular cluster color-magnitude diagrams. One can probe details of interior mixing and nuclear processing with "bumps" in red giant luminosity functions and changes in surface abundances. However, when pushed to the limit, the comparison of theory and observation shows that even in these simple cases there are important unsolved problems. In other circumstances (e.g., Type I supernovae) stellar structure studies are more like much of the rest of astrophysics (e.g., active galactic nuclei)—the first order explanation is still open to debate. This mix of mature and "breaking" subdisciplines makes the subject of stellar structure a timely one.

While many of the basic aspects of stellar structure were known earlier, the subject came into its own in the 1960's with the wide spread availability of large computers. Not only was it possible to make sequences of evolutionary models, but detailed comparisons with the observations became possible. The careers of Icko and some of the conference participants span the epoch of modern stellar structure studies. While we did not plan a "historical" conference, this was an especially appropriate time to reflect on how the subject has evolved, its successes and failures, and what we can anticipate in the next decades.

Just as in the mid-60's the subject is now at a threshold. Space observations allow the exploration of spectral regions not accessible from the ground; photometric samples of unprecedented size allow the study of short phases of evolution; stars in very dense stellar systems can be observed. New detectors make what were once the proprietary projects of the largest telescopes accessible to much more modest observatories. A new generation of 8–10 meter telescopes will allow spectroscopic studies of fainter stars. Multi-fiber spectra systems can produce statistical studies where only a few stars could previously be done. On the theoretical side, evolutionary calculations that 10 years ago challenged University Computer Centers (and those of us grappling to get our share of the CPU) can be routinely done on workstations. Supercomputers make it possible to seriously investigate processes like convection which could previously be only crudely parameterized. It is truly an exciting time.

In this volume the reader will find that the variety of topics covered is a fair, albeit incomplete representation, of the scientific issues to which Icko has considered in the course of his career. Indeed, there are few aspects of single and binary star evolution to

† We discovered that the use of "ecology" caused confusion in the broader community and choose a different title for this book.

which Icko has not contributed. In his "stellar model factory" all sort of model stars have been manufactured, and all sort of processes explored. Pre-main sequence (Pre-MS), MS, red giant branch (RGB), horizontal branch (HB), early asymptotic giant branch (Early-AGB), thermally pulsating AGB (TP-AGB), Post-AGB, planetary nebula nuclei (PNN), R CrB stars, white dwarfs (WD), double degenerates (DDs), blue straggler stars (BSS), novae, supernovae, the three dredge ups, ionic diffusion, semiconvection, liquefaction and crystallization, acoustic and thermal pulsations, gravitational radiation, s-processing, envelope burning, neutrino processes, and we could go on.

From the list above, we see that we owe to Icko much of the current nomenclature in stellar evolution. Indeed, during the last three decades perhaps nobody else has contributed as much as Icko to our understanding of the evolution of stars. So it is not by chance that the two of us felt compelled to organize this meeting. Indeed, we both owe to Icko much of the basis from which to fruitfully start our careers in astronomy. For this reason we like to indulge here in some personal recollection.

(AR) In the summer of 1965 Prof. Livio Gratton passed to me a bunch of "Orange Aid Preprints," from Caltech's Kellogg Lab. They were the famous series of papers that appeared in ApJ later that year, concerning the pre-main sequence and post main sequence evolution of stars up to 15 solar masses. I read them all in just a couple of days, and I enjoyed them immensely, so beautifully the physical processes occurring in the stellar interiors were elucidated and followed in the course of their evolution. At that time I was in Frascati, collaborating with Claudio Firmani on hydrodynamical calculations of the gravitational collapse of 100 million solar mass objects. Optimistically enough, with an IBM 7040, we wanted to make central black holes in galaxies, and—who knows?—perhaps quasars. But after a promising start our hydro-code was running into worse and worse oscillations that resisted all our attempts at controlling them. So I—mesmerized by Icko papers—proposed a deal to Claudio: I would have started developing a stellar evolution code, while waiting for him to get rid of those unwanted oscillations.

In the spring of 1966 I did my "concorso" to become "Aiuto Astronomo" (Astronomer Aid). Italian concorsi are similar to those Chinese had to do to get mandarin tenure, i.e. there is a written dissertation on a particular topic you did not know in advance, and you have to demostrate your proficiency in it. As it turned out, the dissertation was on "The sources of Stellar Energy." Well, I should admit I was very lucky. I used only 4 of the 8 hours allowed for the dissertation, and I carefully described the evolution of a 3 solar mass star, just dumping on the paper all I had learned from Icko's article, including my free hand versions of the celebrated figures that so much contribute to make Icko's papers clear and appealing. I won the concorso, and got my permanent position in astronomy. For the good and the bad Icko bears some responsibility for this, at least an indirect one. From my point of view, I must say have I have a deep gratitude, for both the scientific enlightning ... and the rest.

In 1968 I moved to Bologna, and a couple of years later the Faculty assigned me to teach the course of Theoretical Astrophysics. I taught stellar evolution, being it the only thing I had some knowledge of. Ever since a main part of the course is dedicated to the evolution of the three solar mass star, following 1965 Icko's paper that by now has been studied by 26 generations of Bolognese students in astronomy.

(RTR) In the fall of 1965 I was beginning my second year as a graduate student at MIT. I had been operating under the delusion that I should be a theoretical particle physicist. My research plans for the summer had collapsed when I went to talk to my potential advisor (who had taught me for two semesters), and he greeted my inquiry with, "Are you a student here?" I spent the summer reading long novels and returned to MIT still thinking particle theory. Having no research topic I had to select an additional course. I noticed in the catalog a new course, "The Theory of Stellar Interiors," which sounded rather interesting. Its real virtue though was that it was taught in the afternoon and part of my quest as a graduate student was not having to get up before lunch. So by lucky accident I arrived in Icko's course.

Toward the end of that semester it finally sunk in that (1) I really didn't like particle physics; (2) Icko's course was lots of fun; (3) Icko (having not been properly indoctrinated in MIT protocol) was actually friendly. Still I signed up for a particle physics second semester. However, after one lecture by an eminent physicist who lectured while looking at the ceiling, I said to myself, "this is stupid," and dropped the particle physics stuff adding courses by Alar Toomre and Philip Morrison in addition to Icko's course. The students in that semester along with me were Bill Watson, Saul Rappaport, and Marvin Geller (who is now in atmospheric physics at Stony Brook). I also went to Icko to discuss a research project and became his first PhD student. Bill Watson and Ray Talbot were also Icko's advisees at the same time, but I worked most closely with him.

Icko's fundamental series of papers was an integral part of the course, and, as with Alvio, shaped the way I think about stars (and make graphs). I still use these papers in the course in Stellar Interiors which I now teach. Nobody has done similar papers since then.

After getting my PhD, I continued working at MIT with Icko as a postdoc. I then moved on to Caltech as a postdoc in Kellogg with Willy Fowler. I was still trying to finish up some of my projects with Icko and was not able to get sufficient computing time to complete the calculations. So I begged Icko to let me return to Cambridge over Christmas–New Years to compute red giant tracks. (Icko and many MIT advisors had "suggested" to their graduate students that they become trained computer operators, so that they could do computing on Saturday nights and holidays when the batch computer centers of the era were ordinarily closed.) On a brief trip home to North Carolina I developed an intense muscle spasm in my lower back. By the time I returned to Cambridge I could hardly move. I still remember lying flat on my back on the floor on New Years Eve with the IBM 360-65 (with a 165 kilobyte user partition) merrily chugging away. One of the comments I made in the resulting paper was: "The net effect of the helium flash is to lift the degeneracy of core by burning a small amount of helium and a large amount of computing funds." Little did I know about how effectively the red giants were burning money. My holiday computer binge consumed Icko's entire summer salary for the following year. (Icko related his side of this episode at the conference dinner.)

Other than the actual convergence of the participants in Elba, this was an almost entirely electronic meeting. There were no large scale (snail) mailings. Most communication was by email supplemented by fax where required. All papers (including all but 4 of the figures) in this volume were submitted electronically. With this format it has been possible to update many papers through January of 1997.

We did not feel the necessity to distinguish between a Local and Scientific organizing committee. Many of the details were handled by Federico Ferrini of Pisa University. These included arranging a marvelous birthday concert by Laura Minguzzi (flute) and Stefano Tedesco (oboe). The local arrangements in Marciana Marina were ably done by Antonella Sapere of EIPC.

The conference received financial support from the Physics Department of Pisa University, the Italian CNR, and the University of Virginia Astronomy Department.

From Stellar Evolution to Tera-Flops Gravothermal Dynamics of Stellar Systems

By DAIICHIRO SUGIMOTO

Graduate School of Arts and Science, University of Tokyo, Japan

Theory of stellar structure and evolution is understood in terms of gravothermodynamics which includes the gravitational interaction into thermodynamics. The star can be a gravothermal system or a normal thermal system according to equation of state and its configuration. Similar concept also applies to the stellar systems such as globular cluster. It was successful, I believed, to develop a theory of gravothermal collapse and expansion, and then a non-linear gravothermal oscillation of the core of globular clusters with the approximation of gaseous models for stellar structure. However, the society of stellar dynamics was skeptical about such understanding because the stellar systems consisting of discrete stars could behave differently from the gaseous systems of stellar interior. To answer to this question, we needed large-scale N-body simulations for the discrete stellar systems. So we constructed a special-purpose computer of one-teraflops speed that was dedicated to gravitational N-body problems. It has been in operation since July 1995. Thanks to recent advances in observation, the dense stellar systems are observed with great angular resolution where the star-star collision affects evolution of each star as well as dynamical evolution of the stellar system itself. Now, it is the time to integrate the theory of stellar evolution with the dynamical evolution of the stellar system.

1. Star as a Gravothermal System

The star is a self-gravitating system thermally in steady state. We call such one as a gravothermal system, and the theory to describe it as the gravothermodynamics. The most essential gravothermal nature of a star is described by the Virial theorem as

$$\Omega + 3(\gamma - 1)K = 0 , \tag{1.1}$$

where Ω and K are the gravitational and the thermal (internal) energies of the star, and γ is the ratio of the specific heats approximated as constant throughout the star. The total energy of the star, $E_{\rm tot}$, and its time change are expressed as

$$E_{\rm tot} = \Omega + K = -(3\gamma - 4)K , \tag{1.2}$$

$$L = -\frac{dE_{\rm tot}}{dt} = (3\gamma - 4)\frac{dK}{dt} > 0 , \tag{1.3}$$

where L is the energy loss rate from the star, *i.e.*, the stellar luminosity. When γ is greater than 3/4, the internal energy K increases as the star dumps its energy away to the outer space as its luminosity. It is associated with a rise of the temperature, and we call the star has the *negative specific heat*. Here, we should notice that this negative specific heat is only an apparent, renormalized specific heat where the hydrostatic readjustment of the star is tacitly taken into account.

It can be discussed for different phases of stellar evolution. It is more convenient to use a system of non-dimensional variables for the stellar structure, *i.e.*,

$$P = P_{\rm c}\varpi \ , \quad \rho = \rho_{\rm c}\sigma \ , \quad M_r = M_0\varphi \ , \quad r = r_0\xi \ , \tag{1.4}$$

where the subscript c denotes the value at the center of the star. The parameters M_0 and r_0 are determined so that the coefficients to the right hand side of the equation for hydrostatic equilibrium should vanish as

$$\frac{d\varpi}{d\xi} = -\frac{\varphi\sigma}{\xi^2} , \quad \frac{d\varphi}{d\xi} = r^2\sigma , \tag{1.5}$$

$$M_0 = \frac{1}{(4\pi G^3)^{1/2}} \frac{P_c^{3/2}}{\rho_c^2} , \quad r_0 = \frac{1}{(4\pi G)^{1/2}} \frac{P_c^{1/2}}{\rho_c} . \tag{1.6}$$

The boundary conditions to equation (1.5) are $\varpi = \sigma = 1$ and $\varphi = 0$ at the center ($\xi = 0$), and we can integrate it starting from the center if a relation between ϖ and σ is given by a distribution of the (local) polytropic index or of the specific entropy. Then we obtain a value of φ at the surface ($\varpi = 0$) which we denote as φ_1. The values of φ_1 for the polytrope of index $N = 1.5$ and 3 are not so much different each other, *i.e.*, 10.73 and 16.14, respectively.

If we disregard this relatively small difference, the mass is regarded as proportional to $P_c^{3/2}/\rho_c^2$ according to equation (1.6). Using equation of state, therefore, we can draw curves for constant T_c's on a logarithmic plane of mass versus the central density as can be found in Sugimoto & Nomoto (1980) for instance . In the equation of state we take account of the radiation pressure, the ideal gas for ions, the partially degenerate and partially relativistic electrons. Using only this plane we can easily understand the main behaviors of stellar evolution as discussed by Sugimoto & Nomoto (1980).

In particular, the star in the non-degenerate regime is stable against the perturbation in the nuclear energy generation, *i.e.*, it is a *gravothermal system* having negative specific heat. On the contrary, the star in the degenerate regime is unstable because of its nature of positive specific heat. We shall call such system as a *thermal system* because it behaves along the line of common sense, *i.e.*, with the positive specific heat. Transition from the thermal to the gravothermal system is the phenomenon of the nuclear core flash. Slow transition to the reverse direction limits the minimum mass for nuclear burning.

The discussion above indicated that the distinction between the gravothermal and the thermal systems comes from the difference in the equation of state. However, we should notice that it was on the basis of equation (1.5), *i.e.*, of the spherical geometry. The situation changes, however, if we consider a plane-parallel geometry.

It is related with shell flashes in evolved stars. When a nuclear shell brining is ignited in the helium shell surrounding a carbon-oxygen core, or in the bottom of accreted hydrogen-rich envelope on a white dwarf, for instance, the shell burning becomes unstable if the pressure scale height H_p is small as compared with the radius, *i.e.*, when the value of

$$V = -\frac{d\ln P}{d\ln r} = \frac{r}{H_p} , \tag{1.7}$$

at the bottom of the nuclear burning shell, *i.e.*, $V_{1,e}$, is large. It implies that the configuration of the nuclear burning shell is locally rather flat than spherical. In the flat, *i.e.*, plane-parallel configuration, the pressure at its bottom is determined only by the weight of the overlying layers irrespectively of their thermal states, and it stays at a constant value. Thus, the shell responds to the nuclear energy input with the normal positive specific heat with constant pressure; *i.e.*, the shell burning is unstable.

When this instability grows to a finite amplitude, the nuclear burning shell expands due to the energy input. It makes the pressure scale height larger, the value of $V_{1,e}$ smaller, and, as a result, the configuration closer to a spherical one. This transition from the flat to the spherical configuration changes the local specific heat from the positive to the negative value, and makes the instability cease. Such a non-linear development of the shell flash was discussed by Sugimoto & Fujimoto (1978) in almost analytical fashion.

2. Gravothermal Catastrophe

The problem of negative specific heat appears also in stellar dynamics. It was first discussed by Antonov (1962) and then extended by Lynden-Bell & Wood (1968). We consider a star cluster and ask what is its equilibrium state. For simplicity, the cluster is assumed to be surrounded by an adiabatic wall. It is approximated by a solution of an isothermal gas sphere in the theory of stellar structure. It is in hydrostatic and thermal equilibrium in the sense that there are no flows of matter and heat. It is also at the state of entropy extremum, though not necessarily at its maximum.

We can ask if it is in stable or unstable equilibrium corresponding to the entropy maximum or minimum, respectively. The answer depends on the density contrast of the equilibrium solution that is defined using the density at the center ρ_c and the density just inside the wall ρ_b as $D \equiv \rho_c/\rho_b$. When D is smaller than $D_{cr}(= 709)$, it is stable as in the limiting case of no self-gravity ($D = 1$). When D is greater than D_{cr}, it is unstable to lead a thermodynamic instability. Such situation was shown by Lynden-Bell & Wood (1968) by the method of turning point in a linear series, and the instability was named as gravothermal catastrophe.

The same problem was analyzed by Hachisu & Sugimoto (1978) by means of normal mode method for linearized stability theory. They obtained not only the neutral point of the stability but also eigen-modes of perturbation on the distribution of specific entropy and resultant temperature change for different cases of the density contrast D. In addition they calculated the inverse of the tensor specific heat which relates perturbation in the distribution of the specific entropy (times temperature) and the resultant change in the distribution of temperature. For the case of infinite D such solutions were analytically expressed by Bettwieser & Sugimoto (1984).

Physical process of this thermodynamic instability is as follows. Let us consider that an amount of heat (specific entropy, if we express more exactly) is removed from an inner region and is deposited at an outer region. Then, the temperature and thus the pressure in the inner region drop due to the removal of heat, which leads to the gravitational contraction of the inner region. *Vice versa* in the outer region. This process is called as the hydrostatic readjustment. As a result of this contraction, the gas is adiabatically compressed in the inner region and its temperature rises. If this temperature rise over-compensates the temperature drop due to the removal of heat, the effective (gravothermal) specific heat is negative. The reverse process takes place in the outer region. If this is the case, a temperature gradient decreasing outward appears spontaneously, and it helps further transferring of heat from the inner to the outer region. This instability is associated with the gravitational contraction of the core.

One may ask what happens when this instability grows to a finite amplitude, *i.e.*, to a non-linear regime. Calculation to the finite amplitude is easy for stellar evolutionists because it is nothing but a phase of gravitational contraction of a star bounded by a fixed adiabatic wall. Hachisu *et al.* (1978) did such calculation and showed the followings. When the initial perturbation corresponds to the outward flow of heat, the contraction of the inner core proceeds, the density contrast becomes higher, thus the system is kept unstable, and, finally, the central density becomes infinity in a finite time. This is really the gravothermal catastrophe to core collapse.

When the initial perturbation corresponds, on the other hand, to the inward flow of heat, the temperature becomes decreasing inwards, the expansion of the core proceeds, the density contrast decreases to become lower than D_{cr}, and then the stability resumes. The temperature gradient disappears, and, finally, the gas sphere tends to another isothermal equilibrium with lower density contrast and with a lower temperature than in the initial isothermal equilibrium. This gravothermal instability is not the catastrophe but is simply the gravothermal expansion with an upper bound. In the case of the linearized theory, the contraction and expansion is the same concept each other since the superposition of different modes is possible with different sign and amplitude. However, they are different in the non-linear regime.

For the stellar evolutionists it was evident that the self-gravity, the sphericity and the equation of state of ideal gas are the essential ingredients of this instability. The discussions given in this and in the preceding section can be applied to a stellar system such as a globular cluster when we translate the gas particles to individual stars in the globular cluster. However, stellar dynamicists thought that the granularity of the system due to discrete stars might change the situation. The stellar dynamicists did calculations of the gravothermal contraction by the method of gravitational N-body problem, by Monte Carlo method, and by solving (orbit averaged) Fokker- Planck equation in energy space. Their results proved to be almost the same as the gaseous model of stellar evolutionists.

3. Post-collapse Evolution

The timescale till the complete core collapse was estimated to be of the order of one Giga-year for most globular clusters. Therefore, their cores should have collapsed. However, their ages are older than the collapse time. What happened after the collapse? It should survive even after the collapse. This is the problem of the post-collapse evolution, which was one of the main topics of the IAU Symposium on Dynamics of Star Clusters in 1984 (Goodman and Hut 1985).

The idea is as follows. In the collapsed core the stellar density is so high that close encounter between the stars will take place. In the event of a three-body encounter there is a possibility that one of the three stars gets high energy and the other two result in an orbitally bound state to form a binary star. If we take account of a finite size of the colliding stars, even a two-body encounter can result in the formation of a binary system when enough amount of their orbital energy is transferred to the (internal) energy of deformation of the stars to form an orbitally bound system. The energy transferred to each star will excite non-radial oscillation of the star, but it will soon be dissipated.

From the standpoint of stellar dynamics, these binaries play a role of energy source in the core. When a star collide with a binary, there is a higher possibility to suffer a super-elastic scattering where the binary becomes gravitationally more deeply bound (de-excited) and the colliding star gets orbital energy. In other words, the collision of a star with binary enhances the two-body correlation and the corresponding energy is transferred to the mean field. This energy input is called as *binary heating* or *binary burning* since it corresponds to the nuclear burning in the main-sequence star. So, the stellar evolutionists may ask whether this binary burning is stable or not. When the energy is inputted into the mean field, the core will expand and the stellar density in the core decreases. It will reduce the collision rate between the stars and binaries and thus reduce the energy generation rate.

It was studied by Sugimoto & Bettwieser (1983) and Bettwieser & Sugimoto (1984) using gaseous model and appropriate energy generation rate. They found that it is not stabilized but that a non-linear oscillation takes place with the amplitude as large as several orders of magnitudes in the central density. Moreover, the trajectory of the center of the core follows a refrigeration cycle in the plane of temperature versus specific entropy, *i.e.*, it turns round in the opposite direction to an engine, say, Carnot cycle.

They interpreted this non-linear oscillation as follows. When energy is inputted into the mean field by star-binary collision, the central region expands and the temperature drops. This yields a temperature gradient decreasing toward the center. It triggers the heat flow from outer region to the inner core to yield further expansion. In other words, the energy input from the binary only triggers the gravothermal expansion of the core and further expansion proceeds spontaneously as the gravothermal expansion without any additional input of energy. As the central density becomes low enough, the timescale of the gravothermal expansion, which is proportional to the timescale of heat transport, becomes longer and longer, and at last the temperature profile outside of the core is altered by heat conduction. Then, the gravothermal instability ceases. The energy inputted by the binary burning during one cycle of oscillation was calculated as negligible compared with the change in the gravitational potential energy during the cycle. This fact together with the trajectory of the refrigeration cycle proved that the interpretation above was right.

One might ask why the binary burning in globular cluster leads a non-linear oscillation differently from the stable nuclear burning in the main-sequence star. Most important difference comes from the the timescale of heat transport. In the case of stellar interior, heat is transported by photons and its timescale is shorter in the outer layers than in the central region. Therefore, the thermal change in the central region can be adjusted by the whole star through a quick response of the outer layers. In the case of globular cluster, on the contrary, the heat is transported by collisions between the stars which is faster in the core. Therefore, the outer region does not know anything about thermal change in the central region. This makes the central core behave like a gravothermally unstable *subsystem.*

One more important difference is that the energy generation rate is too small in the case of globular clusters. The reason is that the coupling constant related with the energy generation is the same gravitational constant G as one governing the global structure. In the case of stellar structure, on the other hand, the former is one for the strong interaction (atomic nuclei are also the correlation) but the latter is the gravitational constant.

The stellar dynamicists, however, insisted that the granularity of the system might be important which can not be described by the gaseous models. From a standpoint of the stellar evolutionists it was clear that the gravothermal nature was not affected by the granularity and that the oscillation was gravothermal. However, such physical consideration was not accepted. The stellar dynamicists began computations with the orbit averaged Fokker-Planck equations. After much effort they obtained results which were essentially the same as the gaseous model. However, such calculation corresponded, in principle, to a case of infinite number of particles, though practically determined by the mesh size in numerical computation.

In view of such situations, Goodman (1987) computed gaseous models in great detail. He presented his result as follows: The gravothermal oscillation does not occur for N smaller than $N_{\rm cr} = 7000$ but does occur for N greater than $N_{\rm cr}$. This statement sounded confusing because the gaseous model should have no concept, in itself, of the number of particles N. What he did was that he expressed the model parameters translated to N-body models. In particular, the binary heating rate was expressed by a decreasing function of N, because the timescale of star-star collisions was longer for larger N models. Thus, his result implied that the gravothermal oscillation takes place when the binary heating rate is lower. It is consistent with our consideration given above. We see that the physical understanding and representation are often different even between sub-communities in astronomy.

Despite all the discussions given above, there remained uncertainty which could only be solved by N-body calculations. Though the system is collisional in the collapsed core, it is well collisionless in the outer regions. The star suffered from a super-elastic scattering by a binary has high energy and is easy to escape from the core or even from the cluster because of its smaller cross section for gravitational scattering. Such star will deposit its orbital energy in the outer region after revolving many times its eccentric orbit. Therefore, the energy transport is not so simple as in the gaseous model. Some people including myself tried to do N-body calculations. However, the speed of available computers limited the number of particles N to about 3000 at the largest, for which the statistical noise obscured the results and any definite conclusion could not be derived. The problem was too much computer intensive, and we badly needed much faster computers.

4. Teraflops Computer Dedicated to Many-Body Problems

The requirement of large-scale N-body computation was one of the motivations for me to develop the dedicated computer. We made a prototype machine GRAPE-1 in 1989 (Sugimoto *et al.* 1990) at the cost of 2000 US dollars. Its performance was as high as 200 Mega-flops equivalent though the word length was limited to 12 bits. Nevertheless, it was very powerful to simulate collisionless system such as galaxies.

The idea is as follows. The N-body problem can be nicely separable into two parts; one to calculate gravitational forces and the other to calculate the motion of particles with the force obtained by the former part. We connect the former part, which we shall call as GRAPE, to a host computer. To advance the system by one time step, GRAPE performs floating point operations of the order of N squared, while the host performs only of the order of N. In addition, the algorithm of GRAPE is simple and can be fixed and may be controlled with a small number of parameters, while the host requires complicated programming depending on the problems to solve. Therefore, it is appropriate to design and construct GRAPE by ourselves as a very fast *hardware* of fixed algorithm, and to use a commercially available workstation or computer for the host which may be N times slower than GRAPE. We named such a system consisting of GRAPE and the host as the GRAPE system, where GRAPE works as an accelerator to the host.

The communication between the host and GRAPE requires sending the coordinates of the particles and receiving the forces as seen from the host. The amount of communication for one time step is of the order of N so that no communication bottlenecks take place. From the standpoint of programming, the host calls GRAPE in place of subroutine to calculate the forces. The calculation of the forces can easily be parallelized. Thus, we can avoid hard working on programming even on a massively parallelized GRAPE. The high speed computations in GRAPE can be realized by calculating the forces by pipeline architecture, and by massively parallel connection of the elements of GRAPE. We took the name of GRAPE from GRAvitational PipelinE.

To proceeded further from GRAPE-1 we designed a customized chip which calculates the gravitational force per unit mass as well as its time derivative with the accuracy of 32/64 bits. The time derivatives can be used in the Hermite integration scheme which advances the time step to $t + \Delta t$ only by using the solution at t. This version of the customized chip we named as HARP (Hermite AcceleratoR Pipeline) chip. We connected about 2000 HARP chips in parallel. The hierarchy of this connection is as follows. Eight HARP chips are mounted on one Multi-Chip Module of size 7.5×11.0 cm consisting of 19 routing layers in a ceramic base. Six Multi-Chip Modules are connected in parallel on one Processor Board of size 36×45 cm together with memories to store the coordinates and the velocities of particles. Another customized chip is added which takes care of Aarseth's individual time step method for integrating systems containing vastly different timescales (stiff problems). Nine Processor Boards are put in one box of size 57w $\times$ 65d

$\times$ 109h cm where distribution of data and collection of the resulting forces and their time derivatives are done by a Control Board. We call this box as a Cluster, and four Clusters consist of GRAPE-4 system together with DEC3000/900 as the host machine at the starting time (Taiji *et al.* 1996; Sugimoto *et al.* 1995; Makino *et al.* 1994).

The GRAPE-4 contains 1692 working chips operated at 32 MHz clock cycle. During the manufacturing of the Multi-Chip Module, one chip of 48 became inoperable on the average so that the number of chips in right operation became $47 \times 9 \times 4 = 1692$. One chip contains 20 floating point operations in pipeline so that its speed is 20 flop $\times$ 32 MHz = 0.64 Gflops. Therefore, GRAPE-4 operates at the peak speed of 1.08 Tera-flops. Its sustained speed depends on the problem to compute, in particular, on the number of particles N in the system and on the communication speed of the host machine. When N is relatively small the time used in transferring data between the host and the GRAPE machine become appreciable. When N is large enough, the sustained speed tends to the peak speed though the larger N requires longer computation time for one time step. For typical problems we can secure the sustained speed higher than 50 percent of the peak speed. Such a relatively high ratio of the sustained to the peak speeds is obtained even in the massively parallel computer with a complicated (inhomogeneous) programming. This is another merit of the dedicated machine.

It entered in continuous operation in July 1995. Since then, there have not been any troubles on the machine. The cost for the machine was only 1.6 Million US dollars which includes the designing of the customized chips. Such a high performance to the cost came from the pipeline architecture of the chip where every element is always working without any idle time. Thus, our Tera-flops machine is small; it is accommodated in a laboratory of 20 square meters and consumes only 10 kW of the electric power.

We have many astrophysical objectives to investigate with this machine. In the first place, however, we calculated the post collapse evolution of globular cluster to check the gravothermal oscillation in the N-body system. It requires computation time approximately proportional to the cube of N; the number of forces to calculate is proportional to N squared, and the number of time steps must be proportional to N since the ratio of the thermal relaxation time to the particle crossing time is proportional to N.

We did computations with different N's ranging from 3000 to 32000 (Makino 1996; Makino 1996a). When N was chosen as large as 32000 the statistical noise is small enough and the gravothermal oscillation was clearly seen. Its characteristics such as the time scale (and its dependence on N) of the non-linear oscillation, and the appearance of negative temperature gradient during the expansion phase to drive the gravothermal oscillation, and the refrigeration cycle to prove the gravothermal nature of the oscillation were found to be quite consistent with the theory developed by the gaseous model.

5. Toward Further Interactions with Stellar Dynamicists

From the side of observation, new data are coming out concerning dense stellar systems such as the center of globular clusters and the nuclei of galaxies by means of the Hubble Space Telescope, adaptive optics to resolve the stars, three dimensional spectroscopy to obtain energy distribution in near infrared region, in particular, simultaneously for two dimensionally projected many stars, and VLBA/VLBI observation of water maser spots not only for their velocities but also for time changes of the velocities. In globular clusters the dependence of the mass functions on the distance from the center of the cluster is now being observed. It should be interpreted in terms of mass segregation during the dynamical evolution of the cluster. During the progress of mass segregation the stars evolve and could become neutron stars (Sugimoto 1996). The existence and spatial distribution of X-ray binaries and blue stragglers should have been determined by competition among the stellar evolution, the mass segregation, and the stellar encounters to form binary stars.

In the event of physical collision of the stars a variety of phenomena should occur. The envelope of a red giant star will be dissipated. A part of the envelope will be left un-dissipated and its mass fraction will determine the appearance of the remnant star. The star becomes a red giant again if an appreciable fraction is left, while the star becomes something like a central star of planetary nebula if almost all of the envelope is dissipated away. When two main-sequence stars collide each other, the result will be a merger but some mass should escape carrying orbital angular momentum with it. When a compact star such as white dwarf or neutron star collides or spirals-in another star, different phenomena including additional nuclear reactions will take place because the strongly bound mass of the compact stars will not easily be dissipated.

For the stellar evolutionists these problems are not only of the three dimensional hydrodynamics but also of the hydrostatic stellar structure because the hydrodynamic phase will be followed by thermal evolution of the remnant star. It is easy for the stellar evolutionists to foresee what are those remnant stars. When the evolution of close binary stars became a topics in early 1980's, we got a variety of subjects to be solved in parameter ranges different from ordinary evolution of single stars. Now, the ranges are being vastly extended by introducing evolution in dense stellar systems. The stellar dynamicists are ready to include such stellar evolution into their dynamical models by using super-fast computers.

This work was supported in part by the Grant-in Aid for Specially Promoted Research (04102002) of the Ministry of Education, Science, Sports and Culture.

REFERENCES

Antonov, V.A. 1962, Vest. Leningr. Gos. Univ., 7, 135

Bettwieser, E. & Sugimoto. D. 1984, MNRAS, 208, 493

Goodman, J. 1987, ApJ, 280, 298

Hachisu, I. & Sugimoto, D. 1978, Prog. Theoret. Phys., 60, 123

Hachisu, I., Nakada, Y., Nomoto, K., & Sugimoto, D. 1978, Prog. Theoret. Phys., 60, 393

Lynden-Bell, D. & Wood, R. 1968, MNRAS, 138, 495

Makino, J. 1996, in Proc. IAU Symp. 174, in Dynamical Evolution of Star Clusters, eds. P.Hut and J. Makino (Kluwer) , 151

Makino, J. 1996a, ApJ, 471, 796

Makino, J., Taiji, M., Ebisuzaki, T., & Sugimoto, D. 1994, Proc. Supercomputing '94, IEEE Computer Society Press, Los Alamitos , 429

Sugimoto, D. 1996, in Proc. IAU Symp. 174, Dynamical Evolution of Star Clusters, eds. P.Hut and J. Makino (Kluwer) , 1

Sugimoto, D. & Bettwieser, E. 1983, MNRAS, 204, 19p

Sugimoto, D. & Fujimoto, M.Y. 1978, Publ. Astron. Soc. Japan, 30, 467

Sugimoto, D. & Nomoto, K. 1980, Space Sci. Rev., 25, 155

Sugimoto, D., Chikada, Y., Makino, J., Ito, T., Ebisuzaki, T., & Umemura, M. 1990, Nature, 345, 33

Sugimoto, D., Makino, J., Taiji, M., & Ebisuzaki, T. 1995, Proc. first Aizu Int'l Symp. on Parallel Algorithm/Architecture Synthesis, IEEE Computer Society Press, Tokyo , 38

Taiji, M., Makino, J., Fukushige, T., Ebisuzaki, T., & Sugimoto, D. 1996, in Proc. IAU Symp. 174, Dynamical Evolution of Star Clusters, eds. P.Hut and J. Makino (Kluwer) , 141

Why Stars *Don't* Become Red Giants—and Why They Do!

By JOHN FAULKNER[1]

[1]UCO/Lick Observatory, Board of Studies in Astronomy and Astrophysics, University of California, Santa Cruz, CA 95064, USA

1. Summary

Stars do *not* become red giants because their envelopes develop thermal instabilities. Even though they may expand on thermal timescales, envelopes lag behind, rather than lead, expansion induced and dictated by deep interior developments. Thermal imbalance or even instabilities may in some cases describe *how*, but they do not tell us *why* stars expand to giant dimensions.

Following Hoyle, we show that as dense cores and shell-like energy-generating structures inevitably develop in low-mass stars, it is increasingly difficult to store very much mass outside but "close" to those cores. Yet such a star cannot avoid storing the rest of the mass with which Nature has endowed it. *This is its structural imperative — to store its remaining mass.* It then does what many of us might do with an overwhelming storage problem — *it remodels the attic.*

Mass storage is the clew† — the not so slender thread — on which this story hangs. We follow it to an understanding of low-mass red giants, as dense cores with tenuous but mass-storing envelopes; we see that it applies to other stars, e.g. to horizontal branch stars, particularly so to those with higher $Z_{\rm CNO}$; and finally, by using the insight so gained, we see that the horizontal branch itself acts as a *missing link* to the expanding behaviour of more massive stars. By producing time changing sequences of such analogue HB models with artificially increasing $Z_{\rm CNO}$ (changes clearly confined to the deep interior) we demonstrate the kinds of behaviour attributed by one prominent group of authors to changing external causes such as envelope opacities. We conclude that stars have deep, not shallow reasons for becoming red giants.

On the way to these conclusions, we derive a large number of new analytical results:

• We derive an *explicit* asymptotic theory for the main structural properties of red giants. In that limit the most important internal parameter $\rho_{sh}/\bar{\rho}_c$ (the ratio of shell to mean interior core density) is found in terms of other physical inputs; many other things of interest (e.g. the relative radial extent of the radiative region surrounding a core) are themselves intimately related to that density ratio parameter. The theory is explicitly worked out for the case of electron-scattering opacity, but may be readily extended to other forms for the opacity behaviour.

• In the electron-scattering asymptotic limit we derive a remarkable new "Luminosity, Compactness, Core Mass Relationship," again *completely explicit*, and of the form $L \cdot (\rho_{sh}/\bar{\rho}_c) \propto M_c^3$. Thus it looks like the classical homogenous star's Mass-Luminosity relation, but is both explicit *and* incorporates a $\rho_{sh}/\bar{\rho}_c$ term which gives expression to the relative compactness of the core mass M_c. This result is otherwise independent of the energy generation rate and of the physical size of the core. There is a more complex Kramers opacity analogue.

• A first order theory of the approach to asymptotia is derived. Thus the last parts of

† **clew**, *n.* **3.** (in ancient legend) a ball of thread unwound to serve as a guide through a labyrinth or maze. — *The World Book Dictionary*, 1974

the curving transition of nuclear burning regions to their ultimate lines in the $(\log\rho, \log T)$ plane may be calculated explicitly. An explicit non-linear extension of the method takes representative conditions back even closer to the giant stars' main-sequence antecedents.

- We solve explicitly for the mass contained in the radiative region outside the core, finding that it (and the corresponding quantity in computed models) has a relatively simple power-law dependence on core-mass, and thereby deduce the existence of (and value for) the point of maximum convective zone penetration.

Unfortunately, the extended arguments and analysis required for an adequate comprehension of the theory and its implications do not lend themselves to satisfactory description in the narrow confines here permitted. The full details will appear elsewhere.

The Last Stars

By GREGORY LAUGHLIN

Department Physics, University of Michigan, Ann Arbor, MI 48104, USA

Although the study of evolving M dwarfs involves time scales much longer than the current age of the universe, there are several compelling reasons for examining the future of these low mass stars. Because the low mass stars are the most numerous, one can argue that the theory of stellar evolution is incomplete until the long term evolution of the majority of stars has been calculated. The future of the galaxy will also be determined by the properties of these low mass stars. In a more practical vein, evolving M stars clarify the process by which stars make the transition to red giant structures.

1. Introduction

At first glance, the forthcoming demise of the red dwarfs appears to be one of the least pressing problems in stellar astrophysics. Many Hubble times will pass before the faintest stars even begin to show outward signs of evolution.

Nevertheless, the fate of the M dwarfs holds a certain intrinsic value. The faintest stars comprise the dominant component of the initial mass function. One can hardly argue that the theory of stellar evolution has a balance of emphasis when the future development of the majority of stars is largely unknown. We (Laughlin, Bodenheimer and Adams; 1996) have recently addressed this discrepancy by computing the future course of M dwarf evolution. This contribution is intended to provide an overview of our results. For a fuller treatment, the reader is refered to the full Journal article.

The properties of the present day M stars have received a great deal of attention, and considerable effort has gone into refining both the observations and theoretical models of M dwarfs on the main sequence. For a recent review of the observational status, see Burrows & Liebert (1994). Starting with the efforts of Kumar (1963), successive authors have sought to improve the theoretical main sequence models, with an eye toward providing agreement with the observed properties.

When the classical field of numerical stellar evolution was being established (for a review, see Iben 1967) the uncertainties in the required input physics were simply too large to say much about the evolution of M stars with any degree of confidence. Although Iben, as early as 1963, wrote of his interest in computing the stellar evolution of the M dwarfs, and some unpublished efforts were made by Bodenheimer (1965) to evolve a 0.1 $M_\odot$ star, the thrust of numerical stellar evolution was aimed squarely at the heavy stars. It is something of a curiosity that even as the necessary input physics improved during the 1970s and 80s, almost nothing was published on the long term evolution of M stars, aside from a general (but nevertheless insightful) essay by Salpeter (1982).

2. The Stellar Evolution Code

The code employed in our evolutionary calculations incorporates the Lagrangian Henyey method (Henyey, Forbes & Gould 1964) to evolve the equations of stellar structure. The overall computational package is an updated version of the program described by Laughlin and Bodenheimer (1993). Connoiseurs of the low mass stars will perhaps be the only ones interested in list of specific input physics which went into our main sequence model. The details are given in the full Journal article.

Before describing the computed tracks, it is important to point out that a number of uncertainties come into play as our M stars evolve. The first major source of inaccuracy and uncertainty is associated with the fact that we use grey model atmospheres with the *Rosseland mean* opacities of Alexander et al. (1983) to model energy transport in the outermost layer. The spectra of M stars bear only a rather faint resemblance to blackbodies. They are fraught with thick molecular absorption lines, back-warming, and other serious complications. Significant progress with non-grey model atmospheres for M dwarfs is now being made (cf. Allard & Hauschildt 1995), but we have not yet been able to take full advantage of this work because fresh problems arise when one considers the gradual transition to the helium rich atmospheres which today's red dwarfs will eventually acquire.

The problem of frequency dependent radiative transfer in the atmosphere calls forth a second shortcoming in our models, namely the lack of molecular and grain opacity tables for mixtures with depleted hydrogen fractions. One expects that within a highly evolved yet still completely convective star, a scarcity of hydrogen will lead to increased transparency at a given temperature and density, due to the non-reactivity of helium, and the absence of absorbing molecules such as water and methane. We account in ad-hoc fashion for this general decrease by using an idealized opacity function of the form

$$\kappa_{\rm He} = \kappa_{\rm H}[1 - (0.7 - X_{\rm H})/2]\,, \qquad (2.1)$$

which is invoked in the cool radiative atmospheric layers.

A third source of uncertainty hinges on the question of long term mass loss, a phenomena which our evolutionary models do *not* consider. Low mass stars are known to produce significant coronal displays, as exhibited by their X-ray flares. Stellar winds almost certainly result from this activity, but the amount of mass that they carry away is unknown. The depletion associated with such winds could conceivably be important for evolution that encompasses many Hubble times (Salpeter 1982).

3. The Evolution of M Stars

Figure 1 charts the six trillion year life of a 0.1 $M_\odot$ star in the Hertzsprung Russell Diagram. After condensing out of a molecular cloud core, the star leisurely descends the convective Hayashi track, and (after 2 billion years) arrives on the zero age main sequence with a photospheric temperature of $T_* = 2228$ K, and a luminosity of $\log_{10}[L_*/L_\odot] = -3.38$.

Upon reaching the main sequence, the evolutionary time scale slows dramatically, and the star suffers only extraordinarily slow and gradual signs of aging. For the first trillion years, the mass fraction contained in ^{3}He increases steadily. The mass fraction of hydrogen, and the two helium isotopes are displayed in the inset chart in Figure 1, which plots core composition as a function of time. Because the 0.10 $M_\odot$ star is completely convective, ^{3}He produced in the core is mixed throughout the entire structure. The star thus maintains a homogeneous composition while increasing its average molecular weight, and consequently grows steadily hotter and more luminous. After 1.38 trillion years, the star has a mass fraction in ^{3}He of 9.95%, which constitutes a maximum value. At this moment, the central temperature is 4.8 $\times 10^6$ K. At later times, the ^{3}He is consumed faster than it is produced and the mass fraction declines.

The buildup of ^{3}He has an interesting effect on the star. As ^{3}He accumulates, the nuclear energy generation rate at a particular temperature and density increases. The core is therefore able to expand and steadily reduce its density, even in the face of a slowly increasing total luminosity. As the hydrogen mass fraction of the star drops from

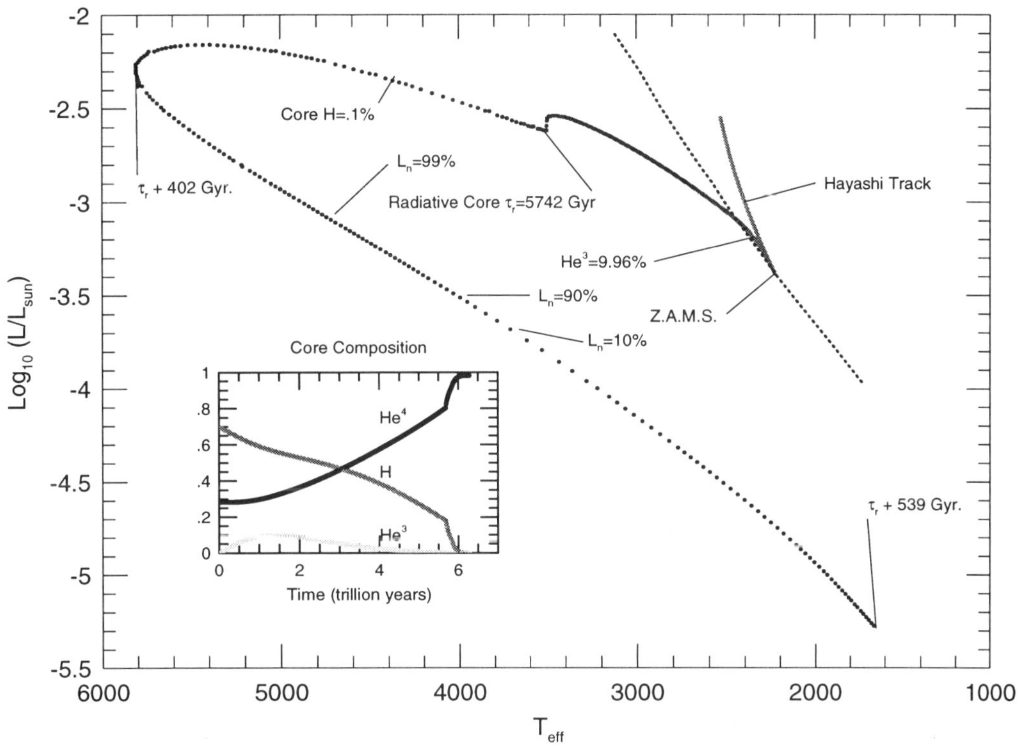

FIGURE 1. Evolution in the Hertzsprung Russell Diagram for a star of mass $M_* = 0.10\ M_\odot$. Each small filled circle delineating the evolutionary track represents a separate converged model. The star begins its evolution on a pre-main sequence Hayashi track and ends up as a helium white dwarf, cooling into obscurity. The inset diagram shows the chemical composition of the star over the course of its evolution; the mass fractions in H, ^{3}He, and ^{4}He are plotted as functions of time.

its initial value of $X_{\rm H} = 0.7$ to $X_{\rm H} = 0.5$, the central density decreases from 309 to 204 gm/cm^3. The temperature in the core during this epoch rises slowly from 4.4 to 5.1 million degrees.

Between 1.5 and 4 trillion years, the mass fraction of ^{3}He declines steadily, eventually reaching the structurally negligible values associated with equilibrium PPI burning. During this broad main sequence phase, the star is methodically turning itself into ^{4}He. After 3.05 trillion years, ^{4}He comes to dominate the mass fraction. At this point the photosphere has heated to a temperature of $T_* = 2500$ K, and the luminosity is just less that 0.1% of the current solar value.

As the amount of available hydrogen per gram diminishes, the energy generation rate at a particular temperature and density declines. The central regions are compelled to grow denser and hotter in order to satisfy the energy demands of the star. After 5.74 trillion years, the star contains only 16% hydrogen by mass, the surface temperature is T_* = 3450 K, and the luminosity is $\log_{10}[L_*/L_\odot] = -2.54$. This point represents a critical juncture in the evolution of the star. The conditions in the core are such that electron conduction is capable of providing the requisite energy transport, and convection ceases in the center. This development of a conductive-radiative core, which soon spreads through the nuclear burning region, almost immediately fixes the composition of the envelope. The unchanging envelope composition in turn allows for the development of a mild composition inhomogeneity, as the remaining hydrogen in the conductive-radiative core is rapidly converted into helium. The development of the conductive core causes the entire star to contract slightly and produces a mild but sudden decline in luminosity. This behavior can be seen on the evolutionary track in the Hertzsprung Russell diagram.

Once the radiative core has developed, the near-eternal youth of the star draws to a close, and the evolutionary time scale accelerates. The hydrogen-exhausted isothermal core steadily increases in mass as a modest nuclear shell source works its way outward through the mass of the star. The shell burning source is located within the conductive, non-convective region of the star, and so the envelope composition continues to remain fixed. The evolution of the star in the H-R diagram during this phase is toward rapidly higher temperatures. The unassuming 0.10 $M_\odot$ red dwarf stars of today will eventually be hotter than the current sun, although they will never be more than $\sim$1% as bright. Evolution to higher temperatures occurs because the whole star is compelled to contract to keep the nuclear energy generation going at a sufficient rate.

By the time the core temperature reaches 12 million degrees, the center of the star is deriving the majority of its support from degeneracy pressure. Further contraction fails to produce higher temperatures, and both the overall luminosity, as well as the central temperature begin to decline. The star persists in being driven to higher surface temperatures over the next several billion years as it contracts, but eventually the surface temperature also reaches a maximum ($T_* = 5807$ K). At this moment, the star "turns the corner" in the Hertzsprung Russell diagram, and begins to fade towards lower luminosities and surface temperatures.

The shell source is still active at the turn-around point. Attenuated nuclear reactions are sufficient to provide nearly all of the steadily declining luminosity during a protracted phase of further contraction. Slowly, the shell source is extinguished, and the star ends its life as a cooling, low mass helium white dwarf with a moderately enriched (15.5%) hydrogen envelope. The overall final hydrogen mass fraction of the remnant is just more than 1%. In keeping with their frugal reputation, the lowest mass M stars will be startlingly efficient in converting their initial allotment of hydrogen into helium. Our evolutionary calculations followed the remnant down to a luminosity of $\log_{10}[L_*/L_\odot] = -5.287$ and a surface temperature of $T_* = 1651$ K, conditions which prevail 539 billion years after the star first develops its radiative-conductive core. The total nuclear burning lifetime of a .1 $M_\odot$ star will be somewhat longer than six trillion years.

The future development of a $M_* = 0.1 M_\odot$ star is placed in perspective by Figure 2, which depicts the evolution of stars ranging from $M_* = 0.06\ M_\odot$ to $M_* = 0.25 M_\odot$. The 0.06 $M_\odot$ object lies beneath the threshold for equilibrium hydrogen burning, and thus rapidly cools and fades away along the hydrogen brown dwarf branch. The figure shows that the general trend of evolution for stars with masses between $M_* = 0.08 M_\odot$ and $M_* = 0.16 M_\odot$ is quite similar to the 0.10 $M_\odot$ case described above. The models are mainly delineated by increasingly dramatic blue-ward excursions following the first appearance of a radiative conductive core. Furthermore, as the masses of the stars increase, radiative cores appear at progressively earlier times in the evolution, endowing the final dwarfs with progressively richer concentrations of unburned hydrogen in their outer layers.

It appears that the mass range in the vicinity of $0.16 \le M_*/M_\odot \le 0.20$ represents a transitional zone within the theory of stellar structure. Upon achieving core degeneracy, the stars in this mass range continue to increase their core temperatures. Rising core

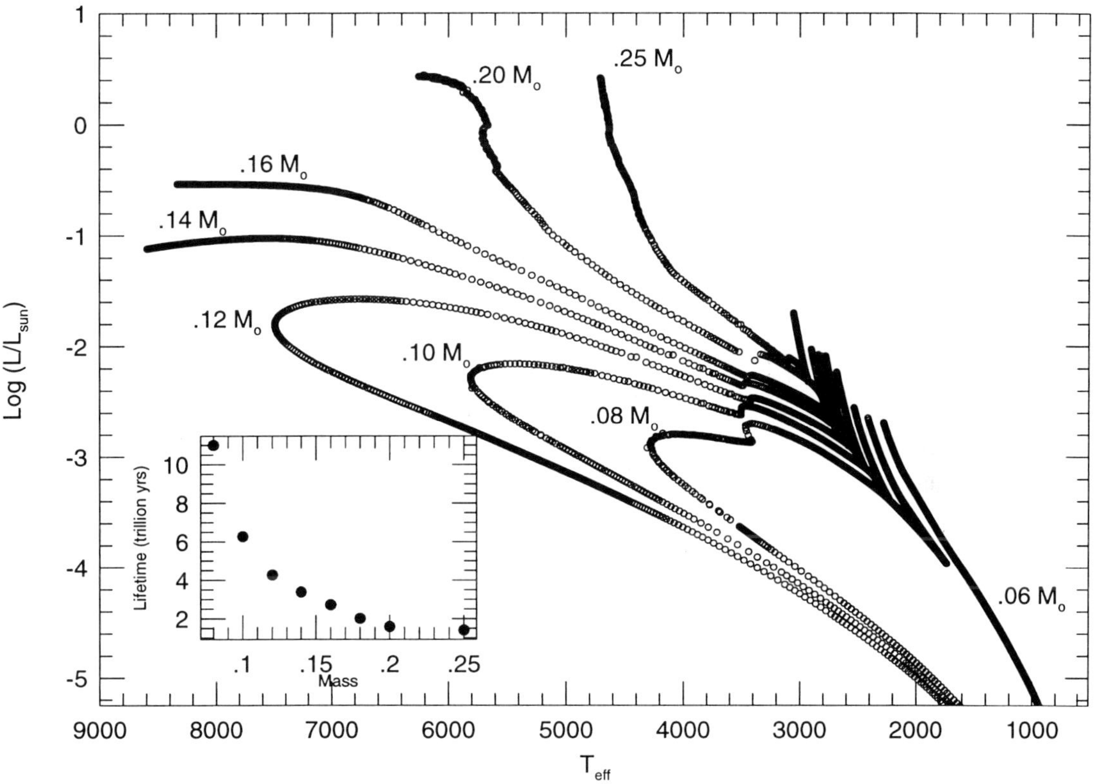

FIGURE 2. Evolution in the Hertzsprung Russell Diagram for stars with masses in the range $0.06\ M_\odot \leq M_* \leq 0.25\ M_\odot$. The inset diagram shows the corresponding main sequence lifetimes as a function of stellar mass.

temperatures are associated with increasing luminosity generation in the overlying shell sources, and hence a mounting luminosity for the entire star as the shell source burns its way outwards. The transitional $0.16 \leq M_*/M_\odot \leq 0.20$ stars are able to produce increasingly larger expansions of the overall stellar radius after the radiative hydrogen-exhausted core has developed. Further numerical experiments, discussed in our Journal article have shown that the maximum expansion factor is closely correlated the final hydrogen fraction present in the envelope.

Our $M_* = 0.25 M_\odot$ model develops a radiative core before its hydrogen fraction has dropped below $X_\mathrm{H} = 0.5$. The 0.20 $M_\odot$ stars will increase their radii by more than a factor of ten, and their tracks in the Hertzsprung Russell diagram (see Figure 2) are very reminiscent of a classic ascent up the giant branch. Indeed, when confronted with the later evolution of the 0.25 $\mathrm{M}_\odot$ model, the Henyey code was plagued with a drastically decreasing time step which put a stop to further computational progress. This particular obstacle is one which methods such as the Eggleton scheme were designed to circumvent. We therefore declare that the lowest mass red giants will have masses on the order of $M_* \approx 0.25\ M_\odot$.

4. Low Mass Red Giants

One conclusion to be drawn from the Stellar Ecology Workshop is that the nature of the exact physical mechanism which drives the development of red giants is the subject of lively, and even heated debate. Many detailed studies of the problem have been given (see, e.g., Iben & Renzini 1984; Whitworth 1989; Eggleton & Cannon 1991; Renzini et al. 1992), but a universally *accepted* explanation of the phenomenon is still lacking. The fact that stars of sufficiently low mass grow neither red nor giant provides a new foothold for attacking the problem. In particular, the behavior of our models tends to suggest that the photospheric opacity is playing at least a partial role in spurring the development of the lowest mass red giants. Our overall scenario (which is closest in spirit to that of Renzini et al. 1992) is outlined here, and developed in more detail in the Journal article.

[**A**] As stars evolve, mass increase and contraction of their isothermal helium cores lead inexorably to higher core temperatures and increasing luminosity from the active shell source. This luminosity is intrinsic to the mass of the helium core, and is largely independent of the envelope composition. More importantly, the total stellar luminosity is an increasing function of time.

[**B**] In equilibrium, the internal luminosity L_* of the star is radiated into space. The surface properties of the star obey the usual relation

$$L_* = 4\pi R_*^2 \sigma_B T_*^4 \,, \tag{4.1}$$

where the symbols have their usual meanings. Faced with the increasing luminosity described above, a star in equilibrium must either increase its radius or increase its temperature (or both). Numerically we found that very low mass stars, $M_* \sim 0.1 M_\odot$, become blue dwarfs, i.e., they increase their temperatures. On the other hand, higher mass stars, $M_* \sim 0.25 M_\odot$ tend to increase their radius and keep the temperature roughly constant.

[**C**] Near the stellar surface, convection must become superadiabatic, and hence increasingly ineffective. Stars have no choice but to radiate from their photospheres. The consequences of this requirement are shown in Figure 3, where we plot the evolution of the $M_* = 0.06$, 0.08, 0.10, 0.12, 0.16, and 0.25 $M_\odot$ models as a function of their *photospheric* densities and temperatures. The size of a point corresponding to a converged model is in direct proportion to the stellar radius. The expansion of the $M_* = 0.25\ M_\odot$ model as it becomes a red giant is dramatically visible. The solar composition opacity table has been plotted as a greyscale backdrop to the diagram; darker regions correspond to larger opacities. The general increase in opacity toward the top of the diagram is due to H^- and hydrogen ionization, whereas the opaque region below $\log_{10} T \approx 3.3$ is due to molecules and grains. This figure demonstrates that in order to achieve efficient radiation, there is a general overall tendency for evolving stars to adjust their envelope structure so as to occupy more transparent regions in the photospheric $\rho - T$ diagram.

[**D**] If the opacity in the stellar photosphere is a sharply increasing function of temperature, then the requirement that the star is both in equilibrium as well as radiating from its photosphere implies an effective upper bound

$$T_* \leq T_{\rm max} \tag{4.2}$$

for the photosphere. Once the star is in a configuration where the photospheric temperature is approaching $T_{\rm max}$, the star has no choice but to increase its radius rather than increase its temperature. In this manner, the opacity controls, in part, when a star becomes a red giant.

[**E**] The maximum temperature $T_{\rm max}$ is determined by the opacity, which in evolving low mass stars is strongly dependent on the relative hydrogen fraction in the atmosphere.

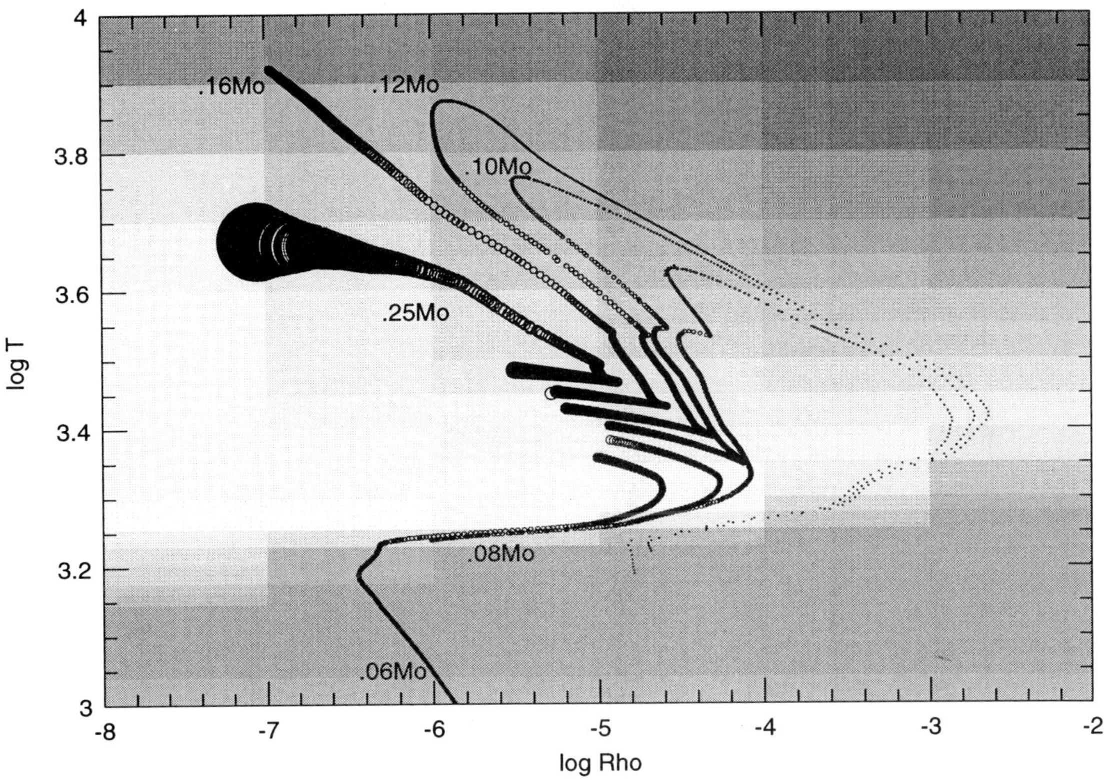

FIGURE 3. Photospheric evolution in the density-temperature plane. The evolutionary sequences are plotted on top of the hydrogen rich opacity table. Darker shades of gray represent higher opacities. The size of the open circles representing each converged model is in proportion to the actual radius of the star. The stars begin their evolution on the Hayashi track clustered fairly closely together near the point $\log_{10} T \approx 3.3$, $\log_{10} \rho \approx -5.5$. Notice that the $M_* =$ 0.08 $M_\odot$ model slides all the way along the grain opacity boundary to a photospheric density $\log_{10} \rho \approx -5.5$ before reaching the main sequence.

Our evolutionary calculations show that as the overall stellar mass increases, radiative cores appear earlier and earlier in the course of stellar evolution, and lock in progressively larger hydrogen fractions in the stellar envelope. The net effect is that the higher mass stars end up with increasingly hydrogen-rich atmospheres, and hence see large photospheric increases in opacity as the temperature rises from 3000 to 10^4 K, sufficient to enhance the development of red-giant structures once the stellar mass is on the order of 0.25 $M_\odot$.

The foregoing line of argument is only a part of the story. We have not yet fully explored the role of the compositional discontinuity between the core and the envelope in determining the overall structure of the smallest red giants. As emphasized by Whitworth (1989) and Faulkner (1996), inhomogeneous core/envelope structures provide structural conditions which dispose stars toward giant-like configurations. The existence of an opacity "wall" of the type described, is a necessary, but not sufficient condition for red giant development.

5. Implications

The evolution of the low mass stars will have important ramifications for the evolution of the galaxy on time scales that greatly exceed the current age of the universe. The overall luminosity of the galaxy is determined by two competing effects. The first is that the luminosity of main sequence stars is an increasing function of time. As discussed above, this feature of stellar evolution is an important part of the process by which stars become red giants. The second effect is that the star formation rate, and hence the total number of stars, is expected to be a decreasing function of time over the long time scales associated with evolving M dwarfs (see Adams & Laughlin 1996 for more details). These two effects tend to cancel and the luminosity of the galaxy will remain *very roughly* comparable to its present day value over the time range $\tau < 10^{13}$ yr. After that time, when the lowest mass stars have completed their post-main sequence evolution and have begun their descent as helium white dwarfs, the luminosity of the galaxy will plummet precipitously.

This work was supported by Fred Adams' NSF Young Investigator Award, by NASA Grant No. NAG 5-2869, and by an AAS International Travel Grant. I would like to thank Peter Bodenheimer and Fred Adams, my co-authors on the main Journal article.

REFERENCES

Adams, F. C., & Laughlin, G. 1996, Rev. Mod. Phys., In Press
Alexander, D. R., Johnson, H. R., & Rypma, R. I. 1983, ApJ, 272, 773
Allard, F., & Hauschildt, P. H. 1995, ApJ, 445, 433
Bodenheimer, P. 1965, Unpublished computer output
Burrows, A., & Liebert, J. 1994, Rev. Mod. Phys., 65, 301
Eggleton, P. P., & Cannon, R. C. 1991, ApJ, 383, 757
Faulkner, J. 1996, This volume
Henyey, L. G., Forbes, J. E., & Gould, N. L. 1964, ApJ, 139, 306
Iben, I., Jr. 1967, ARA&A 5, 571
Iben, I. Jr., & Renzini, A. 1984, Physics Reports, 105, 329
Kumar, S. S. 1963, ApJ, 137, 1121
Laughlin, G., Bodenheimer, P. & Adams F.C. 1996, ApJ, Submitted
Laughlin, G., & Bodenheimer, P. 1993, ApJ, 403, 303
Renzini, A., Greggio, L., Ritossa, C., & Ferrario, L. 1992, ApJ, 400, 280
Salpeter, E. E. 1982, in Essays in Nuclear Astrophysics (Cambridge: Cambridge Univ. Press)
Whitworth, A. P. 1989, MNRAS, 236, 505

Why Stars Become Red Giants

By DAIICHIRO SUGIMOTO

Graduate School of Arts and Science, University of Tokyo, Japan

1. Red Giant Regime and Dwarf Regime

In this Workshop Professor John Faulkner discussed a case of adding mass onto a helium star. Initially the star is on the helium main-sequence. As the hydrogen-rich gas is added on it, the radius of the star becomes larger and larger. However, it does not approach finally to a red giant but to the hydrogen main-sequence star when the mass of the original helium star becomes negligible as compared with the mass of the added hydrogen-rich gas. John's understanding is right. However, it is also well known that a helium white dwarf approaches the red giant when a small amount, *i.e.*, even less than one percent of the white dwarf mass, of hydrogen-rich gas is added. This is the process that is the reverse to the evolution from a red giant to a white dwarf through a phase of planetary nebula. Such phenomenon occurs when the star is *electron degenerate.*

There is a similar situation for a star consisting of a helium core and a hydrogen-rich envelope. In the pre-helium flash stage, the star has an electron degenerate core and lies on the red giant branch. In the post-helium flash stage, it has a non-degenerate core and is not a red giant any more. What are the essential difference between them?

The differential equation of stellar hydrostatic equilibrium does not contain any temperature T, but the pressure P and the density ρ so that the important physical parameter is not T but $\Theta \equiv P/\rho$. The ratio of Θ's between the center (denoted by the subscript c) and the inner edge (denoted by ,i) of the hydrogen-burning shell (denoted by H), *i.e.*, $\Theta_c/\Theta_{H,i}$ is as high as about 200 for the pre-flash stage near the tip of the red giant branch of a globular cluster, while it is as low as about 10 for the post-flash stage at the horizontal branch. Such a great difference comes from the difference in the electron degeneracy. For the non-degenerate core such a high value as 200 can only be realized in a star with iron core, which is well a red giant because of its high central condensation of the mass.

2. Boundary Value Problem versus Initial Value Problem

We should be aware of the distinction between the boundary value problem and the initial value problem. In other words the following two questions are different each other; why a hydrostatic equilibrium solution is a red giant or a dwarf, and why a star can or can not swell up to a red giant.

Let us consider, in the first place, the *boundary value problem* and analyze a change of stellar structure in equilibrium responding to a virtual change in the physical parameter. The star consists of a helium core and a hydrogen-rich envelope of appreciable mass fraction. Let us consider a change due to a virtual gravitational contraction of the core. Because of the strong dependence of the hydrogen burning rate on the temperature, the value of T_H remains practically unchanged, *i.e.*, T_H changes but only slightly. Therefore, the radius r_H also remains practically unchanged so that the volume of the core remains practically unchanged. Nevertheless, the central density increases. This must lead to the decrease in $\rho_{H,i}$ which implies the corresponding decrease in P_H. In order to sustain the envelope against the gravitational attraction of the core with this lowered pressure P_H, the weight of the envelope should be lighter, *i.e.*, the bulk of the envelope mass should lie at larger radii where the gravity is weaker. This implies that a red giant structure is a consistent solution.

The question how realized is such a large radius of the red giant is a different issue: It is a matter of the *initial value problem.* The large radius can be realized as follows. When $P_{\rm H}$ becomes lower, the hydrogen-burning shell contracts slightly with a slight rise in $T_{\rm H}$, and the enhanced nuclear reaction produces the energy necessary to push up the envelope against the gravity.

However, such response as discussed above is not necessarily the case both for the boundary value and the initial value problems, when the envelope is relatively light due to its small mass fraction and when the star is a dwarf. The lowered $P_{\rm H}$ can be compensated by a contraction of $r_{\rm H}$ and a resultant small increase in $T_{\rm H}$. The star may find another hydrostatic equilibrium with this small contraction where the additional energy generation due to hydrogen shell-burning can be transported out in steady state. In this case the star does not swell but remains to be about the same size. Such case is rather along the common sense in usual physical systems. However, the star having a large spherical envelope is different; it is a gravothermal system.

The discussions above indicate that there are two different types of solutions corresponding to the two different types of response to the virtual change. We shall call the former type as the *red giant regime* and the latter type as the *dwarf regime.* In the next section we shall show that such difference in the regimes stems from a difference in topology of stellar structure solutions in U-V plane.

3. Exact Analytical Discussion for the Two Different Regimes

Stellar structure is represented in a U-V plane which is defined by

$$U = \frac{4\pi r^3 \rho}{M_r} = \frac{d\ln M_r}{d\ln r} \,, \tag{3.1}$$

$$V = \frac{GM_r\rho}{rP} = -\frac{d\ln P}{d\ln r} \,. \tag{3.2}$$

By differentiating these equations we obtain

$$d\ln r = -\frac{d\ln U - d\ln V}{D} \,, \tag{3.3}$$

$$D \equiv 2U + V - 4 \,. \tag{3.4}$$

The U-V plane is divided into two regions by the critical curve with $D = 0$. We shall call the regions with $D > 0$ and $D < 0$ as the *dwarf region* and the *red giant region*, respectively. The boundary condition lies in the dwarf region both at the center ($U = 3$, $V = 0$) and for the surface ($U = 0$, $V = \infty$). Therefore, a solution lying only in the dwarf region is possible, which we call as the *dwarf solution.*

The U-V curve can cross the critical curve of $D = 0$ only with the angle of 45 degree in the $\ln U$-$\ln V$ plane because in equation (3.3) both the numerator and the denominator should vanish simultaneously. According to the same equation, the U-V curve of stellar structure extends from the center to the surface to the direction of decreasing $\ln(U/V)$ in the dwarf region and of increasing $\ln(U/V)$ in the red giant region. This implies the followings; when the U-V curve once enters into the red giant region, it makes a loop in order to come out again to the dwarf region and continue to the surface. We call such one as the *red giant solution.*

Such a difference in topologies between the dwarf solution and the red giant solution is associated mainly to the difference in the central condensation of the solution. The solution has the topology of the red giant when the central condensation, *i.e.*, $\Theta_c/\Theta_{H,i}$, is high (Sugimoto & Nomoto 1980) and at the same time the mass fraction of the hydrogen-rich envelope is appreciably large. The critical value for the mass fraction of the envelope depends on the degree of central condensation of the core [see constant mass fraction curves, *e.g.*, in Hayashi *et al.* (1962)]. The crossing of the U-V curve over the $D = 0$ curve from the dwarf to the red giant regions takes place in many cases at the interface from the helium zone to the hydrogen-rich envelope where the U-V curve runs 45 degree in the $\ln U$-$\ln V$ plane only by the change in the mean molecular weight but without any change in the pressure and temperature. It makes Θ larger across the interface and partly compensates the large difference of its value between the center and the interior side of the hydrogen-burning shell. The relation between the evolutionary track in the horizontal branch and the increase in the mean molecular weight of the helium-burning core was discussed in Hayashi *et al.* (1962).

When the central condensation in the core is very high, the value of $U_{H,i}$ is quite small because U is defined as three times of the ratio of the local density to the mean interior density as seen in equation (3.1). Therefore, the value of V is very close to 4 on the line of $D = 0$. On the other hand, the envelope solution exterior to the hydrogen-burning shell runs in the red giant region near the line of $V = 4$ because the electron scattering opacity dominates and thus the polytropic index N as defined by $d\ln\rho/d\ln P \equiv N/(N+1)$ is close to 3. For the collapsed type solution of the polytrope of index N the value of V approaches $N + 1$ where r becomes small [see Chandrasekhar (1939)]. Therefore, in the red giant region of the U-V plane, the envelope solution runs close to the line of small D so that the integral of equation (3.3), *i.e.*, the radius change is very large. This implies that the star is a red giant. It even makes the stellar radius too large to be compatible with its surface boundary condition.

However, the polytropic index in the convection zone can become smaller than 4 when the surface convection zone extends deep in the envelope. Then, the stellar radius can be smaller than the case without any deep convection zone. This is the reason why the star can lie properly on the Hayashi line of red giant branch by deepening the surface convection zone. In other words, the surface boundary condition determines the depth of the surface convection zone.

The transition of the topology from one for the dwarf solution to the other for the red giant solution takes place usually with its thermal state kept in quasi-equilibrium. However, there is also a phase of sudden change of the topology as in the case of the helium flash from one for the red giant to the other for the dwarf. Such a sudden change is realized by relaxing the condition of thermal quasi-equilibrium during the helium flash.

REFERENCES

Chandrasekhar, S. 1939, *An Introduction to the Study of Stellar Structure* (The University of Chicago Press)

Hayashi, C., Hōshi, R., & Sugimoto, D. 1962, Suppl. Prog. Theoret. Phys. No.22 , 1

Sugimoto, D. & Nomoto, K. 1980, Space Sci. Rev., 25, 155

Some Implications of Refinements to the Physics and Chemistry of Lower-Mass Stars

By DON A. VANDENBERG† AND ALAN W. IRWIN

Department of Physics & Astronomy, University of Victoria, Victoria, BC V8W 3P6, Canada

Of the basic input physics (e.g., opacity, nuclear reaction rates) used to construct stellar models, the equation of state probably has the greatest potential to alter our understanding of the structure and evolution of lower-mass stars ($\leq 1.2\mathcal{M}_\odot$, say). In this investigation, the consequences of adopting a number of recent formulations that include a variety of non-ideal effects are examined. In particular, we consider their impact on the properties of a standard solar model, on turnoff age-luminosity relations, and on the predicted core mass at the onset of the Helium Flash. Allowance for non-solar number abundance ratios of the heavy elements is another area in which stellar models are steadily being refined. Here, we report some of the affects of adopting $[\alpha/\mathrm{Fe}] > 0$, where "α" represents such elements as O, Ne, Mg, Si, S, etc. Finally, we discuss the impact of recent revisions to the bolometric correction scale on the ages of the Galactic globular clusters.

1. Introduction

To a very considerable extent, our present understanding of the evolution of lower-mass stars dates back to a series of classic papers written by I. Iben in the mid-to-late 1960's. Certainly, the stellar models that he computed at that time are qualitatively indistinguishable from more modern computations — even in a quantitative sense, the differences are small — and they represented a significant advancement over the pioneering calculations by e.g., Sandage & Schwarzschild (1952) and Hoyle & Schwarzschild (1955). But what really gives those papers a timeless quality are the exquisite physical explanations of how and why stars evolve as they do. For instance, Iben (1965) explained how the detailed morphology of the evolutionary track that is predicted for a star evolving from a fully-convective structure on the Hayashi line to the main sequence could be readily understood in terms of the changing conditions in the stellar interior and the effects of initial nonequilibrium ^{12}C abundances. Similarly, the causes of the morphological differences between the post-main-sequence tracks for 1.25 and $1.0\mathcal{M}_\odot$ stars, such as the variation in the slope of the subgiant branch and the presence or absence of a blueward hook following central hydrogen exhaustion, were elucidated by Iben (1967). Further, the controlling influence of the evolving helium core in driving a star up the red-giant branch was made understandable by Iben (1968a) through a thorough analysis of computed models complemented by simple, yet elegant, analytic arguments.

Besides such fundamental contributions to stellar evolutionary theory, I. Iben (and colleagues) developed and/or employed many of the techniques that are still widely used to analyze observed C-M diagrams. Iben (1968b) devised the so-called "R-method" to infer the helium contents of globular cluster (GC) stars. Iben & Rood (1970) used the predicted turnoff luminosity versus age relations from computed isochrones to deduce the ages of several GCs. Simoda & Iben (1970) constructed theoretical luminosity functions and demonstrated their dependence on the hydrogen abundance. Many more examples could be given: the main point of these brief remarks — besides recalling a few of

† Killam Research Fellow. An N.S.E.R.C. operating grant is also gratefully acknowledged.

the many landmark papers written by the honoree of this meeting — is that the vast majority of the work that has been carried out in the intervening years has simply served to *refine* the basic picture of stellar evolution developed by I. Iben (and several others, most notably M. Schwarzschild and P. Demarque). This is especially true for the pre-horizontal-branch phases of evolution — to which the discussion in the present study is restricted.

In this investigation, we describe some refinements to stellar models for lower-mass stars arising from improvements to the equation of state (§2), to the assumed element mixtures (§3), and to the bolometric corrections that are needed, along with the color transformations, to produce synthetic C-M diagrams (§4).

2. Equation-of-state Considerations

Developed in order to determine the occupation numbers needed for opacity calculations, the OPAL (Rogers 1994, and references therein) and the MHD (Hummer & Mihalas 1988; Mihalas, Däppen, & Hummer 1988) equations of state are two of the most complex and detailed formulations presently available. Because they are so expensive computationally, the only practical way in which they can be used to construct stellar models is through interpolations in pre-calculated tables. But even these very sophisticated treatments have their shortcomings, such as the neglect of relativistic corrections. Consequently, they must be supplemented in some way if the gas under consideration is relativistically degenerate. Moreover, considerable care must be taken with any interpolation procedure to ensure that derivative quantities are accurately calculated and that the errors arising from the coarseness of the grid defined by the independent variables are small. And, although the "additive volume" technique for chemical composition interpolation appears to work well (see Dorman, Irwin, & Pedersen 1991), it is unphysical in ionization or dissociation zones and it may therefore lead to inaccuracies in modelling them. So there is much to be said for an analytic equation of state (EOS), especially if it can mimic to high accuracy the OPAL and/or MHD formulations.

Last Spring, we (Irwin et al. 1996; ISVR) set out to try to develop such an EOS. At about that time, Pols et al. (1995; PTEH) had just completed a major revision of the Eggleton, Faulkner, & Flannery (1973; EFF) code, significantly improving upon the treatment of molecular hydrogen and pressure ionization and including Coulomb corrections. Like its predecessor, the PTEH EOS is extremely fast because it is a completely explicit formulation (by virtue of using the temperature, T, and a degeneracy-related quantity, f, as the input variables) and it is thermodynamically consistent. For efficiency reasons, ISVR also chose to use T and f as the independent variables: the resultant EOS therefore bears many similarities to the PTEH version though, as indicated in Table 1, it contains much more physics and it has the flexibility to treat several non-ideal effects in a variety of ways.

Through an appropriate combination of the available options, ISVR are able to emulate the PTEH, MHD, or OPAL equations of state. An example of how well they can do this is illustrated in Figure 1, which plots the difference in $\ln P$ (where P is the pressure) as determined using the ISVR EOS in the OPAL mode, on the one hand, and from OPAL tables (Rogers, Swenson, & Iglesias 1996), on the other. The same chemical composition is assumed and the entire range of density and temperature spanned by the OPAL tables is considered. The agreement is clearly very good over nearly the entire ρ, T-plane: only in a small region of the parameter space do the differences become noticeable, rising to a maximum of $\approx 1.5\%$, which is still rather small. Such variations can be attributed, in part, to differences in the treatment of the Coulomb interaction. They may also arise

Effect	PTEH	MHD	OPAL	ISVR
H_2	x	x	x	x
H_2^+		x		x
Partial Ionization of the Metals		x	x	x
Non-relativistic, arbitrarily degenerate electrons		x	x	x
Arbitrarily relativistic and degenerate electrons	x			x
Debye-Hückel Coulomb				x
Debye-Hückel Coulomb, τ-corrected		x		x
Full-Order, One-Component Plasma Coulomb	x			x
Low-Order, Multi-Component Plasma Coulomb			x	
Electron Diffraction			x	
Non-Relativistic Electron Exchange			x	x
Electron-Electron Occupation Probability	x			x
Ion-Electron Occupation Probability				x
Ion-Ion Occupation Probability		x		x
Planck-Larkin Occupation Probability			x	x
Neutral-Neutral Occupation Probability		x		x
Neutral-Ion Occupation Probability		x		x

TABLE 1. Effects Treated by Various Equations of State

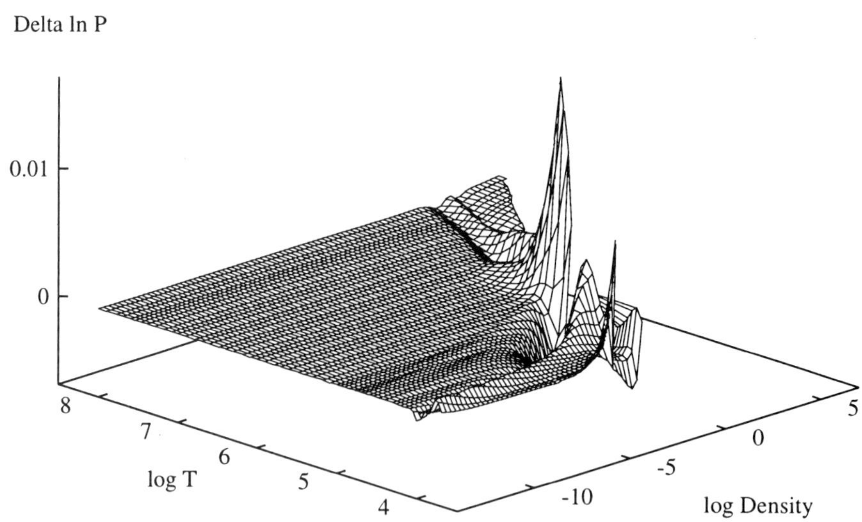

FIGURE 1. Pressure differences: simulated OPAL minus tabulated OPAL

from the fact that, whereas the OPAL (and MHD) codes consider all possible excitation states, ISVR assume ground-state configurations only — which is the main reason why the latter formulation is so fast. It is fortunate that the neglect of excitation has only minor ramifications for thermodynamic quantities and that, mainly through a judicious choice of constants in the pressure-ionization terms, ISVR are able to emulate the OPAL (and MHD) treatments so well. (ISVR can match the MHD EOS to within 2% and the PTEH EOS nearly exactly.)

The relative importance of the various terms in the EOS and the regions on the ρ, T-plane where they are significant can be readily evaluated using the ISVR code. For

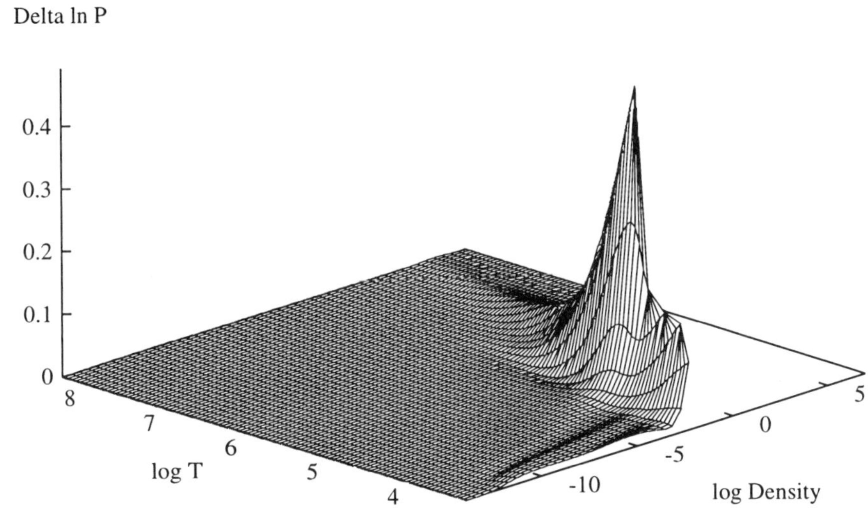

FIGURE 2. Pressure differences: simulated OPAL, including electron exchange, minus simulated OPAL, neglecting electron exchange

instance, Figure 2 shows that electron exchange is a relatively low-temperature and intermediate-density effect and that neglecting it would introduce an error amounting to as much as 0.45 in $\ln P$. As noted in Table 1, neither the MHD nor the PTEH equations of state take this non-ideal effect into account, which is clearly a deficiency of these formulations.

However, it turns out that, as far as predicting the gross properties of most stars is concerned, the Coulomb interaction is by far the dominant non-ideal effect and electron exchange is of secondary importance. This is certainly true for the calculation of a standard solar model, or the structures of lower main-sequence stars down to $0.2\mathcal{M}_\odot$, or the evolutionary tracks of 0.8–1.0 $\mathcal{M}_\odot$ stars up to the tip of the red-giant branch — which represent the extent of the tests that IBVR have made to date. Preliminary results from some of these test calculations are given in Tables 2 and 3. The first of these lists the mass-fraction abundance of helium (Y) and the value of the mixing-length parameter ($\alpha_{\rm MLT}$) that are required to be able to predict the observed solar luminosity and temperature when different EOS formulations are employed in the stellar evolution code. The second tabulates the predicted surface luminosity, effective temperature, and helium core mass at the onset of the Helium Flash for values of [Fe/H] from -2.0 to 0.0 and masses for which the expected lifetimes are $\sim$ 10 Gyr. Results are provided for an EFF-type EOS (the version used by e.g., VandenBerg 1992), which does not treat the Coulomb interaction or electron exchange; for the simulated PTEH EOS, which takes Coulomb interactions into account but not exchange; and for the simulated OPAL EOS, which allows for both of these effects. Electron exchange clearly plays a fairly minor role in such calculations (cf. the differences between the PTEH and OPAL results). However, in much more sensitive tests of the equation of state, such as the modelling of the observed solar oscillation frequencies, it has been shown (e.g., Elliott 1996) that models based on the OPAL EOS do a superior job of matching the observations compared with those based on the MHD formulation. One final point to conclude this section: the net effect of non-ideal corrections in the EOS insofar as the ages of star clusters are concerned is to reduce the age at a given turnoff luminosity by $\approx$ 7% (cf. Chaboyer & Kim 1995).

EOS	Y	α_{MLT}
EFF	0.2816	1.85
PTEH	0.2722	1.95
OPAL	0.2692	1.90

TABLE 2. Properties of Standard Solar Models (Assuming $Z = 0.0188$)

EOS	Mass	[Fe/H]	[α/Fe]	Y	α_{MLT}	$\log L$	$\log T_{\rm eff}$	$\mathcal{M}_c$
EFF	0.8	−2.0	0.3	0.235	1.85	3.294	3.6181	0.4956
	0.9	−1.0	0.3	0.239	1.85	3.366	3.5591	0.4834
	1.0	0.0	0.0	0.282	1.85	3.402	3.4934	0.4695
PTEH	0.8	−2.0	0.3	0.235	1.95	3.293	3.6235	0.4918
	0.9	−1.0	0.3	0.239	1.95	3.366	3.5669	0.4800
	1.0	0.0	0.0	0.272	1.95	3.408	3.5006	0.4673
OPAL	0.8	−2.0	0.3	0.235	1.90	3.297	3.6205	0.4903
	0.9	−1.0	0.3	0.239	1.90	3.369	3.5627	0.4784
	1.0	0.0	0.0	0.269	1.90	3.410	3.4965	0.4657

TABLE 3. Selected Properties of RGB-tip Stars

3. Concerning Stellar Models with [α/Fe] > 0

VandenBerg et al. (1996) have recently completed the computation of a large grid of evolutionary tracks and isochrones for several values of [Fe/H], assuming [α/Fe] $= 0.0, 0.3$, and 0.6 in each case. The α-elements (including O, Ne, Mg, Si, S, Ar, Ca, and Ti) were assumed to scale together (cf. Wheeler, Sneden, & Truran 1989) and opacities were calculated for the precise element mixtures that were considered. Due to the lack of availability of a comprehensive set of model atmospheres, a scaled-solar T–τ relation was used to derive the surface pressures. However, the value of $\alpha_{\rm MLT}$ required to fit the Sun (1.88) was found to be essentially the same as that needed to match the properties of the best-observed of the Population II subdwarfs; namely, HD 103095. Moreover, with this choice of $\alpha_{\rm MLT}$, the predicted temperatures along the giant branches of globular clusters agreed well with those inferred from infrared photometry (e.g., Frogel, Persson, & Cohen 1981). Consequently, the adopted treatment of the surface layers seemed to produce a reasonably realistic $T_{\rm eff}$ scale.

Figure 3 illustrates how the shape and location of a 14 Gyr isochrone on the theoretical H-R diagram, for two different [Fe/H] values, is affected by variations in [α/Fe]. Although the models have been computed to much fainter and brighter magnitudes, we concentrate here on the vicinity of the turnoff to emphasize two main results. First, the turnoff luminosity, or equivalently, the luminosity of the subgiant branch at a fixed $T_{\rm eff}$, decreases as [α/Fe] increases. That is, the age at a given turnoff luminosity is predicted to decrease with increasing [α/Fe] (see Figure 4). This seems to be at odds with the conclusion reached by Chieffi, Straniero, & Salaris (1991) that "the enhancement of the α-elements does not lead to younger ages" (for the Galactic globular clusters). However, they apparently derived the cluster distances from theoretical horizontal-branch (HB) models, which have reduced luminosities if [α/Fe] > 0. But if they had used the main-sequence (MS) fitting procedure to set the distance scale (cf. Bolte & Hogan 1995), then increased distances and significantly younger ages would have resulted. As is evident in Fig. 3, the main sequence tends to be redder at a fixed magnitude or brighter at a

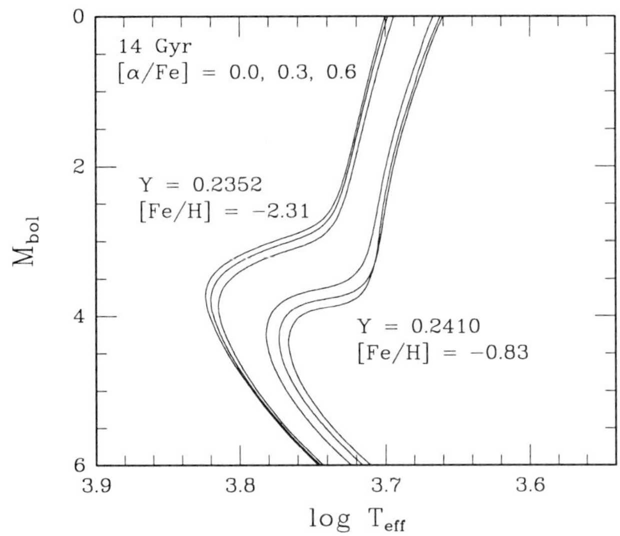

FIGURE 3. VandenBerg et al. (1996) isochrones for two values of [Fe/H] assuming, in each case, $[\alpha/\mathrm{Fe}] = 0.0$, 0.3, and 0.6.

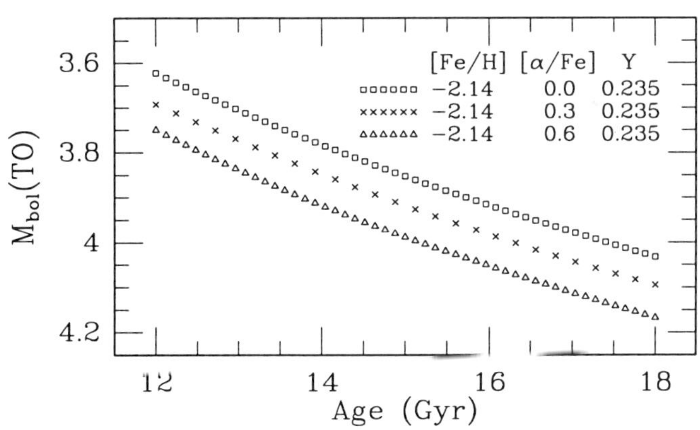

FIGURE 4. Turnoff luminosity versus age relations as predicted by VandenBerg et al. (1996).

fixed temperature/color when enhancements in the α-element abundances are assumed, especially at the higher [Fe/H] values. Thus the Chieffi et al. conclusion is contingent upon their assumptions concerning the cluster distances. A more accurate statement of the impact of non-solar α/Fe number abundance ratios on ages is that, at a fixed turnoff luminosity, the age is reduced by about 7% ($\sim$ 1 Gyr) for each 0.3 dex increase in [α/Fe] (see, e.g., Fig. 4).

The second result that we wish to draw attention to in Fig. 3 is the behavior of the red-giant branch (RGB) location with [α/Fe] in the most metal-rich case of the two that have been plotted. Whereas the giant branch is predicted to become cooler when [α/Fe] is increased from 0.0 to 0.3, a further enhancement to $[\alpha/\mathrm{Fe}] = 0.6$ has no additional impact on RGB temperatures. The reason for this is that the giant branch has become sufficiently cool that the α-elements cease to be effective as electron donors and the surface opacity no longer changes with increasing [α/Fe]. (For even higher [Fe/H] values,

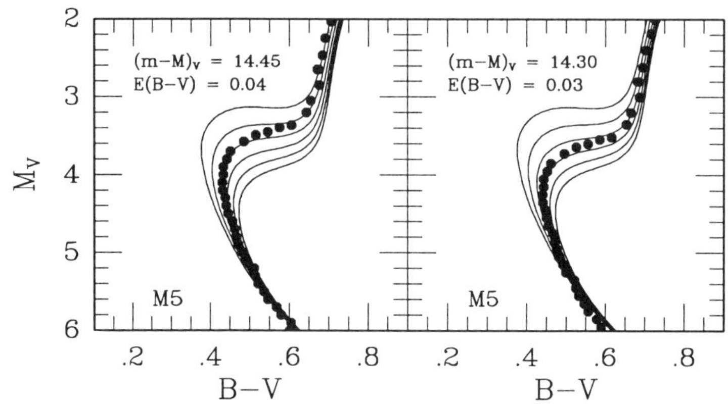

FIGURE 5. Fit of isochrones for [Fe/H] = −1.31, [α/Fe] = 0.3, and ages from 8 to 18 Gyr to the M5 C-M diagram (Sandquist et al. 1996) for the indicated distances and reddenings.

the position of the giant branch (at fixed mass) is completely independent of [α/Fe], so the case [Fe/H] = −0.8 represents something of a "transition metallicity".) These results are clearly in conflict with the suggestion by Chieffi et al. (1991) that it is possible to mimic isochrones for non-solar α/Fe number abundance ratios by those for scaled-solar mixes with a suitably chosen Z. Their approach may work for very metal-poor compositions, but not in the more metal-rich regime.

4. Revised Bolometric Corrections and the Ages of Globular Clusters

There has been some suggestion (cf. Salaris, Degl'Innocenti, & Weiss 1996) that current uncertainties in the bolometric correction (BC) scale are such that they alone contribute an ≈ 2 Gyr uncertainty in the estimated ages of the most metal-poor globular clusters. If correct, this would be quite a surprising result since one would have expected the bolometric corrections appropriate to 6000–7000 K turnoff stars to be especially well determined. Certainly it is an issue that warrants further investigation.

Since the BC scale involves a constant that can be set to satisfy the solar constraint, there should be little disagreement concerning the BC values at near-solar gravities, temperatures, and metallicities. But how do the available predictions vary with [Fe/H]? Considering $T_{\rm eff} = 6000$ K and $\log g = 4.5$, VandenBerg & Bell (1985; VB85) predict that the bolometric correction in V changes from −0.094 at [Fe/H] = 0.0 to −0.118 at [Fe/H] = −2.0, for a difference of −0.024. This agrees well with the value of −0.019 obtained by Buser & Kurucz (1992) based on the same model atmospheres as used by VB85. However, for the same input parameters, a difference of −0.091 is found from Kurucz (1979) model atmospheres and −0.113 from Kurucz (1992) atmospheres. Thus, the choice of which BC scale is used to transform a set of isochrones can, by itself, lead to differences in the predicted turnoff M_V at a given age by up to ∼ 0.1 mag, with consequent ramifications for cluster ages. For instance, the Bolte & Hogan (1995) estimate of 15.8 Gyr for M92, using isochrones that adopted the VB85 bolometric corrections, would have been closer to 14.3 Gyr, had those isochrones used the Kurucz (1992) BC's instead. In fact, the lower age estimate is arguably the preferred one since R. Bell's (1996, private communication) latest model atmospheres yield a BC scale that agrees extremely well with the one by Kurucz (1992). However, as is widely appreciated, distance uncertainties represent a much more serious concern and ages as high as 16 Gyr cannot be ruled out yet.

One constraint on cluster distances/ages that has not been given much attention since it was discussed at some length by Iben (1967) is the slope of the subgiant branch (SGB). Because stars evolve rapidly through the shell-narrowing phase, the subgiant segment of an observed C-M diagram will be similar in appearance to the subgiant track of the most massive stars present in the cluster. Considering Figure 5, for instance, the observed SGB slope in M5 clearly favors $(m - M)_V \leq 14.30$. (In making such comparisons, it is important that the isochrones provide a reasonable match to both the observed main-sequence and RGB locations so that the effects of convection theory on the morphology of the SGB are minimized.) Based on this argument, ages ≤ 12 Gyr (or so) appear to be precluded. Iben's insights concerning the SGB (and the many other phases in a star's life) are well worth keeping in mind. A tip of the hat to you, Icko!!

REFERENCES

Bolte, M., & Hogan, C.J. 1995, Nature, 376, 399

Buser, R., & Kurucz, R.L. 1992, A&A, 264, 557

Chaboyer, B., & Kim, Y.-C. 1995, ApJ, 454, 767

Chieffi, A., Straniero, O., & Salaris, M. 1991, in The Formation and Evolution of Star Clusters, ed. K. Janes, ASP Conf. Ser., 13, 219

Dorman, B., Irwin, A.W., & Pedersen, B.B. 1991, ApJ, 381, 228

Elliott, J.R. 1996, MNRAS, 280, 1244

Eggleton, P.P., Faulkner, J., & Flannery, B.P. 1973, A&A, 23, 261 (EFF)

Frogel, J.A., Persson, S.E., & Cohen, J.G. 1981, ApJ, 246, 865

Hoyle, F., & Schwarzschild, M. 1955, ApJS, 2, 1

Hummer, D.G., & Mihalas, D. 1988, ApJ, 331, 794

Iben, I., Jr. 1965, ApJ, 141, 993

Iben, I., Jr. 1967, ApJ, 147, 624

Iben, I., Jr. 1968a, ApJ, 154, 581

Iben, I., Jr. 1968b, Nature, 220, 143

Iben, I., Jr., & Rood, R.T. 1970, ApJ, 159, 605

Irwin, A.W., Swenson, F.J., VandenBerg, D.A., & Rogers, F.J. 1996, ApJ, submitted (ISVR)

Kurucz, R.L. 1979, ApJS, 40, 1

Kurucz, R.L. 1992, Solar Abundance Model Atmospheres for 0, 1, 2, 4, 8 km/s (CD-ROM 19)

Mihalas, D., Däppen, W.D., & Hummer, D.G. 1988, ApJ, 331, 815

Pols, O.R., Tout, C.A., Eggleton, P.P., & Han, Z. 1995, MNRAS, 274, 964 (PTEH)

Rogers, F.J. 1994, in IAU Colloq. 147, The Equation of State in Astrophysics, ed. G. Chabrier & E. Schatzman (Cambridge: Cambridge Univ. Press), p. 16

Rogers, F.J., Swenson, F.J., & Iglesias, C.A. 1996, ApJ, 456, 902

Salaris, M., Degl'Innocenti, S., & Weiss, A. 1996, preprint

Sandage, A., & Schwarzschild, M. 1952, ApJ, 116, 463

Sandquist, E., Bolte, M., & Stetson, P.B. 1996, ApJ, in press

Simoda, M., & Iben, I., Jr. 1970, ApJS, 22, 81

VandenBerg, D.A. 1992, ApJ, 391, 685

VandenBerg, D.A., & Bell, R.A. 1985, ApJS, 58, 561 (VB85)

VandenBerg, D.A., Swenson, F.J., Rogers, F.J., Iglesias, C.A., & Alexander, D.R. 1996, in preparation

Wheeler, J.C., Sneden, C., & Truran, J.W., Jr. 1989, ARAA, 27, 279

Globular-Cluster Luminosity and Mass Functions from HST Data†

By IVAN R. KING[1], ADRIENNE M. COOL[1], GIAMPAOLO PIOTTO[2], JAY ANDERSON[1] AND CRAIG SOSIN[1]

[1]Astronomy Dept., University of California, Berkeley, CA 92720-3411, USA

[2]Dipartimento di Astronomia, Università di Padova, Vicolo dell' Osservatorio 5, I-35122 Padova, Italy

Hubble Space Telescope observations carry the luminosity functions of globular clusters several magnitudes fainter than ground-based observations have been able to do reliably. In the course of deriving each LF we produce a color–magnitude diagram that is itself of great interest and can provide an important check on stellar-structure theories. From the LFs—after correction from the local to the global LF of a cluster—we derive mass functions, but only by using mass–luminosity relations that are themselves dependent on the correctness of the stellar models. For four globulars of very similar metallicity we have bypassed the MLR issue by comparing LFs directly (since they should all convert to MFs via the same MLR). The LFs of M15, M30, and M92 are virtually identical, while NGC 6397 is strikingly deficient in faint stars. We tentatively attribute this difference to the much greater vulnerability of NGC 6397 to Galactic tidal shocks.

1. Introduction

The Hubble Space Telescope (*HST*) gives us two special advantages in studying globular clusters: (1) It reaches faint limiting magnitudes. (2) Its superb resolving power allows us to study the crowded centers of the clusters, with all of the fascinating anomalies that occur there.

The *HST* limit, for reasonable exposure times, is around 27 or 27.5 in V, which is about three magnitudes fainter than one usually achieves with ground-based telescopes of the 4-meter class. (We note in passing, however, that in principle the 10-meter Keck telescopes should be able to go a little fainter than *HST*, since for a faint color–magnitude diagram resolution is no problem; one simply moves farther out in the cluster. For the photometry itself, calculations of theoretical Keck performance predict a fainter limiting magnitude than that of *HST*—but unfortunately the present complement of Keck instruments does not include a camera that gives good images over the field size that is needed.)

In the cluster centers, of course, *HST* has no competition. It became immediately evident that we can see anomalous populations there, such as blue stragglers; but that is not within the topic of this paper. Similarly for the sharp density peaks that we encounter in the collapsed cores of some clusters. But what *is* relevant to the subject of this paper is that the mass function is a strong function of position in the cluster. In an earlier paper (King, Sosin, & Cool 1995) we showed that the relative numbers of high- and low-mass stars differed by nearly two orders of magnitude between the center of NGC 6397 and an outer field.

The reason that this is important to know is that we tend to observe a luminosity function in one small part of a cluster, and need to convert it to a global LF. In Figure 1

† Based on observations with the NASA/ESA *Hubble Space Telescope*, obtained at the Space Telescope Science Institute, which is operated by AURA, Inc., under NASA contract NAS5-26555

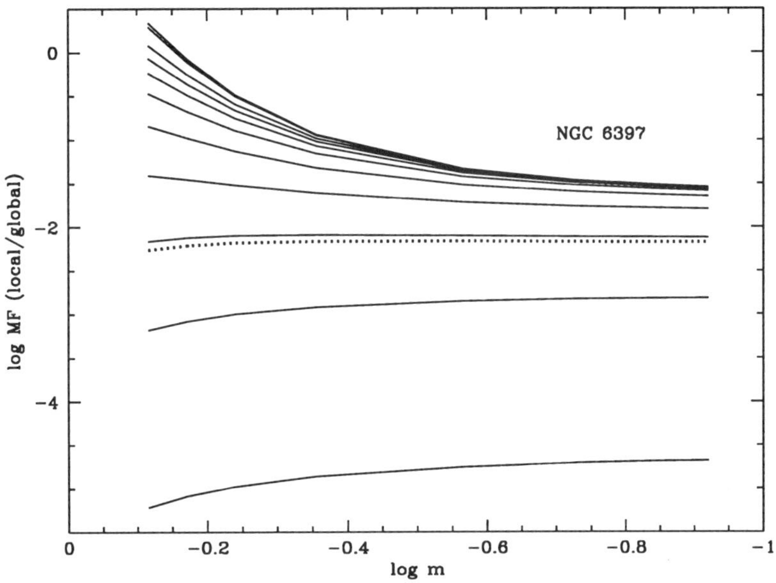

FIGURE 1. Differences between local mass functions at different radii (in projection) and the global mass function, in our model of NGC 6397. The dotted line represents the radius from which our observed MF came, while the solid lines refer to radii from the core, at the top, to the periphery, at the bottom. The relative vertical positions of the curves reflect radial differences in density, but the zero point is arbitrary, since local and global MFs have different dimensions.

we illustrate in a more general way what dynamical modeling says about the correction from local to global, as a function of stellar mass and of position in the cluster. The dotted line refers to the radius at which we derived the LF of NGC 6397, a radius at which the correction is fortuitously small.

What we observe, unfortunately, is not the mass function, which is really what is of interest astrophysically, but rather the luminosity function, which then has to be converted to an MF, with difficulties that we will discuss below. In practice the LF usually comes from a color–magnitude diagram. (In principle an LF could be derived from observations in a single color band, but in practice one band is usually not enough, because of the correction for field stars, which in some cases requires a color–magnitude diagram. And one usually wants to have the CMD for its own sake anyway.)

2. HST photometry

With sufficient care, photometry of quite high precision can be done with HST. Figure 2 shows our CMD of NGC 6397 (see Cool, Piotto, & King 1996); to emphasize the faintness of *HST* photometry, we note that these measurements do not reach *up* as high as the turnoff.

This CMD has several interesting features. The main sequence is quite narrow, with a width determined only by our photometric errors; down to $V = 24$ its FWHM is only $0^{\rm m}.04$ in color. It extends nearly to the lower limit of hydrogen burning, but field-star contamination keeps us from reaching that exciting region. We hope to solve this problem in 1997 by measuring proper motions that will allow us to separate cluster stars from the field down to the limits of our photometry. This is another area in which *HST* offers unprecedented accuracy; we were able to demonstrate, over the two-month span of

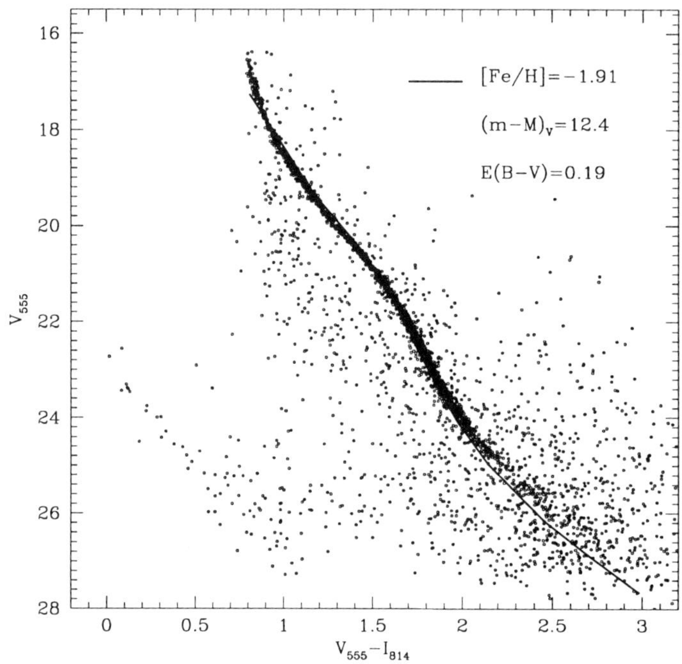

FIGURE 2. Color–magnitude diagram of NGC 6397. Note the lower main sequence, the white-dwarf sequence, and the field contamination. The solid line is a theoretical sequence from Alexander *et al.* (1996).

our observations, the 2 1/2 milliarcsecond displacement of the cluster due to its proper motion.

Another interesting aspect of the main sequence is its shape. In the "straight" part below the turnoff are two striking bends. It is not our place to explain them; we merely provide the data to the theoreticians of stellar structure and atmospheres (the latter because of the problem of converting between $\log T_{\rm eff}$ and an observational color). This turns out to be an important test for the theories. Alexander *et al.* (1996) have succeeded in fitting our MS rather well (Fig. 2), but not all groups whom we have contacted have done so well, and data such as these will help them to improve their theories.

Also of great interest is the white-dwarf sequence that slopes across the lower left of Fig. 2. Similar and more richly populated white-dwarf sequences have been observed in M4 by Richer *et al.* (1995) and by Renzini *et al.* (1996) in NGC 6752, and we now have further observations of NGC 6397 that show a much better populated sequence than the one in Fig. 2. White-dwarf sequences allow a study of the cooling rates of stars of that type, and Renzini *et al.* have shown how they can also be used to improve our knowledge of the distances of the clusters.

We return now to the problem of determining luminosity functions and mass functions. We have already mentioned the problem of correcting for field stars, which is clearly a more serious problem for some clusters than for others. But even after field-star removal, getting a LF is not simply a matter of counting. We need completeness corrections, which have to be done quite carefully. The methods are discussed elsewhere (e.g., Cool *et al.* 1996, Anderson, King, & Sosin 1995), and we will not go into them here.

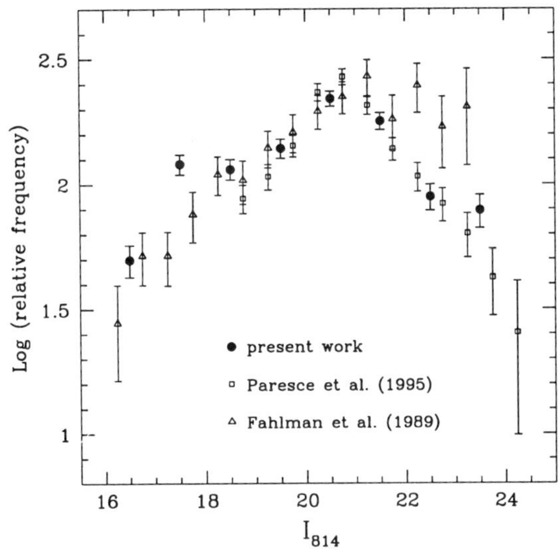

FIGURE 3. *HST* and ground-based luminosity functions of NGC 6397 (Cool *et al.* 1996). The two *HST* determinations (dots and open squares) were done by different observers in different fields but agree well with each other. The ground-based determination (triangles), which we believe to suffer from inadequate field-star corrections, would suggest significantly more faint stars.

3. The luminosity function of NGC 6397

Although *HST* does very well at the faint end of the LF, it does not provide us with adequate information at the bright end, because the images of the bright stars saturate, and also because there are too few of them in the small area covered by an *HST* field. It is therefore essential to tie to additional ground-based data. One needs the bright end of the LF in order to know how many of the more massive stars to include in a dynamical model of the cluster, and also because it can be dangerous to compare the LF's of clusters over too short a range in magnitude (see an example, below).

Figure 3, taken from our recent paper on NGC 6397 (Cool *et al.* 1996), shows an example of an *HST* LF. Here we illustrate two independent *HST* results that are in substantial agreement (except that we consider the faintest point given by Paresce *et al.* to be unreliable, because it falls in a magnitude range for which field-star corrections are not yet adequate). The ground-based LF of Fahlman *et al.* we believe to be in error at the faint end, not because it is ground-based but because their study was done in a in single color band, with field-star corrections estimated from a comparison field; as we have said, this method is less reliable than our recommended approach of making the separation by means of a CMD.

4. Conversion from luminosity function to mass function

As we have emphasized, it is the luminosity function that we observe, but it is the mass function that one really wants. Converting an LF to an MF requires knowing the mass–luminosity relation (MLR), which is poorly known for low metal abundances. And the conversion is particularly sensitive to the MLR, because it depends not only on the value but also on the slope of the relation. As we have already shown (Fig. 2) the models of

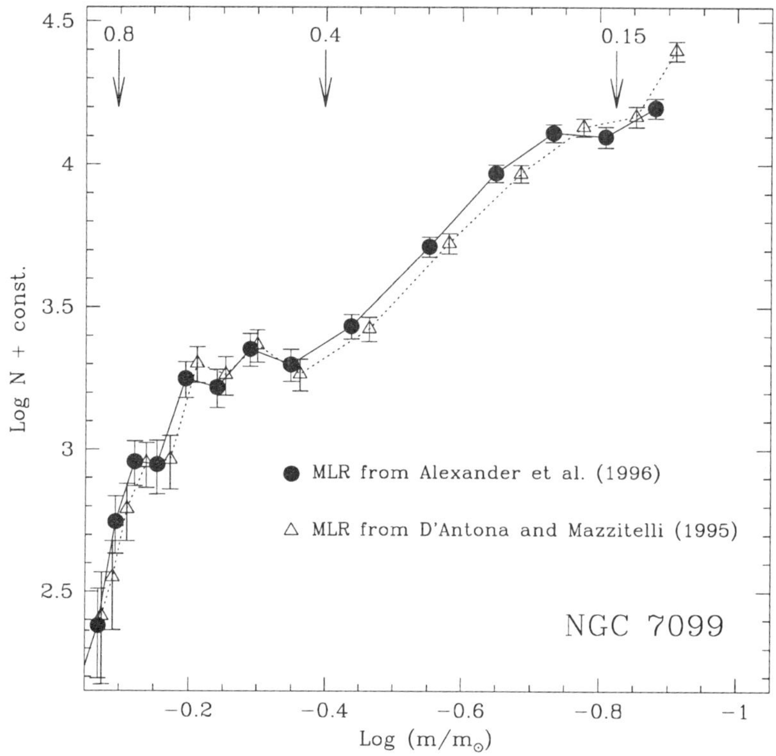

FIGURE 4. The mass function of M30 (NGC 7099), as calculated from our LF for that cluster with two different mass–luminosity relations. For convenience, three particular mass values are identified at the top on a linear scale.

Alexander *et al.* (1996) fit our MS shape in the CMD rather well. (Those models are also discussed in the present volume in the paper of Cassisi *et al.*, which immediately follows this one.) It is reasonable to hope that if models fit the luminosities, temperatures, and radii so well, they will also do well at predicting the masses. For this reason we use the MLR of Alexander *et al.* in Figure 4 to present a MF for the globular cluster M30. (We choose to illustrate M30 because it has a steeper LF—and therefore MF—than NGC 6397, a characteristic that is important for the discussion that ends this section. We also note that M30 has nearly the same metallicity as NGC 6397, so that the same MLR should apply.)

To illustrate typical differences that arise from the use of a different MLR, we also show in Figure 4 the MF that results from the same LF data if we instead use the MLR of D'Antona & Mazzitelli (1995). The difference at the low-mass end is quite significant. Such differences are particularly important because of how they impinge on the question of how much of the mass of a globular cluster (and perhaps of galaxies and of the universe?) resides in low-mass stars. The criterion for domination of the total mass by the low-mass stars is that the logarithmic slope in such a figure be steeper than −1. Even for the MLR of D'Antona & Mazzitelli, which gives the steepest MFs among all the MLRs that we have tried, the slope of the MF is somewhat short of meeting this criterion.

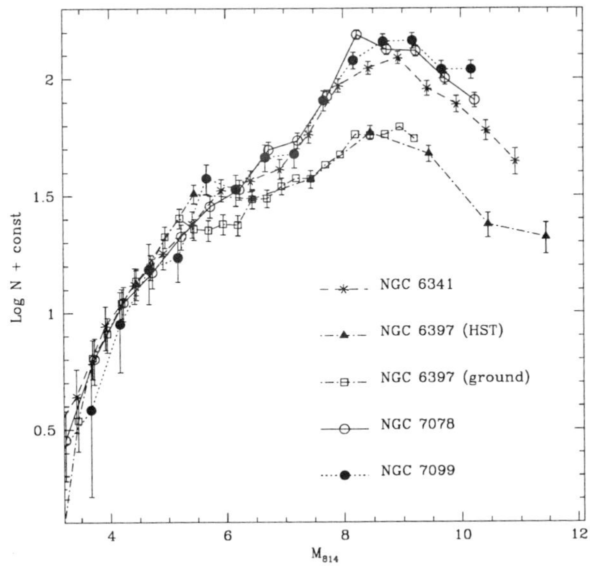

FIGURE 5. M_I LFs for NGC 6397, M15 (NGC 7078), M30 (NGC 7099), and M92 (NGC 6341). The LF for NGC 6397 has been extended up to the turnoff using ground-based data. Note the similarity between the LFs of M15, M30, and M92, and the significantly shallower slope of the LF of NGC 6397 relative to those of the other three clusters.

5. Differences between mass functions

There is one case, however, where we can bypass the MLR entirely. If clusters have the same metallicity, we can avoid MLR problems by comparing LFs directly, since the conversion would be the same for all of them. In Figure 5 is a comparison of the LFs of four clusters, all of which have metallicities between −1.9 and −2.2. It is striking how similar the LFs of M15, M30 and M92 are, and how different that of NGC 6397 is. We will discuss the probable cause of this difference, but first we return to our earlier remark about how important it is to include the brighter part of the LF. In a recent paper on the mass function of M15, De Marchi & Paresce (1995) compare its LF with that of NGC 6397 and conclude that "The remarkable similarity of the two LFs stands out even at a superficial inspection." But Figure 5 shows how different they really are. How could De Marchi and Paresce have made such an error? The answer is that their data did not go brighter than $M_I = 6.5$, and over their limited range of coverage it is possible to force some similarity between the two curves. But when we go another 3 magnitudes brighter, in Fig. 5, the difference becomes unmistakable.

6. The case of NGC 6397

But what about NGC 6397—why is it different from the other clusters? The process that suggests itself immediately is depletion of low-mass stars from the outer parts of the cluster as a result of tidal shocks. In Figure 6 is shown the Galactic orbit of NGC 6397, compared with that of M15, each followed for the same length of time. N6397 is hit much harder and more frequently by the tidal shocks of the Galactic plane, in a much denser part of the plane.

As for the other two clusters, the orbit of M92 (Dauphole *et al.* 1996) does carry it into

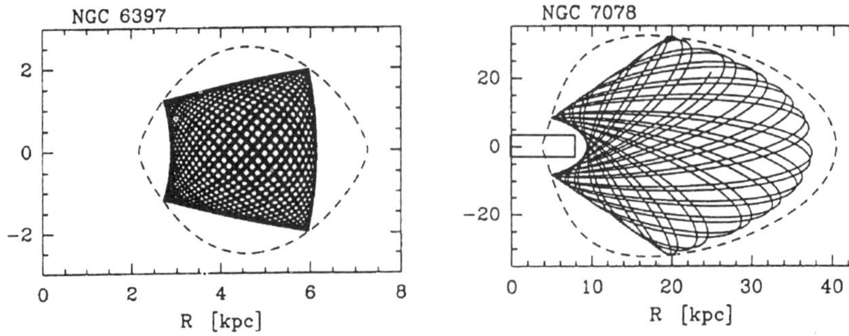

FIGURE 6. Galactic orbits of NGC 6397 and M15 (NGC 7078), in a meridional plane (the unlabeled vertical coordinate is Z), as calculated by Dauphole *et al.* (1996). Note that the orbit of NGC 6397 is plotted on a greatly expanded scale and remains within the inner part of the Galaxy; the small rectangle in the M15 figure marks the entire area that is shown in the NGC 6397 figure.

a region of strong shocking, but much less often than NGC 6397. Also, the relaxation of M92 is so much slower than that of NGC 6397 that the former cluster may be unable to feed low-mass stars out to its periphery for detachment by tidal shocks. As for M30, its orbit is not known, but its present position and radial velocity make it unlikely that M30 experiences strong tidal shocks.

The comparison of the LFs of these four clusters is discussed in more detail elsewhere (Piotto, Cool, & King 1996).

7. Piotto's correlations

This comparison raises an interesting question: could tidal shocking be the main cause of differences in the MFs of clusters? Piotto and his collaborators (Piotto 1991, 1993; Djorgovski, Piotto, & Capaccioli 1993) have been pointing out for some years that the slopes of mass functions depend more on position in the Galaxy than on any other characteristics, and one of the main consequences of position in the Galaxy is to experience strong or weak tidal shocks.

Finally, an even more interesting speculation suggests itself: Here we have three globular clusters with mass functions that are practically identical. Is it possible that all clusters are born with the same mass function, and that any differences that now exist are due to their later dynamical history? Only further study—probably with *HST*—will show.

8. Conclusion

In this paper we have shown some of the things that we are doing with regard to luminosity functions and mass functions of globular clusters. Another paper in this volume, by Piotto *et al.*, deals with color–magnitude diagrams, especially some very strange horizontal branches. And other groups are exploiting *HST* in other ways for clusters. There is no question that *HST* is now a major player in the area of stellar populations in globular clusters.

REFERENCES

Alexander, D. R., Brocato, E., Cassisi, S., Castellani, V., Ciacio, F., & Degl'Innocenti, S. 1996, A&A, submitted

Anderson, J., King, I. R., & Sosin, C. 1995, in Calibrating HST: Post Servicing Mission, eds. A. Koratkar & C. Leitherer (Baltimore: STScI), p. 300

Cool, A. M., Piotto, G., & King, I. R. 1996, ApJ, 468, 655

D'Antona, F., & Mazzitelli, I. 1995, ApJ, 456, 329

Dauphole, B., Geffert, M., Colin, J., Ducourant, C., Odenkirchen, M., & Tucholke, H.-J. 1996, A&A, in press

De Marchi, G., & Paresce, F., 1995, A&A, 304, 202

Djorgovski, S., Piotto, G., & Capaccioli, M. 1993, AJ 105, 2148

King, I. R., Sosin, C., & Cool, A. M. 1995, ApJL, 452, L33

Piotto, G. 1991, in The Formation and Evolution of Star Clusters, ed. K. Janes (San Francisco: ASP), p. 200

Piotto, G. 1993, in Structure and Dynamics of Globular Clusters, ed. S. G. Djorgovski and G. Meylan (San Francisco: ASP), p. 233

Piotto, G., Cool, A. M., & King, I. R. 1996, AJ, submitted

Renzini, A., et al. 1996, ApJL, 465, L23

Richer, H. B., et al. 1995, ApJL, 451, L17

Very Low Mass Stars: From Observations to Theory and Back

By E. BROCATO[1,2], S. CASSISI[1,3], AND V. CASTELLANI[1,2,4]

[1]Osservatorio Astronomico Collurania, Via M. Maggini, 64100 Teramo, Italy

[2]Istituto Nazionale di Fisica Nucleare, LNGS, 67100 L'Aquila, Italy

[3]Dipartimento di Fisica, Universitá de L'Aquila, Via Vetoio, 67100 L'Aquila, Italy

[4]Dipartimento di Fisica, Universitá di Pisa, Piazza Torricelli, 56100 Pisa, Italy

We briefly review the observational and theoretical investigations on Very Low Mass (VLM) stars. New evolutionary computations, extending to lower main sequence masses than previous computations already given for globular cluster stars, are presented. Comparisons with updated observational data both for globular cluster stars and VLM stars with known parallaxes are also shown. We discuss observational evidences supporting theoretical predictions concerning the Color - Magnitude (C-M) location of faint Main Sequence stars as far as concern the dependence of the lower main sequence on the metallicity.

1. Introduction

As it has been early recognized, the largest amount of stars in our Galaxy is formed by low mass stars still burning H at the center of their rather unevolved main sequence structures. However, investigating the lower main sequence has been a tantalizing goal for a long time. As a matter of the fact, for a long time observations have been hampered because the intrinsic faintness of these structures. In the mean time, it was early understood that similar simple stellar structures do require a particular sophisticated physics to describe the state of the gas at low temperatures and relatively high densities as well as the radiative opacity affecting their cool atmospheres.

The first light on that problem came from observation. After some pioneering works (see, e.g., Spinrad 1973, Upgren & Weis 1975, Veeder 1974, Mould & Hyland 1976) we were progressively facing an increasing body of observational data, as finally given by the rather large sample of stars with known parallaxes (Monet *et al.* 1992, Dahn *et al.* 1995). This sample already disclosed the occurrence of a well populated and, rather well defined sequence of stars to be interpreted as solar metallicity Low Main Sequence (LMS) objects, with the additional evidence for metal poor subdwarfs spreaded toward higher effective temperatures. Such an evidence obviously stimulated the theoretical competition to fit the data. The first attempts (Grossman 1970, Hoxie 1970, Copeland *et al.* 1970, Grossman & Graboske 1971, Hoxie 1973, Grossman *et al.* 1974) were revisited and largely improved (VandenBerg *et al.* 1983, D'Antona 1987, Dorman *et al.* 1989, Burrows *et al.* 1993, Nelson *et al.* 1993) to address the problem of deriving as more realible as possible theoretical models of VLM stars. Although the agreement with the observations improved significantly within the past decade or so, a discrepancy still exists in all these models, which usually predict too large effective temperatures for a given luminosity.

Whereas theory is with some difficulty reaching observations, following the improved efficiency of the Hubble Space Telescope one is now facing a tremendous improvement in the observation of stars at the faint end of the globular cluster main sequence. The recent HST observations (NGC6397 - Paresce *et al.* (1994) & Cool *et al.* (1996), M30 - Piotto & King 1996) have disclosed a well defined and well populated sequence which should put tight observational constraints on the evolutionary behavior of metal poor, low main sequence (LMS) stars. Thus again observation is stimulating new theoretical work.

However, the suggestion for revisiting the theoretical evolutionary scenario for VLM stars follows the evidence for recent and significant improvements in the physical inputs or, more precisely, in both the opacity and the EOS. As for the EOS, Saumon *et al.* (1995 hereinafter SCH) have recently provided an accurate EOS appropriate for dense and cool stellar objects, which should represent a relevant improvement in comparison with other EOSs suitable for dense and cool objects. It is interesting to notice that SCH EOS provides, when used in computing stellar models, results in good agreement with the ones obtained adopting the EOS by Rogers *et al.* (1996, OPAL EOS) for computing more massive stellar models. In Figure 1, we compare isochrones obtained from stellar models computed adopting the SCH EOS in the VLM stars' range and alternatively the new OPAL EOS or the 'old' Straniero (1988) EOS for the models with mass equal or greater than $0.6M_{\odot}$. The good match obtained between the OPAL EOS models and the SCH EOS ones largely improves previous evaluations.

As far as opacity is concerned, Alexander (1994) and Alexander & Ferguson (1994) computed Rosseland mean opacities, which account accurately for both atomic and molecular absorbers. Relying on such 'fresh' physical inputs, it is obviously interesting to explore whether or not the current evolutionary scenario is changed and it provides a better agreement with the new observational data. However, it is worth noting that another further complication in computing models of VLM objects is related to the treatment of the atmosphere (Burrows *et al.* 1993, Dorman *et al.* 1989, Baraffe *et al.* 1995). In the atmosphere of these cool stars one expects that convection can deeply affects atmospheric layers for two reason: i) the adiabatic gradients are small due to the dissociation of H_2 and, ii) the opacity can became large due to the presence of molecules. Therefore, in structures less massive than about $0.5\,M_{\odot}$ the atmosphere can become unstable against convection at very low optical depth ($\tau \approx 0.01$) and the use of normal $T(\tau)$ relation is not more valid. The solution of this problem relies on the possibility of incorporating model atmospheres (Allard & Hauschildt 1995, Brett 1995, Brett & Plez 1993) in stellar evolutionary codes (see Baraffe *et al.* 1995 and Chabrier *et al.* 1996). However, significative discrepancies do still exist between different sets of model atmosphere (Bessel 1995) due mainly to the different techniques adopted in computing the opacity. So we choose for the moment to keep on the Eddington approximation, i.e. to adopt a $T(\tau)$ relation (in particular the Krishna Swamy (1966) one) which describes the temperature stratification in the outermost layers. As it will be shown in the following, we think that this assumption has to be regarded as a first order, of course, but not to bad approximation.

2. Theoretical VLM stellar structures.

Selected sequences of stellar models have been computed to cover the range of hydrogen burning stars below $0.8\,M_{\odot}$. Computations have been performed adopting the following metallicities: Z=0.0001 - 0.0003 - 0.0006 - 0.001 - 0.006, and a Helium abundance (Y) equal to 0.23 everywhere. A sequence of models with solar metallicity and $Y = 0.28$ has been also computed. According to the calibration of solar models (Salaris & Cassisi 1996), in all computations it has been adopted a mixing length parameter as given by $\alpha = 2.2$. The models have been transferred from the theoretical plane to the observational one by adopting bolometric corrections and color-temperature relation by Allard & Hauschildt (1995) and by Kurucz (1992). Figure 2 reports the location of 10 Gyr sequences in the observational ($M_V, V - I$) plane for the labelled assumptions on the metallicity. Further

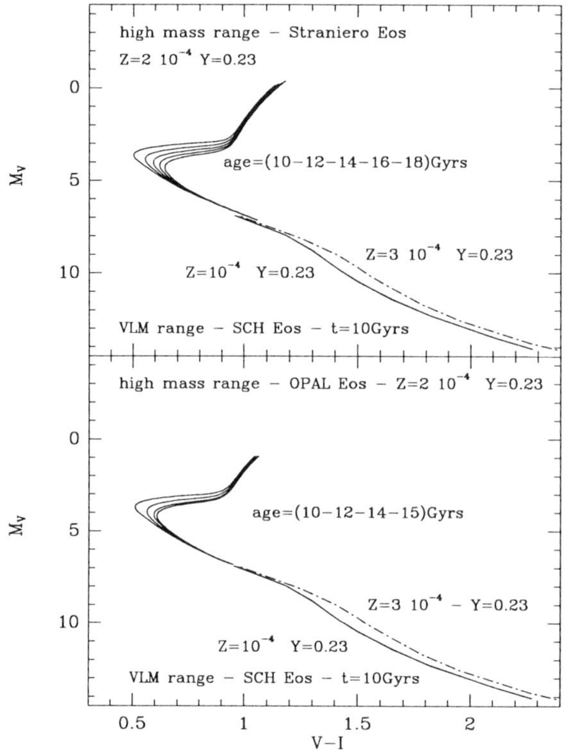

FIGURE 1. Comparison between isochrones obtained using the OPAL Eos and the SCH Eos (*Bottom panel*) and, isochrones computed adopting the Straniero Eos and the one of SCH (*Top panel*).

details on these models, as well as a detailed description of VLM evolutionary features can be found in Alexander *et al.* (1996). However in the following section we wish to address the attention of the reader toward two unexpected evolutionary properties of VLM stellar models.

2.1. *Some interesting features....*

As it is already known, during the H burning phase stars less massive than about $0.3\,M_\odot$ are fully convective. Analizing in detail the behavior of similar stellar structures one finds a peculiar behaviour of the density against the time. As a matter of fact, in "normal" models (more massive than $0.3\,M_\odot$) the consumption of central H drives a continuos increase of both central temperature and central density. However, in fully convective structures the central density keeps decreasing all along the major phase of H burning. It has been found (Alexander et al 1996) that such 'anomalous' behavior is related to the increase of the molecular weight in the whole structure due to both the convective mixing and the production of Helium by nuclear reactions. It may be worth noting that such behavior of the central density is quite similar to the one of central density during the first phases of He burning in more massive stars.

By discussing the effect of the age on the HR diagram location of VLM stars, one finds the well known feature that age plays a negligible role for structures older than about 10 Gyr. But one can notice the further curious property for which the $0.3M_\odot$ model appears as the less affected by the variation of the age, while less massive models show a progressively increasing sensitivity to the age. One understands such an occurrence

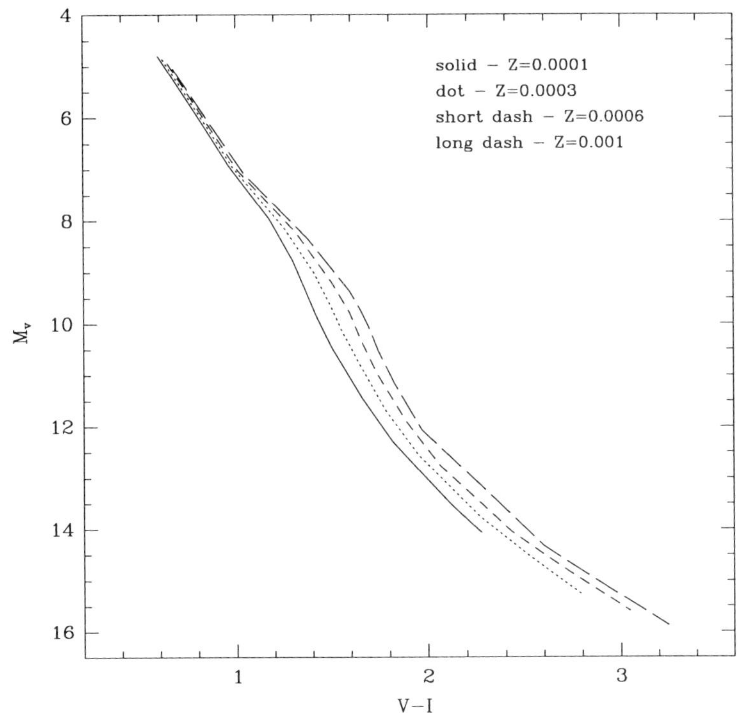

FIGURE 2. Theoretical expectations for the (V,$V-I$) color - magnitude diagrams of faint MS stars for the labeled assumtions on the metallicity.

when accounting for the role played by the time evolution of the ^3He abundance in these stars (for more details see Alexander *et al.* 1996).

3. Comparison with the observational data.

3.1. *NGC6397 and metal poor VLM models.*

In figure 3, we show the comparison between the HST data for NGC6397 by Cool *et al.* (1996) and our theoretical prescriptions for two selected metallicity. The fine agreement obtained between the observational data and the theoretical sequence for [Fe/H] = −1.91 appears largely satisfactory. Note that the agreement has been achieved for values of both reddening and distance modulus well compatible with the current estimates.

3.2. *VLM with known parallaxes and solar VLM models.*

The sample of stars with known parallaxes has been for a long time a fundamental tool to check the reability of the theoretical scenario for VLM stellar objects. In Figure 4, we show the most complete, presently available, color - magnitude diagram for such stars from the extensive parallax survey being carried out by the US Naval Observatory (Monet *et al.* 1992, Dahn *et al.* 1995). In the same figure, we report also our theoretical predictions for solar composition VLM stars together with similar data as presented by by D'Antona & Mazzitelli (1994, 1996) and by Baraffe *et al.* (1995). These last Authors use the same EOS and opacity evaluations as in our computations, but adopt model atmosphere as outer condition.

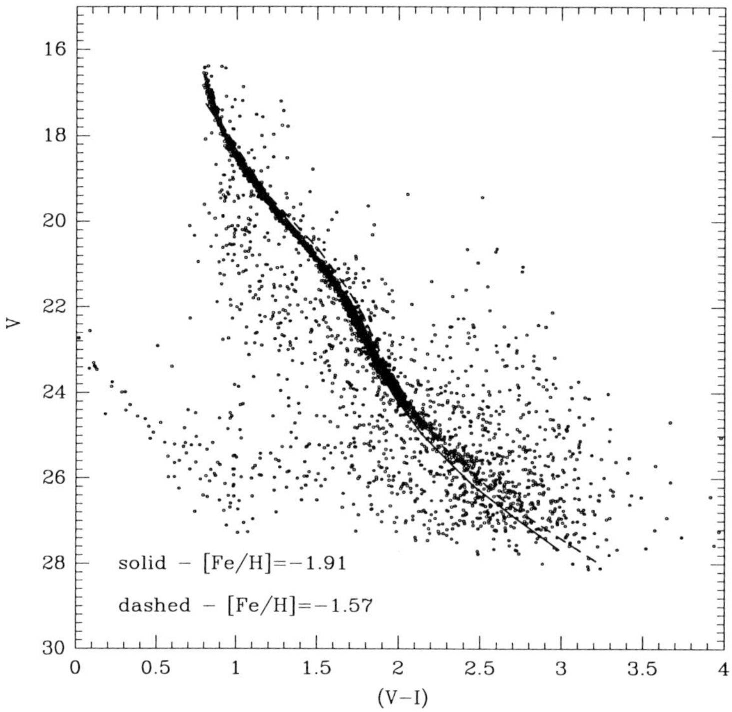

FIGURE 3. (V, $V-I$) CM diagram for lower main sequence stars in the globular cluster NGC6397. The comparison with theoretical sequences for $t = 10\,\mathrm{Gyr}$ and for the two labeled assumptions about the star metallicity has been performed adopting for the cluster $(m-M)_V = 12.4$, $E(V-I) = 0.19$.

All models fail in accurate reproducing the location of the lowest MS masses. This occurrence is probably due to the difficult task of taking into the right account the proper influence of metals on both the opacity and the EOS. This occurrence suggests that the knowledge of the evolutionary scenario for VLM stars is not yet complete, so remaining a tantalizing goal for stellar theoricists.

4. The ranking of low luminosity Main Sequence stars with the metallicity.

As it is well known, a fundamental feature of the Red Giant Branch (RGB) in Galactic globular is its sensitive dependence on the metallicity of the stars in the cluster: the cluster RGBs become bluer and steeper when the metallicity of the cluster decreases. However, all recent theoretical works on VLM stars (Baraffe *et al.* 1995, D'Antona & Mazzitelli 1996, Alexander *et al.* 1996) have shown that the location in the H-R diagram of such stars is also expected to depend sensitively on the metallicity, again becoming bluer and steeper when the metallicity is decreased (see Figure 2).

In the following, we will report the evidence that such a prediction on the Low Luminosity Main Sequences is eventually supported by recent accurate photometries of galactic globular cluster (for more details see Brocato *et al.* 1996). We have selected a little sample of clusters with accurate photometries for the Main Sequence, namely: NGC6397 (Cool *et al.* 1996), M30 (Piotto & King 1996), M4 and M55 (Fahlman *et al.* 1996). In

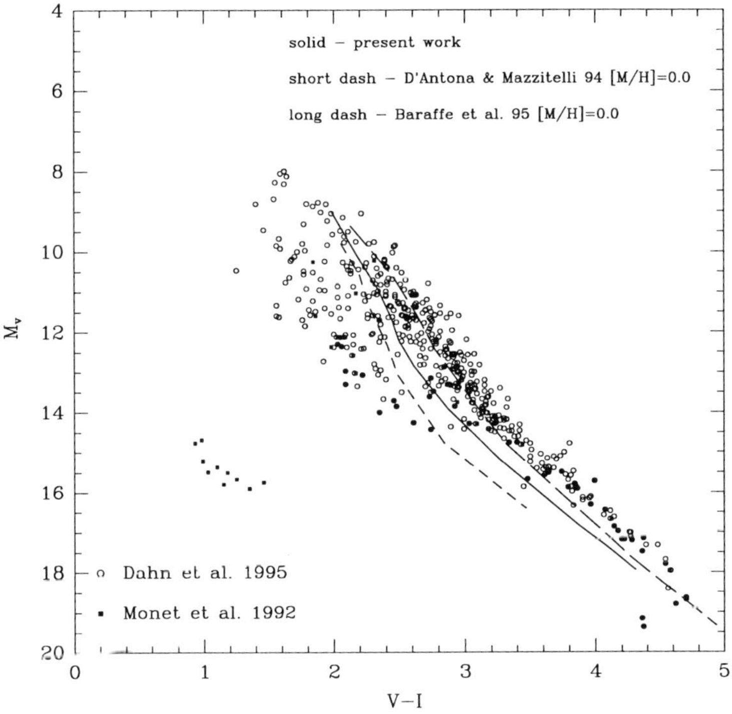

FIGURE 4. (M_V,$V-I$) CM diagram for faint stars with known parallaxes from Monet et al. (1992) and Dahn et al. (1995) with superimposed theoretical distributions for solar metallicity models.

Cluster	[Fe/H]	$(m-M)_V$	$E(B-V)$
M4	-1.19	12.65	0.40
NGC6397	-1.82	12.50	0.16
M55	-1.82	14.10	0.14
M30	-1.91	14.85	0.01

TABLE 1. Values of the metallicity, distance modulus and reddening for the selected globulars

table 1 we report some important parameters for these clusters. Metallicities have been taken from the compilation provided by Carretta & Gratton (1996) for all cluster, but M55, for which we adopt the value reported in the work by Chaboyer & Kim (1995). As far as the distance modulus and reddening, we adopt the current values in the literature. According to the data in Table 1 and accounting for a possible indetermination on the metallicity of about 0.10dex, the clusters NGC6397, M55 and M30 may well have the same metallicity. As a matter of the fact, by overplotting the C-M diagrams of these clusters one finds a nice global ovelap between the three stellar distributions, once the differential differences in both reddenings and distance moduli have been accounted for.

As a most interesting point, let us compare the metal poor cluster NGC6397 with the more metal rich M4. Figure 5 shows that at the fainter luminosities the M4 Main Sequence clearly develops toward much redder color than NGC6397. The amount of such effect is so large that it can not be related to observational errors and/or indeterminations

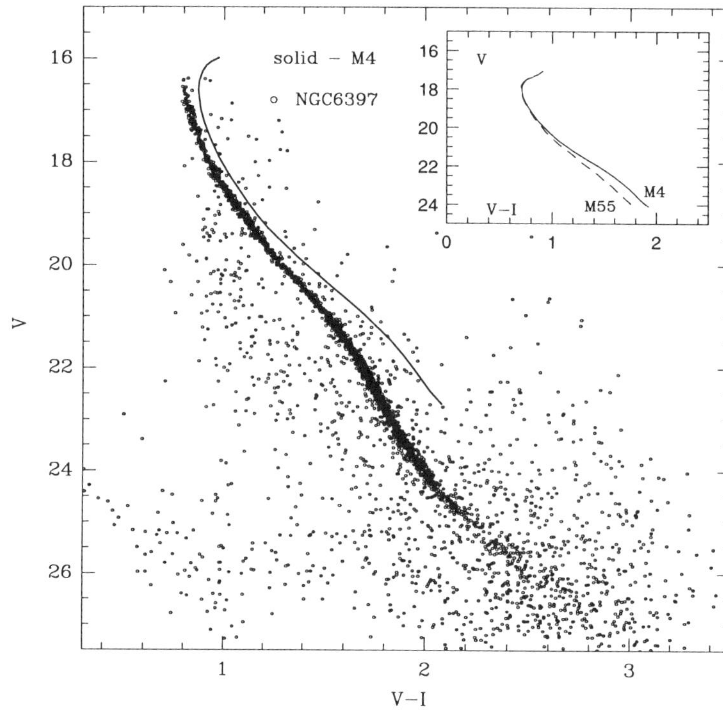

FIGURE 5. Comparison between the C-M diagrams of NGC6397 and M4 with M4 reduced to the same reddening and distance modulus of NGC6397. In the box, we present the comparison between the mean lines of M4 and M55 (see text for more details).

in the trasformation of the HST colors. To go deeper in this investigation, in the same figure we compare the two clusters M4 and M55 simply applying a shift to the stellar distributions for obtaining a close overlap in the Turn Off region. Once again we find that the MS slope of M4 is lower than the M55 one.

So, as expected on theoretical grounds, we conclude that the Low MS of globular clusters is a sensitive indicator of the cluster metallicity as the Red Giant Branch.

REFERENCES

Alexander, D. R. 1994, private comunication

Alexander, D. R., Brocato, E., Cassisi, S., Castellani, V., Ciacio, F. & Degl'Innocenti, S. 1996, A&A, in press

Alexander D.R. & Ferguson J.W. 1994, ApJ 437, 879

Allard, F. & Hauschildt, P. H. 1995, ApJ, 445, 433

Baraffe, I., Chabrier, G., Allard, F. & Hauschildt, P. H. 1995, ApJL, 446, L35

Bessel, M. S. 1995, in Proceeding of the ESO workshop "The bottom of the Main Sequence and Beyond", ed. C.G. Tinney, p. 123

Brett, J. M. 1995, A&A, 295, 736

Brett, J. M. & Plez, B. 1993, ProcASA, 10, 250

Brocato, E., Cassisi, S. & Castellani, V. 1996, ApJL, submitted to

Burrows, A., Hubbard, W. B., Saumon, D. & Lunine J. I. 1993, ApJ, 406, 158

Carretta, E. & Gratton, R.G. 1996, A&ASS, in press

Chaboyer, B. & Kim, Y-C. 1995, ApJ, 454, 767

Chabrier, G., Baraffe, I. & Plez, B. 1996, ApJL, 459, L91

Cool, A. C., Piotto, G. & King, I. R. 1996, ApJ, in press

Copeland, H., Jensen, J. O. & Jorgensen, H. E. 1970, A&A, 5, 12

Dahn, C. C., Liebert, J., Harris, H. C. & Guetter, H. H. 1995, in Proceeding of the ESO workshop "The bottom of the Main Sequence and Beyond", ed. C.G. Tinney, p. 239

D'Antona, F. 1987, ApJ, 320, 653

D'Antona, F. & Mazzitelli, I. 1994, ApJS, 90, 467

D'Antona, F. & Mazzitelli, I. 1996, ApJ, 456, 329

Dorman, B., Nelson, L. A. & Chau, W. Y. 1989, ApJ, 342, 1003

Fahlman, G.G., Mandushev, G., Richer, h.B., Thompson, I.B. & Sivaramakrishnan, A. 1996, ApJ, 459, L65

Grossman, A. S. 1970, ApJ, 161, 619

Grossman, A. S. & Graboske, H. C. 1971, ApJ, 164, 475

Grossman, A. S., Hays, D. & Graboske, H. C. 1974, A&A, 30, 95

Hoxie, D. T. 1970, ApJ, 161, 1083

Hoxie, D. T. 1973, A&A, 26, 437

Krishna-Swamy, K. S. 1966, ApJ, 145, 174

Kurucz R.L. 1992, in Barbuy B., Renzini A. (eds.), IAU Symp. n. 149, "The Stellar Populations of Galaxies", Kluwer, Dordrecht, p. 225

Monet, D. G., Dahn, C.C., Vrba, F. J., Harris, H. C., Pier, J. R., Luginbuhl, C. B. & Ables, H. D. 1992, AJ, 103, 638

Mould, J. R. & Hyland, A. R. 1976, ApJ, 208, 399

Nelson, A. L., Rappaport, S. & Chiang, E. 1993, ApJ, 413, 364

Paresce, F., De Marchi, G. & Romaniello, M. 1995, ApJ, 440, 216

Piotto, G. & King, I. R. 1996, private comunication

Rogers F.J., Swenson F.J. & Iglesias C.A. 1996, ApJ 456, 902

Salaris, M. & Cassisi, S. 1996, A&A, 305, 858

Saumon, D., Chabrier, G. & Van Horn, H. M. 1995, ApJS, 99, 713

Straniero O. 1988, A&AS 76, 157

Upgren, A. R. & Weis, E. W. 1975, ApJL, 197, L53

VandenBerg, D. A., Hartwick, F. D. A., Dawson, P. & Alexander, D. R. 1983, ApJ, 266, 747

Veeder, G. J. 1974, AJ, 79, 1056

Three Steps on the Age Ladder

By ALVIO RENZINI

European Southern Observatory,
D-85748 Garching b. München, Germany, and
Department of Astronomy, University of Bologna, Italy

This paper reports the main results of three projects that combine optical and near-IR data obtained partly with the *Hubble Space Telescope* (HST) and partly with ESO telescopes. The first project deals with the determination of an accurate distance and absolute age of the nearby, metal poor globular cluster NGC 6752. This was achieved with a new method based on fitting the cluster white dwarf sequence to a local sequence of white dwarfs with known trigonometric parallax. An age of $15 \pm 1.5 \pm 0.5$ Gyr is derived, with values towards the minimum in this range being obtained when models including helium diffusion, Livermore opacities and equation of state are used. The second project deals with the dating of the Galactic bulge. Here the luminosity difference between the horizontal branch and the turnoff is used to get the age of two metal rich clusters in the bulge, compared to the age of NGC 6752. Within current uncertainties of the HB luminosity-metallicity relation the age of the two clusters, and of the bulge as a whole, is found to be the same as that of NGC 6752, indicating that the bulk of the Galactic spheroid formed some 15 Gyr ago, i.e. very early in the evolution of the Galaxy. The third project moves from IR observations of one of the same bulge clusters, and shows that its brightest AGB stars are as bright as the brightest stars observed in M32, from which one concludes that a population component significantly younger than ~ 15 Gyr is no longer required to account for the presence of such bright stars in M32.

1. Introduction

In the summer of 1970 Icko published a seminal paper in *Scientific American.* It was condensing his efforts towards the determination of the age and the helium content of galactic globular clusters (GC), describing with colorful figures the R-method to get the helium, and the $\Delta V_{\rm TO}^{\rm HB}$ method to get the age. I was fascinated by the beauty of such a fully theoretical approach, which is formally independent of distance and reddening, hence protected against the corresponding errors. While Icko left globular clusters soon after 1970, in successive versions his two methods have been applied and refined by at least three generations of astronomers in the following 26 years, both on the theoretical side and on the observational front. Here I present some further development in the same direction.

Of course, for the result to be correct the theoretical models must be "right", while we know that they are based on several assumptions and approximations. Occasionally, however, tables of stellar models are used as if they were like mathematical tables of, say hyperbolic functions: i.e. taking them at face value, without paying much attention to the (systematic) errors they may propagate. In the specific case, it is worth instead being aware of how uncertainties and approximations in the model input physics may affect the two functions involved in the age determination. These are the theoretically predicted turnoff (TO) luminosity-age relation $[t = t(M_{\rm V}^{\rm TO}, Y, Z)]$, and the ZAHB luminosity at nearly the TO temperature $[(M_{\rm V}^{\rm HB}(Y, Z)]$.

The age-TO luminosity relation is affected by errors in opacity, reaction rates, equation of state, and by the possible occurrence of processes such as deep mixing, diffusion, core cooling by WIMPS, etc ... Rotation can instead be excluded from this list of items, because the observed tightness of the TO locus implies that – at least around TO – all stars evolve at the same pace. Other physical uncertainties affect instead M_V^{HB}, which in turn has a strong dependence on the core mass at the central helium flash (M_c). Now core rotation might become important, along with neutrino losses, and possible axion cooling, while radiative opacity plays a minor role in determining M_c. Not so thermal conductivity, as the core is highly degenerate prior to the helium flash.

The determination of accurate ages plays a crucial role in modern observational cosmology. It offers an independent estimate of the age of the universe, and provides a dating for the fossil records that help to reconstruct the past history of galaxies back to their formation epoch. We are then interested in reducing as much as possible both the observational and the systematic errors affecting the dating procedures. This latter requirement translates into a simple direction: to get age estimates as cogent as possible one should make the least possible use of stellar models! Of course, they cannot be entirely eliminated, as after all the use of the age-TO luminosity relation cannot be avoided, because it ultimately represents the only *clock* in the dating procedure. The use of theoretical HB models – which in fact played the role of a standard candle in Icko's 1970 paper – is however not essential, as other (empirical) standard candles can be used instead.

In this presentation to the meeting celebrating Icko's 65th birthday I will report some results of a wide collaborative effort to improve upon absolute dating of Galactic GCs, the Galactic bulge, and the stellar population of M32. I will emphasize the methodological connections that link – indeed as in a ladder (cf. Renzini, 1992) – the three corresponding steps.

2. The Absolute Age of the Halo from a Precise Measurment

The age of Galactic GCs must be estimated by using theoretical relation between age and the TO luminosity. A simple relation that fits the 'old' isochrones of VandenBerg & Bell (1985):

$$\mathrm{Log}\, t_9 \simeq -0.41 + 0.37\, M_V^{TO} - 0.43\, Y - 0.13\, [\mathrm{Fe/H}], \qquad (1)$$

allows to estimates the contribution of the uncertainty in each input quantity in establishing the final uncertainty in age. Here t_9 is the age in Gyr units, Y the helium abundance, [Fe/H] the iron abundance in standard notations, and M_V^{TO} the absolute visual magnitude of the main sequence turnoff. In turn, $M_V^{TO} = V^{TO} -$ mod, where V^{TO} (the TO apparent magnitude) is the directly *observable* quantity, and mod is the cluster distance modulus. For a long time distances to GCs have been estimated using two kinds of standard candles: RR Lyrae and subdwarfs. So far these methods have delivered distances to clusters that are uncertain by about 0.25 mag in the modulus, which implies a $\sim$ 22% error in the derived cluster age ($\sim$ 3 Gyr for an age of 15 Gyr). All other input quantities convey substantially smaller errors. The current accuracy in determining V^{TO} is now better than 0.1 mag, which implies a less than $\sim$ 9% error in age. The primordial helium is known to within $\sim$ 2%, which generates a negligible $\sim$ 2% error in age. The metal content of the best studied clusters is uncertain by perhaps 0.3 dex (most of it being systematic), which translates into a $\sim$ 9% uncertainty in age. One has also to pay attention to the *composition* of metallicity (e.g. enhanced [O/Fe], or [α/Fe]), a point to which I shall return later.

Hence, the first concern when deriving accurate cluster ages is to reduce the uncertainty in the distance. In the traditional methods (using RR Lyrae and subdwarfs) the metallicity of both the standard candle stars and of the clusters need to be measured, and the resulting distance is affected by errors in metallicity which can dominate the age error budget (Renzini 1991). Before Hipparcos only five subdwarfs could be used to get GC distances by the main sequence fitting to local subdwarfs with good trig parallaxes. After Hipparcos this number is expected to increaase enormously, thus giving more weight to this method. However, the systematic uncertainty in the metallicity scale is difficult to quantify, and will continue to inprint itselp in the derived GC distances, hence in their inferred ages. This problem can be overrun if cluster and subdwarf abundances are measured with strictly the same method. Systematic effects may also arise when using RR Lyrae with a distance scale calibration based on the Magellanic Clouds (e.g., Walker 1992), as there is no guarantee that the α-element enhancement in the MC RR Lyrae is the same as in the Galactic spheroid.

The white dwarf (WD) method presented here is instead free from these limitations. It consists of fitting the WD cooling sequence of a GC to an appropriate empirical cooling sequence constructed using local WDs with well determined trig parallax, and observed with the same instrument (here WFPC2) used for the cluster WDs. The method does not involve metallicity determinations, as WDs have virtually metal free atmospheres. Moreover, their local space density of is much higher than that of subdwarfs, and therefore accurate trig parallaxes can be obtained for a potentially much larger sample of calibrators. However, cluster WDs are very faint, with $V \gtrsim 24$ even in the closest GCs. HST was therefore required to detect them and to obtain photometric data of adequate accuracy.

The result is shown in Fig. 1, reproduced from Renzini *et al.* (1996, hereafter Paper I) to which I refer for all technical details. The resulting distance modulus, corrected for reddening, is 13.05±0.05. For the cluster parameters we adopted from the literature $V_{\rm TO} = 17.4 \pm 0.07$ and $A_{\rm V} = 0.12 \pm 0.06$, [Fe/H]=–1.54±0.3, and $Y = 0.23 \pm 0.02$. Thus, $M_{\rm V}^{\rm TO} = 4.23$, and entering equation (1) one gets a cluster age of 18.0 Gyr. This assumes solar proportions for the cluster heavy elements. Under the same assumption, the models of Salaris et al. (1993) [see their equation (6)] yield an age of 17.8 Gyr. However, there is now ample evidence that the α-elements (i.e., O, Ne, Mg, Si) are enhanced relative to iron in metal poor halo stars and clusters, with $[\alpha/\rm Fe] = 0.4$–0.6 at the metallicity of NGC 6752 (e.g., Bessell, Sutherland, & Ruan 1991).

Salaris et al. have shown that what matters in the age-$M_{\rm V}^{\rm TO}$ relation is the overall heavy element abundance $[\rm M/H] = [\rm Fe/H] + \rm Log\,(0.638 f_\alpha + 0.362)$, where $f_\alpha = \rm dex\,[\alpha/Fe]$, rather than the detailed distribution. Thus, for $[\alpha/\rm Fe] = 0.4$ and 0.6 the models of Salaris et al. give a cluster age of 16.1 and 15.3 Gyr, respectively. For $[\alpha/\rm Fe] = 0.6$ (i.e., $[\rm M/H] = -1.1$) we estimate an age of 15.2 Gyr from Fig. 7 in Bergbush & VandenBerg (1992), once more showing the good agreement between different sets of stellar models when identical cluster parameters and assumptions are adopted. This latter estimate assumes no helium diffusion inside stars during their main sequence lifetime. When helium diffusion is allowed, Bergbush & VandenBerg models yield an age of 14.0 Gyr for $[\rm M/H] = -1.1$. The error to attach to these age determinations can be estimated from the propagation of the errors in the various input paramenters $V_\circ^{\rm TO}$, $(m - M)_\circ$, [M/H], and Y. We estimate the overall error to be $\sim 10\%$, with a systematic uncertainty related to helium diffusion of $\sim \pm 0.5$ Gyr. In summary, adopting the central value [α/Fe]=0.5, one obtains an age of $15 \pm 1.5 \pm 0.5$ Gyr (random plus systematic uncertainties).

The wide interest in getting accurate ages has continued to prompt efforts to improve the input physics to evolutionary models, and to use it to run new evolutionary sequences. These efforts include the isochrones computed by Chaboyer & Kim (1995) with the most recent *Livermore* opacities and equation of state (EOS), and Chaboyer (private communication) informs me that the inclusion of Coulomb-interaction effects on the

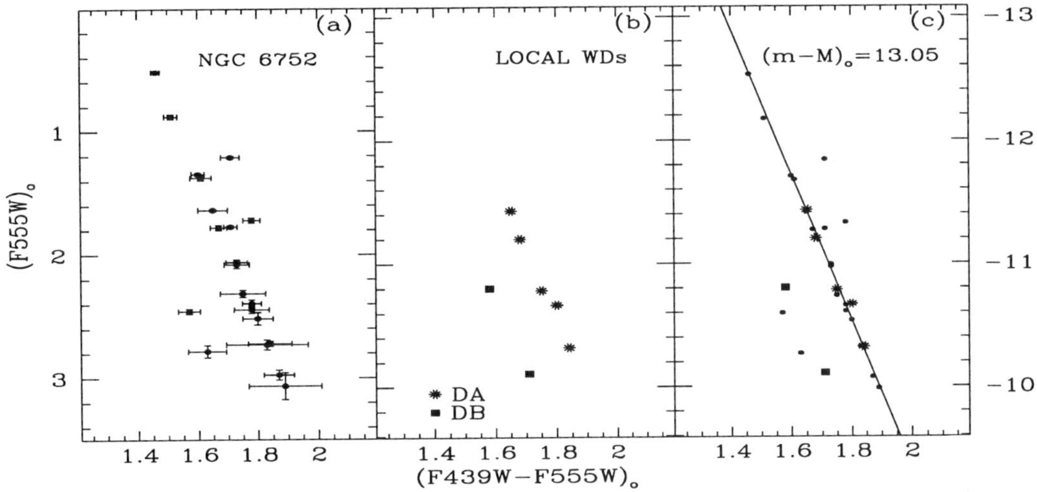

FIGURE 1. (a) The instrumental color-magnitude diagram for the cluster white dwarfs detected on the CCD chip #4 of the WFPC2; (b): the instrumental *absolute* color-magnitude diagram for the local, calibrating white dwarfs (from WF4 data only) of known trig parallax that are listed in Table 1 (in order of decreasing F555W luminosity to allow cross identification). WDs of the DA and DB varieties are represented by different symbols; (c): the instrumental color-magnitude diagram of the cluster and local WDs, with the former ones having been shifted in magnitude to match the local sequence. This operation delivers the distance modulus of the cluster: $(m - M)_\circ = 13.05$. The straight line is a linear fit to the cluster WD sequence (From Paper I).

EOS has the effect of reducing the estimated ages by about 1 Gyr. Salaris *et al.* models did include such effects, so there remains to be seen how such models compare with those of Chaboyer & Kim. In an updated version of Salaris *et al.* models, Chieffi & Straniero (1996) have run new sequences with with the most recent Livermore opacities and EOS (but without diffusion). They give

$$M_{\rm V}^{\rm TO} = 2.0550 + 2.2707\,{\rm Log}\,t + 0.4968\,[{\rm M/H}] + 0.0510\,[{\rm M/H}]^2 \qquad (2)$$

when only the new opacities are incorporated, and

$$M_{\rm V}^{\rm TO} = 2.0756 + 2.3191\,{\rm Log}\,t + 0.5253\,[{\rm M/H}] + 0.0568\,[{\rm M/H}]^2 \qquad (3)$$

when including also the new EOS. With $M_{\rm V}^{\rm TO} = 4.23$, $Y = 0.23$, and $[{\rm M/H}] = -1.1$ the first relation gives an age of 14.8 Gyr (0.5 Gyr younger than Salaris *et al.*), and the second relation gives 14.1 Gyr. When allowing for diffusion, a further ~ 1 Gyr reduction could be obtained, reducing the age estimate of NGC 6752 near 13 Gyr. With halo globulars having formed at least 1–2 Gyr after Big Bang, this places a lower limit of 14–15 Gyr for the age of the Universe, with obvious implications.

We are now at an exciting time on the quest for accurate GC ages. Besides the addition of the WD method, the traditional methods for GC distances are being revitalized, with HST providing a direct calibration of the $M_{\rm v}^{\rm HB}$-[Fe/H]relation from the observation of GCs in M31 (e.g., Fusi Pecci *et al.* 1996), and Hipparcos greatly expanding the sample of subdwarfs with good trig parallax. The comparison of the results of the three methods will be most interesting. In parallel, a detailed comparison of the $M_{\rm V}^{\rm TO}(t, Y, [{\rm M/H}])$ relations from different sets of isochrones will allow us to better understand the possible systematic effects brought in by the theoretical models.

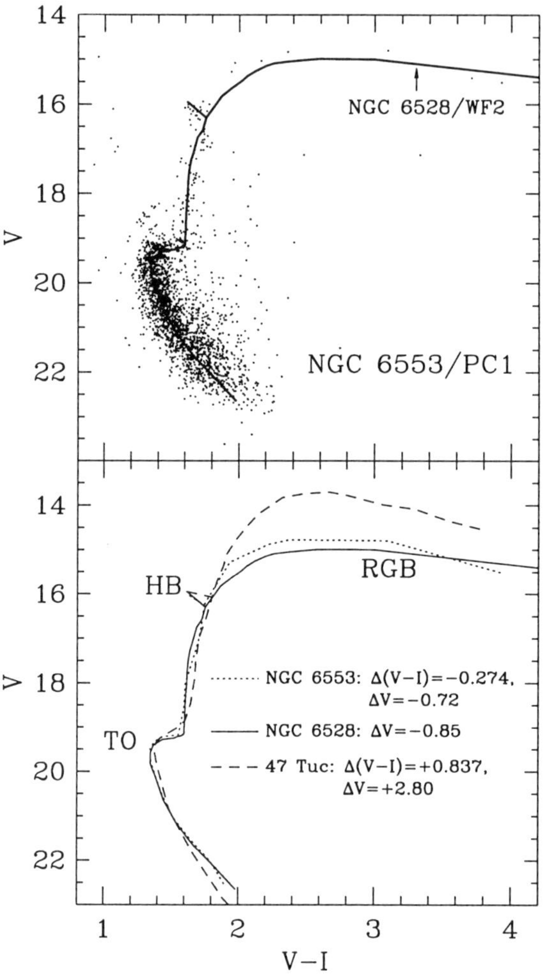

FIGURE 2. Upper panel: the color-magnitude diagram of the bulge globular NGC 6553 for stars in the PC field of WFPC2, with superimposed the mean locus of the cluster NGC 5528, as sampled by chip #2 of WFPC2. Lower panel: The mean loci of NGC NGC 6528, NGC 6553, and 47 Tuc. Each locus has been shifted as indicated, in order to bring into coincidence the end of the HB (from Paper II).

3. The Age of the Bulge Relative to the Halo

Having established the absolute age of one cluster, by using Icko's $\Delta V_{\rm TO}^{\rm HB}$ method one is now in the position of getting relative ages to it for other clusters, and in a fashion that avoids depending on HB models. This approach was used by Ortolani *et al.* (1995, Paper II) to date two metal rich GCs in the Galactic bulge and the Galactic bulge field population itself. Sufficiently deep HST photometry was then obtained for two metal rich clusters of the bulge ([Fe/H] $\simeq$ 0.0), namely NGC 6528 and NGC 6553. Fig. 2 shows that the color-magnitude diagrams of the two clusters are remarkably identical, from the MS all the way to the tip of the RGB. Moreover, their $\Delta V_{\rm TO}^{\rm HB}$ values are the same, or even slightly larger, than that of the more metal poor cluster 47 Tuc ([Fe/H] = −0.7). Fig. 3 compares the $\Delta V_{\rm TO}^{\rm HB}$'s of the two bulge clusters with those of several halo clusters, including NGC 6752 and 47 Tuc. This comparison implies that the two bulge clusters are nearly as old as the halo clusters, and demonstrates that the bulge underwent rapid chemical enrichment to solar abundance and beyond, very early in the evolution of our Galaxy. As shown in Fig. 3, it is not yet possible to assess whether the two bulge clusters are slightly older or younger than the halo GCs, because of the still surving uncertainty in the $M_{\rm v}^{\rm HB}$ – [Fe/H] relation.

Moreover, Fig. 4 shows that the MS luminosity function of the two clusters is indistinguishable to that of the stars in Baade's Window (the low-reddening bulge field at $\sim 4^{\circ}$ from the Galaxy center), that was obtained from observations with the ESO NTT with superb seeing ($0\rlap{.}''4$). From this Ortolani *et al.* infer that the whole bulge formed quickly, some 15 Gyr ago, and set an upper limit of $\sim$ 10% by number to any intermediate age population in the bulge. Indeed, a larger proportion of intermediate age stars would have resulted in a shallower fall off of the bulge luminosity function around TO (i.e., for $20.5 \gtrsim V \gtrsim 19.5$), where instead it coincides with that of the cluster. In conclusion, the *fossil* evidence tells us that the whole Galactic spheroid is pretty old indeed.

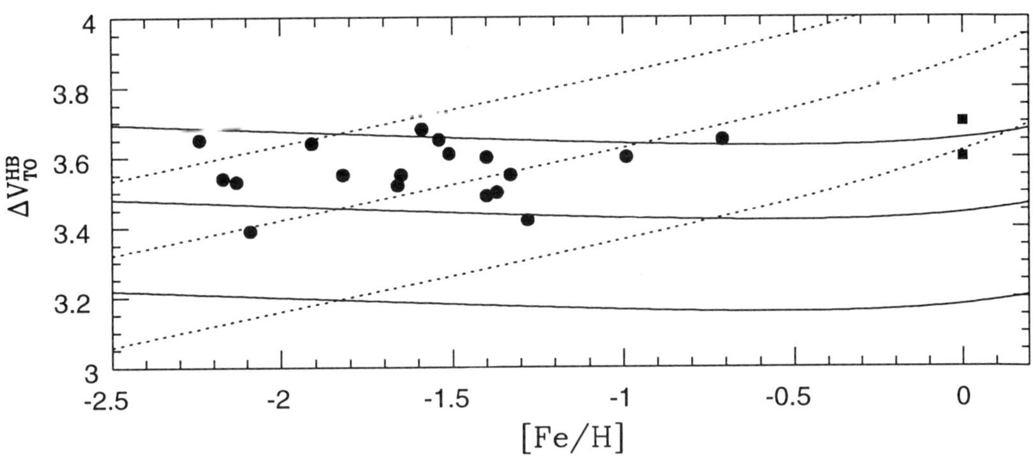

FIGURE 3. The luminosity difference between the HB and the turnoff $\Delta V_{\rm TO}^{\rm HB}$ for a representative set of halo globular clusters (filled circles) and for NGC 6528 and NGC 6553 (filled squares). For this display [Fe/H]=0.0 and 0.1 has been adopted for these two clusters. The lines refer to ages of 18, 15, and 12 Gyr (upper, middle, and lower lines, respectively), with these assumptions: 1) turnoff luminosity-age relation from Eq. (1); 2) helium abundance $Y = 0.23 + 3.0 \times Z$; 3) no enhancement of α-elements (i.e., solar proportions); 4) HB luminosity-metallicity relations $M_{\rm v}^{\rm HB} = 1.17 + 0.39$[Fe/H] from Sandage & Cacciari (1990, solid lines), and $M_{\rm v}^{\rm HB} = 0.73 + 0.15$[Fe/H] from Walker (1992, dotted lines). (From Paper II).

There are important lessons to draw from these conclusions. Our Milky Way is a spiral galaxy in a very loose group that is located rather away from major density peaks in the distribution of galaxies. Nevertheless, her whole spheroidal component looks $\sim$ one Hubble time old, from the halo globular clusters all the way to the inner bulge. Along with the well known homogeneity of elliptical galaxies, this appears to require a cosmological framework allowing a very quick build up of spheroids – no matter whether in clusters or in the field – early in the evolution of the universe. With a mass of $\sim 2 \times 10^{10} M_{\odot}$, the old age for the bulk of the spheroidal population implies a typical star formation rate $\sim 20\, M_{\odot}{\rm yr}^{-1}$ at the epoch of spheroid formation, some 14-15 Gyr ago (having assumed $\sim 10^9$ yr for the duration of the star formation process). This is well in the range of star formation rates inferred by Steidel *et al.* (1996) for their sample of $z \gtrsim 3$ galaxies. When considering that such galaxies have also effective radii of 1-3 kpc (typical of galactic bulges, cf. Giavalisco *et al.* 1996), we can conclude that the fossil evidence is indeed in fine agreement with what is seen directly at very high redshift.

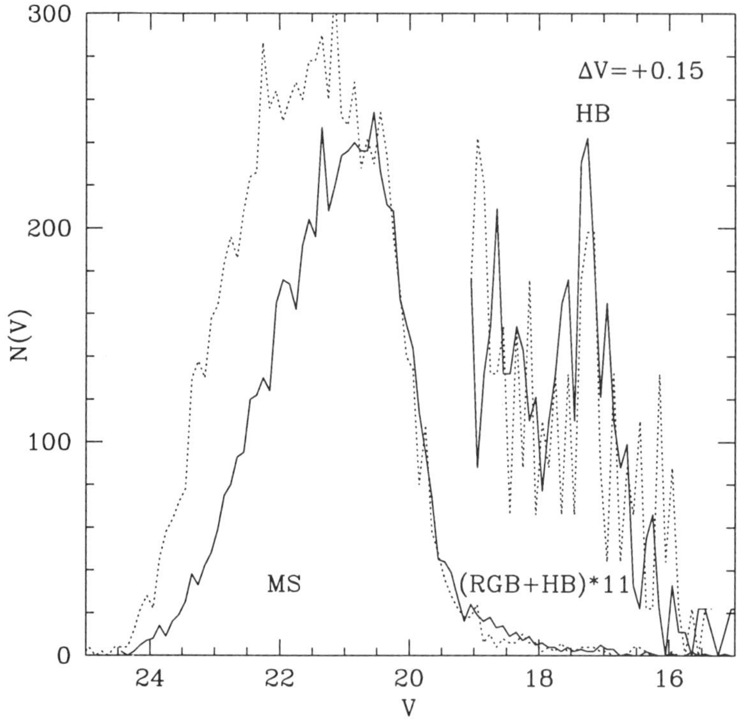

FIGURE 4. The luminosity function (LF) of main sequence (MS) and red giant (RGB+HB) stars in NGC 6528 (WF2 field, dotted line) and in Baade's Window (BW, solid lines). The cluster LF has been shifted by $\Delta V = +0.15$ so as to bring into coincidence its HB peak (marked on the figure) with that of BW, and multiplied by a factor of 2 so as to normalize the two distributions at $V = 20.45$, where both are reasonably complete, or to the same number of RGB+HB stars brighter than V=19.45 in this figure. Note that below $V \simeq 21$ the bulge LF is progressively more incomplete compared to that of the cluster. The cluster LF has been suitably broadened with a Montecarlo simulation to mimic the depth effect present in the BW field. For this display, the RGB+HB LFs have been multiplied by a factor 11, in order to avoid overlap with the LF of the disk foreground stars (From Paper II).

4. The Age of M32 Relative to the Galactic Bulge

Having obtained the age of the metal rich clusters in the bulge, this allows us to climb yet another step along the age ladder. The main sequence TO is indeed unreachable – even with HST – for galaxies beyond the dwarfs orbiting the outskirts of our Galaxy. To extend dating of old populations to other members of the Local Group one has to use other age indicators. One opportunity is offered by brightest AGB stars, which luminosity is theoretically predicted to correlate with age (e.g., Iben & Renzini 1983), and indeed does so among Magellanic Cloud GCs. However, as pointed out by Frogel & Elias (1988), GCs as metal rich as 47 Tuc exhibit a high luminosity extension of the AGB, contrary to metal poor clusters. The two clusters in the bulge allow to check in quantitative terms whether this remains true even at higher metallicities, and to assess the validity of bright AGB stars as age indicators. To this end, Guarnieri, Renzini, & Ortolani (1997, Paper III) have obtained near-IR array data for NGC 6553 using the ESO/MPIA 2.2m telescope at La Silla, then combining near-IR and HST data for this cluster.

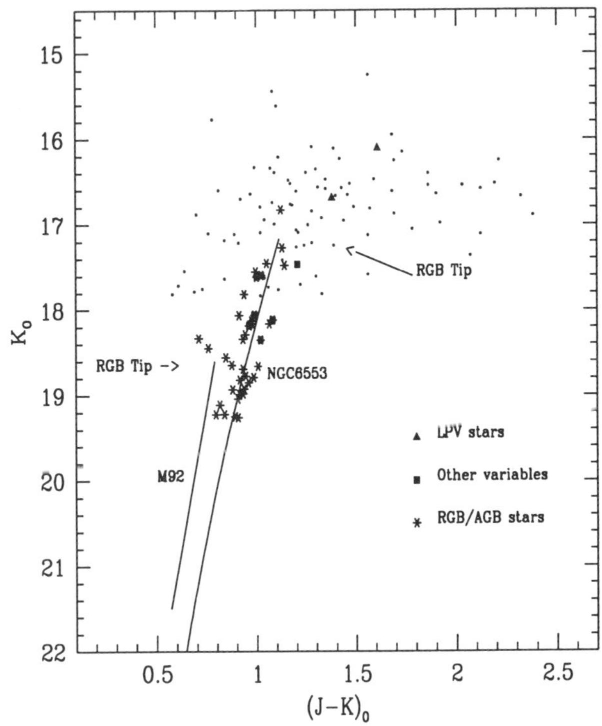

FIGURE 5. The $K_\circ - (J-K)_\circ$ color magnitude diagram of stars in M32 (Freedman 1992), with superimposed the brightest variable and non-variable stars in NGC 6553, as they would appear if at the same distance of M32 (From Paper III).

Fig. 5 shows the IR color-magnitude diagram for the stars in the dwarf elliptical M32 (from Freedman 1992, her Table 1) with superimposed the bright stars in NGC 6553, adjusted to the distance and reddening of M32. Many of the stars in M32 appear to have a luminosity which is consistent with them being still in the upper RGB, prior to the core helium flash, while several are definitely brighter than the RGB tip. Given the relatively low surface brightness of the observed M32 field, perhaps only a fraction of such bright objects may be blends of fainter stars. In fact, although the number of bright RGB stars per resolution element is relatively small, some ambiguity remains about the total luminosity sampled by Freedman's observations, which might allow the number of the brightest stars in the sample to be consistent with a blue straggler origin (cf. Renzini 1993). In any event, Fig. 5 shows that the two long period variables (LPV) in NGC 6553 are nearly as bright as the brightest objects in M32, demonstrating that a metal rich population, nearly as old as the galactic halo GCs, is indeed able to generate stars as bright as the brightest ones observed in M32. Guarnieri *et al.* then conclude that an intermediate age component (few to several Gyr) is no longer required to account for the brightest stars in M32.

Much of this conclusion depends on the adopted distance modulus for NGC 6553, that was based on the HB luminosity. In support of the adopted distance in Paper III we note that a shorter distance modulus would worsen the match between the brightest RGB stars and the theoretical RGB tip luminosity, and would place the cluster physically outside the bulge. This seems unlikely given that the color-magnitude diagram of NGC 6553 is indistinguishable from that of the cluster NGC 6528, which bulge membership is out of question.

In Paper III it was also noted that the comparison with M32 depends crucially on just two LPVs V4 and V5, while their membership in the cluster has not yet been proven spectroscopically. However, their location respectively 16 and 72 arcsec from the cluster center (respectively ~ 0.4 and ~ 2 core radii) leaves little doubt that they are indeed cluster member. This is especially the case for V4, which is the brightest star. Note also that the magnitudes of V4 and V5 reported in the figures are from single-epoch observations, and therefore their time averaged luminosity may be significantly lower (typical LPV amplitudes are ~ 1 mag in K). However, this is also the case for the M32 stars in Fig. 5.

Acknowledgements: I would like to conclude this brief review by paying once more my tribute of gratitude to Icko for having inspired – a long time ago – much of this work. I would also like to thank all those with whom I have collaborated in carrying out the three projects, i.e., all my co-authors in Paper I, II, and III.

REFERENCES

Bergbush, P.A., & VandenBerg, D.A. 1992, ApJS, 81, 163

Bessell, M.S., Sutherland, R.S., & Ruan, K. 1991, ApJ, 383, L71

Chaboyer, B., & Kim, Y.-C. 1995, ApJ, 454, 767

Chieffi, A., & Straniero, O. 1996 (Private communication)

Freedman, W.L. 1992, AJ, 104, 1349

Frogel, J.A., & Elias, J.H. 1988, ApJ, 324, 823

Fusi Pecci, F. *et al.* 1996, AJ, 112, 1461

Giavalisco, M., Steidel, C.C., & Macchetto, F.D. 1996, ApJ, 470, 189

Guarnieri, M.D., Renzini, A., & Ortolani, S. 1997, ApJ, in press (Paper III)

Iben, I.Jr. 1970, Scientific American, July, p. 27

Iben, I.Jr., & Renzini, A. 1983, ARA&A, 21, 271

Ortolani, S. Renzini, A., Gilmozzi, R., Marconi, G., Barbuy, B., Bica, E., & Rich, R.M. 1995, Nature, 377, 701 (Paper II)

Renzini, A. 1991. Observational Tests of Cosmological Inflation, ed. T. Shanks *et al.* (Dordrecht: Kluwer), p. 131

Renzini, A. 1992, in The Stellar Populations of Galaxies, ed. B. Barbuy & A. Renzini, (Dordrecht: Kluwer), 325

Renzini, A. 1993, Galactic Bulges, ed. H. Dejonghe & H.J. Habing (Dordrecht: Kluwer), 151

Renzini, A., Bragaglia, A., Ferraro, F., Gilmozzi, R., Ortolani, S., Holberg, J.B., Liebert, J., Wesemael, F., Bohlin, R.C. 1996, ApJ, 465, L23 (Paper I)

Sandage, A. & Cacciari, C. 1990, ApJ, 350, 645

Salaris, M., Chieffi, A., & Straniero, O. 1993. ApJ, 414, 580

Steidel, C.C., Giavalisco, M., Pettini, M., Dickinson, M., & Adelberger, K.L. 1996, ApJ, 462, L17

VandenBerg, D.A., & Bell, R.A. 1985, ApJS, 58, 561

Walker, A.R. 1992, ApJ, 390, L81

The Stellar Population in the Galactic Bulge

By SERGIO ORTOLANI[1], B. BARBUY[2], E. BICA[3]

[1]Department of Astronomy, University of Padua, Italy

[2]Department of Astronomy, University of São Paulo, Brazil

[3]Department of Astronomy, University Rio Grande do Sul, Porto Alegre, Brazil

In the recent years we have concentrated efforts to collect Colour-Magnitude Diagrams (CMDs) of the globular clusters, and some adjacent fields, projected in the central parts of the Galaxy. So far we were able to determine reliable reddening, distances and metallicity estimates from direct measurements on the CMD morphologies for 29 known cluster candidates, extended up to 36 galactic longitude degrees within 10 degrees from the galactic plane. Most of them are projected in the direction of the central 5°, where the survey is essentially completed.

The results favour a flattened bulge extending to 4.5 kpc from the Galactic center populated mainly by very old, metal rich clusters. Some peculiar clusters have also been identified.

---

1. Introduction

The formation and spatial distribution of the Galactic bulge require observational constraints to establish whether its stellar population is very old (Larson 1990) or is a younger, disk-like component (Raha et al. 1992), and if its shape is spherical or extended, or perhaps a bar (Spergel et al. 1996). Yet other possibilities are a flattened bulge or a disk-like system (Zinn 1985; Armandroff 1989; Ortolani et al. 1993; Minniti 1995a).

The results coming from radio observations of neutral hydrogen, molecular lines, supernova remnants and H II region distribution (see e.g. Georgelin & Georgelin 1976), show that all these typically disk population indicators are sharply dropping at a distance of about 4–4.5 Kpc from the Sun in the direction of the galactic bulge, leaving a wide, 3–4 kpc radius region centered on the galactic center, free from population I indicators (with the exception of a very limited region concentrated on the galactic center of about 100 pc cf. King 1989). A natural question is then to see if the properties and distributions of the star clusters present a corresponding change.

Recently Ortolani et al. (1995), using *Hubble Space Telescope* (HST) data of NGC 6528 and NGC 6553, two metal-rich globular clusters located at low galactic latitude, within the bulge region, with a CMD morphology identical to that of the Baade Window (see their Fig. 3), reached the conclusion that the age of these clusters is similar to that of the globular cluster 47 Tuc, and favour then an old age for the bulge.

In this paper we address the question of the spatial and metallicity distribution. Another key point is to compare the spatial distribution of these central globulars with determinations of the distance to the Galactic center (Reid 1993 and references therein).

2. The Survey

Until recently no reliable data were available for most clusters in the direction of the galactic bulge (Racine & Harris 1989). We now have gathered CCD CMD data for 16 out of 17 known globular clusters within 5°, and for the first time it is possible to study the properties of this inner system from such kind of data. 13 more known cluster candidates at galactic latitudes below 5° have been observed, at galactic longitudes extending up to 36 degrees. Most of the observations have been carried out in two steps, in a first step with a relatively wide field, using the $7' \times 7'$ field available at ESO NTT 3.5 m equipped with EMMI or with the 1.5 m Danish telescope and the direct CCD camera. In a second step the high resolution SUSI camera at NTT (with a sampling of 0.15 ″/pixel) has been used to resolve the stars in the cluster cores under very good seeing conditions (with a FWHM from 0.9 to 0.3 arcsec). The first results showed that 23 objects were genuine star clusters, while 6 were different objects (one galaxy, one supernova remnant, two planetary nebulae, one open nearby cluster, and one just non-existing; see Bica et al. 1996 and Ortolani et al. 1997a).

3. Results

The 23 observed, genuine globular clusters, are unexpectly divided into two very sharply defined classes according to their horizontal branch (HB) morphology: red horizontal branch and blue extended horizontal branch. No intermediate HB have been detected. Most of them (13) are red HB, against 8 blue HB.

The discovery of an extended blue HB is much more difficult than a red HB, because the points are spread in a wider area, in the color magnitude diagrams, and the blue tail is often blended with the contaminating foreground filed disk main sequence. Only very good seeing and photometry selected close to the cluster core allows the detection of a blue HB in very low galactic latitude clusters. In many cases we had to re-observe them in excellent seeing conditions to confirm the existence of the blue extended HB. We are not then surprised that very few blue HB clusters have been so far detected in the galactic bulge.

Red HB clusters have also quite tilted red giant branches (RGB), recalling metal rich morphologies with [Fe/H] around −0.7 to solar. On the contrary, blue HB clusters show very steep, almost vertical RGBs, with a morphology very similar to those of the most metal poor clusters of the halo such as M 30. A considerable fraction of these bulge clusters is core collapsed (10 clusters, for a fraction of 44 percent). This core collapsed sample, in turn, is dominated by blue HB clusters (8 blue against only 2 red HB).

As concerning the age, we detected the main sequence turnoff for only 5 clusters. Following the method originally proposed by Iben & Renzini (1984) based on the horizontal branch - main sequence turnoff luminosity difference, we found that four of them, NGC 6553, 6528 (Ortolani et al. 1995), NGC 6652 (Ortolani et al. 1994), and Terzan 1 are clearly old, comparable to the halo clusters, while a remaining one, still under study, appears relatively younger.

Distance determination is a key problem in any space distribution analysis of globular clusters. In the case of low latitude clusters the high reddening makes the problem even more complicate because of the high dependence of the derived total extinction from the reddening vs extinction adopted coefficients. It is easy to see that the uncertainty on this coefficient is the driver on the distance measurement error when the reddening is greater than $E(B-V) = 0.3$–0.4. Current estimates of the ratio $R = A_v/E(B-V)$, are from 3.1 and 3.6 (Olson 1975; Cardelli et al. 1989; Savage & Mathis 1979). It is not even clear if the ratio R is constant (Cardelli et al. 1989). Applied to a reddening of $E(B-V) = 0.3$ mag., the extinctions derived from the two extreme values of the coefficient R (3.1 and 3.6) differ by about 0.15 magnitudes, comparable, or greater than the photometric errors and HB absolute magnitude uncertainty used for the distance calibration. A test made on the globular cluster Terzan 2 (Ortolani et al. 1997b), which was observed also in the infrared by Kuchinski et al. (1995), shows a better agreement between optical and infrared distances if the higher value $R = 3.6$ is adopted, but its surrounding bulge field diagram indicates a distance from the Sun around 8.0 kpc (the current Sun to the Galactic Center distance) only for an intermediate coefficient value of about $R = 3.3$. Evidence that the ratio $R = 3.1$ is too low is independently given in Guarnieri et al. (1996, see also Guarnieri et al. 1997 for IR results), comparing again visible and infrared photometry.

From a selected sample of the 14 best data clusters located within 5 degrees from the galactic center, we derived the average distance of the clusters as a function of the ratio R. The distance distribution of these clusters is strongly peaked at a distance from the Sun of $D = 7.0$ kpc for $R = 3.6$, but it increases to $D = 7.7$ kpc for an intermediate ratio R=3.3, and reaches $D = 8.5$ kpc for the often adopted $R = 3.1$.

With the caution due to these distance uncertainties, we found only one of these 14 globular clusters closer to the Sun than 3.5 kpc, while all the remaining are located within about 4 kpc (with $R = 3.3$) from the galactic center. This region coincides with the HII avoidance zone and with the radius of the Galactic Molecular Ring (Burton 1976). More details on the bulge cluster distribution are presented in Barbuy et al. (1996).

Metal abundance analysis from high resolution spectra has been obtained so far only for NGC 6553 (Barbuy et al., 1992). The result show that its metallicity is near solar with [Fe/H] $= -0.2$, in excellent agreement with the recent findings from McWilliam & Rich (1994) on Baade's Window field stars, and comparable to the average metallicity available in the literature for the other red HB bulge clusters. We compared also the metallicities derived mostly from integrated or single star low dispersion spectroscopy and integrated photometry, with the CMD morphologies obtained from our photometry. Whith no exceptions all the red HB clusters have also high metallicities even if sometimes differences in the metallicities do not perfectly correspond to an equivalent CMD morphology change, possibly due to measurement errors of the metallicities or contamination. It is instead more difficult the interpretation of the discrepancy in some blue HB clusters. We have at least two relevant cases (HP1 and NGC 6749, Ortolani et al. 1997a; Rosino et al. 1996) for which a high metallicity between [Fe/H] $= -0.30$ and -0.56 (Armandroff & Zinn 1988; Minniti 1995b; Webbink 1985) is given in the recent literature. These two clusters have also a relatively steep RGB, comparable to that of NGC 6752 and M 30. The discrepancy does not appear simply due to contamination problems. Most of the other blue HB clusters have not been deeply investigated for metallicity through spectroscopic observations. If the low metallicity of these blue HB globular clusters is confirmed, then they can be candidates to be among the oldest in our Galaxy.

REFERENCES

Armandroff, T. 1989 AJ, 97, 375.

Armandroff, T., Zinn, R. 1988, A.J. 96, 92.

Barbuy, B., Bica, E., Ortolani, S. 1996, in prep.

Bica, E., Clariá, J.J., Bonatto, C., Piatti, A.E., Ortolani, S., Barbuy, B. 1995, A&A, 303, 747.

Barbuy, B., Castro, S., Ortolani, S., Bica, E. 1992, A&A 259, 607.

Burton, J.B. 1976, Ann. Rev. A&A 14, 275.

Buzzoni, A., Fusi Pecci, F., Buonanno, R., Corsi, C.E. 1983, A&A 123, 94.

Cardelli, J.A., Clayton, G.C., Mathis, J.S. 1989, Ap. J. 345, 245.

Dean, J.F., Warpen, P.R., Cousins, A.J. 1978, MNRAS, 183, 569.

Geogelin, Y.M., Georgelin, Y.P. 1976, A&A 49, 57.

Guarnieri, M.D., Ortolani, S., Renzini, A., Montegriffo, P., Barbuy, B., Bica, E. 1996, in *Proceedings of Science with the Hubble Space Telescope, II*, Paris, France, December 4-8, 1995, eds. P. Benvenuti, F.D. Macchetto & E.I. Schreier, p. 338.

Guarnieri, M.D., Ortolani, S., Montegriffo, P., Renzini, A., Barbuy, B., Bica, E., Moneti, A.

1997, A&A, in press.

Iben, I.Jr., Renzini, A. 1984, Phys. Rep. 105, 329.

King, I. 1989, in *The Milky Way as a Galaxy*, Sauverny-Versoix, Switzerland, by Gilmore G., King I., van der Kruit P., editor R. Buser, p. 41.

Kuchinski, L.E., Frogel, J.A., Terndrup, D.M., Persson, S.E. 1995, AJ 109, 1131.

Larson, R.B. 1990, PASP, 102, 709.

McWilliam, A., Rich, M.R. 1994, ApJS, 91, 749.

Minniti, D. 1995a, AJ, 109, 1663.

Minniti, D. 1995b, A&A 300, 109.

Olson, B.I. 1975, PASP 87, 349.

Ortolani, S., Bica, E., Barbuy, B. 1993, A&A 273, 415.

Ortolani, S., Bica, E., Barbuy, B. 1994, A&A 286, 444.

Ortolani, S., Renzini, A., Gilmozzi, R., Marconi, G., Barbuy, E., Bica, E., Rich, R.M. 1995, Nature 377, 701.

Ortolani, S., Bica, E., Barbuy, B. 1996, A&A 306, 134.

Ortolani, S., Bica, E., Barbuy, B. 1996a, MNRAS, in press.

Ortolani, S., Bica, E., Barbuy, B. 1996b, A&A, in press.

Racine, R., Harris, W.E. 1989, AJ, 98, 1609.

Raha, N., Sellwood, J.A., James, R.A., Kahn, F.D. 1991, Nature 352, 411.

Reid, M. 1993, Ann. Rev. A&A 31, 345.

Rosino, L., Ortolani, S., Barbuy, B., Bica, E. 1996, MNRAS, submitted.

Savage, B.D., Mathis, J.S. 1979, Ann. Rev. A&A 17, 73.

Spergel, D.N., Malhotra, S., Blitz, L. 1996, in *Spiral Galaxies in the Near-IR*, eds. D. Minniti, H.W. Rix, Springer-Verlag,p. 128

Webbink, R.F. 1985, in *Dynamics of Star Clusters* IAU Symp. 113, eds. J. Goodman & P. Hut, (Dordrecht: Reidel), p. 541.

Zinn, R. 1985, ApJ 293, 424

Tests of Evolutionary Sequences of Solar Metallicity Stars. HST Observations of the Bulge Globular Clusters NGC6528 and NGC6553

By FRANCESCA LANTERI CRAVET[1], M.D.GUARNIERI[2], A.RENZINI[2], AND S. ORTOLANI[3]

[1]Department of Astronomy, University of Bologna, Italy

[2]European Southern Observatory, Garching bei Munchen, Germany
[3]Osservatorio Astrofisico di Padova, Padova, Italy,

The colour-magnitude diagrams (CMDs) of stellar clusters, in particular those of globular clusters, can be used for testing and calibrating theoretical evolutionary sequences. For the galactic Bulge metal-rich clusters, in particular, they can be useful to test the evolutionary models of stars having nearly solar metallicity (*i.e.* $Z = Z_{\odot}$).

The optical (V, $V-I$) and (I, $V-I$) CMDs study of the Bulge globular clusters NGC 6553 and NGC 6528, both observed with the Hubble Space Telescope (HST), has allowed us to compare the experimental results with the theoretical ones, with regard to: **i)** the contribution of the evolutionary phases to the total luminosity in the V band and **ii)** the duration of the RGB-bump phase relative to the lifetime of the star during the Horizontal Branch phase.

The theoretical data we used, come from models having: **i)** t = 10–15 Gyr, $Y = 0.27$, $Z = 0.02$ (Maraston 1996, private communication; see also Maraston poster) and **ii)** t = 10 Gyr, $Y = 0.28$, $Z = 0.02$ (Catelan & de Freitas Pacheco 1996). In this frame, NGC 6528 and NGC 6553 show an age of $t >$ 10 Gyr (as already found by Ortolani *et al.* 1995), $Y \sim 0.28$ and $Z \sim 0.02$. The agreement between our data and the theoretical predictions is very good.

The high photometric accuracy of the HST permits the separation of single evolutionary phases in the CMDs of the bulge clusters. This can be accomplished despite the peculiarity of their CMDs due to their chemical composition, the high differential reddening, and large field contamination from background objects.

1. Differences between Halo and Bulge clusters.

In each field of science, the interaction between theory and observation occurs following two possible directions: 1) the theory tries to explain and organize the observed phenomena into, as much as possible, simple models; 2) the observations are, at the same time, the *test-bed* of the related theory. For this reason, it is up to the theory to indicate which observations are more suitable. In Renzini & Fusi Pecci (1988) study, is exhaustively shown how the globular cluster studies can be useful to test and calibrate the stellar evolution theoretical models.

The most famous galactic globular clusters (GCs), that is the halo clusters, have enabled us to verify the evolutionary sequences of stars having the following characteristics: **i)** $M \lesssim M_{\odot}$ and age of about 12–18 Gyr; **ii)** metallicity range of $-2.3 \leq \mathrm{[Fe/H]} \leq -0.7$; **iii)** primordial helium abundance ($Y \sim 0.23$).

It's a long time since the halo clusters have been studied and they are very familiar to us. Their observations are not very difficult, due to their position in regions of the Galaxy that are quite far from us, but not very crowded and reddened. But the Milky Way also contains clusters of stars which are located in the inner regions called the *Bulge*. This region presents greater problems when observing: **i)** a greater density of matter in the shape of stars and clouds of gas and dust; **ii)** our line of sight is along the galactic disk and **iii)** this last region is more populated than the rarefied halo.

The field disk and/or Bulge stars produce a strong contamination on the cluster CMDs. The interstellar clouds weaken and *alter*, through the reddening phenomenon, the luminous information coming from the objects. The presence of the differential reddening, variable inside the observed field, further complicates the scene.

The stellar fields located in the Bulge direction and belonging mainly to the disk + Bulge population, contain a multitude of objects located at different distances from us, having different age and chemical compositions. For these reasons, we don't know much about the Bulge population and, above all, we don't know if and how much, it is homogeneous or composite.

Going back to the evolutionary theory, if the Bulge clusters were very similar to those of halo, i.e., if they had the same range in chemical composition and age, it would not be worth studying them, as their *halo brothers* are much simpler to observe.

However the Bulge GCs, (i.e. those which formed *in loco* and not those which are halo intruders (Minniti 1994), simply crossing the region because of their elliptical orbits) show very different characteristics with respect to the halo GCs, because their environment of formation is very different.

The *halo family* formed in an metal-poor environment, peripheral and not very much evolved, however the *Bulge family* grew up in the central part of the Milky Way, which is a much more lively environment, with a very high density of matter and so much more evolved. From this point of view this region is more difficult and less quiet than the halo.

The most recent theories regarding the galactic formation (for complete references see Ortolani et al. 1995) support, on observational basis (see paragraph n.2), the scenario in which the Bulge built up during the first formation of the global galactic structure. For this reason it has undergone a rapid evolution with a subsequent chemical enrichment due to the local high density conditions. The age of the two families should be nevertheless more or less the same (Ortolani et al. 1995).

The consequences of the two different formation environments are as the following:
• Halo GCs: a) the low probability of interactions with other objects has allowed the formation and maintenance of clusters of very large dimension (i.e., ω Cen); b) they are metal-poor and their helium abundances are primordial. • Bulge GCs: a) generally they have small dimensions as they were born and grown (better say: remained) in a crowded place, so they had to undergone the gravitational effects of the surrounding objects, suffering mass loss. Probably the same clouds, from which they formed, for the reasons had small dimensions. Moreover the higher density, maybe favoured by external pressures, secured that the Jeans limiting mass was lower, allowing the collapse of smaller clouds; b) they are, from the chemical point of view, much more evolved because they formed from gas already enriched by chemical processes.

Since such chemical differences exist between the two group of clusters, the Bulge globulars extend the empirical tests of the theoretical evolutionary sequences. So it is worth to study them, despite of the observational difficulties which have been anyway recently reduced thanks to the use of the Hubble Space Telescope.

Quantitatively, the chemical abundances of the Bulge population, and then of the Bulge GCs, are:
• $Y \gtrsim 0.28$ (Terndrup 1988; Renzini 1994; Minniti 1995a);
• $-1.0 \leq$ [Fe/H] $\leq +0.5$ (Minniti et al. 1995b).

2. Observations with HST of the Bulge globular clusters NGC 6528 and NGC 6553: data and main results

On February 1994, the instrument WFPC2 mounted on board of HST has observed the two Bulge GCs NGC 6528 and NGC 6553. Two filters (F555W amd F814W) corresponding to the V and I bands, have been used.

From the images of the first cluster, 19479 stars have been detected (including the background stars) while the final sample of the second object contains 26709 stars. All the images have been reduced using DaophotII (Stetson, 1987).

Ortolani et al. (1995), from the analysis of the CMDs and luminosity functions, gave the following conclusions:

i) Metallicity (Z): The two clusters have a nearly identical metallicity, the difference is less than 10%, quantitatively this is nearly solar: [Fe/H] $\sim 0.0 \pm 0.3$ ($Z \sim 0.02$) ;

ii) Age (t): also the age is very similar with an error less than 10% . Moreover this age is more or less the same of the halo clusters, with an error of 20%. The absolute age determination depends on the calibration of the Age - $V(TO - HB)$ relation and could vary between ~ 12 and ~ 18 Gyr.

Being consolidated the membership of the two objects to the Bulge region (based on considerations as chemical composition, kinematics and spatial collocation), this result has allowed us to establish that Bulge and Halo are coeval, the first one has been enriched to solar metallicity and beyond in less than few Gyrs.

iii) Helium content (Y): Lanteri Cravet (1996) has used the HST data to obtain the helium abundance of the two clusters via the R and R' method (Buzzoni et al., 1983):

- for NGC 6528 $Y = 0.29 \pm 0.02$;
- for NGC 6553 $Y = 0.26 \pm 0.02$.

These values are in agreement with those obtained by near-infrared data by Guarnieri et al. (1996): $Y \sim 0.29 \pm 0.03$ for NGC 6528 and $Y \sim 0.28 \pm 0.03$ for NGC 6553.

3. Test of theoretical results for stars having solar metallicity

NGC 6528 and NGC 6553 have the suitable chemical composition to verify the predictions of the evolutionary theory regarding stars having $Z = Z_{\odot}$. We have carried on the comparison between the empirical and theoretical results relatively to the items:

- Contributions of the evolutionary phases to the total luminosity in the V band;
- Ratio between the duration of the RGB-bump phase and duration of the HB phase.

The next two subsections describe the single items.

3.1. *Contributions of the evolutionary phases to the total V luminosity*

We have calculated the empirical contributions related to the evolutionary phases MS + SGB, RGB, HB, AGB to the total V luminosity for both clusters. A similar comparison have been done on 1988 by Renzini & Fusi Pecci on the halo globular cluster M3.

The table summarizes our results and reportes also the theoretical results obtained by Maraston (1996, private communication, see also Maraston poster) for two population models having: $Y = 0.27$, $Z = 0.02$ and $t = 10^{10}$ yr or $t = 1.5 \times 10^{10}$ yr, respectively.

The agreement between theory and observations is good, particularly for what it concerns the RGB and AGB (to be noted that the AGB phase is the one more subject to uncertainties, both from the theoretical and observational point of view). The observational MS + SGB values are affected by uncompleteness at the lower end while the theoretical HB values are probably the result of a branch which is too red.

Evolutionary Stages	NGC 6553	NGC 6528	Theoretical values, $Y = 0.27$, $Z = 0.02$ $t = 10$ Gyr	$t = 15$ Gyr
MS + SGB	0.36	0.38	0.41	0.43
RGB	0.36	0.32	0.35	0.35
HB	0.20	0.22	0.15	0.14
AGB	0.09	0.08	0.09	0.08

TABLE 1. Contribution of evolutionary phases to the V total luminosity

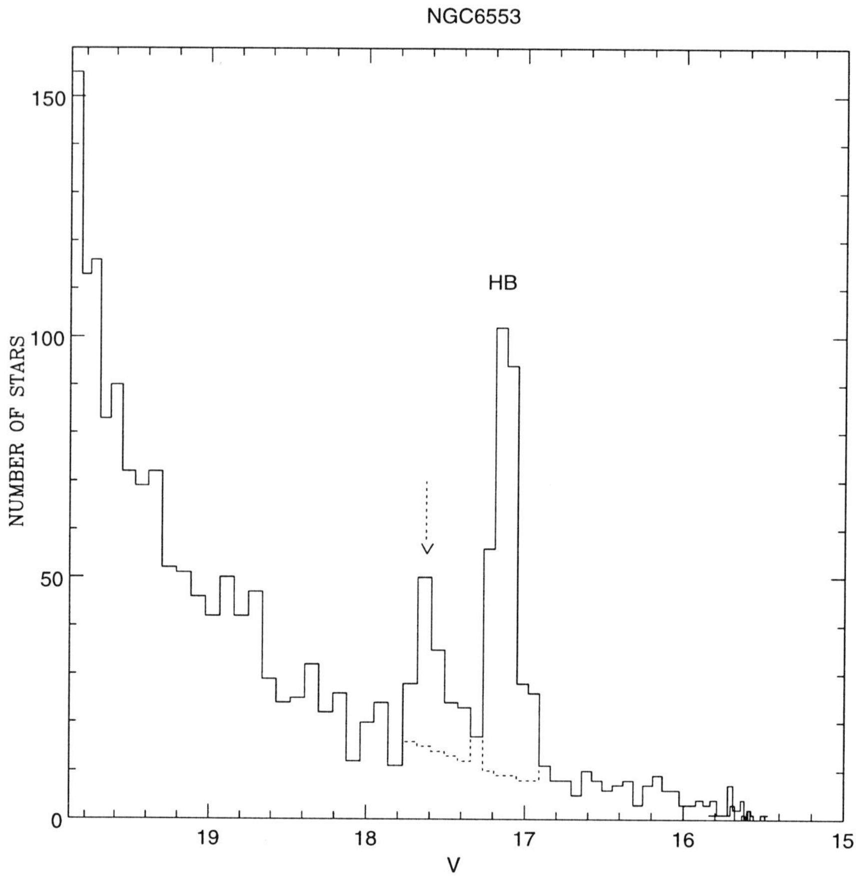

FIGURE 1. NGC 6553 GB differential luminosity function. The arrow indicates the location of the RGB-bump

3.2. *Duration of the RGB-bump phase*

The so-called RGB-bump is a clump of points along the RGB evolutionary phase. This clump is due to a temporary reversal in the star path that is ascending the RGB: the star stops and goes toward slight fainter magnitudes for some time and then starts againg ascending the RGB. This phenomenon is due to the shell approaching and then burning

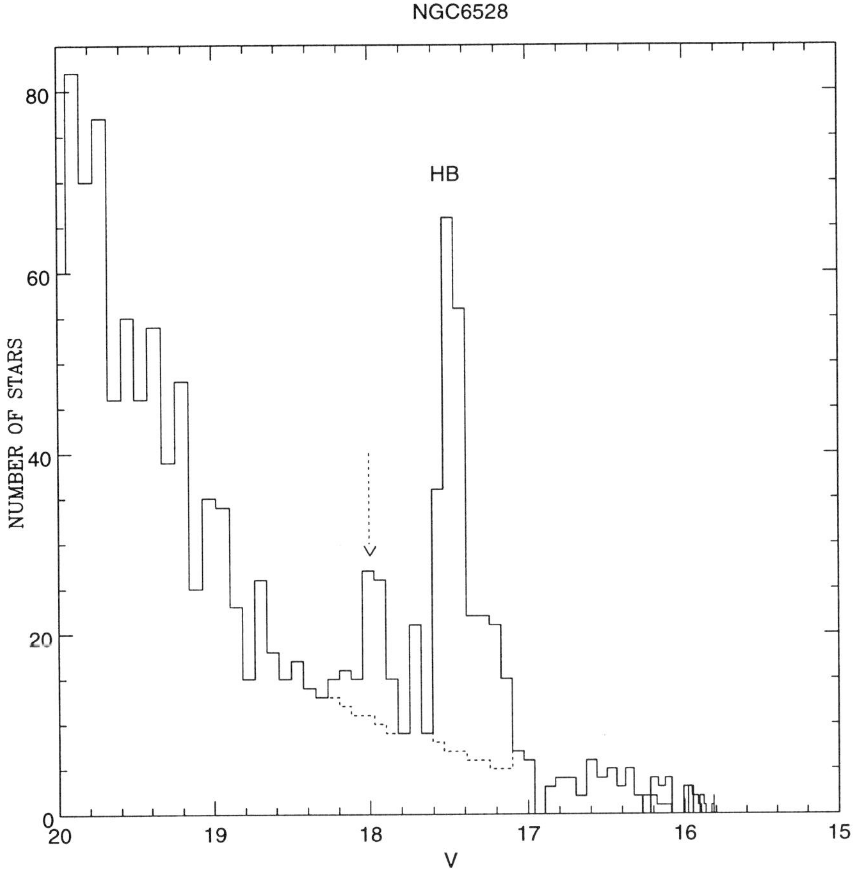

FIGURE 2. NGC 6528 GB differential luminosity function. The arrow indicates the location of the RGB-bump

through the composition discontinuity left by the maximum penetration of the convective envelope (Renzini & Fusi Pecci 1988, Sweigart 1978, see also King et al. 1985)

On the (V, V-I) CMDs of NGC 6528 and NGC 6553, which have nearly solar metallicity, the RGB-bump is located below the HB (in agreement with Fusi Pecci et al., 1990) and is easily identified. Figs.1 and 2 show the differential V luminosity functions (LFs) of the giant branch of the two clusters. In each plot we notice two peaks: the higher contains the horizontal branch stars while the smaller (indicated by an arrow) contains the stars belonging to the bump. To get out of the problems due to the differential reddening, both LFs have been build by counting the stars in interval parallels to the reddening line, rather than in interval of magnitude (Lanteri Cravet 1996).

Catelan & de Freitas Pacheco (1996), basing on the theoretical relations of Buzzoni et al. (1983), have estimated the ratio between the RGB-bump and HB evolutionary phases for clusters having $t = 10^{10}$ yr, $Y = 0.28$, $Z = 0.02$, that is with similar characteristics of the Bulge clusters. They found $t(bump)/t(HB) = 0.2$. From the Renzini & Buzzoni (1986) relation: $N(j) = B(t) \times L_{TOT} \times t(j)$, follows that $t(bump)/t(HB) = N(bump)/N(HB)$.

To carry out the star counts in the two evolutionary phases, we have applied a fit (dotted histogram) to the RGB LF represented in Figs. 1 and 2. We obtained the following results:

• for NGC 6528: $t(bump)/t(HB) = 0.23 \pm 0.04$

• for NGC 6553: $t(bump)/t(HB) = 0.19 \pm 0.03$

that, within the errors, is in good agreement with the theoretical predictions.

REFERENCES

Buzzoni, A., Fusi Pecci, F., Buonanno, R., Corsi, C.E. 1983, A&A 123, 94.

Catelan, M., De Freitas Pacheco, J.A. 1996, PASP 108, 166.

Fusi Pecci,F., Ferraro,F.R., Crocker,D.A., Rood,R.T., Buonanno,R. 1990, A&A 238,95.

Guarnieri, M.D., Ortolani, S., Renzini, A., Montegriffo, P., Barbuy, B., Bica, E. 1996, *Proceedings of Science with the Hubble Space Telescope, II* - UNESCO Buildings, Paris, France, December 4-8, 1995, 338.

King,C.R., Da Costa,G., Demarque,P. 1985, ApJ 299, 674.

Lanteri Cravet, F. 1996, Master Thesis, University of Bologna, Italy.

Maraston, C., 1996 This congress.

Minniti, D. 1994, PASP 106, 813.

Minniti, D. 1995a, A&A 300, 109.

Minniti, D., Olszewski, E.W., Liebert, J.W., White, S.D.M., Hill, J.H., Irwin, M.J. 1995b, MNRAS 227, 1293.

Ortolani, S., Renzini, A., Gilmozzi, R., Marconi, G., Barbuy, B., Bica, E.,& Rich, R.M. 1995, *Nature*, 377, 701.

Renzini, A., Buzzoni, A. 1986, *Spectral Evolution of Galaxies*, eds. Chiosi,C., Renzini,A., Dordrecht:Reichel, p. 195.

Renzini, A., Fusi Pecci, F. 1988, ARA&A, 26, 199.

Renzini, A. 1994 A&A, 285, L5.

Stetson P.B. 1987, PASP 99, 191.

Sweigart, A.V., 1978 *The HR Diagram*, eds. A.G.D. Philip, D.S. Hayes, Dordrect: Reidelp, p.333.

Terndrup, D.M. 1988, AJ 96, 884.

The Relative Ages of Galactic Globular Clusters: A Multi-color Approach

By IVO SAVIANE[1], ALFRED ROSENBERG[2,3], AND GIAMPAOLO PIOTTO[1]

[1]Department of Astronomy, University of Padova, Padova, I 35122, Italy

[2]Osservatorio Astronomico, Padova, I 35122, Italy

[3]Telescopio Nazionale Galileo , Padova, I 35122, Italy

A large sample of color–magnitude diagrams of galactic globular clusters with good photometry in the turn-off region has been collected, both in the BV and in the VI colors. A distance in color between the turn-off and the red giant branch has been defined and measured on our sample, and the experimental trend with the cluster metallicity has been determined for 31 clusters in BV and 12 clusters in VI. It is shown that the bulk of the globular's shares a common region in the diagrams, although some peculiar clusters exist. The dispersion in color around the fitted linear relation is entirely compatible with the experimental errors, suggesting a negligible age dispersion for most of the globulars. From the (albeit small) sample in the VI colors, we infer that **an almost null dependence on [Fe/H] is obtained for the $(V - I)$ color difference**. Finally, a comparison with the theoretical predictions is offered, which leads to an estimated formal age dispersion of 0.1 Gyr (from both color dispersions).

---

1. Introduction

Introduction The measurement of the *relative* ages of galactic globular clusters (GGC) is of fundamental importance in order to reconstruct the history of the Galaxy formation and of its chemical enrichment. For this task, we must be able to distinguish age differences $\leq 10^9$ yr, and to investigate the age dependence upon clusters metallicities and positions within the Galaxy. Furthermore, there have been claims that age differences could also be responsible for the long-standing *second parameter problem* (e.g. Lee, Demarque & Zinn (1994), LDZ), which is still demanding a theoretical explanation within standard stellar evolution.

Relative age determinations *at a fixed metallicity* have considerable advantage over absolute determinations, since only 2 error sources enter into the procedure, namely (a) the conversion from the theoretical plane to the observational plane and (b) the conversion from the instrumental color-magnitude diagram (CMD) to the CMD in some standard system. In other words, we only need to measure *differential* quantities in the CMD, which makes these methods independent from distance or reddening determination.

For these reasons, considerable attention has been given in the past years to relative age investigations. Among these studies there are those of VandenBerg, Bolte & Stetson (1990), Sarajedini and Demarque (1990), VandenBerg and Stetson (1991), and Stetson, VandenBerg and Bolte (1996). These studies have revealed that (a) most of the GGCs are coeval and (b) there exists a small group of objects which have younger ages (in an interval of a few Gyrs). The main drawback is that they are all based on short wavelength photometry (BV) where the effects of metallicity are strong: indeed, only the metal–poor GCs ([Fe/H] < -1.3 dex) have been widely investigated, and the relative age determination have been restricted to subgroups of clusters with similar metal content (within ± 0.1 dex in[Fe/H]).

This problem has motivated our investigation of the relative GGC ages, based on the best available photometry, with the purpose of extending the set of colors used for the analysis. We will give here a brief progress report.

2. The Method

Differential studies are based either on the so-called 'horizontal' method or on the 'vertical' method. Both methods exploit the fact that some CMD features are little dependent on the age of the population, so the position of the TO with respect to these features can be used as an age indicator. The horizontal method is based on the color difference between the TO and some point along the red-giant branch (RGB), while the vertical method uses the magnitude difference between the TO and the horizontal branch (HB), at some fixed temperature.

SVB extensively analyse the advantages and disadvantages of using either the vertical or the horizontal method; in short, while theoretical arguments would favour the vertical method, the observational requirements are better met with the horizontal method. Furthermore, when theoretical models are used in a differential sense, then also most of the uncertainties connected with the latter method disappear. In fact, SVB demonstrate that, although the absolute temperatures are affected by the adopted mixing-length parameter, metallicity, α enhancement or He content, this does not affect the relative age scale, particularly if we restrict our comparisons to clusters with similar metal content. We decided consequently to use the horizontal method, in a way explained as follows.

We measure the width in color between the TO an the RGB with respect to the point on the RGB which is 2.2 magnitudes brighter in V than the TO, and define this parameter as as $\delta(B-V)_{2.2}$ or $\delta(V-I)_{2.2}$. This choice is justified by Fig. 1 where a set of isochrones from Bertelli et al. 1994 are plotted: each panel displays a set of models for a fixed age, and isochrones for different metallicities have been registered so that their TO coincide. The upper and the lower rows display the theoretical loci in the $(M_V,\ B-V)$ plane, and in the $(M_V,\ V-I)$ plane respectively. From left to right the ages are 16, 12.5, 10 and 8 Gyr; the metallicities are $Z = 0.0004$, 0.001, 0.004, 0.008, and in each box a solid vertical dash marks a fixed point in color. The reference RGB point is marked by a dashed line in Fig. 1, and it can be seen that this point is in a CMD region where the color difference TO-RGB is large enough to reduce the effect of the measurement errors, and at the same time its magnitude is faint enough to ensure a linear dependence on [Fe/H]; this allows an easy interpolation of the theoretical models to be compared with the observations. At the moment, the Bertelli et al. 1994 models are the only ones covering the required range of metallicities, ages, and colors.

Fig. 1 shows a feature which, if confirmed, would be of extremely high relevance: the $(V-I)$ color difference is, in practice, **independent of the cluster metallicity**. That is to say, the dispersion of the theoretical $\delta(V-I)_{2.2}$ parameter is within the typical measurement errors. This would make $\delta(V-I)_{2.2}$ an age indicator without any further assumption, and would make the relative dating of GGCs a simple matter of collecting the relevant data.

3. The Observational Data Set

The observational data set The selected clusters include the VBS database plus objects whose photometry has appeared more recently. Whenever possible, we made our measurements on tables published in electronic format. In the other cases, hardcopies of the published CMDs have been used. The TO and the RGB points have been estimated

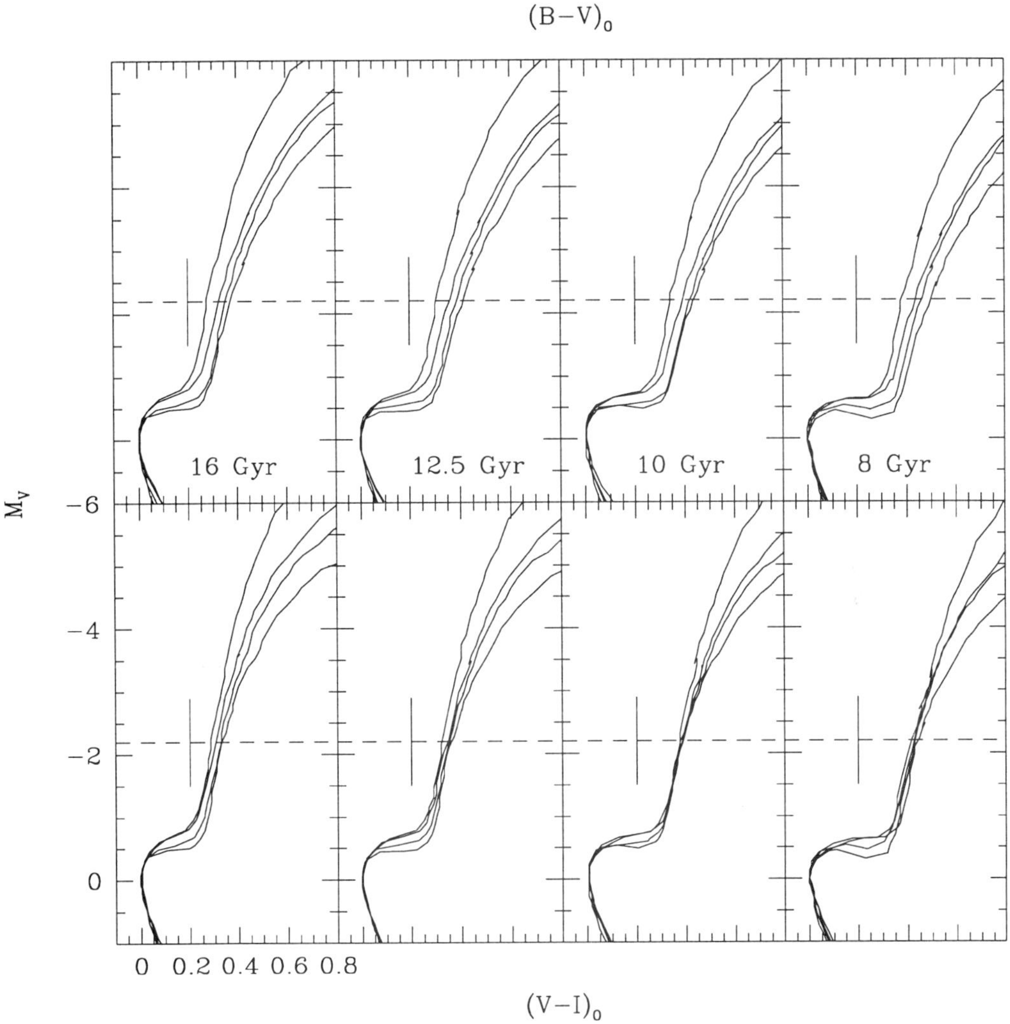

FIGURE 1. Each panel displays a set of isochrones from Bertelli et al. 1994 for a fixed age and varying metallicity, registered so that their TO coincide. The upper and the lower rows display the theoretical loci in the $(M_V,\ B-V)$ plane, and in the $(M_V,\ V-I)$ plane respectively. From left to right the ages are 16, 12.5, 10 and 8 Gyr; the metallicities are $Z = 0.0004$, 0.001, 0.004, 0.008, and in each box a solid vertical dash marks a fixed point in color. The reference RGB point is marked by a dashed line. The color difference between the TO and the RGB gets larger when the age gets lower, for both colors, but its dependence on metallicity is much smaller for the $(V-I)$ color.

by eye, and the associated error depends on the width and the 'sharpness' of the CMD branches. The measurements have been carried out independently by two of us (IS and AR), in order to check for possible systematic errors: in most cases, the discrepancies were within the observational uncertainties. This procedure allowed (a) a check of the influence of different calibrations for the same object and (b) a check of the influence of the uncertainty on the TO position: in both cases, these proved to be negligible. In fact, the color difference is a differential quantity, so a zero-point error in the calibration curve does not affect its value; furthermore, the effect of color terms is not important on

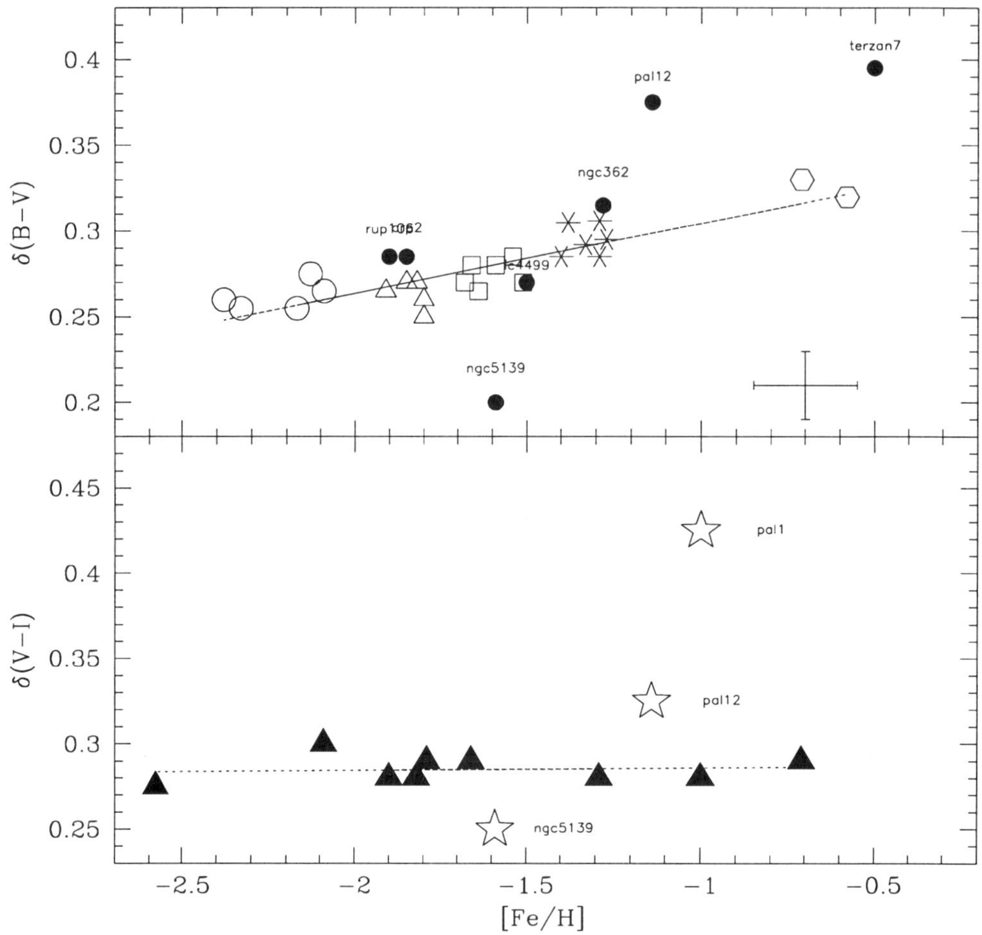

FIGURE 2. The measured $\delta(B-V)$ (upper panel) and $\delta(V-I)$ (lower panel) color differences as a function of the cluster metallicity (see text for the definition of the parameters). In the upper panel, different symbols identify different metallicity groups: open circles, open triangles, open squares, asterisks and open hexagons refer, respectively, to clusters in the groups $[Fe/H] \leq -2.00$, $-2.00 \leq [Fe/H] \leq -1.75$, $-1.75 \leq [Fe/H] \leq -1.45$, $-1.45 \leq [Fe/H] \leq -1.20$, $-1.20 \leq [Fe/H]$. A typical error bar is sketched in the lower right corner of the upper panel. Notice the almost null sensitivity of $\delta(V-I)$.

a ~ 0.3 magnitude scale. Finally, the near verticality of the lower RGB makes an error of even 0.1 mag in the TO position of no importance in the evaluation of the RGB color.

Fig. 2 shows the observed color differences as a function of the metallicity[Fe/H], both for the $(B-V)$ (upper panel) and the $(V-I)$ (lower panel) colors. It is clear that the available sample for the $(B-V)$ color ('blue' sample) is much larger than that for the $(V-I)$ color ('red' sample). We measured 31 clusters for the blue sample and 11 clusters for the red sample: the richness of the former data set allowed a separation into 5 metallicity groups, which are shown with different symbols in Fig. 2. For the blue sample, the clusters marked with filled circles represent objects which are reported as young clusters in the literature, or generically 'anomalous'; the name of these clusters is also shown in the figure. These objects will be discussed in more detail in a forthcoming paper. For the red sample, a star marks the suspect younger or anomalous objects; the VI photometry for these objects has been obtained by our group, together with NGC 6809 and NGC 1851.

A linear regression has been computed on the data, excluding the anomalous ones; for the two colors, the result is $\delta(B-V) = 0.04\,[\mathrm{Fe/H}]+0.35$ and $\delta(V-I) = 0.0014\,[\mathrm{Fe/H}]+0.29$ with rms dispersions of 0.009 and 0.007 mag respectively.

4. Discussion

Three main points are unveiled by Fig. 2: (a) the bulk of GGCs is distributed in a narrow (linear) region; (b) objects outside the 3σ region are very few and (c) apparently at longer wavelengths the metallicity dependence is very small. We must recall that the position of an object in Fig. 1 is determined only by its age and metallicity, so our interpretation is that (a) the bulk of GGCs share a common age; (b) some GCs are younger and (c) the VI color difference has a dependence on[Fe/H] that is $\sim$ 30 times smaller than BV. This strongly suggests to use the latter color baseline for the horizontal method to get globular clusters relative ages.

To have some insight into the age dispersion, we use the models plotted in Fig. 1 to translate the color spread into an age spread: at 16 Gyr we find $\Delta(\delta(B-V))/\Delta(\log t_9) \simeq 0.3$ and $\Delta(\delta(V-I))/\Delta(\log t_9) \simeq 0.4$. Thus, the quoted observed dispersions correspond to an age dispersion of $\sim$ 0.0027 in log, or $\sim$ 0.1 Gyr.

The lower panel of Fig. 2 suggests that further VI data must be obtained in order to (a) find the dependence of the color difference vs. [Fe/H] with an enlarged and more significant sample of objects; (b) independently assess (or discard) a common age for most of the globular clusters; (c) cross-check the age of the suspected young objects and finally (d) possibly identify new young GC candidates.

REFERENCES

Bertelli, G., Bressan, A., Chiosi, C., Fagotto, F., Nasi, E., 1994, A&AS, 106, 275

Lee Y., Demarque P., Zinn R., 1994, ApJ, 423, 248

Sarajedini A., Demarque P., 1990, ApJ, 365, 219

Stetson P.B., VandenBerg D.A., Bolte M., 1996, PASP, 108, 560

VandenBerg DA, Stetson PB, 1991, AJ, 102, 1043

VandenBerg D.A., Bolte M., Stetson P.B., 1990, AJ, 100, 445

A Photometric Study of NGC 458

By VINCENZO RIPEPI[1], AND ENZO BROCATO[2]

[1]Dipartimento di Fisica, Università della Calabria, Arcavacata di Rende, 87036 Rende (CS), IT

[2] Osservatorio Astronomico di Collurania, 64100 Teramo, IT

We present a CCD investigation of the poorly known SMC young Globular Cluster NGC 458. The NTT data, presented here, allowed us to study in detail the more internal regions of the cluster which are less contaminated by the field. On the basis of theoretical isochrones, a preliminary evaluation of the age is also given.

---

1. Introduction

The observation of stellar clusters is a relevant tool in testing the goodness of stellar evolution theories. In particular when intermediate-mass stars (about 2–9 $M_\odot$) are investigated, one has to observe young clusters. Unfortunately such a clusters in our galaxy, in general, contains too few stars to allow significant tests to stellar theoretical models, so that it is useful to study young populous clusters in near galaxies; in particular in SMC one of the richest cluster is NGC 458, which appears poorly investigated. In fact there is no direct measurement of his metallicity or reddening, and the estimations of the age (see Hodge (1983), Elson & Fall (1985) and Stothers & Chin (1992)) ranges from 50 to 130 Myr. The only previous photometric work on this cluster is due to Arp (1959) who obtained a photographic *BV* CM diagram either of the cluster core and of the surrounding field. For this reasons, NGC 458 was observed and a new CMD for this cluster was derived. In the following we will present our preliminary results on NGC 458, giving estimates for metallicity, reddening and age of this cluster.

2. Observations, data reduction and the resulting CM diagram

The CM diagram is based on three *B* and three *V* frames obtained with ESO NTT telescope equipped with SUSI (La Silla, Chile) in december 1995. These data was reduced using the ROMAFOT package (Buonanno et al. 1979, 1983) for crowded field photometry and BV instrumental magnitudes for 1056 stars within a radius of 65$''$ were measured.

The small size of the SUSI field of view, covering only the NGC458 core, was compensated by using a set of observations based on *B V* frames obtained at the 0.9 CTIO Telescope (Cerro Tololo, Chile), and kindly provided by A. R. Walker. The relationships for calibration were derived by using 9 photoelectric standards by Walker (1987) and Alvarado et al. (1995) located in the field covered by CTIO measurements. Common stars between NTT and CTIO data were then used to calibrate the NTT photometry.

The NTT-based CM diagram of NGC 458 core ($r < 65''$, 1056 stars) is presented in Fig. 2. It shows a well defined MS extending over about four magnitude in V, V=18.2 mag is the estimated visual magnitude of the TO and a consistent population of evolved stars (34 stars) can be recognized from $(B-V)_0 \simeq -0.1$ mag to $(B-V)_0 \simeq 1$ mag.

The CTIO-based CM diagrams, reported in Fig.s 2, resemble the main features of the NTT ones. However, it can be recognized that, selecting the external region ($r > 120''$) of the cluster, a clump of field stars appears at $(B-V)_0 \simeq 1.0$ mag and $V_0 \simeq 19$ mag.

This clump has been already recognized to be populated of helium burning intermediate mass stars belonging to the field (2–3 $M_\odot$, see Castellani et al. (1990), Bencivenni et al. (1991)).

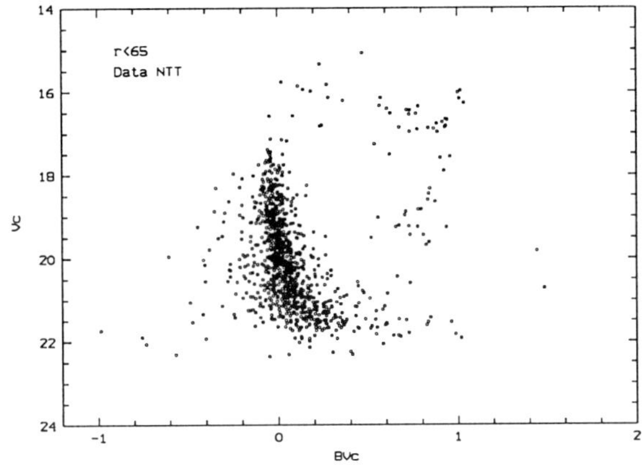

FIGURE 1. NTT-data; CM diagram for the central zone of NGC 458: 1056 stars in $r < 65''$)

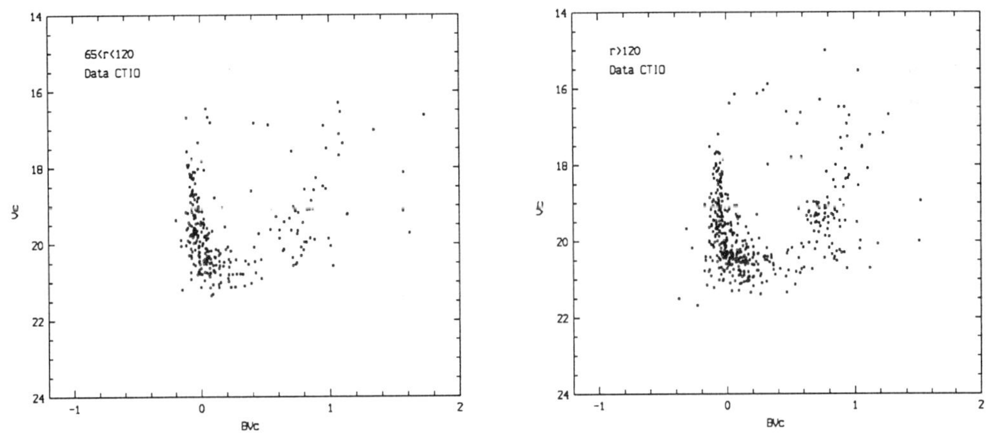

FIGURE 2. CTIO data; left: CM diagram for $65'' < r < 120''$; right: $r > 120''$

It is interesting to note that the field of SMC and LMC appears to be very similar according to theoretical previsions discussed in Bencivenni et al. (1991).

3. Results

Since for NGC 458 there is no indication in literature about reddening we first estimated this quantity.

To obtain preliminary indication, we compared the CM diagram of NGC 458 with that of the LMC cluster NGC 1866 which has known reddening and distance modulus and presents a quite similar morphology (Brocato et al. (1989)). Concerning NGC 458 we assumed the distance modulus of SMC, i.e. $DM = 18.9$ mag (Westerlund (1990)). Then, by assuming the quoted values of distance modulus the two Main Sequences fairly overlap if a reddening of $E(B-V) = 0.11$ is adopted for NGC 458 (Fig. 3 left). This figure also discloses that the evolved stars of the two clusters are located at about the same luminosity, but NGC 458 evolved stars extend to much bluer colors (temperatures). Recalling that the temperature extension of the He-burning blue loop in intermediate mass stars depends on metallicity (Brocato et al. (1993)), a possible explanation is that NGC 458 and NGC 1866 have quite similar ages (about 110 Myr) but different metallicity (i.e. NGC 1866 more metal-rich than NGC 458). This suggestion is supported by Fig. 3 right, where we overplot two isochrones (Cassisi et al. (1994)) of different metallicities ($Z = 0.02$ and $Z = 0.006$) but same ages (110 Myr) on the observational data.

We also checked the consistency of the age evaluation by plotting (Fig. 4 left) the observed LF of NGC 458 to compare the observed bright MS termination end to the theoretical one obtained by the quoted isochrone of $t = 110$ Myr.

As a conclusion we find that a fairly good isochrone fitting for NGC 458 (Fig. 4 right) is reached with the following values: $DM = 18.9$, $E(B-V) = 0.11$ mag, $t = 110$ Myr and $Z = 0.006$.

Independent evaluation of metallicity and reddening would certainly improve the reliability of the values derived in this work for the first time and which should be regarded as temptative estimates.

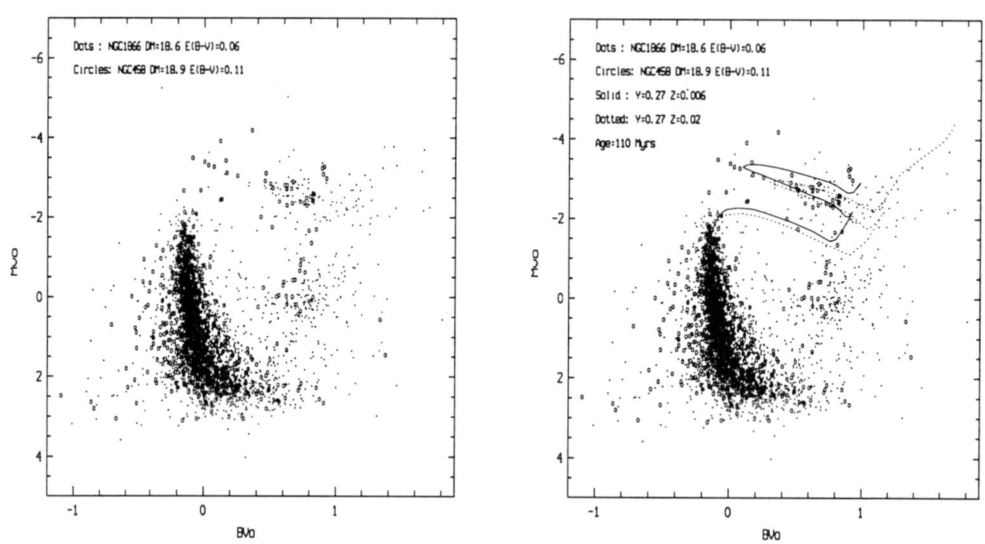

FIGURE 3. Comparison between CM diagram of NGC 458 and NGC 1866; Left: observational data only; Right:as in Left, but with superimposed isochrones for 100 Myrbut different metallicity

4. Conclusions

We presented a CM diagram for NGC 458 mainly based on NTT data; we are able to fairly identify the various stages of stellar evolution present in this cluster; in particular the MS and the blue loop of the evolved stars are clearly defined.

On the basis of our diagram and by comparison with observational data of NGC 1866 in LMC we estimated the reddening ($E(B-V) = 0.11$ mag) of NGC 458 and we suggest that these two cluster have very similar age (110 Myr) but different metallicity, being NGC 458 less metal-rich.

Moreover, theoretical isochrones provide a fairly good fit assuming the following quantities: $Z = 0.006$, $E(B-V) = 0.11$ mag, age $= 110$ Myr.

As a final remark we have to note that more work is still necessary in order to im-

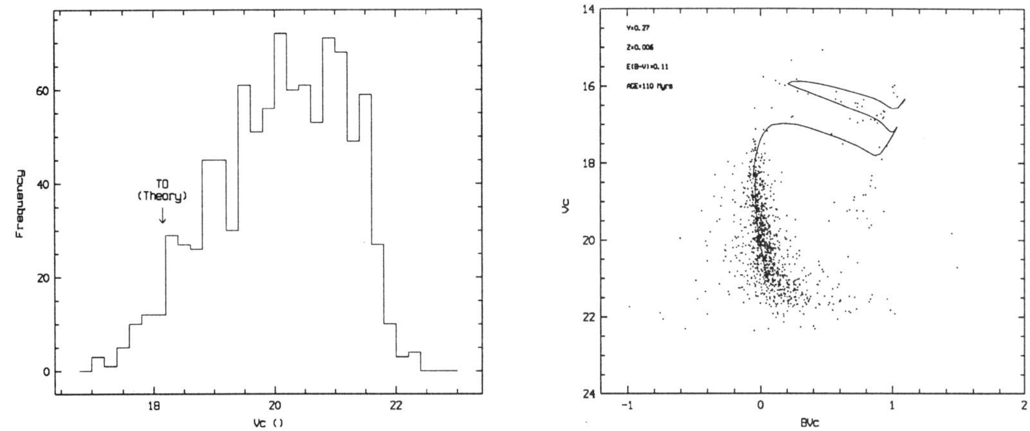

FIGURE 4. Left: Observed LF for NGC 458 with superimposed the location of the theoretical TO for an age of 110 Myr; Right: the adopted isochrone fitting

prove and refine the interpretation of NGC 458: in particular independent evaluation of reddening and metallicity would greatly improve the reliability of the age determination.

5. Acknowledgments

We are in debt with A.R. Walker that kindly provided us the CTIO data.

REFERENCES

Alvarado F., Wenderoth E., Alcaino G. and Liller W., 1995, AJ, 110, 646

Arp H., 1959, AJ, 64, 175

Bencivenni, D., Brocato, E., Buonanno, R. & Castellani, V., 1991, AJ, 102, 137

Brocato E., & Castellani V., 1993, ApJ, 410, 99

Brocato E., Buonanno R., Castellani V. & Walker A.R., 1989, ApJS, 71, 25

Buonanno R., Corsi, C.E., De Biase G.A. & Ferraro, I., 1979, in: International Workshop on Image Processing in Astronomy, p. 354, eds, G.Sedmak, M. Capaccioli & R.J. Allen, ICTP, Trieste, Italy

Buonanno R., Buscema, G., Corsi, C.E., Ferraro, I. & Iannicola, G., 1983, A&A, 126, 278

Cassisi S., Castellani, V. & Straniero, O., 1994, A&A, 282, 753

Castellani, V., Chieffi, A. & Straniero, O., 1990, ApJS, 74, 463

Elson, R.A.W. & Fall, S.M., 1985, ApJ, 299, 211

Hodge, P.W., 1983, ApJ, 264, 470

Stothers R.B. & Chin C., 19992, ApJ, 390, 136

Walker M. F., 1987, PASP, 99, 179

Westerlund, B.E. 1990, A&AR, Vol. 2, n.1, 29

Opisthodromism on the Horizontal Branch?

By ROBERT T. ROOD, JON WHITNEY, AND NOELLA D'CRUZ

Department of Astronomy, University of Virginia

Since about 1970 the standard model of the horizontal branch (HB) has assumed that on the average stars lose about 25% of their mass on the first ascent of the RGB and that some parameter varies star-to-star leading to roughly a 10% dispersion in this mean mass loss. With two mass loss parameters, basically a mean and a dispersion, Rood (1973) was able to make simulated HBs which resembled observed HBs. Unfortunately one could draw few conclusions based on HB color morphology. With perfectly plausible assumptions about how mass loss varies (or does not vary) with parameters like composition, one could make HBs jump all over the color magnitude diagram (CMD).

How has the situation has changed? There have been many improvements in the input physics, but the major uncertainty confronting the models—mass loss—remains. We only have a dimensional parameterization for mean mass loss, and the parameter which varies from star-to-star leading to the dispersion has never been identified much less understanding in a quantitative way how it affects the mass loss. On the observational side, many more HB CMDs have been obtained with larger samples and higher precision. It is not clear whether this has helped or not. HBs show a distressing diversity which defy simple classification schemes. For example, do truly bimodal distributions require even more free parameters? We hope not. Indeed, we suggest turning the problem around, i.e. using pathological HB morphology to study mass loss.

---

1. Retrospective 1996—RTR

I began my dissertation project on horizontal branch (HB) stars almost exactly 30 years ago. Several times in that period I have taken advantage of a conference proceedings volume (Rood & Seitzer 1982, Rood & Crocker 1985, Rood & Crocker 1989) to reflect on my progress or lack thereof. This time Icko has provided me with a word which captures my feelings on all of these occasions—opisthodromism. Although it appears in no dictionary which I have checked, the etymology of opisthodromism gives it a clear meaning—being carried backward while trying to run forward.† On this occasion I add another chapter to the series, starting by returning to the very beginning of the HB in an undoubtedly spotty history.

The HB was first observed in globular clusters by Shapley in the second decade of this century, (e.g. M13—Shapley 1915 & M3—Shapley & Davis 1919). In his book *Star Clusters* Shapley (1930) presents tabular arrays barely recognizable to my late 20th century mind-set. By the late 30's HB photometry was displayed much as it is now (e.g. Hachenberg 1939—M92 & Greenstein 1939—M4). As far as I can find HB stars were not referred to as "horizontal branch" before Baade's (1944) landmark paper on stellar populations. Baade used the HB as one of the identifying characteristics for Population II. By the early 50's the nomenclature had become standard (Johnson & Schwarzschild 1950—M15, Arp, Baum, & Sandage 1953—M92, Sandage 1953—M3). In particular, the color magnitude diagram (CMD) of M3 presented by Sandage (1953) clearly showed all of the principal sequences and became the textbook example for a globular cluster CMD. Sandage (1953) also gave a detailed comparison of the M3 & M92 HBs in what might be the earliest discussion of HB morphology.

† One can hardly have a conference on Elba without the word "palindrome" coming to mind. The etymologically curious wander naturally to "dromedary" and "opisthodromism."

Within a few years Arp (1955) showed the great diversity in the morphologies of the observed cluster HBs. It was recognized that the dominant or first parameter driving HB morphology was metallicity (Sandage & Wallerstein 1960), and Faulkner (1966) showed how this arose naturally in core helium burning stars. However, data as early as that compiled by Sandage & Wallerstein (1960) showed that more than one parameter appeared to be driving HB morphology. This was firmly established by Sandage & Wildey (1966) and van den Bergh (1967), and the first theoretical attempt to identify the "second parameter" (Iben & Rood 1968) soon followed. The next dramatic observational evidence for the diversity of HB morphology was the CMD of NGC 6752 presented by Russell Cannon at the 1973 Frascati Globular Cluster workshop. The CMD was not published for many years (Cannon 1985), but I, and I suspect many other participants, kept a tattered thermal copier copy, showing the horrid vertical HB and its gap. The ensuing years have given even more surprises. Strongly bimodal HBs are found in NGC 2808 (Ferraro *et al.* 1990) and NGC 1851 (Walker 1992). Even at this meeting we are confronted with additional gaps in NGC 2808 and prominent blue extensions to the HBs of the metal rich clusters NGC 6388 and NGC 6441 (Piotto *et al.* 1997, Sosin *et al.* 1997).

Our earliest attempts to explain observed HBs (e.g. Iben & Rood 1968) were made without invoking mass loss—Icko's contention was that "when I got things right" the unpalatable necessity of introducing a free parameter would vanish. Unfortunately as the helium abundance settled toward that produced in primordial nucleosynthesis and opacities improved, HBs with no mass loss became "red spots." An alternate hypothesis was that the HB was a mass sequence rather than an evolutionary sequence. Because the mass of the helium core at the time of the helium flash was insensitive to total stellar mass, it was plausible to model the HB as sequence of stars with constant core mass ($M_{\rm core}$) and varying total mass (M). The zero-age HB (ZAHB) of stars just beginning core He burning is populated with stars which have undergone some mean mass loss (ΔM) with dispersion around that mean [$\sigma(M)$](Iben & Rood 1970; Faulkner 1972). Subsequent evolution carries stars away from their ZAHB location, but under most circumstances the primary factor affecting HB morphology is what part of the ZAHB is populated. Thus four mass parameters were required to model HBs: $M_{\rm RG}$, $M_{\rm core}$, ΔM, & $\sigma(M)$. Two of these, $M_{\rm RG}$ & $M_{\rm core}$, can be determined with evolutionary calculations. Adopting a truncated gaussian mass distribution, Rood (1973) showed $\Delta M \sim 0.25 M_{\rm RG}$ and a $\sigma(M) \sim 0.1\Delta M$ could mimic the observed CMDs of many clusters. There was no reason to adopt a gaussian mass distribution other than that it worked and that random variables often have such distributions. HBs are still modeled in this fashion (e.g. Lee, Demarque, & Zinn 1990, 1994; Catelan 1993; Catelan & de Freitas Pacheco 1993)

Adopting this model age, helium abundance (Y), iron abundance ([Fe/H]), and the summed abundance of the CNO elements with respect to iron ([(C + N + O)/Fe]) all affect where stars fall on the HB. One could fix [Fe/H] and vary one the others to explore the second parameter problem, but this requires an assumption about mass loss, e.g. that the mean mass loss is a constant or that the mass loss was given by the Reimers' formula (Reimers 1977). Making this assumption Lee, Demarque, & Zinn (1990, 94) have argued that dominant second parameter is age. However, there is evidence that stellar rotation (Peterson 1983, 85a, 85b) and stellar density within a cluster (Fusi Pecci *et al.* 1993) affect HB morphology, most plausibly through mass loss. Neither blue HB gaps as in NGC 6752 (Crocker, Rood, & O'Connell 1988) nor bimodal HBs as in NGC 2808 (Rood *et al.* 1993) can easily be accommodated into a simple HB scheme. In addition, recent evidence has shown that several second parameter pairs have essentially the same age (M3/M13—Catelan & de Freitas Pacheco 1995, Ferraro *et al.* 1997; M3/NGC 6752—Richer *et al.* 1996; NGC 288/NGC 362—Stetson, VandenBerg & Bolte 1996).

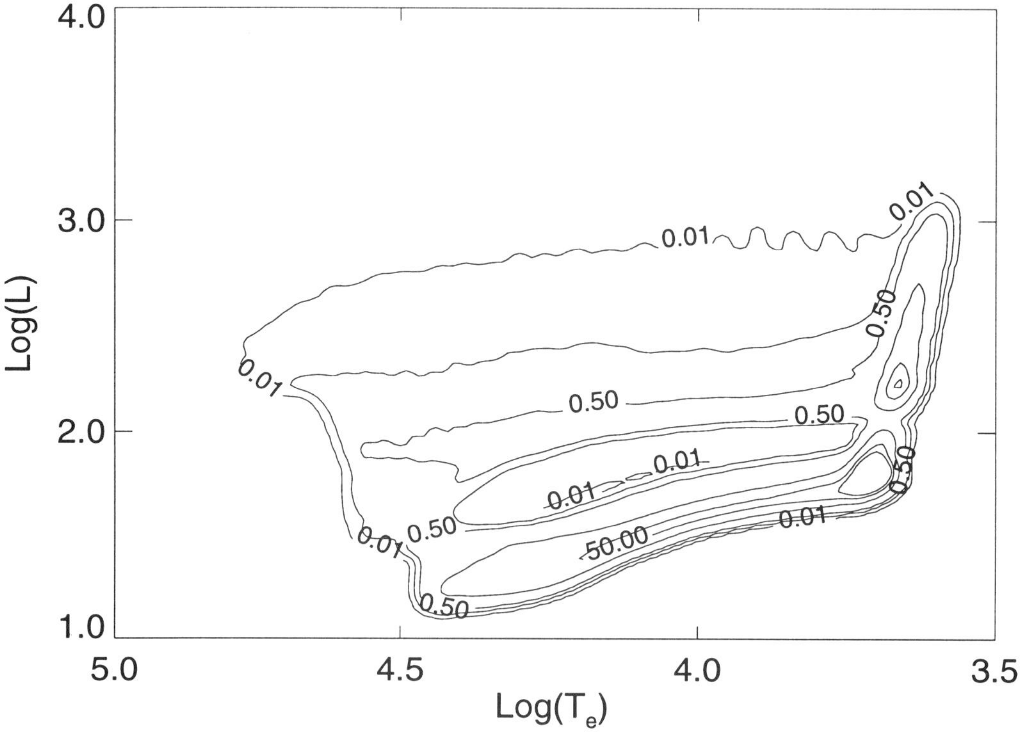

FIGURE 1. Contour plot Hess Diagram based on the Dorman *et al.* (1993) HB and post-HB tracks for [Fe/H] = −1.5, [O/Fe] = 0.6. This image has been scaled logarithmically to bring out the densities at lower levels.

The situation has become so complex that any assumption concerning mass loss is suspect. Indeed, it seems appropriate to turn the process around and use observed HB morphology to deduce something about mass loss which is what we attempt here.

2. Hess Diagrams of the HB

Several years ago we (see Whitney, Rood, & O'Connell 1997) began to explore new ways to compare theory with observations and developed a technique which represents the CMD as an "image" rather than a collection of points. The "intensity" of each pixel is the observed or theoretical density of stars in the CMD. Such a representation combines the advantages of a CMD with the temporal information contained in number counts and luminosity functions (LFs). A further advantage is that the full range of modern image processing software can be utilized in analysis. As it turns out our "new" technique is not new at all but was developed by Hess (1924) and described in Trumpler & Weaver (1953). Hess Diagrams have received little attention in recent decades (e.g. Kraft & Hiltner 1961), with but one application to globular cluster HBs (Bono *et al.* 1991).

Theoretical HB Hess Diagrams are equivalent to discrete point synthetic HBs with essentially infinite HB and post-HB population. They are constructed by weighting each pixel of the CMD image according to the amount of time evolving stars spend in that region of the CMD. The resulting image is a population density plot, indicating where evolving HB and post-HB stars spend their lifetimes, and thus, where they are most

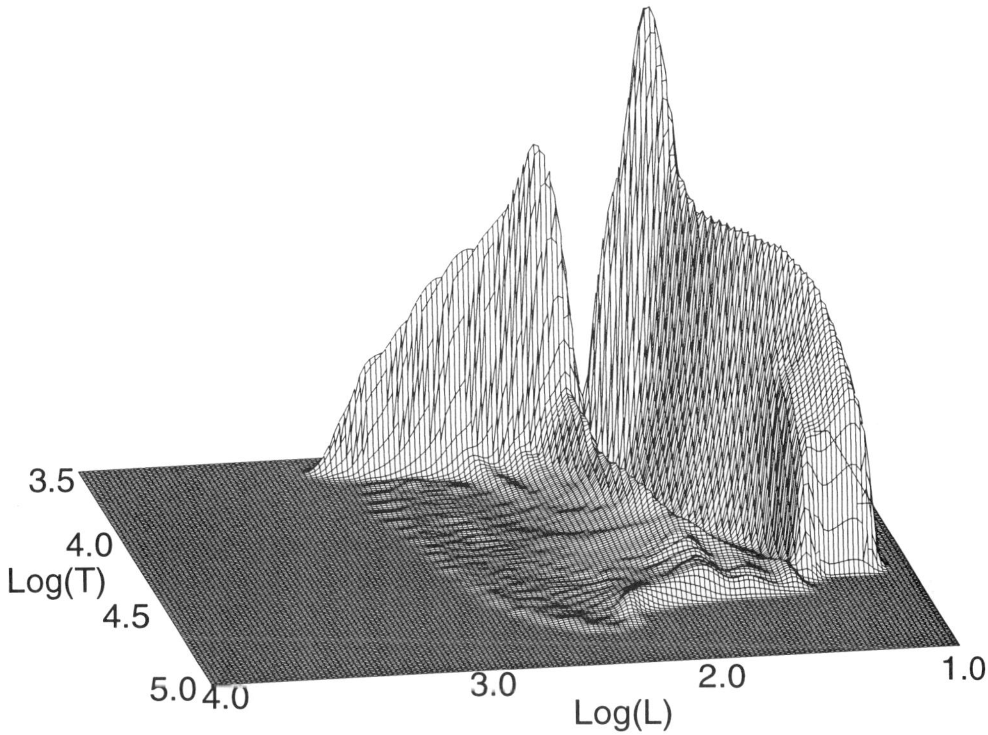

FIGURE 2. Surface plot representation of the Hess Diagram shown in Fig. 1

likely to be observed. Observational Hess diagrams are constructed by summing over stars with each star spread over its error ellipse.

In Fig. 1 we show a theoretical Hess Diagram based on the Dorman *et al.* (1993) HB and post-HB tracks for [Fe/H] = −1.5, [O/Fe] = 0.6. Fig. 2 shows a different representation, rather like viewing the HB from above the planetary nebula nuclei region. We have taken a flat ZAHB mass distribution—$P(M)$ = constant for $0.488\,M_\odot < M < 0.900\,M_\odot$. While such a distribution is unphysical it is still instructive. Parts of the diagram not populated in this case will never be populated no matter what the real $P(M)$ is. Further, it can be seen whether "observed" features like gaps have some origin in the underlying evolution or whether they arise from "structure" in $P(M)$ or in the transformation to the observed quantities.

Basically Fig. 2 is the block of modeling-clay from which an observed HB must be sculpted. The unsculpted HB shows several features: (1) the heavily populated HB, evenly populated except at the extreme ends; (2) the AGB "clump" separated from the HB in luminosity by a gap; (3) a somewhat less populated, but still significant band above the HB. This band is separated from the main HB by a gap at low log $T_{\rm eff}$ but merges with the HB the highest log $T_{\rm eff}$. *Notice that there are no gaps along the HB.*

The first step toward sculpting an HB is the conversion to colors and magnitudes. This is analogous to bending and stretching the clay block. The result of converting to the V, $B-V$ plane is shown in Fig. 3. The CM image retains the major features of the HR image, with some modifications. The most obvious difference is at the blue end of the image where the HB drops in V and becomes shortened in $B-V$. This occurs due to the large bolometric correction for the hotter stars, and the relative insensitivity

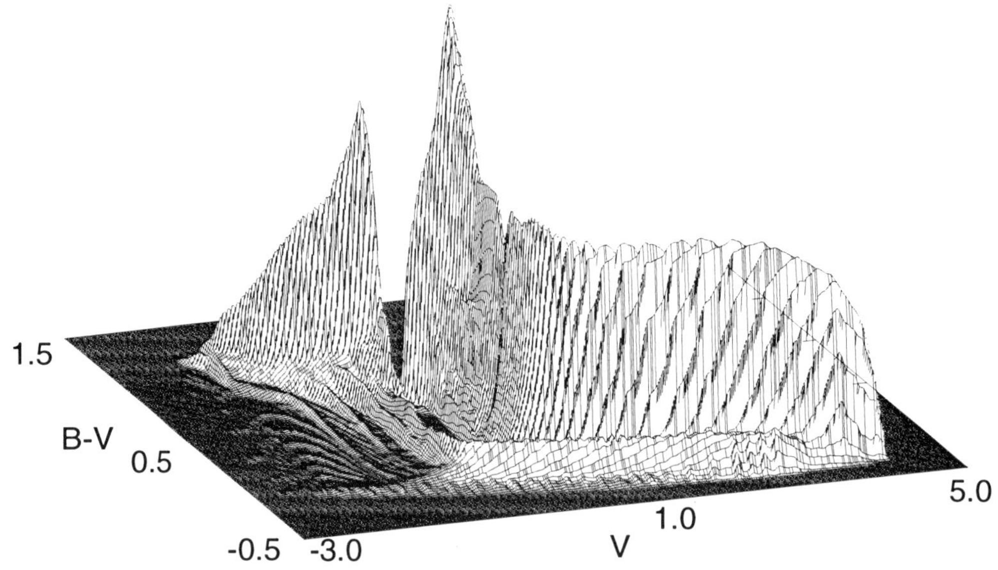

FIGURE 3. The Hess Diagram shown in Fig. 2 transformed to the V, $B-V$ plane.

of $B-V$ to effective temperature for $T_{\rm eff} > 10000\,{\rm K}$. Another interesting feature is the first appearance of a saddle or "gap" in the HB near $B-V = 0.2$ to 0.3. The saddle appears where $B-V$ is most sensitive to effective temperature. The HB is effectively stretched in $B-V$, so the population density decreases compared to the adjoining regions. This gap may be the origin of the mild mid-HB deficiencies seen in clusters like M4 (Cudworth & Rees 1990). However, it account for the gaps, essentially devoid of stars, like that observed in NGC 2808, nor does it not explain the gaps which occur in the "blue tail" portion of the HB of a number of clusters (Fusi Pecci *et al.* 1993). As can be seen in Fig. 3, the blue tail is populated quite smoothly. From this we conclude that *BHB gaps do not naturally arise from HB track morphology or from transformations from (*$\log L$, $\log T_{\rm eff}$*) to (*V, $B-V$*), hence these gaps are a product of the mass loss process.*

The final step to observed HBs is to fold in a realistic mass distribution along the HB. Basically one now carves away clay from "raw HB" as shown in Fig. 3. The result of "carving" with the truncated gaussian distribution of Rood (1973) ($\sigma(M) = 0.03\,M_\odot$) is shown in Fig. 4. The mean HB mass in the upper left corner is $0.74\,M_\odot$ and decreases in steps of $0.02\,M_\odot$ clockwise. If so inclined one could interpret this as an age progression with steps of 1.6 Gyr. While there is no structure along the ZAHB, one does see gaps in the distribution orthogonal to the ZAHB. We anticipate that this sort of structure may eventually provide mass-loss-independent tests for HB models.

3. One Natural Way to Produce a BHB Gap

Extreme horizontal branch stars (EHB) are stars with such small envelope masses that the ensuing helium shell burning phase occurs at high $\log T_{\rm eff}$—the so-called AGB-manqué and P-EAGB stars (Greggio & Renzini 1990, Dorman, Rood & O'Connell 1993).

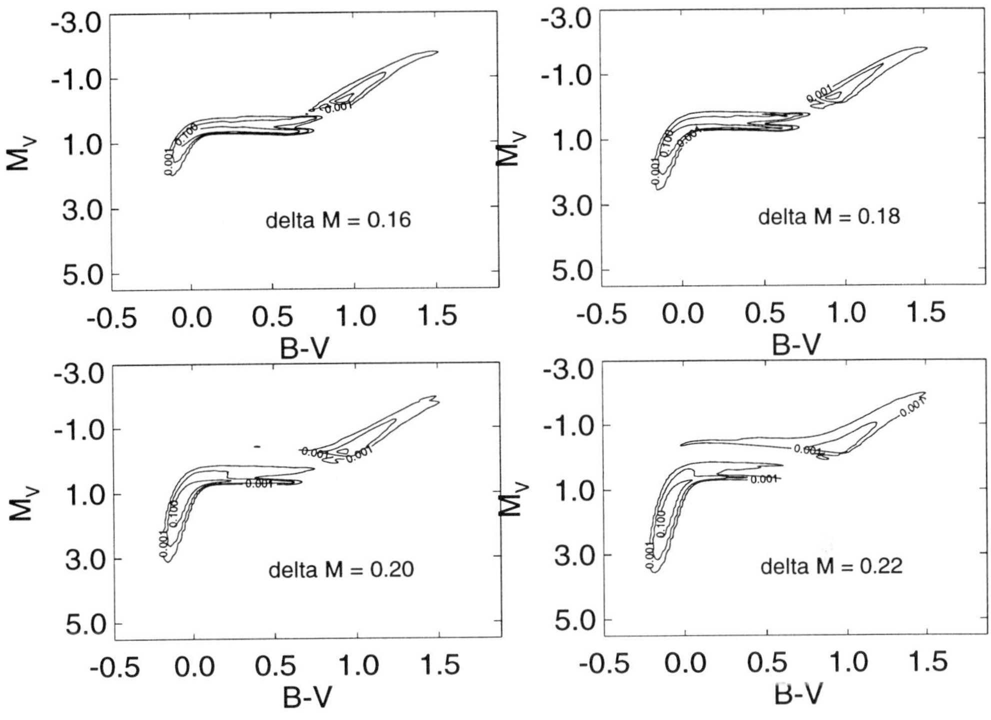

FIGURE 4. Contour plots of theoretical Hess diagrams with the mean HB mass decreasing from 0.74 $M_\odot$ in steps of 0.02 $M_\odot$. These mass steps are equivalent to age differences of $\sim$ 1.6 Gyr.

Some clusters have HB blue tails which extend into the EHB region. There is sometimes a gap at or near the beginning of EHB regions (e.g. M13—Ferraro *et al.* 1997 and NGC 2808 (Piotto *et al.* 1997). One especially puzzling aspect of clusters with extensive EHB populations is that the EHB is populated by a very narrow range of HB masses. Stars with very large RGB mass loss rates peel-off the RGB before the helium flash never reaching the HB. Smaller mass loss rates produce ordinary HB stars. Surely some fine tuning is required for stars to have lost almost all of their envelopes just as they reach the helium flash. But the large observed EHB populations suggest otherwise. Maybe this is a clue to the nature of the mass loss mechanism.

Castellani & Castellani (1993) found that some stars that peeled-off the RGB underwent a He flash at high $T_{\rm eff}$ and became EHB stars. Intrigued by this result, we (D'Cruz *et al.* 1996) set-out to find a mass loss "law" which would produce such "hot-flashers" without fine tuning. As a prelude we planned to evolve stars using the Reimers' formula to find out how fine the tuning had to be. Much to our surprise we found that no fining tuning was required. Our results suggest that any mass loss formulation which leads to rapidly increasing $\dot{M}$ while going up the RGB will give a similar result, so long as the mass loss turns off at some point.

The commonly held view is that the core temperature rises because of the advance of the H burning shell. In actuality the core heating is due to the contraction of the outer core. The increasing core size leads to a situation where the outer core can contract, but core contraction continues for some time even if the core mass is fixed. Consequently any star that gets to within ~ 0.4 mag of the RGB-tip will ignite He whether it reaches the tip or not.

Our results do not depend on the stars' being hot-flashers. We terminated mass loss when the star peeled-off the RGB. Since the Reimers' formula is a surface description, peel-off did not occur until the envelope mass ($M_{\rm env}$) was very small. Our resulting EHB stars formed a tight blue hook at the end of the HB. Almost certainly the mass loss phenomenon depends on some bulk property of the convective envelope, so $\dot{M}$ will go to 0 at some $M_{\rm env}^{\rm min}$, perhaps $\lesssim$ few 0.01 $M_\odot$. In this case the flash may might still occur on the RGB and the blue hook will be spread toward lower log $T_{\rm eff}$.

We explored the range of the Reimers' parameter $\eta_{\rm R}$ which led to EHB stars and the results are shown in Fig. 5. The range of $\eta_{\rm R}$ producing different types of ZAHB stars are indicated as a function of metallicity. We have split the BHB (normally defined as all stars hotter than the RR Lyrae) into 3 groups: BHB1—the traditional "horizontal" HB (in the V, $B-V$ plane); BHB2—the blue tails or essentially vertical HB (excluding the EHB); and the EHB. Thinking in $\eta_{\rm R}$ (or some more generalized mass loss parameter $\eta_{\rm ML}$) space we see that the magnitude and range of $\eta_{\rm ML}$ feeding the EHB is almost independent of [Fe/H](assuming constant age). It is no more difficult to make EHB stars at high metallicity than at low. Indeed, it is the mid-HB which is hard to populate at high metallicity. Further, EHB stars (at least at low metallicity) are funneled into a narrow range of log $T_{\rm eff}$ at the hot end of the HB enhancing bimodality. Simply by modeling the HB using $P(\eta_{\rm ML})$ rather than $P(M_{\rm HB})$ one can produce EHB stars without fine tuning and produce a gap at the BHB2/EHB transition. We conjecture that the lower BHB gap observed in many clusters is this EHB gap.

Observations of the location of the EHB gap and the distribution of stars on the hot side of the gap will yield information on the mass loss mechanism. For example, the larger $M_{\rm env}^{\rm min}$ is, the more spread out in log $T_{\rm eff}$ the EHB stars will be.

To get EHB stars the mass loss efficiency distribution must extend to values roughly double that required for normal HBs, ($\eta_{\rm ML}^{\rm EHB} \sim 2\eta_{\rm ML}^{\rm norm}$). This occurs in some clusters but not in others; we do not know why—the simple correlations seem to be vanishing (see Ferraro *et al.* 1997). There might be two mass loss mechanisms operating—one present in all clusters, the other only in some. If the second mechanism is very effective, all of the stars in which it operates could be separated from the main HB leading to the observed "upper BHB" gaps in some clusters.

4. Summary

- Gaps along the HB do not arise naturally from the models. $\Rightarrow$ HB gaps probably tell us something about mass loss.
- There are many strange HBs. There are no simple one-to-one correlations between HB morphology and other parameters. $\Rightarrow$ It is dangerous to assume HB morphology can be a measure of age.
- If one thinks in terms of $P(\eta_{\rm ML})$ rather than $P(M_{\rm HB})$, then
  - Fine tuning is not required to produce stars with very low envelope masses.
  - A gap (the lower or EHB gap in M13 and NGC 2808) can arise naturally.
  - As metallicity increases the mid-HB becomes difficult to populate. Bimodality is easy to explain in high metallicity systems.
  - It is just as easy to make EHB stars at high metallicity as at low.
  - In *some* globular clusters over entire range of cluster metallicity, $P(\eta_{\rm ML}) > 0$ for values of $\eta_{ML} \sim 2 \times \eta_{\rm ML}^{\rm norm}$. I.e., globular clusters know how to make EHB stars.
  - There may be two or more mass loss mechanisms.

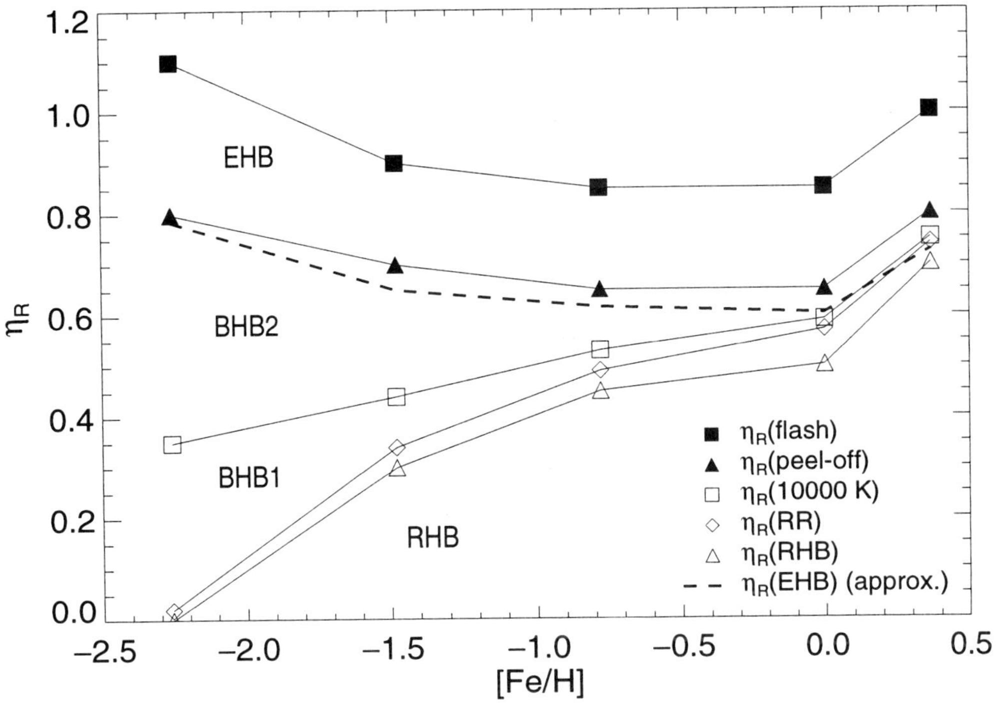

FIGURE 5. This plot of η_R vs [Fe/H] shows η_R values that correspond to (i) the blue end of RHB — η_{RHB}, (ii) RR Lyraes — η_{RR}, (iii) effective temperature of 10000 K — η_{BHB}, (iv) η_{peel} — the transition between classic tip flashers and hot He flashers, (v) η_{fl} — the largest value of η_R for which a flash occurs. The dashed line is the approximate location of η_{EHB}, the lowest value of η_R for which EHB stars form.

Even though there is no room for a complete discussion here, we note that the blue tail and EHB stars are probably not the result of binary mergers or collisions because:

- The observed stars have parameters more or less like those of single star models. One would expect that mergers and collisions produce a more diverse population. Basically, if it looks like a duck and quacks like a duck, there is a pretty good chance it is a duck.
- There is no variation (at least in a gross sense) of BHB/EHB population with position in a cluster. This is in contrast to the blue stragglers, generally considered to be certifiable interaction products. BSS are generally strongly concentrated toward the cluster center (M3 BSS are bimodal)

This research was supported by NASA Long Term Astrophysics Grant NAGW-2596.

REFERENCES

Arp, H. C. 1955, AJ, 60, 317

Arp, H. C., Baum, W. A., & Sandage, A. R. 1953, AJ, 58, 4

Baade, W. 1944, ApJ, 100, 137

Bono, G., Fulle, M., & Pulone, L. 1991, in The Formation and Evolution of Star Clusters, edited by K. Janes (PASP, San Francisco), p. 234

Cannon, R. D. 1985, in Observational Tests of Stellar Evolution Theory, ed. A. Maeder & A. Renzini (Dordrect: Reidel), 123

Castellani, M., & Castellani, V. 1993, ApJ, 407, 649

Catelan, M. 1993, A&AS, 98, 547

Catelan, M. & de Freitas Pacheco, J. A. 1993, AJ, 106, 1858

Catelan, M., & de Freitas Pacheco, J. A. 1995, A&A, 297, 345

Cudworth, K. M., & Rees, R. 1990, AJ, 99, 1491

Crocker, D. A., Rood, R. T., & O'Connell, R. W. 1988 ApJ, 332, 236

D'Cruz, N. L., Dorman, B., Rood, R. T., & O'Connell, R. W. 1996, ApJ, 466, 359

Dorman, B., Rood, R. T., & O'Connell, R. W. 1993, ApJ, 419, 596

Faulkner, J. 1966, ApJ, 144, 978

Faulkner, J. 1972, ApJ, 173, 401

Ferraro, F. R., Clementini, G., Fusi Pecci, F., Buonanno, R., & Alcaino, G. 1990, A&AS, 84, 59

Ferraro, F. R., Paltrinieri, B., Fusi Pecci, F., Cacciari, C., Dorman, B., Rood, R. T. 1997, submitted to ApJ Letters

Fusi Pecci, F., Ferraro, F.R., Bellazzini, M., Djorgovski, D.S., Piotto, G., Buonanno, R., 1993, AJ, 105, 1145

Greggio, L. & Renzini, A., 1990, ApJ, 364, 35

Greenstein, J. L. 1939, ApJ, 90, 387

Hachenberg, O. 1939, Zs. f. Ap., 18, 49

Hess, R. 1924, in Probleme der Astronomie: Seeliger Festscrift, ed. H. Kienle (Springer: Berlin) p. 265

Kraft, R. P., & Hiltner, W. A. 1961, ApJ, 134, 850

Iben, I., Jr., & Rood, R. T. 1968, ApJ, 154, 215

Iben, I., Jr., & Rood, R. T. 1970, ApJ, 161, 587

Johnson, H. L., & Schwarzschild, M. 1950, ApJ, 113, 630

Peterson, R. C. 1983, ApJ, 275, 737

Peterson, R. C. 1985a, ApJ, 289, 320

Peterson, R. C. 1985b, ApJ, 294, L35

Reimers, D. 1977, A&A, 57, 395

Richer, H. B., Harris, W. E., Fahlman, G. G., Bell, R. A., Bond, H. E., Hesser, J. E., Holland, S., Pryor, C., Stetson, P. B., Vandenberg, D. A., Vandenbergh, S. 1996, ApJ, 463, 602

Piotto, G., Sosin, C., Djorgovski, S. G., King, I. R., Rich, R. M., Dorman, B., Liebert, J., & Renzini, A. 1997, in this volume

Rood, R.T. 1973, ApJ, 184, 815

Rood, R. T., & Crocker, D. A. 1985, in The Production and Distribution of C, N, O Elements, ed. J. Danziger, F. Matteucci, & K. Ajär (Garching: ESO), 61

Rood, R. T., & Crocker, D. A. 1989, in The Use of Pulsating Stars in Fundamental Problems of Astronomy, ed. E. G. Schmidt (Cambridge: Cambridge University Press), 103

Rood, R. T., Crocker, D. A., Fusi Pecci, F., Ferraro, F. R., Clementini, G., & Buonanno, R. 1993, in The Globular Cluster–Galaxy Connection, ed. G. H. Smith, J. P. Brodie (San Francisco: ASP), 218

Rood, R. T., & Seitzer, P. O. 1982, in Astrophysical Parameters of Globular Clusters, ed. A.G.D. Philip & D. S. Hayes (Schenectady: L. Davis), 369

Sandage, A.R. 1953, AJ, 58, 61

Sandage, A. R. 1969, ApJ, 157, 515

Sandage, A. R., & Wallerstein, G. 1960, ApJ, 131, 598

Sandage, A. R., & Wildey, R. 1966, ApJ, 150, 169

Shapley, H. 1915, Contri. Mt. Wilson Obs., #116

Shapley, H., & Davis, H. N. 1919, Contri. Mt. Wilson Obs., #176

Shapley, H. 1930, Star Clusters, (Cambridge: Harvard Coll. Obs.), p26ff

Sosin, C., Piotto, G., Djorgovski, S. G., King, I. R., Rich, R. M., Dorman, B., Liebert, J., & Renzini, A. 1997, in this volume

Stetson, P.B., Vandenberg, D.A., Bolte, M., 1996, PASP, 108,560

Trumpler, R. J. & Weaver, H. F. 1953, Statistical Astronomy (Berkeley/Los Angeles: University of California Press) p. 392

van den Bergh, S. 1967, AJ, 72, 70

Walker, A. R. 1992, PASP, 94, 1063

Whitney, J., Rood, R. T., & O'Connell, R. W. 1997, in preparation

HST Observations of High-Density Globular Clusters†

By G. PIOTTO[1], C. SOSIN[2], I. R. KING[2], S. G. DJORGOVSKI[3], R. M. RICH[4], B. DORMAN[5], A. RENZINI[6], S. PHINNEY[3], AND J. LIEBERT[7]

[1]Dipartimento di Astronomia, Università di Padova, Vicolo dell' Osservatorio 5, I-35122 Padova, Italy

[2]Astronomy Dept., University of California, Berkeley, CA 92720-3411, USA

[3]Astronomy Dept., MS 105-24, California Institute of Technology, Pasadena, CA 91125, USA

[4]Astronomy Dept., Columbia University, 538 W. 120th St., Box 43 Pupin, New York, NY 10027, USA

[5]Laboratory for Astronomy and Solar Physics, Code 681, NASA Goddard Space Flight Center, Greenbelt, MD 20771, USA

[6]Dipartimento di Astronomia, Università di Bologna, Cp. 596, I-40100 Bologna, Italy

[7]Steward Observatory, University of Arizona, Tucson, AZ 85721, USA

In this paper we present preliminary results from our HST project aimed at exploring the connection between stellar dynamics and stellar evolution in the cores of high-density globular clusters.

1. Introduction

Until a few years ago Galactic globular clusters (GC) were regarded as ideal laboratories for testing stellar evolution theories and for studying stellar dynamics in *simple* stellar systems, and as important relics of the formation of the Galaxy. For many years most of the research proceeded as if the problems in the single fields mentioned above could be understood and solved independently of each other.

Now, however, each branch of GC studies is realizing that further progress depends on viewing each cluster as a kind of *ecosystem* of interrelated species. We have a growing body of observational evidence that dynamical interactions among stars in high-density clusters can modify their stellar content. Color gradients have been observed from the ground in the central regions of some post-core-collapse (PCC) or high-concentration clusters (*cf.* Djorgovski & Piotto 1993 for a review and references). The gradients are always in the sense of a bluer center, and extend even to the far-UV wavelengths. The gradients reflect radial changes in the stellar population. In addition, Fusi Pecci, Ferraro, & Cacciari (1993) have shown that the length of the horizontal branch (HB), and the presence and the extent of blue tails in particular, are correlated with the cluster density and concentration, in the sense of more concentrated or denser clusters having bluer and longer HB morphologies. The theoretical understanding of these phenomena remains unclear, though it is very likely that binaries and stellar interactions are involved in modifying stars located on the evolved branches of the color–magnitude diagram (CMD).

The exceptional resolving power of HST is of fundamental importance in this kind of study, as it allows observing faint stars in the center, down to the main sequence.

† Based on observations with the NASA/ESA *Hubble Space Telescope*, obtained at the Space Telescope Science Institute, which is operated by AURA, Inc., under NASA contract NAS5-26555

As an example, in Figure 1 we compare one of the best ground-based images on which our previous investigation (Djorgovski & Piotto 1993) was based with the corresponding HST frame.

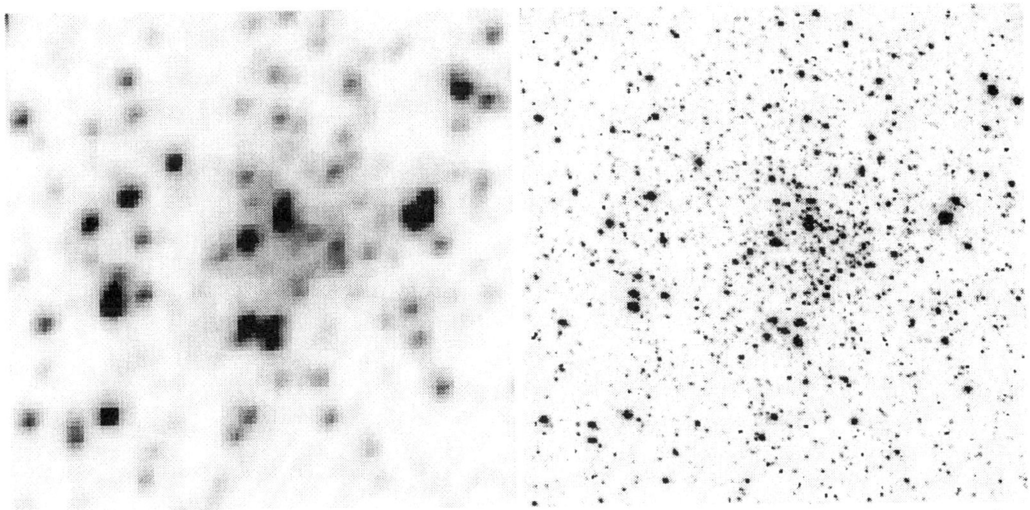

Figure 1. The *left panel* shows a ground-based V-band image of the inner 21×21 arcsec^2 of the GC M30, taken at the ESO NTT telescope (seeing FWHM = 0.6 arcsec). The *right panel* shows the same field from the PC frame of the WFPC2 camera in the F555W band.

In this paper we present some preliminary results from our HST project specifically devised to explore the connection between stellar dynamics and stellar evolution by investigating the GC star population in the very inner regions of 10 high-density clusters.

2. Observations and data reduction

The central regions of 10 GCs were observed with the WFPC2 camera on HST in the F218W, F439W, and F555W bands, during Cycle 6. The targets were chosen on the basis of either their high density and/or concentration or because of the strong UV flux of unknown origin detected by IUE. In our analysis, we used also the archive data (from an HST program by Yanny *et al.*) on 3 additional PCC clusters: NGC 6624, NGC 7078, and NGC 7099. Our cluster sample spans a factor of 100 in metallicity, covering almost the entire GC metallicity range. In this paper we focus on the F439W and F555W images.

The stellar photometry has been obtained with DAOPHOT, setting the parameters as discussed in Cool & King (1995). The F439W and F555W instrumental magnitudes have been transformed into standard B and V magnitudes following Holtzman *et al.* (1995). There is a general agreement in the photometric zero point with the previous ground-based investigations, with the noticeable exception of NGC 2808, which seems ~ 0.1 mag fainter in V and redder than in Ferraro *et al.* (1990).

3. The color–magnitude diagrams

The CMDs for the 13 clusters are shown in the following paper by Sosin *et al.*.

From a minimum of ~ 3000 (NGC 6652) to more than 27000 (NGC 7078) stars have been identified in each cluster, from ~ 2 magnitudes below the TO to the tip of the giant branch (GB). Such a large sample of stars is of particular importance to check the stellar clock (Renzini & Fusi Pecci 1988), and a comparison with the theoretical luminosity functions (LF) is in progress.

All the evolved branches of the CMD are clearly identifiable in the 13 clusters. In particular, we note that all the 13 clusters show a well-defined blue straggler (BS) sequence, extending from the TO to about 2.5 magnitudes brighter. In only a couple of cases (NGC 6441 and NGC 6522) does the strong field contamination make it difficult to extract the true cluster BS population. A population of supra-HB stars is visible in at least 7 clusters: NGC 1851, 1904, 2808, 6388, 6441, 6522, and 7078.

The most exciting features in the present set of CMDs concern the HB. Six clusters (NGC 1904, 2808, 6388, 6441, 6552, 7078), plus probably also NGC 362, show extended blue HB tails. In all cases the blue tail seems to extend down to the photometric limit of our data. In particular, the HB of NGC 2808 is densely populated down to at least two magnitudes below the TO. There are at least two unexpected and, at the moment, inexplicable findings:

- the blue horizontal branches, with extended blue tails, in the two metal-rich clusters NGC 6388 ([Fe/H] = −0.60) and NGC 6441 ([Fe/H] = −0.53);
- the clumpy nature of the extended blue HB tail of NGC 2808.

3.1. *NGC 6388 and NGC 6441*

A more detailed discussion of NGC 6388 and NGC 6441 will appear in Rich *et al.* (1997).

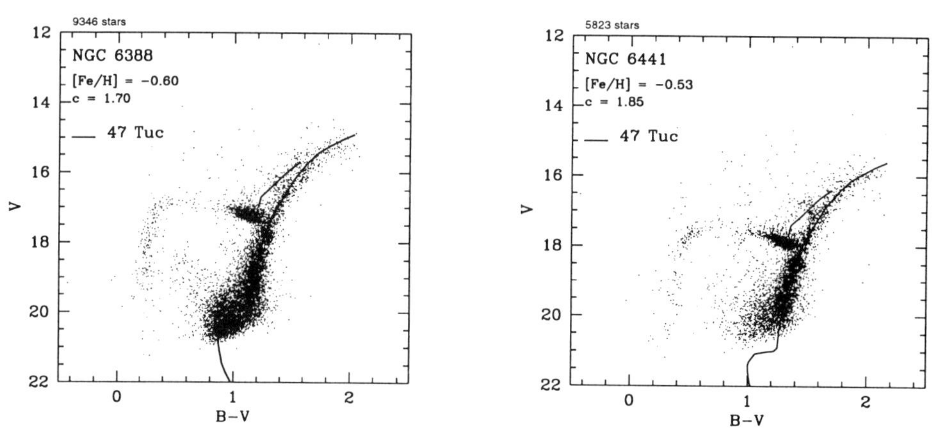

Figure 2. (left) The $(B, B-V)$ CMD of NGC 6388 is compared with the fiducial sequence of 47 Tuc. The slopes of the giant branches of the two clusters are similar (the NGC 6388 giant branch is marginally flatter), suggesting similar metal content. (right) As in the left panel, but for NGC 6441. Again, the metallicity of NGC 6441 cannot differ much from that of 47 Tuc. The color extent of the red HBs of NGC 6388 and NGC 6441 is greater than that of the HB of 47 Tuc.

Regarding the average metal content of NGC 6388 and NGC 6441 there are very few doubts. Figure 2 reproduces their CMDs. For comparison, we have over-plotted the fiducial points of the CMD of 47 Tuc ([Fe/H] = −0.71). This comparison clearly shows that the giant branches of NGC 6388 and NGC 6441 have similar slopes to that of—or are marginally flatter than—the giant branch of 47 Tuc. This means that, on average, they have the same—or slightly higher—metallicity than 47 Tuc. On the basis of their metal content we would expect to see only a red stub of the HB. Instead, we see also a prolonged HB tail, which extends to the limit of our photometry in both of them, with a clear presence of at least one and possibly two gaps, which remind us of the gaps in NGC 2808. No standard evolutionary model can reproduce these blue tails and the red HB at the same time. We note here that in view of the presence of both the (unexpected) blue tail and of the red HB stub, NGC 6388 and NGC 6441 might be regarded as two further examples of clusters with a bimodal HB (like NGC 1851 and NGC 2808), or better, as the most metal rich among the known bimodal-HB GCs. Whatever the origin of the blue tail is, NGC 6388 and NGC 6441 tell us that a truly metal-rich old population can also create hot blue stars, and this result might be related to the ultraviolet flux increase towards shorter wavelengths discovered in elliptical galaxies (Bertola *et al.* 1980).

A close inspection of the CMDs in Figure 2 (*cf.* also Figure 1 in Sosin *et al.*) shows that the red HB is peculiar, as is in general the entire CMD. Indeed, we can compare the CMDs of NGC 6388 and NGC 6441 with the CMDs of the other metal-rich clusters in the present sample, i.e., 47 Tuc, NGC 5927, and NGC 6624. In these three clusters, the red HB (RHB) is very well defined, well confined to a restricted color interval, almost parallel to the color axis, and very well separated from the red GB (RGB). NGC 6388 and NGC 6441 have a completely different RHB, spread out in a larger color interval, inclined with respect to the color axis, merging with the RGB on the red side. These features might be thought to be due to differential reddening or to crowding effects. We are presently running artificial-star tests, but it is unlikely that internal photometric errors can be the explanation, as the effect is perfectly visible in a similar way also in the CMDs from the WF3 and WF4 chips alone, which are much less crowded. There might be some differential reddening, though the average reddening is not very high (< 0.4 mag) and the covered field is very small, particularly with the PC camera, while the effect is visible in every single WFPC2 chip. A direct comparison of the CMDs in different regions of NGC 6388 shows that there is no average zero-point difference at scales larger than about 8 arcsec. So if there is some differential reddening, it must be present at scales smaller than 8 arcsec, still leaving the average reddening the same at large scales in the chip area of about 160×160 arcsec^2. On the contrary, there might be some differential reddening in NGC 6441, at the level of a few hundredths of a magnitude. Differential reddening might be a possibility, though it is does not seem likely, at least for NGC 6388.

There is another possibility, which might be an interesting working hypothesis. At a first glance, the CMDs of NGC 6388 and NGC 6441 strongly recall the CMD of ω Cen, apart from the RHB, raising the suspicion that also in the former two clusters there might be some spread in metallicity. At least qualitatively, a spread in metallicity might be an appealing explanation for the anomalous HB and for the spread in the RGB. Of course, this is just a possibility, suggesting that it might be of some interest to study the metal content of a few single stars in these clusters.

Another working hypothesis is that there might be two stellar populations in the cluster, with two different ages. This possibility has been excluded for all the other known bimodal GCs, and it seems unlikely that it can work for NGC 6388 and NGC 6441. Unfortunately, the present material does not allow testing this hypothesis.

There is a further possible explanation for the BHB tail in NGC 6388 and NGC 6441: it might well be possible that their BHBs originate from the same phenomenon (whatever it is) of core dynamics that is responsible for the bluer and longer HB tails in the clusters with denser cores (Fusi Pecci *et al.* 1993, Buonanno *et al.* 1997). Again, NGC 6388 and NGC 6441 would be the most metal-rich clusters in which the phenomenon has been discovered. Tidal stripping of red-giant envelopes during close stellar encounters has been suggested as a possible cause (see, however, Djorgovski *et al.* 1991). Indeed, Rich *et al.* (1997) have calculated that NGC 6388 and NGC 6441 have among the highest collision rates for any globular cluster in the Galaxy. It is therefore possible that these high collision rates are responsible for the extended BHB. But this scenario presents a problem: if tidal collisions are responsible for the production of the hot HB stars, we would expect them to have a different radial distribution compared to other evolved stars in the cluster. However, we find that blue and red HB stars, subgiants, red giants, and asymptotic GB stars all have the same radial distribution (in these and in all the other 11 clusters), and this might be a serious problem for the tidal-stripping model.

3.2. *NGC 2808*

A detailed discussion of the results on NGC 2808 will be presented by Sosin *et al.* (1997).

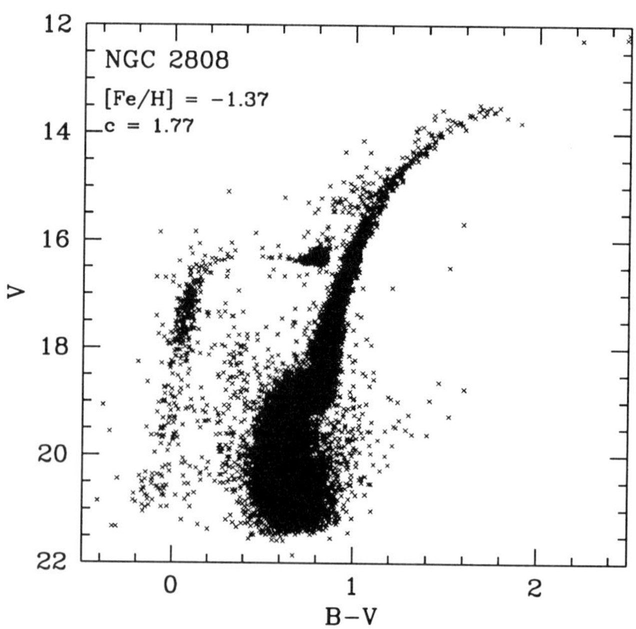

Figure 3. The V vs. $(B-V)$ CMD for 27286 stars in NGC 2808. Note that the HB population is divided into four groups, separated by three gaps.

The $(V, B-V)$ CMD of NGC 2808 is presented in Figure 3. This cluster was already known to have a bimodal HB with a blue and red HB more or less evenly populated. What is new here is that the long blue tail of the HB extends down to $V = 21$, at the limit of our photometry. The blue tail is not evenly populated: there are two significant gaps (*cf.* also Figure 2 of Sosin *et al.*, in the following paper) at $V = 18$ and at $V = 20$ and colors F218W$-B = -1.3$ and -2.0†, corresponding to effective temperatures $\log T_{\rm eff} = 4.23$ and 4.40, respectively. As shown in Sosin *et al.* (1997), the two gaps in the blue tail correspond to masses near 0.54 and 0.495 $m_\odot$, and the widths of the depleted regions in mass are very narrow, $\sim 0.01\ m_\odot$. To these two gaps we have to add also a third gap,

† Note that while the F439W and F555W magnitudes have been converted to Johnson B and V magnitudes, the F218W magnitudes have been calibrated to the STMAG instrumental system, using the zero points in the tables of Holtzman *et al.* (1995).

between the red and blue HB. As far as the discontinuities in the HB distribution are concerned, NGC 2808 is not unique. If we look at the 13 CMDs in the present paper, we will see gaps which recall the gaps in the blue tail of NGC 2808 (but are not necessarily similar or of similar origin) also in NGC 1904, NGC 6441, NGC 7078, and perhaps NGC 6388. Moreover, two significant gaps at more or less the same temperature as the gaps in NGC 2808 are also present in the blue HB of high-latitude halo subdwarfs (Newell 1973). So gaps might be not an exceptional feature in the blue HB.

It is not clear what the origin of these gaps is. In any case, either the BHB stars have a particular set of properties which make them land in well-defined regions of the HB after the helium flash, or the stars which happen to fall on the gap regions rapidly evolve off them. While this last possibility might (at least partially) explain the intermediate gap (close to the mass where the HB evolution changes from predominantly redward to blueward), no obvious reason can be found for the other two gaps (*cf.* Sosin *et al.* 1997).

As in the case of NGC 6388 and NGC 6441, at the moment we have no satisfactory explanation for the multi-modality of the HB of NGC 2808. It might be due to dynamical effects, mass-loss processes, or a combination of factors. For sure, the peculiar HBs of these clusters strongly remind us how poor is our knowledge of the mass-loss processes, which are the main cause of the HB morphology, as they set the final stellar envelope mass, i.e., where the stars should move after the helium flash.

4. The blue stragglers

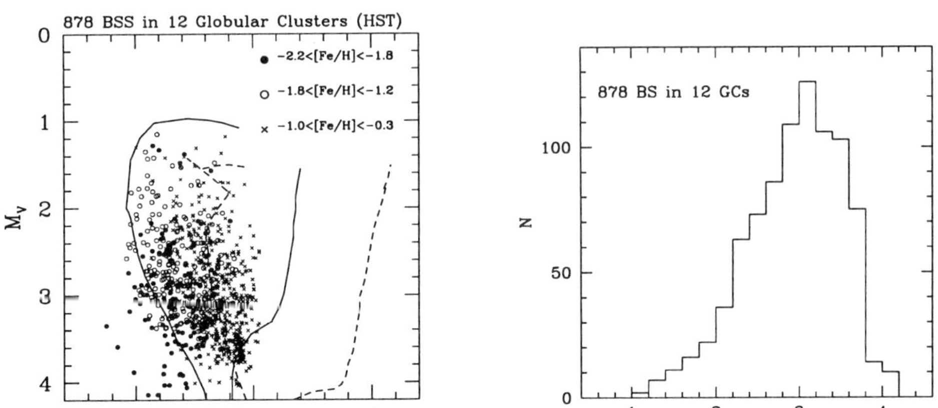

Figure 4. The *left panel* shows The M_V, $(B-V)_0$ CMD for the 878 BSs selected from 12 clusters in our sample. The BSs have been divided into 3 metallicity groups. The isochrones for a 1.6 $m_\odot$ and a 0.8 $m_\odot$ star with metallicities $Z = 10^{-4}$ and 4×10^{-3} are also plotted. The *right panel* shows the V luminosity function for the same BSs.

As already noticed in the previous Sections, all of the 13 clusters in the present sample have a population of blue stragglers. We will discuss the properties of the BSs in a forthcoming paper (Piotto *et al.* 1997).

Here we want to present only a few preliminary results. A sample of 878 BSs have been extracted from 12 of the 13 clusters in our sample (even though NGC 6522 appears to have a BS population, the contamination by field stars prevented us from extracting its BSs). This is the largest sample of BSs in GCs available so far (*cf.* Fusi Pecci *et al.* 1993 for a comparison). Moreover, all the BS magnitudes are in a single, photometrically homogeneous system.

Figure 4, *left panel*, shows the M_V,$(B\text{-}V)_0$ CMD for the 878 BSs. Distance moduli and reddenings have been extracted from the compilation by Djorgovski (1993). The BSs of Figure 4 have been divided into three metallicity groups (see labels). As expected, the BSs become bluer and bluer as the metallicity decreases. In the same figure we have plotted the isochrones (from Bertelli *et al.* 1994) for a 1.6 $m_\odot$ star for the two metallicities approximately corresponding to the extreme metallicities of our sample of GCs (the full line is for $Z = 10^{-4}$ and the dashed line is for $Z = 4 \times 10^{-3}$, respectively). On the red side are the isochrones for a 0.8 $m_\odot$ star for the same metallicities. The *right panel* of Figure 4 shows the LF from the 878 BSs.

Here we briefly summarize the main results on the BS population:

- The brightest BSs have magnitudes as bright as, but not exceeding the magnitude of the most metal-poor isochrone for a 1.6 $m_\odot$ star. In other words, the masses of the BSs in our sample do not exceed twice the mass of a normal TO star (note that their masses could be smaller, according to the BS evolutionary models by Baylin & Pinsonneault 1995).
- There are many BSs redder than the $Z = 4 \times 10^{-3}$ isochrone. This might be due both to photometric errors and to the evolution of these stars off the main sequence.
- It might be harder to explain the presence of a group of metal-poor BSs significantly bluer than the $Z = 10^{-4}$ isochrones. We need to run artificial-star experiments, but it seems unlikely that all those objects come from photometric errors.
- In all the 12 clusters, a Kolmogorov–Smirnov test has shown that the BSs are more concentrated than the subgiants of the same magnitudes, with a confidence level that is always greater than 99.99%. This result is of particular interest because this is the only group of stars whose radial distribution differs significantly from the others.
- The BSs closer to the center are marginally bluer and significantly brighter than the BSs farther from the center.
- The LF shows that there are BSs up to 3 magnitudes brighter than their corresponding TO. The LF rises steeply up to $M_V \sim 3.1$, while the significance of the apparent decline at fainter magnitude is uncertain, due to the possible bias in selecting the BSs. No significant gaps can be found in the present LF.

REFERENCES

Baylin, C.D., & Pinsonneault, M.H. 1995, ApJ, 439, 705

Bertelli, G., Bressan, A., Chiosi, C., Fagotto, F., Nasi, E., 1994, A&AS, 106, 275

Bertola, F., Capaccioli, M., Holm, A.V., Oke, J.B. 1980, ApJ, 237, L65

Buonanno, R., Corsi, C.E., Bellazzini, M., Ferraro, F.R., & Fusi Pecci, F. 1997, to appear in AJ, February 1997

Cool, A.C., & King, I.R. 1995, in *Calibrating HST: Post Servicing Mission*, eds. A. Koratkar & C. Leitherer (Baltimore: STScI), p. 290

Ferraro, F.R., Clementini, G., Fusi Pecci, F., Buonanno, R., & Alcaino, G. 1990, A&AS, 84, 59

Fusi Pecci F., Ferraro, F.R., & Cacciari, C. 1993, in *Blue Stragglers*, ed. R.A. Saffer, ASPCS 53 (San Francisco: ASP), p. 97

Djorgovski, S. 1993, in *Structure and Dynamics of Globular Clusters*, eds. S.G. Djorgovski & G. Meylan, ASPCS 50 (San Francisco: ASP), p. 37

Djorgovski, S., Piotto, G., Phinney, E.S., & Chernoff, D.F. 1991, ApJ, 372, L4

Djorgovski, S., & Piotto, G. 1993, in *Structure and Dynamics of Globular Clusters*, eds. S.G. Djorgovski & G. Meylan, ASPCS 50 (San Francisco: ASP), p. 203

Holtzman, J.A., Burrows, C.J., Casertano, S., Hester, J.J., Trauger, J.T., Watson, A.M., Worthey, G. 1995, PASP, 107, 1065

Newell, E.B., 1973, ApJS, 26, 37

Piotto, G., et al. 1997, in preparation

Renzini, A., & Fusi Pecci, F. 1988, ARAA 26, 199

Rich, R.M., et al. 1997, in preparation

Sosin, C., et al. 1997, in preparation

Globular-Cluster Color–Magnitude Diagrams with HST†

By C. SOSIN[1], G. PIOTTO[2], S. G. DJORGOVSKI[3], I. R. KING[1], R. M. RICH[4], B. DORMAN[5], J. LIEBERT[6], AND A. RENZINI[7]

[1]Astronomy Dept., University of California, Berkeley, CA 92720-3411, USA

[2]Dipartimento di Astronomia, Università di Padova, Vicolo dell' Osservatorio 5, I-35122 Padova, Italy

[3]Astronomy Dept., MS 105-24, California Institute of Technology, Pasadena, CA 91125, USA

[4]Astronomy Dept., Columbia University, 538 W. 120th St., Box 43 Pupin, New York, NY 10027, USA

[5]Laboratory for Astronomy and Solar Physics, Code 681, NASA Goddard Space Flight Center, Greenbelt, MD 20771, USA

[6]Steward Observatory, University of Arizona, Tucson, AZ 85721, USA

[7]Dipartimento di Astronomia, Università di Bologna, Cp. 596, I-40100 Bologna, Italy

This poster paper illustrates the color–magnitude diagrams discussed by Piotto *et al.* in the preceding paper. We present CMDs for 13 clusters; and we emphasize the discovery of additional blue horizontal branches in two metal-rich clusters, and the four-mode HB of NGC 2808.

---

1. Introduction

The factors that determine the morphology of the horizontal branch in globular clusters are still not well understood. Metal abundance plays an important role, with the most metal-rich clusters usually having the reddest HBs. However, a number of "second parameters," such as age, have been proposed as explanations of the clusters that deviate from this rule.

Here we present preliminary results from a *Hubble Space Telescope* program that aims to explore connections between stellar evolution and cluster dynamics, and to investigate the CMD morphology of some clusters that are difficult to observe from the ground. The central regions of ten clusters were observed with the WFPC2 on *HST*. Some of the clusters chosen had a high central density and/or concentration; others had ultraviolet flux of unknown origin detected by *IUE*. We also include here three additional CMDs that we have derived from archival data originally taken by Yanny in another program.

We present two results: the discovery of additional blue horizontal branches in two metal-rich globular clusters, and the intriguing "clumpy" nature of the blue horizontal-branch tail of NGC 2808.

2. Blue horizontal branches in metal-rich clusters

An exciting result from this program has been the discovery of additional *blue* horizontal branches in two *metal-rich* clusters, NGC 6388 and NGC 6441. Both are extremely crowded from the ground, and their HBs have not been seen previously.

† Based on observations with the NASA/ESA *Hubble Space Telescope*, obtained at the Space Telescope Science Institute, which is operated by AURA, Inc., under NASA contract NAS5-26555

Both clusters were also detected in the ultraviolet by *IUE*, and it was suggested that their UV brightness arose from blue HB stars. While a handful of blue HB stars have indeed been detected in metal-rich populations, blue populations as large as those discovered here have not previously been seen. (For discussion of these points, and references, see the preceding paper by Piotto *et al.*.) What is puzzling, however, is that some other clusters with nearly the same metal abundance have no blue HB stars (see our CMDs of 47 Tuc and NGC 5927). It appears that another "second-parameter" problem may be at hand!

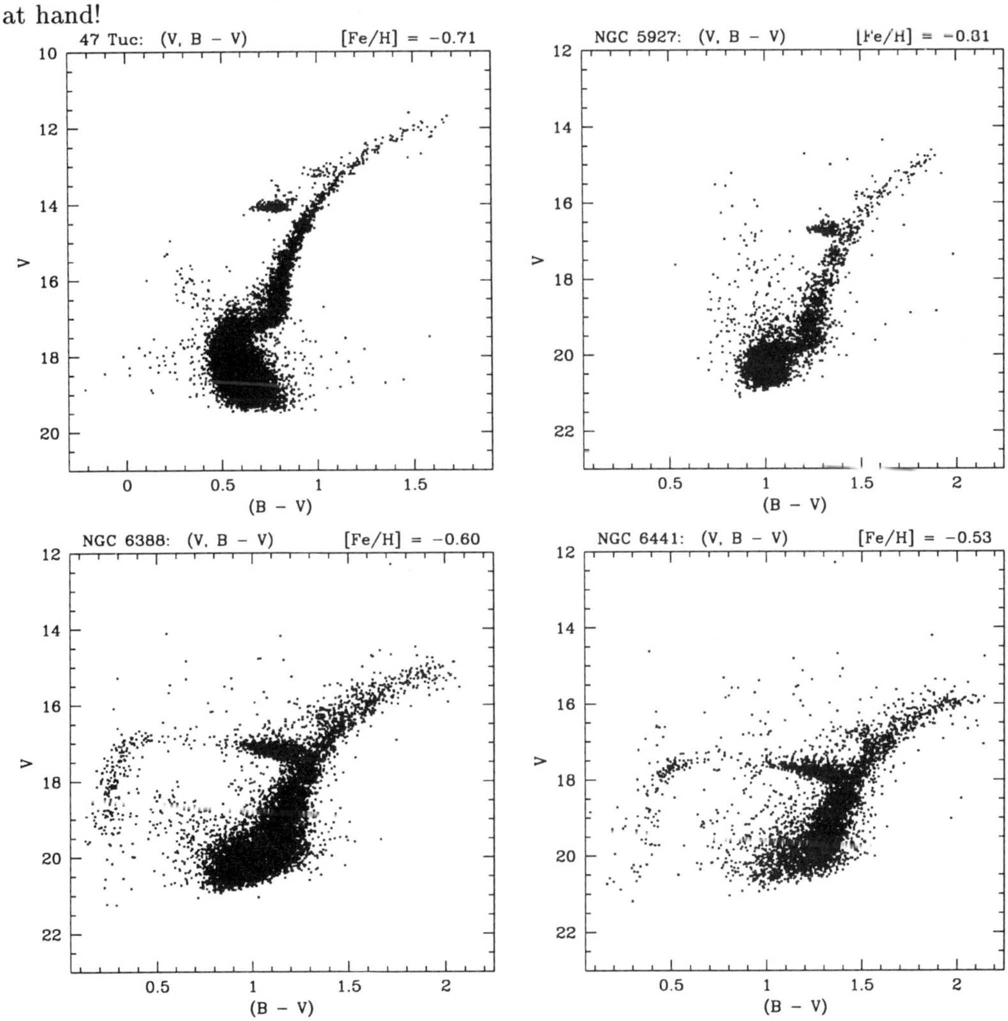

Figure 1. The upper two high-metal-abundance clusters have only a red stub of a horizontal branch, while the lower two have an additional blue portion.

3. The Long, Clumpy HB of NGC 2808

The horizontal branch of NGC 2808 has provided another fascinating result. Already known to be separated from the red stub, its blue end is now seen to extend many magnitudes fainter in V than previously known, down to $V \simeq 21$.

A CMD taken with an ultraviolet filter (F218W, $\lambda_{\rm eff}$ = 2189Å) spreads out the HB tail in color, allowing better separation of stars of different envelope mass. A histogram of this "blue vertical branch" shows that it is made up of at least two distinct groups, and possibly a third at its extreme blue end (in addition to the red stub that does not appear in the UV diagram). No mass-loss mechanism is known that could lead to such well-defined HB clumps.

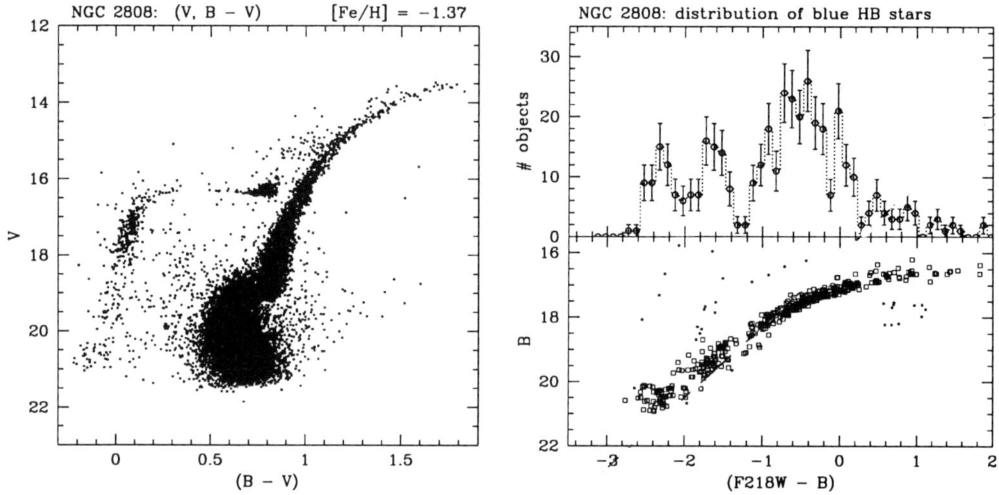

Figure 2. (left) The (B, V) CMD of NGC 2808. (right) The (FUV, B) CMD of NGC 2808, with a graph above it showing the distribution of the stars in color. Note that the red-stub part of the HB is too weak in the FUV to appear in this diagram at all.

4. Other clusters

The analysis of these data continues. Color–magnitude diagrams for our other clusters are shown in Figure 3; a number of interesting features in them are being investigated.

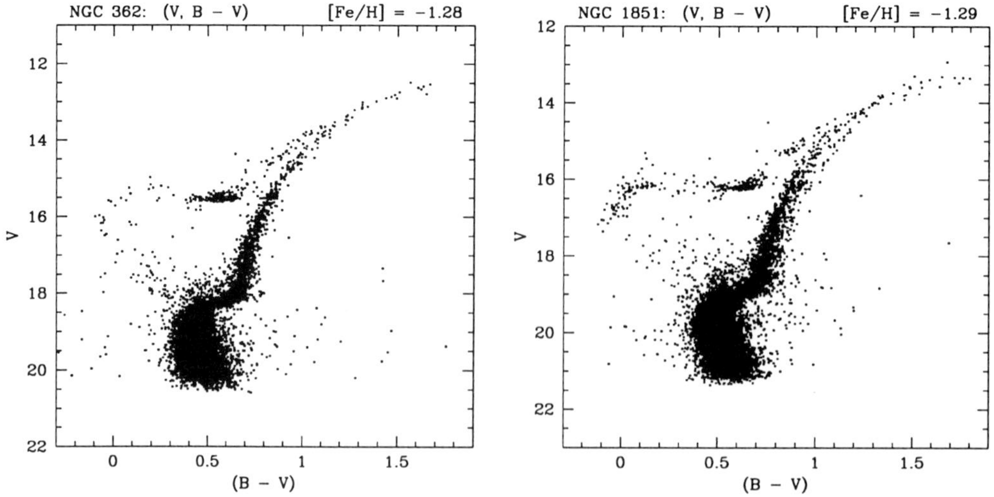

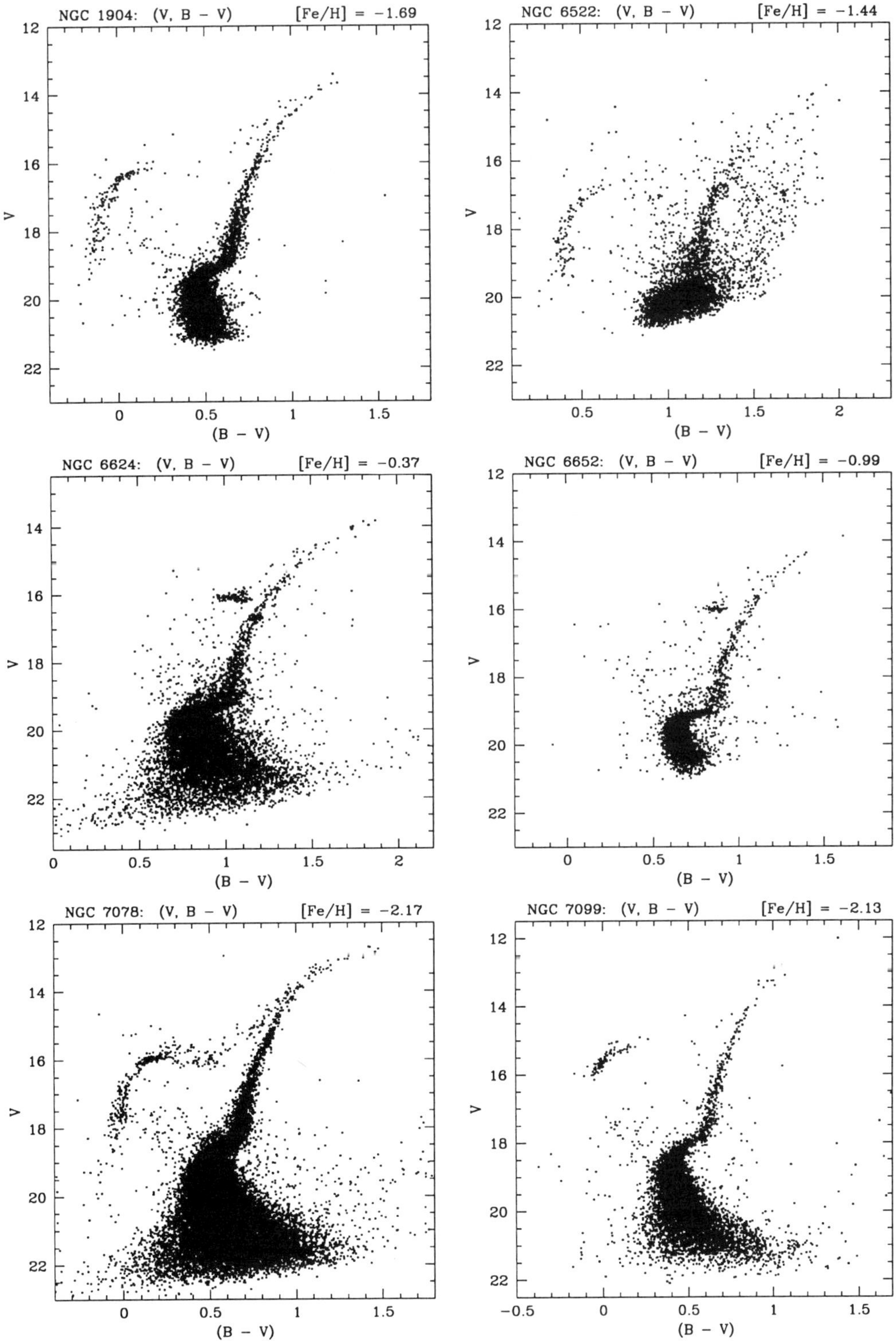

Figure 3. CMDs of eight more globular clusters.

The Horizontal-Branch Morphology of the Outer-Halo Globular Cluster NGC 6229

By M. CATELAN[1], J. BORISSOVA[2], A. V. SWEIGART[1], AND N. SPASSOVA[2]

[1]NASA/Goddard Space Flight Center, Code 681, Greenbelt, MD 20771, USA

[2]Institute of Astronomy, Bulgarian Academy of Sciences, 72 Tsarigradsko chausèe, BG–1784, Sofia, Bulgaria

We present CCD photometry for stars in the central region of the outer-halo globular star cluster NGC 6229. The horizontal-branch (HB) morphology of this cluster is found to be very peculiar. In particular, there is a clear indication of a strongly bimodal distribution along the HB, similar to those found in NGC 1851 and NGC 2808: a well-populated red HB and an extended blue HB tail are clearly present, but the instability strip at the RR Lyrae level seems to be more scarcely populated. A gap along the blue HB has also been positively identified, and there is some (weak) evidence for the presence of radial gradients in the stellar populations. The derived R ratio is extremely small, possibly suggesting a low helium abundance *or* the presence of a very well populated, but as yet undetected, extreme HB reaching down to the main-sequence turnoff level in V. A period-shift analysis for NGC 6229 and NGC 1851 RR Lyrae variables reveals interesting properties: NGC 6229 seems to possess a *negative* mean period shift at a fixed effective temperature in comparison with M3, while NGC 1851 shows the opposite behavior. New, extensive grids of synthetic HB models have been calculated with the purpose of better understanding the origin of the HB bimodality in NGC 6229, as well as the nature of the gap on the blue HB.

1. Introduction

NGC 6229 is a relatively poorly studied globular cluster (GC) in the outer Galactic halo ($R_{GC} \simeq 28$ kpc). It is quite dense ($c \simeq 1.61$) and only mildly metal-poor ([Fe/H] $\simeq -1.44$). The most extensive CMD available to date is from the work by Carney, Fullton, & Trammell (1991, hereafter CFT91), where the outer regions of the cluster were investigated down to a limiting V magnitude just above the turnoff (TO) level. Their study strongly suggested that the HB of NGC 6229 is peculiar for its position in the Galaxy. The consequent need to obtain more extensive samples of HB stars provided the major motivation for us to obtain photometric data for the *inner* parts of the cluster.

2. Observations and data analysis

The observations of NGC 6229 were made using the 2m Ritchey-Chrétien telescope of the Bulgarian National Astronomical Observatory. Both B and V CCD frames were obtained for the cluster under stable seeing conditions.

The reduction of the data for the cluster central regions proved to be challenging, since NGC 6229 is a quite distant and concentrated GC. By applying a wavelet transform analysis (e.g., Murtagh, Starck, & Bijaoui 1995), however, we were able to obtain improved photometric data for the cluster stars. The derived CMD is shown in figure 1. A more detailed description of the data reduction will be presented elsewhere (Borissova *et al.* 1996).

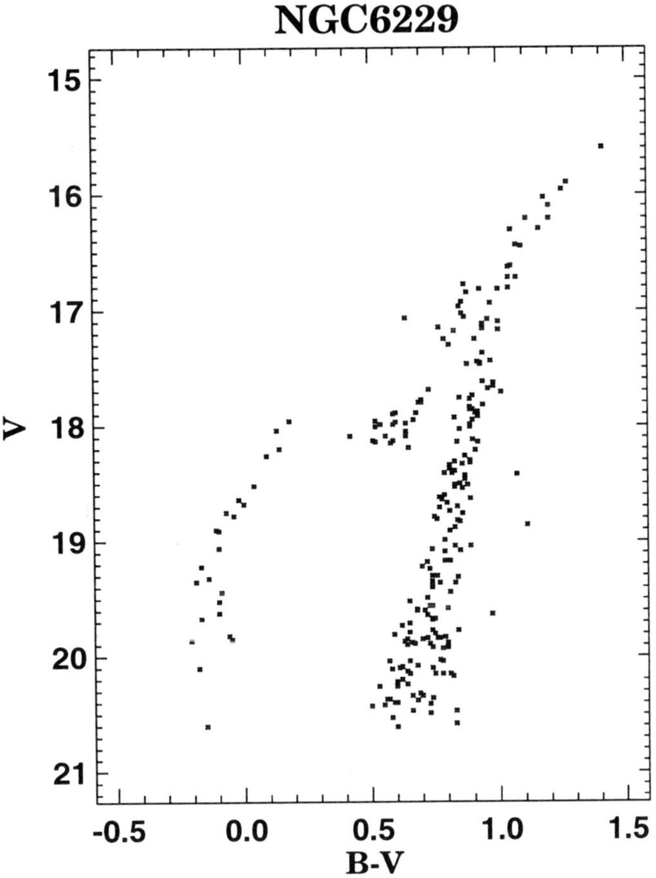

FIGURE 1. The derived CMD for NGC 6229 from our CCD data for the inner parts of the cluster (crosses), superimposed on the CFT91 data for the outer regions (plus signs).

3. The RGB "bump," number ratios, and radial gradients

Combining our increased sample of stars with the CFT91 sample, we have attempted to search for the "bump" in the red giant branch (RGB) luminosity function, following the Fusi Pecci *et al.* (1990) procedure. We do find a subtle feature in the cluster's luminosity function, at approximately the same location as expected on the basis of the Fusi Pecci *et al.* results for other clusters. Curiously, however, this possible "bump" is much less pronounced than found in other clusters of comparable metallicity, but is quite similar to the bump that has been found in M3 (cf. Fusi Pecci *et al.* 1990).

We have also recomputed the relevant R-method (Iben 1968) number ratios and several HB morphology parameters, some of which are given in Table 1. In this table, $R = N_{\rm HB}/N_{\rm RGB}$, $R' = N_{\rm HB}/N_{\rm RGB+AGB}$ (where "AGB" stands for the asymptotic giant branch), and B, V, and R are the numbers of HB stars bluer, within, and redder than the instability strip, respectively. The number of RGB stars considered includes only those giants which are brighter than the HB level. $B2$ is the number of blue HB stars bluer than $(B-V)_0 = -0.02$ (cf. Buonanno *et al.* 1996). The R and R' ratios are significantly lower than the mean for the Galactic GCs, possibly indicating a low primordial

R	R'	$(B-R)/(B+V+R)$	$(B2-R)/(B+V+R)$	DT
1.08 ± 0.10	0.88 ± 0.08	$+0.26 \pm 0.02$	$+0.09 \pm 0.01$	$(1 \pm 1)/(7 \pm 1)$

TABLE 1. R ratio and HB morphology parameters for NGC 6229 (see text)

helium abundance (see also CFT91; note that the total number of HB stars considered has increased by a factor > 2.5 between CFT91 and the present study). Another possible interpretation is that a substantial number of extended blue-HB tail stars has been missed in the present analysis and in CFT91.

We find some (weak) evidence for the presence of radial population gradients in NGC 6229, in the sense that red HB stars may be more centrally concentrated than both RGB and blue HB stars.

4. The HB morphology of NGC 6229

The HB morphology of NGC 6229 is extremely interesting for a variety of reasons:

• The HB appears to be bimodal, in the sense that the instability strip is more scarcely populated than both the red HB and the blue HB. This is indicated in Table 1 by the derived Dickens (1972) type for the cluster. A similar behavior is known to be present in NGC 1851 (Walker 1992) and NGC 2808 (Djorgovski *et al.* 1996);

• A gap is clearly present in the blue HB of NGC 6229, at $B - V \approx 0$. Applying a χ^2 test similar to the one described by Newell (1973) and Crocker, Rood, & O'Connell (1988), we find a probability of 99.8% that the gap is "real;"

• The blue HB tail of NGC 6229 extends to the very limit of our photometry. The existence of an extension reaching the vicinity of the cluster's TO level (as in NGC 2808; cf. Djorgovski *et al.* 1996) is entirely possible, and may be supported by the derived R-method ratios (cf. § 3). The presence of such an extended blue HB would be consistent with the relatively high central concentration of the cluster (e.g., Fusi Pecci *et al.* 1993; Buonanno *et al.* 1996);

• The RR Lyrae variables seem to be peculiar, in the sense that a period shift analysis indicates they are shifted towards *lower periods* at a given temperature in comparison with the variables in M3 [i.e., $\langle \Delta \log P(T_{\rm eff}) \rangle_{\rm N6229-M3} \simeq -0.014$, where the shifts are measured with respect to the lower envelope of the M3 distribution]. Just an opposite behavior appears to be present in NGC 1851 [$\langle \Delta \log P(T_{\rm eff}) \rangle_{\rm N1851-M3} \simeq +0.068$], where a substantial fraction of the RRab Lyraes have anomalously large amplitudes ($A_B \approx 2.0$ mag). More modern light curves are needed to put these results on a firm basis. These points will be discussed more fully elsewhere (Catelan *et al.* 1996).

These peculiarities make NGC 6229 a unique GC in the outer halo. In particular, it is the *only* known outer-halo globular with a bimodal HB. NGC 6229 also has the longest blue HB tail yet detected among the outer-halo clusters. Explaining the HB morphology of NGC 6229 is thus expected to be of the utmost importance for the explanation of the second-parameter problem (Borissova *et al.* 1996), and will require both accurate photometry reaching the cluster's TO level and careful comparison with the evolutionary models.

Finally, we have discovered 12 possible new RR Lyrae stars in the cluster central regions.

5. Synthetic HB models for NGC 6229

In an effort to understand the peculiarities of NGC 6229's HB, we have set out to derive a new, extensive grid of synthetic HB models for the cluster, based on evolutionary tracks specifically computed for the purpose. The complete analysis will be published elsewhere (Catelan *et al.* 1996). Our preliminary analysis of these models shows that:

• Walker's (1992) suggestion that bimodal, NGC 1851-like HBs *may* be obtained from mass distributions derived from anomalously "wide" gaussian deviates is correct. In the case of NGC 6229, we are able to reproduce the main HB morphology parameters derived from the CMD number counts, including the HB bimodality, with a gaussian deviate with a mass dispersion of $\sigma_M \approx 0.05\ M_\odot$;

• However, Kolmogorov-Smirnov statistical tests suggest that genuinely bimodal HB mass distributions are somewhat more satisfactory at reproducing the cluster's overall HB morphology than are unimodal distributions;

• Gaps on the blue HB *may*, in some cases, be the result of statistical fluctuations, combined with the change in track morphology from one that is entirely redward-evolving to one that is blueward loop-dominated;

• The difference in brightness between the blue HB at the instability strip edge and the red HB in bimodal-HB clusters (Stetson, VandenBerg, & Bolte 1996), if real, can be better accounted for by models in which the blue HB originates from helium mixing on the RGB (Sweigart 1996).

This research was supported in part by the Bulgarian National Science Foundation under contract No. F–604/1996 with the Bulgarian Ministry of Education and Sciences. This work was performed while M. C. held a National Research Council–NASA/GSFC Research Associateship. A. V. S. acknowledges support through NASA grant NAG5–3028.

REFERENCES

Borissova, J., Catelan, M., Spassova, N., & Sweigart, A. V., 1996, AJ, submitted

Buonanno, R., Corsi, C. E., Bellazzini, M., Ferraro, F., & Fusi Pecci, F., 1996, AJ, submitted

Carney, B., Fullton, L., & Trammell, S., 1991, AJ, 101, 1699 (CFT91)

Catelan, M., Borissova, J., Sweigart, A. V., & Spassova, N., 1996, in preparation

Crocker, D. A., Rood, R. T., & O'Connell, R. W., 1988, ApJ, 332, 236

Dickens, R. J., 1972, MNRAS, 157, 281

Djorgovski, S., *et al.*, 1996, these proceedings

Fusi Pecci, F., Ferraro, F. R., Bellazzini, M., Djorgovski, S., Piotto, G., & Buonanno, R., 1993, AJ, 105, 1145

Fusi Pecci, F., Ferraro, F. R., Crocker, D. A., Rood, R. T., & Buonanno, R., 1990, A&A, 238, 95

Iben, I., Jr., 1968, Nature, 220, 143

Murtagh, F., Starck, J.-L., & Bijaoui, A., 1995, A&AS, 112, 179

Newell, E. B., 1973, ApJS, 26, 37

Stetson, P., VandenBerg, D. A., & Bolte, M., 1996, PASP, 108, 560

Sweigart, A. V., 1996, ApJL, in press

Walker, A., 1992, PASP, 104, 1063

Calcium Abundances of Blue Horizontal Branch Stars in ω Centauri

By N. L. D'CRUZ[1], R. J. DICKENS[2], D. HATZIDIMITRIOU[3], R. T. ROOD[1], R. W. O'CONNELL[1]

[1]Astronomy Dept, UVa, P.O.Box 3818 Charlottesville VA 22903, USA
[2]Department of Physics, University of Bristol, Tyndall Avenue, Bristol, BS8 1TL, UK.
[3]Physics Dept, University of Crete, Vassilika Vouton, Heraklion, Crete, Greece.

Spectra of 480 Blue Horizontal Branch (BHB) stars in ω Centauri have been obtained using the multiobject fiber spectroscopic facility, Autofib, at the 3.9 m Anglo-Australian Telescope. Preliminary analysis in the temperature range 7500 – 8200 K indicates that a large fraction of BHB stars are at lower metallicities compared to the RR Lyraes and the giants. This could possibly be used to elucidate how mass loss efficiency varies with metallicity. The data can also be used together with simulated HBs to derive the variation of HB luminosity with metallicity.

1. Introduction

The main motivation for obtaining spectra of a large number of blue horizontal branch (BHB) stars in the globular cluster ω Centauri was to derive the variation of HB luminosity with metallicity and to study the analog of the Oosterhoff Effect/Sandage Period Shift Effect in stable stars. Another reason for studying ω Cen's BHB stars is to learn about how mass loss efficiency varies with metallicity, which we address in more detail here.

We use ω Cen because of its unique properties — the stars show a large spread in abundance ($-2.2 \lesssim$ [Fe/H] $\lesssim -0.8$), they have essentially constant reddening, they are at the same distance and the large number of stars allow for adequate sampling of the BHB over its entire length. The spectra are being used to obtain log $T_{\rm eff}$, log g and [Ca/H]. It should be possible to separate out the stars according to metallicity to see how they populate the HB. Or if the stars are divided up in temperature bins, comparing the metallicity distribution between temperature bins will also give an idea of how the HB population varies with metallicity. In this way we hope to get some information on how mass loss efficiency depends on metallicity. We can also compare the abundance distribution to that of the RR Lyraes and the giants.

Dickens et al. (1993; DHR) gave a progress report on this project at "The Globular Cluster-Galaxy Connection" meeting.

2. Observations:

Stars were chosen from "The Catalogue of Magnitudes and Colors in the Globular Cluster ω Cen" (Dickens et al. 1988) within a range in V magnitude of 13.5 – 16.1 and in $B-V$ color of -0.2 to 0.6. The spectra were obtained at the 3.9 m Anglo-Australian Telescope with the multi-object spectroscopic fiber facility *Autofib* and IPCS detector. The spectral range covered was 3600 – 4439 Å with a resolution of 1.8 Å. Standard data reduction and sky subtraction techniques were applied to the spectra using the STARLINK Figaro software. For further details on the observations and for a sample spectrum, see DHR.

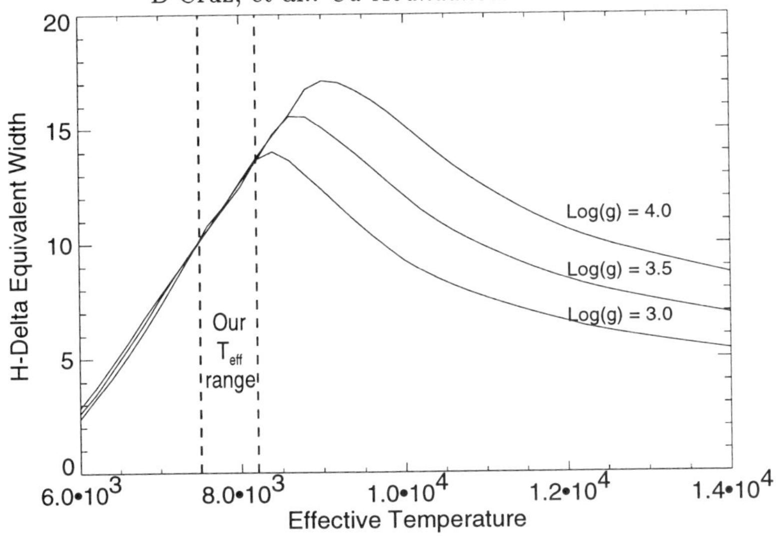

FIGURE 1. Hδ Synthetic Equivalent Widths are [Ca/H]= −1.5

3. Analysis:

Effective temperatures and gravities can be obtained by comparing observed Balmer-line profiles (primarily Hδ) to synthetic ones. Ca abundances are obtained similarly using the Ca II K line at 3634 Å. Model atmospheres are created from the Kurucz (1979) ATLAS 6 program for stars hotter than 7600 K and the MARCS program for cooler stars. Bell's SSG is used to generate the synthetic spectra.

3.1. *Effective Temperature and Gravity*

Least squares fitting between observed and synthetic Balmer line profiles is carried out over the appropriate range in T_{eff} and gravity (R. B. L. Greenlaw's software) to obtain T_{eff} and log g where the residuals are minimum. However, multiple solutions of T_{eff} and log g can be obtained for a given Balmer equivalent width (EQW) as shown in Figure 1.

Hence we have to rely on independent estimates of these quantities to constrain the range. Independent estimates of T_{eff} are found from V, $B-V$ (Dickens et al. 1988); more recent CCD V, $V-I$ photometry (Folgheraiter & Dickens 1997) will also be used. Gravity will be estimated from a modified Balmer Jump parameter and the last visible Balmer line method as detailed in DHR. However, log g is not needed for the results presented here.

We tested our analysis method using flux calibrated, non-fiber spectra of BHB stars in other globular clusters kindly supplied to us by Crocker (1996) and by "analyzing" synthetic spectra to which Gaussian noise was added. From these tests we find that T_{eff} in the range 7500 − 8200 K can be reliably determined as shown in Figure 1. On the other hand, it is obvious that HB luminosity in this temperature range must be obtained from photometry.

3.2. *Ca Abundance*

The Ca abundance can be determined for stars cooler than ∼ 9000 K from the Ca K line. For hotter stars the line is too weak to measure. Since the line is reasonably

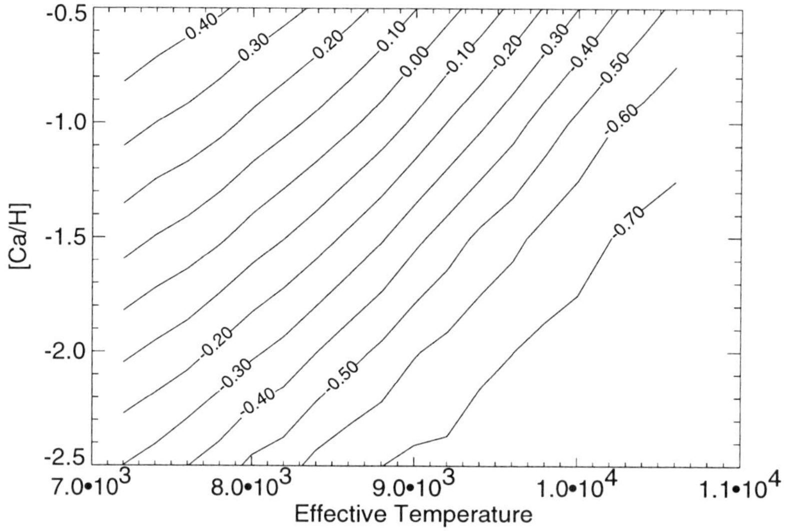

FIGURE 2. Contours of constant log(equivalent width) for the Ca K 3934 Å line as a function of metallicity and $T_{\rm eff}$ at log $g = 3.5$.

gravity independent, we determined [Ca/H] assuming log $g = 3.5$. The cluster has a heliocentric velocity of about 230 km s^{-1}, which partially separates out the interstellar Ca K component at our resolution. We obtained an average interstellar Ca K profile from the hottest stars and subtracted it from normalized cooler spectra to more reliably carry out the profile fitting.

The behavior of the Ca K EQW with $T_{\rm eff}$ and [Ca/H] is shown in Figure 2. It can be seen that if $T_{\rm eff}$ is underestimated (overestimated), [Ca/H] will also be underestimated (overestimated). [Ca/H] is approximately a bi-linear function in $T_{\rm eff}$ and log($EQW_{\rm CaK}$). This implies that the shape of the [Ca/H] distribution depends directly on the shape of the $T_{\rm eff}$ distribution, so systematic offsets between photometric and spectral $T_{\rm eff}$s (which could exist) will not affect the shape of the [Ca/H] distribution much.

4. Results

Our preliminary analysis has allowed us to determine the shape of the [Ca/H] distribution for the cooler stars as shown in Figure 3.

There is a larger fraction of BHB stars at lower metallicity compared to the Ca abundance distribution of the giants (Norris et al. 1996) and the RR Lyraes in ω Centauri (Greenlaw, 1993). The distribution does not show any hint of the bimodality seen in the giants and which possibly exists in the variables.

We feel that this result is valid because (1) The coolest BHB $T_{\rm eff}$s agree well with the blue edge of the RR Lyrae strip so the chance of the temperatures, and hence the abundances, being systematically low is small. (2) We carried out Monte Carlo simulations using the Ca distribution of Norris et al. (1996) over our temperature regime to see how temperature errors affect abundance. We fail to produce the above distribution for a wide range in $T_{\rm eff}$ errors. (3) Kinman (1996) independently analyzed of some of our spectra and finds a similar trend in abundance.

We do not yet understand why there is a difference between the abundance distribution in BHB stars and that in giants and RR Lyrae variables. It is possible that diffusion

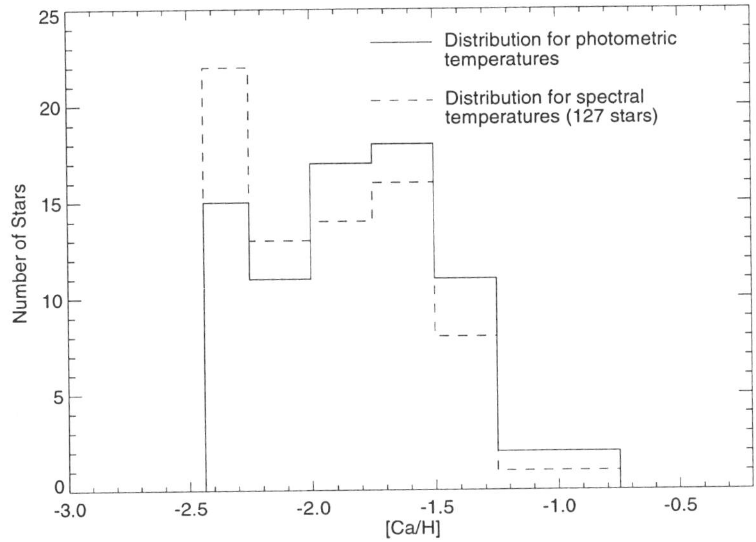

FIGURE 3. [Ca/H] distribution for BHB stars in the 7500 – 8200 K $T_{\rm eff}$ range.

of Ca could be affecting the BHB abundances (Michaud, et al. 1983). BHB stars in other clusters, with similar $T_{\rm eff}$ do not show any evidence for diffusion (Glaspey 1989; Peterson, Rood & Crocker 1995; Green, Liebert & Peterson 1996; Cohen & McCarthy 1997). Alternatively, it could be that the distribution of stars on the BHB is such that the lower metallicity stars dominate in the 7500 – 8200 K range and the hotter stars have a higher fraction of higher metal abundance. This would be somewhat akin to combining the very hot BHB of the intermediate metallicity cluster NGC 6752 with the comparatively cooler BHB of the metal poor cluster M92. We might be able to more easily gain some insight on mass loss efficiency as a function of metallicity in ω Cen if this is the case.

Acknowledgements:
We thank R.A.Bell for the use of his models and code, R.B.L.Greenlaw for making his software available, Debe Crocker for providing us with non-fiber spectra and Tom Kinman for running his analysis method on our spectra. NASA grant NAGW-2596 supported this research. NLD thanks the AAS for an International Travel Grant which partially supported her attendance at this meeting.

REFERENCES

Cohen, J. G. & McCarthy, J. K. 1997, AJ, in Press

Crocker, D. A. 1996, private communication

Dickens, R. J., Brodie, I. R., Bingham, E. A. & Caldwell, S. P. 1988, in R.A.L. Report no. 88-04

Dickens, R. J., Hatzidimitriou, D., & Rood, R. T. 1993, in The Globular Cluster-Galaxy Connection, ed. G. H. Smith & J. P. Brodie (San Francisco: ASP), 194 (DHR)

Folgheraiter, E. & Dickens, R. J. 1997 in preparation

Glaspey, J. W., Michaud, G., Moffat, A. F. J., & Demers, S. 1989, ApJ, 339, 926

Green, E. M., Liebert, J., & Peterson, R. C. 1996, in Formation of the Galactic Halo...Inside and Out, ed. H. L. Morrison & A. Sarajedini, (San Francisco: ASP), 184

Greenlaw, R. B. L. 1993, Ph.D. thesis, University of Leeds

Kinman, T. D. 1996, private communication

Kurucz, R. L. 1979, ApJS, 40, 1

Michaud, G., Vauclair, G, & Vauclair, S. 1983, ApJ, 267, 256

Norris, J. E., Freeman, K.C., & Mighell, K. J. 1996, ApJ, 462, 241

Peterson, R. C., Rood, R. T. & Crocker, D. A. 1995, ApJ, 453, 214

The Cepheid Instability Strip in External Galaxies

By NORMAN R. SIMON AND TODD S. YOUNG

Department of Physics & Astronomy, University of Nebraska, Lincoln NE 68588-0111

Preliminary evidence is presented which indicates the existence of two somewhat different types of distribution of long-period Cepheids in external galaxies. Prototypes are the Large and Small Magellanic Clouds. Linear pulsation calculations are performed to model these distributions. The sturdy nature of the Cepheid P-L relation is affirmed, but both analytic arguments and the linear model grids point to potential systematic errors reaching up to a few tenths of a magnitude if the Cepheids in calibrating and target galaxies have different distributions. We also point out some difficulties posed for stellar pulsation and evolution theory by the long-period Cepheids we have studied: The theoretical blue edges seem too hot and/or the inferred masses too large to account for the observed stars.

1. A Cepheid Dichotomy?

Recent HST measurements for two spiral galaxies, both supposedly near the Virgo Cluster core, show a substantial difference in distance, with NGC 4639 (Sandage, *et al.* 1996) a full magnitude farther than M100 (Ferrarese, *et al.* 1996). Since both of these are primary distances, based on Cepheids — what could be the source of the discrepancy? Is it possible that some part of it might be due to the Cepheid calibration itself? Or, more generally, how might galaxy-to-galaxy differences in the Cepheid distribution affect the distance scale? We shall examine this question in what follows.

Let us begin with the 22 long-period Cepheids from the Large Magellanic Cloud (LMC) used by the HST Key Project team to calibrate the distance to M81 (Freedman, *et al.* 1994a) and for the initial calibration in the case of M100 (Freedman, *et al.* 1994b). These Cepheids span the period range, $10 \leq P \leq 50d$. In the period-magnitude diagram, a free fit to these stars, with the form,

$$M_V = s \log P + q, \tag{1.1}$$

yields a slope $s = -3.3$.

Next, we consider a comparable sample of Cepheids from the Small Magellanic Cloud (SMC) in the same period range. This consists of 39 stars with V-I colors listed by Madore (1985). For these stars a period-magnitude fit according to Eq. (1.1) gives a slope $s = -2.5$. To the Cepheids in the Magellanic Clouds we may add two additional datasets, both from HST observations. These are M81 (Freedman, *et al.* 1994a) and IC 4182 (Saha, *et al.* 1994). Interestingly enough, when Eq. (1.1) is applied to these two samples, the results are $s = -2.5$ and $s = -3.3$, respectively.

Thus, we see some hint of a dichotomy in the period-magnitude diagrams of the four galaxies treated, — namely, a sharp slope, $s = -3.3$, for IC 4182 and the LMC and a shallow slope, $s = -2.5$, for M81 and the SMC. Formal statistical analysis shows that these differences exist at a level of two sigma. Is there reason to believe that this dichotomy is real? If so, what might be its cause?

Figure 1 shows a color-magnitude diagram for the 22-star LMC sample. These points do not seem to display the traditional positive slope dictated by the blue edge of the instability strip (e.g., Chiosi 1990). By constrast, the long-period Cepheid sample in the Small Magellanic Cloud (SMC) does show a positive slope. The latter stars are portrayed

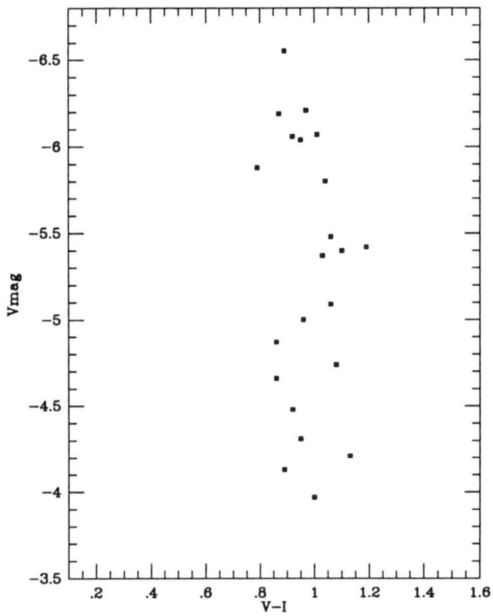

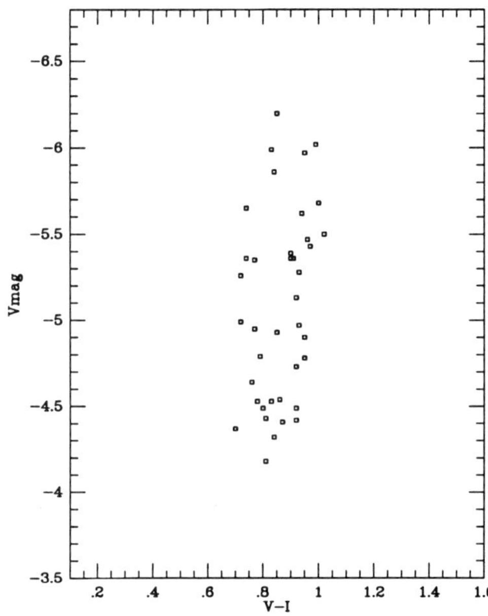

FIGURE 1. (left) Color-magnitude diagram for long-period Cepheids from the LMC. See text.
FIGURE 2. (right) Same for the SMC. See text.

in Figure 2. It turns out that the same difference appears to the eye for IC 4182 versus M81 — the former slows a sharp, nearly vertical, or even negative slope, while the slope of the latter seems more mild. However, here the differences in all cases occur at the one sigma level and thus lack statistical significance.

Finally, one may plot period-color diagrams for the various galaxies. These diagrams (not shown here) display similar characteristics — steeper slopes for IC 4182 and the LMC, shallower slopes for the SMC and M81. In this case, the statistical difference lies between one and two sigma — suggestive to the eye but not statistically strong.

2. Analytic Theory

Does it make any sense theoretically to talk about a dichotomy? We treat this question first with simple analytic theory beginning with the period/mean-density law from pulsation theory, viz.,

$$\log P = 0.9339 \log L - 0.7678 \log M - 3.540 \log T_e + 11.50, \tag{2.2}$$

where P is in days, and L and M in solar units. The coefficients in Eq. (2.2) represent a fit to a few hundred linear nonadiabatic (LNA) Cepheid pulsation models constructed for the present study. The model parameters will be described later.

We now supplement Eq. (2.2) with a generic mass-luminosity relation for Cepheids evolving through the instability strip. This relation has the form

$$\log L = a \log M + b \tag{2.3}$$

Using (2.3) in (2.2), we get

$$\log P = a_1 \log L - 3.540 \log T_e + c_1 \tag{2.4}$$

where

$$a_1 = \left(0.9339 - \frac{0.7678}{a}\right), \qquad c_1 = 11.50 + 0.7678\frac{b}{a}. \tag{2.5}$$

Suppose now that the distribution of Cepheids in the H-R diagram follows the relation

$$\log L = a_2 \log T_e + c_2. \tag{2.6}$$

Inserting (2.6) in 2.4), we obtain

$$\log P = (a_1 a_2 - 3.540) \log T_e + c_1 + a_1 c_2. \tag{2.7}$$

Now the coefficient a in Eq. (2.3) is positive and has a value of about 4.0 (Simon 1995). Thus, the value of a_1 (Eq. 2.5) is about 0.74. On the other hand, the typical value one might assign to the coefficient a_2 in Eq. (2.6) is -20 or -30; this is the slope of the theoretical blue edge (Chiosi 1990). Thus in Eq. (2.7), the term $a_1 a_2$ is negative and dominates the slope, which gets steeper as a_2 increases in magnitude. Hence, as already seen in the four galaxies we have studied, the steeper the slope in the color-magnitude diagram (the observational counterpart of Eq. 2.6), the steeper will be the slope in the color-period diagram (the observational counterpart of Eq. 2.7).

Finally, we use Eq. (2.6) in Eq. (2.4) to eliminate $\log T_e$. The result is

$$\log L = \frac{1}{a_3} \log P + c_3, \tag{2.8}$$

where

$$a_3 = a_1 = \frac{3.540}{a_2}, \tag{2.9}$$

or

$$M_V = \frac{-2.5}{a_3} \log P + c_4. \tag{2.10}$$

If a_2 is large and negative as indicated above, then a_3 is positive, giving the observed negative slope in the period-magnitude diagram. At the same time one sees from Eqs. (2.8) and (2.9) that the larger the magnitude of a_2, the smaller a_3 and thus the steeper the slope in Eq. (2.10). That is, a steep slope in the color-magnitude diagram gives rise to a steep slope in the period-magnitude diagram. Thus we have demonstrated on general grounds that steep slopes in all the diagrams go together, as do shallower slopes.

Eqs. (2.9) and (2.10) can be used further in the following manner. Suppose one has a very steep slope in the color-magnitude diagram as in the LMC. This implies a_2 very large, and therefore $a_3 \simeq a_1 \simeq 0.74$ from Eq. (2.9). Putting this value of a_3 into Eq. (2.10), we obtain a slope of -3.4, very close to the slope actually observed in the LMC. On the other hand, the observed slope in the period-magnitude diagram for the SMC is -2.5. From Eq. (2.10) this implies $a_3 \simeq 1$, which in turn yields, via Eq. (2.9), $a_2 \simeq -14$. This value is somewhat smaller in magnitude than the slope of the theoretical blue edge, but still of the same order. We conclude that elementary pulsation theory gives us a good semi-quantitative picture of the relations among the ridge lines in the diagrams that have been displayed. It remains now to flesh out this picture with an appropriate series of pulsational models.

			Range	
Case	a	b	Te/100K	$M/M_\odot$
A	3.5	0.40	48 - 54	9 - 15
B	4.0	0.15	50 - 55	7.5 - 12.5
C	4.0	0.40	51 - 57	6.5 - 10
D	4.0	0.65	52 - 58	5.5 - 8.5
E	4.5	0.40	53 - 59	5 - 7.5

TABLE 1. Models According to Equation (2.3)

3. Theoretical Models

To aid in investigating the Cepheid samples in the various galaxies, we constructed a large number of linear pulsation models, spanning a wide range of masses, luminosities and temperatures. Five series of models were created, labeled A through E, in accordance with the generic M-L relation given in Eq. (2.3). These cases are listed in Table 1, where the progression is generally from lower to higher luminosity at given mass. The last two columns give for each case the approximate ranges in temperature and mass needed to accomodate the 10 to 50d period domain.

Our next task will be to convert the extragalactic Cepheid data for plotting on an H-R diagram. Given published distances and reddenings for the various galaxies (see Table 2), the absolute visual magnitude of each Cepheid is known. Then, the $\log L$ value for each star becomes a function of the temperature, via the bolometric correction formula given by Simon and Kanbur (1995). (Note that the very small gravity dependence in this formula may be eliminated with virtually no error by taking $\log g = 1.0$.) Thus

$$\log L = f(\log T_e). \qquad (3.11)$$

Now choose one of the cases in Table 2; this fixes a_1 and c_1 (Eq. 2.5). Use of these values in Eq. (2.4), along with Eq. (3.11) and the observed period of the given Cepheid, results in a polynomial equation for $\log T_e$ which may be solved iteratively; then Eq. (3.11) gives log L. Thus for any specified choice of an M-L relation, all of our observed sample of extragalactic Cepheids may be rendered on an H-R diagram.

Figure 3 shows such a diagram for case C of Table 1, while Figure 4 gives it for case A. Cepheids in the different galaxies are represented by different symbols as indicated. In each case the blue edge has also been drawn in as a straight line. We see that if we take the Cepheids to belong to case C, a large number of them are found to lie blueward of the theoretical instability strip, whereas the choice of case A accomodates many more stars. Comparison of Figures 3 and 4 shows that the blue edge changes very little between the two cases. Rather, it is the cooler inferred temperatures corresponding to case A that are responsible for moving most of the stars into the strip. These cooler temperatures are compensated by significantly higher masses, as can be seen from the last two columns in Table 1. While the stars with low L/M, corresponding to case A, greatly ammeliorate the the blue-edge problem, other difficulties are associated with them, as discussed briefly in Section 5, below.

4. The P-L Relation

We now wish to use the linear models to estimate how much error in distance might be incurred by differences in the Cepheid distribution between calibrating and target galaxies. These differences are of two sorts, which might or might not be linked. First

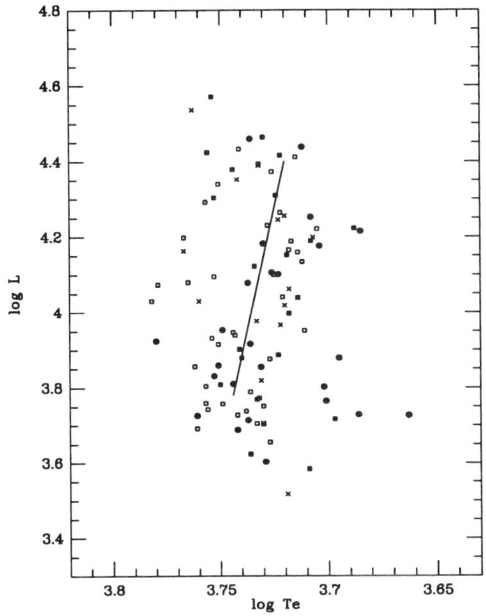

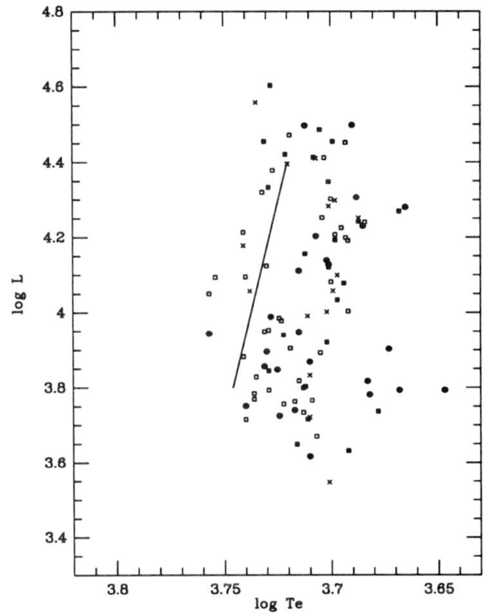

FIGURE 3. (left) H-R diagram for Cepheids in four galaxies, based upon case C. Filled squares: LMC; open squares: SMC; dots: M81; crosses: IC 4182. Solid line: blue edge.

FIGURE 4. (right) Same as Fig. 3, but based on case A.

Galaxy	(m-M)$_{true}$	E(B–V)	3.3E(B–V)	References†
LMC	18.50	0.10	0.33	1
SMC	18.90	0.09	0.30	2, 3
M81	27.80	0.03	0.10	4
IC 4182	28.36	0.00	0.00	5

† 1: Madore & Freedman (1991); 2: Smith *et al.* (1992); 3: Grieve & Madore (1986); 4: Freedman *et al.* (1994a); 5: Saha *et al.* (1994)

TABLE 2. Distances and Reddenings for Four Galaxies

there is the question of different distributions in the H-R diagram, as exemplified by the four galaxies we have treated; and, second, is the matter of different M-L relations such as those presented in Table 1. Of course, in any given case a combination of these complications could be operating.

Figure 5 shows a plot of all the case C models in the magnitude vs. temperature plane. To get at the first question above, the straight lines on the diagram select two subsets of the models, one with a negative, nearly vertical, slope reminiscent of the LMC (dashed lines) and the other with a positive slope more like that of the SMC (solid lines). We shall call these two cuts of the data "steep" and "shallow", respectively. Expanding the analysis to the other M-L relations, we have made similar cuts through the model spaces of all cases in Table 1. Since our goal is to study the P-L relation we focus on the period-magnitude diagram for each of the cuts, fitting the points in each case with a

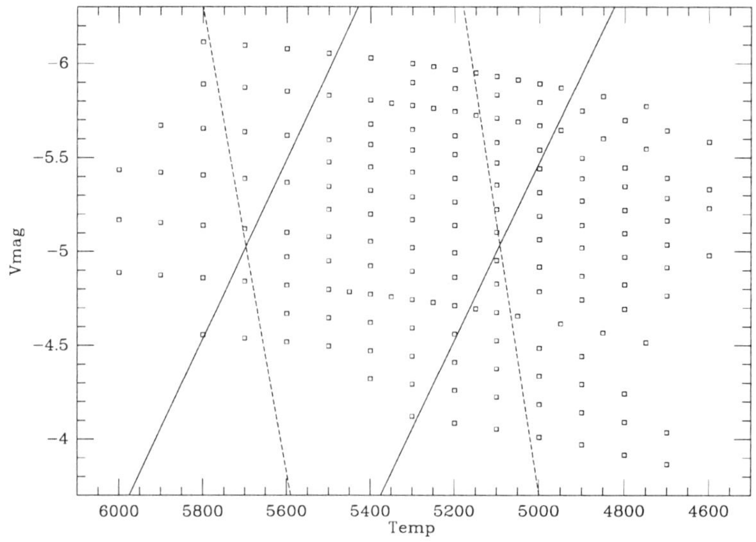

FIGURE 5. Magnitude vs. temperature for the full set of linear pulsation models characterized by case C. Dashed lines contain the "steep" subset; solid lines contain the "shallow" subset.

straight line that has the form of Eq. (1.1). Now we can choose any given distribution as a target (say, case C, "steep cut") and another as calibrator (say, case E, "shallow cut"), forcing the calibrator's slope onto the target and noting the resulting value of the zero point q, which would determine the distance if the exercise were real.

After subjecting each of the distributions to calibration by all of the others, we report the following results. Concerning the circumstance where target and calibrator have different slopes in the H-R diagram, the errors in the distance are small, ranging from 0.03 to 0.10 mag. When the target and calibrator differ in their M-L relation, the errors incurred become larger, typically one or two tenths of a magnitude. The largest error we found in performing this exercise was 0.25 mag.

That the range of uncertainty is so small is remarkable considering, particularly, the large differences among the cases in Table 2. Although the Cepheids in a galaxy governed by case A would be twice as massive as those in a case E galaxy, the estimated error incurred by comparing these two galaxies as if they were alike is only a quarter of a magnitude. On the one hand, this argues *the essential, impervious nature of the Cepheid P-L relation:* No matter how distinct (within reason) the Cepheid distribution in a given galaxy, the standard P-L relation will yield a reliable ballpark distance. On the other hand, our analysis shows that if the Cepheids in a calibrating and a target galaxy have different distributions the systematic error in the distance could conceivably be as large as the entire error budget assigned by Freedman, *et al.* (1994a; b).

5. Discussion

We have presented circumstantial evidence for the existence of various galaxies whose long-period Cepheid distributions differ from one another in the color-magnitude, color-period and period-magnitude diagrams. To study these differences LNA pulsation models were constructed to produce synthetic diagrams which closely resemble the observed ones. Using these models we verify the extremely sturdy nature of the Cepheid P-L relation, but also estimate that reasonable differences in the Cepheid distributions of calibrating and target galaxies could result in distance scale errors of a few tenths of a magnitude. However, since the evidence for a dichotomy in the Cepheid distribution remains shaky statistically, more observations will be necessary, particularly on the long-period samples in the Magellanic Clouds.

In addition to the problem of distance scale, the current investigation also raises questions concerning both pulsation theory and Cepheid evolution. Figure 3 shows that if the middling evolutionary scenario represented by case C is correct, then the current theoretical blue edge must be too cool by 300 to 400 K! On the other hand, if one chooses case A to ameliorate this problem, then other potential difficulties arise. In the first place, we know of no published evolutionary tracks with luminosity-to-mass ratio as small as that of case A. Secondly, case A requires that the masses of the long-period Cepheids be very large (9-15 $M_\odot$). This would have to be jibed in each galaxy with the putative age of the Cepheid sample and the presence or absence of other kinds of stars, including the resonant "bump Cepheids" with periods near 10d. These and other problems with Cepheid evolution have been discussed by Simon (1995).

Ultimately, we must ask of the theory of stellar evolution that it account for the Cepheid sample in a galaxy of known distance on a star-by-star basis, locating each object along an evolutionary track nested in the instability strip and unfolding on a nuclear timescale. It seems that considerable work will be required before this can be accomplished.

REFERENCES

Chiosi, C., 1990, in Confrontation Between Stellar Pulsation and Evolution, ed. C. Cacciari & G. Clementini (San Francisco: ASP), 158.

Ferrarese, L., *et al.*, 1996, preprint.

Freedman, W. L., *et al.*, 1994a, ApJ, 427, 628.

Freedman, W. L., *et al.*, 1994b, Nature, 371, 757.

Grieve, G. R. & Madore, B. F., 1986, ApJS, 62, 427.

Madore, B. F., 1985, in Cepheids: Theory and Observations, ed. B. F. Madore (Cambridge: Cambridge University Press), p. 166.

Madore, B. F. & Freedman, W. L., 1991, PASP, 103, 933.

Saha, A. *et al.*, 1994, ApJ, 425, 14.

Sandage, A. *et al.*, 1996, ApJ, 460, L15.

Simon, N. R., 1995, in Astrophysical Applications of Powerful New Data Bases, ed. S. J. Adelman & W. L. Wiese (San Francisco: ASP), 211.

Simon, N. R. & Kanbur, S. M., 1995, ApJ, 451, 703.

Smith, H. A., Silberman, N. A., Baird, S. R. & Graham, J. A., 1992, AJ, 104, 1430.

Theory and Observation of Cepheids and LPVs in the Magellanic Clouds

By PETER R. WOOD

Mount Stromlo and Siding Spring Observatories, Private Bag, Weston Creek PO, ACT 2611, Australia

Observations of Cepheids and long-period variables (LPVs) in the Magellanic Clouds are discussed. Recent Cepheid observations indicate that evolution and pulsation masses can be bought into agreement if the amount of main-sequence convective core overshoot increases with mass between 2 and 8 $M_{\odot}$. It is shown that new opacities allow precise modelling of the nonlinear light curves of bump Cepheids with a consequent very accurate determination of stellar properties. New observations of LPVs in the Magellanic Clouds reveal the existence of multiple pulsation modes. The multiple $K-logP$ relations seen in the LMC are most easily explained if the Mira variables are fundamental mode, rather than first overtone mode, pulsators.

---

1. Introduction

Stars in the late stages of stellar evolution, such as Cepheids and LPVs, are so bright that they can be observed relatively easily in the Magellanic Clouds. In addition, since the distances to the LMC and SMC are well known, absolute magnitudes can be accurately derived. Finally, the large population of stars in the Magellanic Clouds mean that large samples of Cepheids and LPVs are known there with a wide range in intrinsic properties such as luminosity and pulsation period. The wide range of properties is very useful for comparison of the observations with theory.

Since the 1970's, observations of stars in the Magellanic Clouds have been used widely for comparison with theoretical computations. Many of these computations were done more or less simultaneously with the observations, often by Icko Iben and collaborators. Special mention should be made of the interplay between the theory of dredge-up and carbon star formation, first found by Iben (1975) and Sugimoto & Nomoto (1975), and observations of carbon stars in the Magellanic Clouds. The latter provided valuable data on the luminosities and masses (via cluster membership) of these stars (e.g. Frogel *et al.* 1990). Similarly, Cepheids in the Magellanic Clouds were used by Iben & Tuggle (1975) and Becker, Iben & Tuggle (1977) to confront theory. The LPVs were important because they provided an identifiable sample of AGB stars with masses and luminosities going right up to the AGB limit (corresponding to the core mass reaching the Chandrasekhar limiting mass). This sample of stars has been useful for abundance studies related to third dredge-up and hot-bottom burning (Wood, Bessell & Fox 1983; Smith *et al.* 1995). In the following sections, some recent observational and theoretical results related to Cepheids and LPVs in the Magellanic Clouds are discussed.

2. Magellanic Cloud Cepheids and the Comparison of Evolution, Pulsation and Bump Masses

The evolution mass for a Cepheid variable is obtained from the observed luminosity of the Cepheid together with the theoretical relation between mass and luminosity for stars on helium core burning loops, where the latter relation is obtained from stellar evolution computations. However, the $M-L$ relation is poorly known for intermediate mass stars because of the uncertainty in the amount of convective core overshoot that occurs during the main-sequence phase of evolution (*e.g.* Bertelli, Bressan & Chiosi 1985). It should be possible to calibrate the amount of core overshoot if the masses and luminosities of Cepheids can be measured. Luminosities of Magellanic Cloud Cepheids are easily observed and the (pulsation) mass can be computed from the relation between M, L, $T_{\rm eff}$ and P obtained from pulsation theory alone. A comparison of evolution and pulsation masses has long indicated that evolution masses derived from standard theory without any convective core overshoot were larger than pulsation masses (eg. Iben and Tuggle 1972), the so-called "Cepheid mass discrepancy". However, until recently, a major concern with this procedure was that the various stellar masses derivable from pulsation theory, namely, the pulsation mass, the bump mass and the beat mass, did not agree well (Cox 1980) so that it was not clear that the standard evolution mass was in error rather than the pulsation mass.

The new OPAL opacities seem to have largely removed this discrepancy (Moskalik, Buchler & Marom 1992) so that, hopefully, one can now use the pulsation masses and luminosities of Magellanic Cloud Cepheids to calibrate the evolutionary $M-L$ relation and hence the amount of convective core overshoot during main sequence evolution of intermediate mass stars.

2.1. *Cepheid pulsation masses*

Sebo and Wood (1994; 1995) have recently obtained time sequences of accurate V and I CCD photometry for fields around the SMC cluster NGC 330 and the LMC cluster NGC 1850. The data has been used to compute the pulsation and evolution masses of Cepheids found around the clusters.

In the case of NGC 330, the luminosities and masses were found to be remarkably low (~2-5 $M_\odot$), a reflection of the fact that at low metallicity the blue loops can cross the instability strip at low masses. One of the effects of main-sequence convective core overshoot is to shorten the blue loops and Sebo & Wood (1994) argued that the amount of overshoot in the Cepheids of low (~2 $M_\odot$) would have to be modest, less than $\lesssim 0.25$ pressure scale heights, in order that the loops actually pass through the instability strip.

In and around NGC 1850, Cepheids with a wide range of periods and masses (~4-8 $M_\odot$) were found by Sebo & Wood (1995), with the lower mass limit being larger than around NGC 330 because of the higher metallicity. Sebo & Wood (1995) compared the evolution and pulsation masses of Cepheids near NGC 1850, assuming a distance modulus of 18.50, and found that relatively large amounts of overshoot ($\gtrsim 0.25$ pressure scale heights) were required to bring the evolution and pulsation masses into agreement. The combined results for NGC 330 and NGC 1850 indicate an increasing amount of main-sequence convective core overshoot with mass. This is perhaps not surprising as the many of the Cepheids near NGC 330 have masses barely larger than the minimum mass of ~1.5 $M_\odot$ required for a main-sequence star to develop a convective core.

3. A Bump Cepheid in the LMC

One of the Cepheids studied by Sebo & Wood (1995) in the vicinity of NGC 1850 turned out to be a bump Cepheid near the blue edge of the instability strip. Because of the proximity of this star to the blue edge, theoretical models can be made for it without convection being a dominant factor in controlling the pulsation: hopefully the models will be reliable as a result.

Using linear non-adiabatic pulsation theory and the resonance condition P_0/P_2=2.0 which is thought to pertain in bump Cepheids (Simon & Schmidt 1976), Sebo & Wood (1995) derived a distance modulus to HV 905 of 18.48-18.54 for a reddening of $E(B-$

V)=0.15, or 18.39-18.63 if the reddening was allowed to vary from $E(B-V)$=0.12-0.18. This represents a purely theoretical calibration of the LMC distance modulus which agrees well with other distance modulus estimates such as that for SN1987A (Panagia *et al.* 1991). Using the mass, luminosity and T_{eff} derived from linear theory, a nonlinear light curve was computed which gave a fair reproduction of the light curve of HV 905.

Recent models of bump Cepheids assume a smooth variation of properties such as bump phase with pulsation period. However, a glance at the compendium of LMC Cepheid light curves by Gaposhkin (1970) shows that, at a given period, there is a wide range in light curve amplitudes and shapes. This presumably arises from the finite width of the instability strip. By modelling individual light curves, precise values of T_{eff}, L and M should be derivable for Cepheids. Such an attempt at detailed fitting of a theoretical to an observed light curve has been made for HV 905. It was found that, for a star such as HV 905 near the blue edge of the instability strip, the overall pulsation amplitude is very sensitive to T_{eff}, while the ratio of P_0/P_2 sensitively defines the phase of the bump relative to the main part of the light curve. These two conditions, along with the requirement that the primary (fundamental mode) period equal 11.858 days, are sufficient to totally constrain the intrinsic properties M, L and T_{eff}. The only assumed quantity is the composition, which is taken to be Y=0.25 and Z=0.008, to match that observed for young stars and HII regions in the LMC (Russell & Bessell 1989; Dufour 1984). The light and colour curves resulting from fitting the models to the observed curves are shown in Figure 1. From the offset between the observed and theoretical $V-I$ colour curves, the reddening is obtained directly, while the apparent distance modulus is obtained from the offset between the observed and theoretical V light curves. As a result of these calculations (Wood, Arnold and Sebo 1996), the following properties are derived for HV 905: $M/M_{\odot} = 5.20 \pm 0.2$, $L/L_{\odot} = 4897 \pm 150$, $DM = 18.53 \pm 0.04$ and $E(B-V)$=0.10. The LMC distance modulus obtained here is purely theoretical and is possibly one of the most accurate distance modulii known.

4. New samples of low amplitude LPVs in the Magellanic Clouds

During the last decade, large numbers of LPVs have been discovered in the Magellanic Clouds from photographic searches of time series of I plates take by the UK Schmidt telescope (Wood, Bessell & Paltoglou 1985; Reid, Glass & Catchpole 1988; Hughes 1989). Infrared photometry of these variables has been used to derive their bolometric luminosities. These LPVs, of relatively large amplitude $\Delta I \gtrsim 0.5$, seem to fall into two distinct groups. The first group are the lower luminosity ($M_{bol} \gtrsim -5$) and shorter period ($P \lesssim$ 450 days) LPVs that presumably come from the old and intermediate age population that dominates in the LMC. These LPVs, which are the analogs of the Mira variables in our Galaxy, obey a tight $K-logP$ relation (Feast *et al.* 1989; Hughes & Wood 1990). The longer period, more luminous LPVs are young stars ($M \sim 3-8\,M_{\odot}$) in advanced stages of AGB evolution and they show enhanced s-process abundances indicating that the 3rd dredge-up has been occurring during helium shell flashes (Wood *et al.* 1983). These stars are important as they are the only known examples of massive, upper AGB stars. Observations by Smith *et al.* (1995) have shown that the bright, upper AGB stars have strongly enhanced Li abundances as expected in stars undergoing hot-bottom burning, which occurs when a deep convective envelope cycles material through the hydrogen-burning shell during quiescent AGB evolution (e.g. Iben 1975). It is presumably this hot-bottom burning that prevents the massive AGB stars from becoming carbon stars (Wood *et al.* 1983).

The CCD surveys of Sebo & Wood (1994; 1995) were designed to find LPVs as well

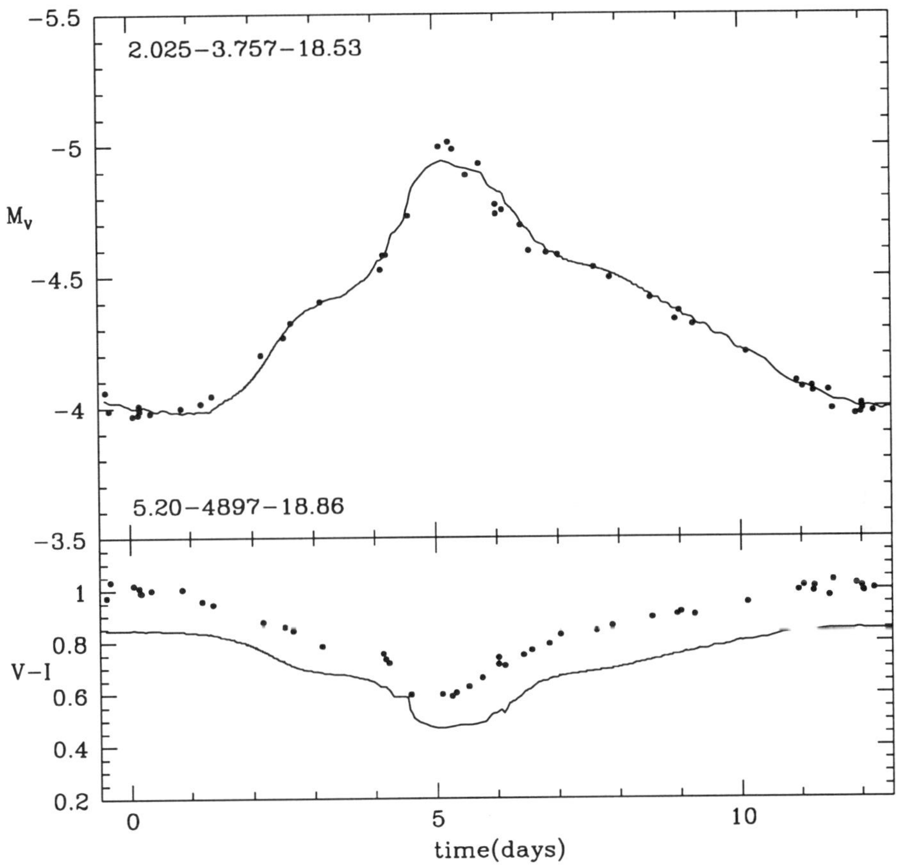

FIGURE 1. Light and colour curves for the LMC bump Cepheid HV 905 (dots) and the best fit theoretical model (continuous curves). In the top panel, the observed light curve has been shifted by the apparent V distance modulus 18.86. The three number at the top of the top panel are P_0/P_2, $T_{\rm eff}$ and true distance modulus, while the three numbers at the bottom of the top panel are M, L and apparent V distance modulus.

as Cepheids and they led to the discovery of a number of LPVs with amplitudes much smaller than those of the previously known Magellanic Cloud LPVs. Wood & Sebo (1996) obtained infrared JK photometry of these small amplitude stars and showed that roughly half of them did not fall on the well-defined $K{-}logP$ sequence occupied by the Mira variables in the LMC. Instead, many of the low amplitude LPVs fell on a sequence parallel to the Mira sequence but separated from it by $\Delta logP \sim 0.35$. Wood & Sebo (1996) argued that this situation is exactly what would be expected if the Mira variables were fundamental mode pulsators while the low amplitude LPVs were first overtone pulsators, since the fundamental and first overtone periods in a typical Mira have $\log P_0/P_1 \sim 0.35$. Furthermore, the theoretical $K{-}logP$ relation for fundamental mode pulsators was found to fit the observed Mira sequence when $T_{\rm eff}$ for the LPVs was consistent with a fit of model evolutionary tracks to the observed old/intermediate age AGB in the LMC.

These results, favouring fundamental mode as the pulsation mode of the Miras, stand in contrast to recent measures of the angular diameters of the Miras (Haniff *et al.* 1995) which suggest that the Miras have such large radii that their pulsation mode would seem to be first overtone. However, recent infrared angular diameter measurement (van Belle *et al.* 1996) suggest that the Haniff *et al.* (1995) results may not be definitive. The solution to this long-standing problem probably lies in obtaining large samples of LPVs of small amplitude so that the $K-logP$ sequences for different modes, and the gaps between them, are defined with more statistical certainty than for the small sample of Wood & Sebo (1996). Data from the MACHO experiment should provide such a sample.

REFERENCES

Becker, S.A., Iben, I., Tuggle, R.S. 1977, ApJ, 218, 633

Bertelli, G., Bressan, A.G., Chiosi, C. 1985, A&A, 150, 33

Cox, A.N. 1980, ARA&A, 18 15

Dufour, R.J. 1984, in IAU Symp. 108, Structure and Evolution of the Magellanic Clouds, ed. S. van den Bergh & K.S. de Boer (Dordrech: Reidel), p. 353

Feast, M.W., Glass, I.S., Whitelock, P.A., Catchpole, R.M. 1989, MNRAS, 241, 375

Frogel, J.A., Mould, J., Blanco, V.M. 1990, ApJ, 352, 96

Gaposhkin, S. 1970, Smithsonian Astrophysical Observatory, Special Report 310.

Haniff, C.A., Scholz, M., Tuthill, P.G. 1995, MNRAS, 276, 640

Hughes, S.M.G. 1989, AJ, 97, 1634

Hughes, S.M.G., Wood, P.R. 1990, AJ, 99, 784

Iben, I. 1975, ApJ, 196, 525

Iben, I., Tuggle, R.S. 1972, ApJ, 178, 441

Iben, I., Tuggle, R.S. 1975, ApJ, 197, 39

Moskalik, P., Buchler, J.R., Marom, A. 1992, ApJ, 385, 685

Panagia, N., Gilmozzi, R., Macchetto, F., Adorf, H.-M., Kirshner, R.P. 1991, ApJ, 380, L23

Reid, N., Glass, I.S., Catchpole, R.M. 1988, MNRAS, 232, 53

Russell, S.C., Bessell, M.S. 1989, ApJS, 70, 865

Sebo, K.M., Wood, P.R. 1994, AJ, 108, 932

Sebo, K.M., Wood, P.R. 1995, ApJ, 449, 164

Simon, N.R., Schmidt, E.G. 1976, ApJ, 205, 162

Smith, V.V., Plez, B., Lambert, D.L., Lubowich, D.A. 1995, ApJ, 441, 735

Sugimoto, D., Nomoto, K. 1975, PASJ, 27, 197

van Belle, G.T., Dyck, H.M., Benson, J.A., Lacasse, M.G. 1996, preprint

Wood, P.R., Arnold, A., Sebo, K.M. 1996, in preparation

Wood, P.R., Bessell, M.S., Fox, M.W. 1983, ApJ, 272, 99

Wood, P.R., Bessell, M.S., Paltoglou, G. 1985, ApJ, 290, 477

RR Lyrae Pulsational Masses Revisited

By G. BONO[1], C. CAPUTO[2], V. CASTELLANI[3] AND M. MARCONI[3]

[1]Osservatorio Astronomico di Trieste, Via G.B. Tiepolo 11, Trieste, 34131, ITALY

[2]Osservatorio Astronomico di Capodimonte, Via Moiariello 16, Napoli, 80131, ITALY

[3]Dipartimento di Fisica, Università di Pisa, Piazza Torricelli 2, Pisa, 56100, ITALY

We discuss mass estimates for double mode RR pulsators, relying on a detailed analysis of physical and numerical ingredients which affect the periods and the modal stability of RR Lyrae linear and nonlinear models. In particular we call attention to the role played by spatial resolution, opacities and full amplitude nonlinear models for firmly removing the discrepancy between pulsational and evolutionary masses. We provide a comparison between theoretical nonlinear models and double-mode RR Lyrae in Galactic globular clusters, showing that, according to the new theoretical scenario, not only pulsational masses but also pulsator luminosities appear in good agreement with the evolutionary prescriptions.

1. Introduction

The study of radial pulsating variables can provide an independent estimation of stellar masses based only on the ratio between the fundamental (P_0) and the first overtone periods (P_1). Petersen (1973) introduced the diagram P_1/P_0 vs. P_0 (hereinafter referred to as PD) as a suitable tool for estimating the actual mass value of double-mode pulsators (RRd).

For HB pulsators the evolutionary theories foresee masses of the order of $M/M_{\odot} \simeq 0.8$ and $\simeq 0.65$ for Oo II and Oo I clusters respectively, but for a long time pulsational masses remained below such evolutionary expectations.

Cox (1991) showed that pulsational computations with updated opacity evaluations reconcile pulsational and evolutionary predictions.

We present a reinvestigation of the problem which discloses some unexpected results and sheds new light on the matter.

2. Results

In Fig. 1 we present the PD for linear, nonadiabatic and radiative RR Lyare models.

These models were computed by adopting a fixed chemical composition ($Y = 0.24$, $Z = 0.0001$), the labeled values of stellar masses and luminosity levels and the analytical opacity approximation provided by Stellingwerf (1975). RRd variables observed in Oo I clusters are marked by *open squares* (IC4499, Clement et al. 1986) and by *stars* (M3, Cox, Hodson & Clancy 1983), whereas those observed in the Oo II clusters are displayed by: *open triangles* (M68, Walker 1994), *cross* (NGC6426, Clement & Nemec 1990), *diamond* (NGC2419, Clement & Nemec 1990) and *open circles* (M15, Nemec 1985 and other authors; for greater details the interested reader is referred to Bono et al. 1996).

The comparison between theoretical models and observational data clearly supports previous results given in the literature under similar theoretical assumptions and disclose the occurrence of the "mass discrepancy problem".

In order to test the dependence of the PD on spatial resolution we have then computed linear RR Lyrae models characterized by a finer spatial resolution. The thoretical periods obtained from this analysis are plotted in Fig. 2.

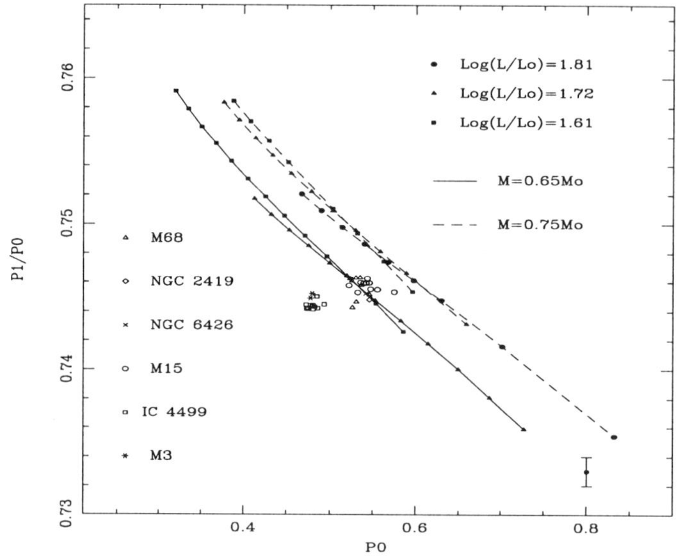

FIGURE 1. PD for linear nonadiabatic models.

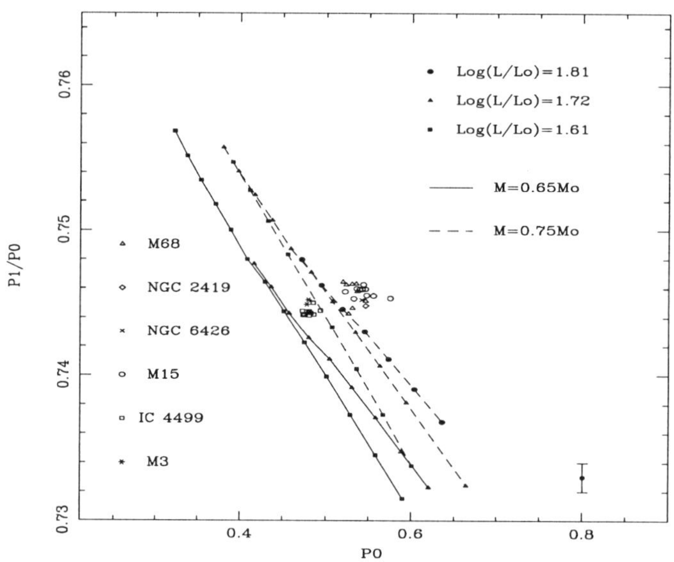

FIGURE 2. PD for linear nonadiabatic models with finer zoning.

We notice that a finer spatial resolution partially removes the degeneracy of the luminosity levels. Moreover, as a most relevant point, these calculations now suggest that the mass value of Oo II RRd variables should be of the order of $M/M_{\odot} = 0.8$, while for Oo I RRd variables mass should increase to about $M/M_{\odot} = 0.70$, in much better agreement with evolutionary prescriptions (see Bono et al. 1996).

But, as discussed in Bono & Stellingwerf (1994), linear periods are only a good approximation of the true nonlinear ones.

In Fig. 3 we show the PD for full amplitude, nonlinear, nonlocal and time-dependent convective models. Here the theoretical points are fewer with respect to the previous plots

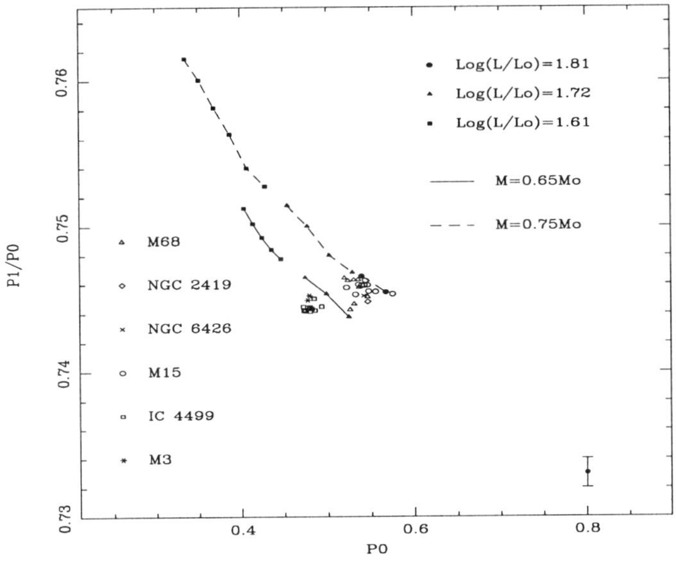

FIGURE 3. PD for nonlinear models.

and there is no more degeneracy of the luminosity level. The reason for this change is that now only models which present a stable nonlinear limit cycle both in the fundamental and in the first overtone are displayed, while the spurious theoretical points connected with models which present a unique stable limit cycle (fundamental or first overtone) have disappeared. The comparison between nonlinear periods and observational data can now give simultaneous useful informations on both stellar masses and luminosities of the pulsators. We obtain a stellar mass of $M/M_{\odot} \simeq 0.7$ for Oo II RRd pulsators, whereas for the RRd variables in IC4499 we estimate a mass of the order of $M/M_{\odot} = 0.60$. Moreover, for M15 and M68 pulsators we find a luminosity close to $\log L/L_{\odot} \simeq 1.8$, which appears somewhat larger than the currently accepted evolutionary predictions.

Finally, in order to account for the effects of opacity assumptions on the PD, in Fig. 4 theoretical periods are referred to nonlinear RR Lyrae models computed by adopting the new opacity tables provided by Roger & Iglesias (1992) for temperatures higher than 10^4 K, and by Alexander & Ferguson (1994) for lower temperatures.

The comparison with observational data is now pointing out a promising theoretical scenario since it predicts a stellar mass slightly greater than $M/M_{\odot} = 0.8$ for RRd pulsators in Oo II clusters and a mass value around $M/M_{\odot} = 0.65$ for RRd variables in IC4499. Both results are now in excellent agreement, within the error bar, with canonical evolutionary predictions. Data plotted in Fig. 4 also suggest a luminosity level of the order of $\log L/L_{\odot} \simeq 1.7$ for Oo II RRd variables, whereas the corresponding luminosity level for RRd variables in IC4499 falls between the computed luminosity levels at $\log L/L_{\odot} = 1.61$ and 1.72, in good agreement with the evolutionary estimates.

3. Conclusions

We have shown that the pulsator masses evaluated through the comparison between linear periods and observational data might be affected by substantial systematic errors. The calculations carried out in this investigation point out that in a linear, nonadiabatic regime the periods, and in turn the masses, present a non negligible dependence on the spatial resolution adopted in computing the envelope models.

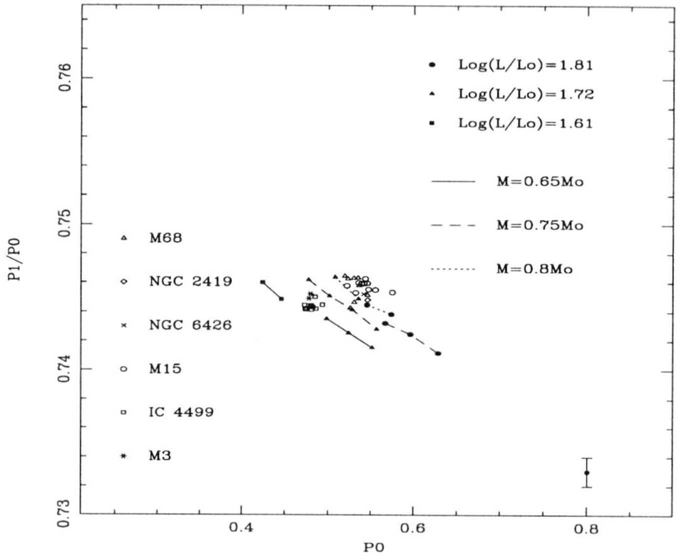

FIGURE 4. PD for nonlinear models with new updated opacity tables.

Even though the discrepancy between linear and nonlinear periods has often been considered negligible, it plays a key role in the case of double-mode pulsators. As a most relevant point, we found that a nonlinear P_1/P_0 vs. P_0 diagram constructed taking into account nonlinear models provides valuable constraints not only on the stellar masses but also on the luminosities of RRd variables.

If we adopt new and updated opacity evaluations, nonlinear computations provide pulsational estimates of both masses and luminosities which are in good agreement with canonical evolutionary predictions.

REFERENCES

Alexander, D. R. & Ferguson, J. W. 1994, ApJ, 437, 879

Bono, G., Caputo, F., Castellani, V., & Marconi, M. 1996, ApJ Lett., accepted

Bono, G. & Stellingwerf, R. F. 1994, ApJS, 93, 233

Clement, C. M., Nemec, J. M., Robert, N., Wells, T., Dickens, R. J., & Bingham, E. A 1986, AJ, 92, 825

Cox, A. N. 1991, ApJ Lett., 381, L71

Cox, A. N., Hodson, S. W., & Clancy, S. P. 1983, ApJ, 266, 94

Jorgesen, H. E., & Petersen, J. O. 1967, Zs. Astrophys., 67, 377

Nemec, J. M. 1985, AJ, 90, 240

Petersen, J. O. 1973, A&A, 27, 89

Rogers, F. J. & Iglesias, C. A. 1992, ApJS, 79, 507

Stellingwerf, R. F. 1975, ApJ, 195, 441

Walker, A. R. 1994, AJ, 108, 555

Pulsational Models for Young Stars of Intermediate Mass

By M. MARCONI[1], AND F. PALLA[2]

[1],Dipartimento di Fisica, Università di Pisa, Piazza Torricelli 2, Pisa, 56100, ITALY

[2]Osservatorio Astrofisico di Arcetri, Largo E. Fermi 5, Firenze, 50125, ITALY

Recently, Kurtz & Marang (1995 MNRAS 276, 191) have discovered δ Scuti pulsation with a period of 4.99 h in the well known pre-main-sequence (PMS) Herbig Ae star HR 5999.

This finding represents the first observational evidence of pulsation in young stars of intermediate mass (2–5 $M_\odot$). As such, it offers useful constraints on the inner structure of these stars and on theoretical models.

We have studied the stability properties of PMS stars of intermediate mass for a wide range of luminosity and effective temperature, using nonlinear, nonlocal time-dependent convective models. The period of the pulsation is successfully reproduced both in the fundamental and in the first overtone modes. This restricts the range of mass, luminosity and effective temperature to a narrow region of the parameter space: $M = 3 \div 3.5\, M_\odot$ for the fundamental mode.

Moreover the growth rates of the selected models are quite small ($\eta \sim 10^{-4}$), thus requiring very long computations.

We present preliminary results for full amplitude pulsators in order to reproduce the amplitude level detected on HR 5999 ($\sim$ 0.013 *mag.*). The best case is for a model with $M = 3.5\, M_\odot$, $L = 120 L_\odot$, $T_{\rm eff} = 8300$ K.

---

1. Introduction

Most of known δ Scuti variables are main-sequence or post-main-sequence stars. However there are a few pre-main sequence stars whose surface boundary conditions (temperature and luminosity) appear to lie in the δ scuti instability strip. One of these is HR5999, a pre-main-sequence Herbig A7III-IVe star that has been intensively studied at many wavelengths (Baade & Stahl (1989), Tjin a Djie et al. (1989), Praderie et al. (1991), Perez, Grady & The (1993)). Such photometric studies have shown that the brightness of this star varies from a maximum of $V \approx 6.8$ down to (occasionally) $V > 8$.

The stellar surface properties are not very accurately known.

The values found in the literature are: $T_{\rm eff} \simeq 7400 - 7800$ K, $L \simeq 30 - 130\ L_\odot$, $M \simeq 2 - 3\ M_\odot$, $\log g \simeq 3.5 - 4$.

Kurtz & Marang (1995) have observed HR 5999 for pulsational photometric light variations characteristic of δ Scuti stars and have found a peak-to-peak pulsation amplitude of about 0.013 mag in Johnson V and a period of 4.99 h, superimposed on a non-periodic photometric variation of 0.35 mag.

Such an observational evidence obviously requires theoretical support in order to verify if a young pre-main-sequence star of intermediate mass can pulsate as a δ Scuti variable.

From the point of view of the theory of stellar pulsation the modelization of a pre-main-sequence star is a quite difficult task. In fact young stars of mass between 2 and 4 $M_\odot$ begin their PMS phase in a *thermally unrelaxed* state, which is very different from the standard PMS initial conditions (Palla & Stahler 1993).

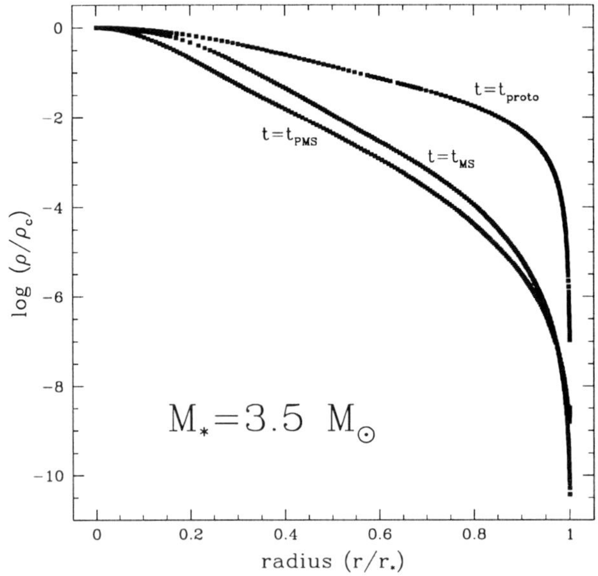

FIGURE 1. Inner density versus percentual radius for a $3.5 M_{\odot}$.

2. Models

In order to reproduce the observed period of HR 5999 we calculated linear non-adiabatic models at varying stellar mass, luminosity and effective temperature, since for δ Scuti as well as for RR Lyrae and Cepheids variables, pulsation depends on gravity and periods are then functions of the stellar parameters M, L and $T_{\rm eff}$.

On the basis of the period selection performed by linear computations we evaluated fundamental and first overtone modal stability through a nonlinear hydrodynamical code implemented by a nonlocal time-dependent treatment of convection.

Both linear and nonlinear codes were originally developed by Stellingwerf (1975) and since then improved and successfully applied to RR Lyrae variable stars (see for more detail Bono & Stellingwerf 1994; Bono et al. 1996).

For our computations of the pulsational instability we have considered the following models: $M = 2,\ 2.5,\ 3,\ 3.5\, M_{\odot}$, $L = 20,\ 40,\ 60,\ 80,\ 100\, L_{\odot}$, $7400\,{\rm K} < T_{\rm eff} < 8800\,{\rm K}$, all with solar composition.

From the evolutionary pre-main-sequence models of Palla & Stahler (1993) we know that not all the combinations of the above selected parameters correspond to physical models.

Fig. 1 shows the inner density distribution for a $3.5\, M_{\odot}$ at three evolutionary times: t_{proto} represents the time at the end of the protostellar accretion phase and thus the beginning of the PMS phase; $t_{\rm PMS} = 2.5 \times 10^5$ yr corresponds to the time when the $3.5\, M_{\odot}$ has luminosities and temperatures equal to those of HR 5999; $t_{MS} = 1.2 \times 10^6$ yr is the end of the PMS phase at the ZAMS.

From this figure we notice that the structure of a pre-main-sequence star with the surface properties of HR 5999 is not very different from that of a main-sequence star. As a consequence we adopted envelope models that contain only a few percent of the total mass and a large fraction (80 %) of the total radius.

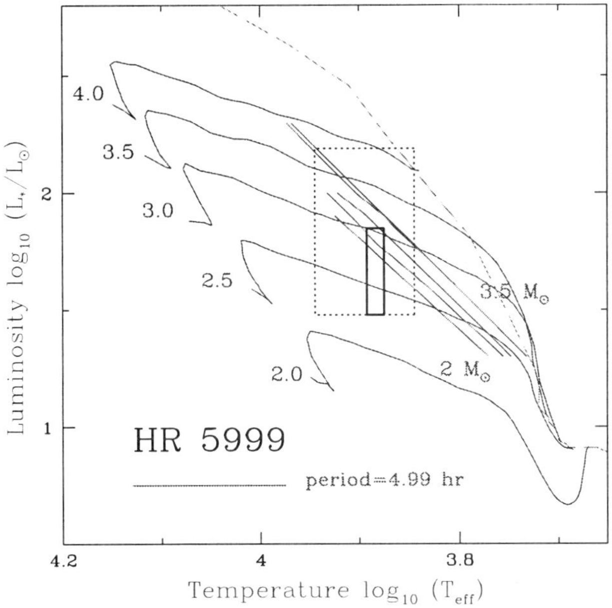

FIGURE 2. Location in the HR diagram of the selected fundamental models

3. Results: linear stability analysis

We have first computed linear nonadiabatic models for all the stellar parameters discussed above.

We have reproduced HR 5999 period of 4.99 h. within an error of 0.12 h. (0.205 d $\leq \Pi \leq$ 0.215 d.) both in the fundamental and firts overtone modes for several combinations of M, L, $T_{\rm eff}$.

As far as the fundamental mode is concerned, in Fig. 2 we show the location in HR diagram of the models which satisfy the period condition specified above. The position of HR 5999 is shown by the asterisk, with the associated uncertainty in luminosity and temperature (solid box) as given by Molster & The (1994) and by Zinnecker & Preibisch (1994). The larger box (dotted) shows the maximum range of uncertainty in L and $T_{\rm eff}$ (from various authors). The lines of constant pulsation period (0.205 d $\leq \Pi \leq$ 0.215 d) are also plotted together with the PMS evolutionary tracks from Palla & Stahler (1993) for the labeled values of stellar mass.

From this figure we note that there are only two cases where the line of constant period intersects the evolutionary track of the appropriate mass within the boundaries of L and $T_{\rm eff}$ given by the solid and dotted boxes. One model has $M = 3.0\,M_\odot$, $L = 120\,L_\odot$ and $T_{\rm eff} = 8300\,$K.

Thus, this analysis allows us to restrict the possible parameter space to a much narrower region.

4. Results: nonlinear stability analysis

Using the results of the linear study, we can analyze two stellar models to search for the long term pulsational behavior with the characteristics observed in HR 5999 (amplitude(V) = 0.013 mag).

Then we apply a perturbation surface velocity and integrate the nonlinear dynamical and convective equations until spurious fluctuations due to the superposition of higher

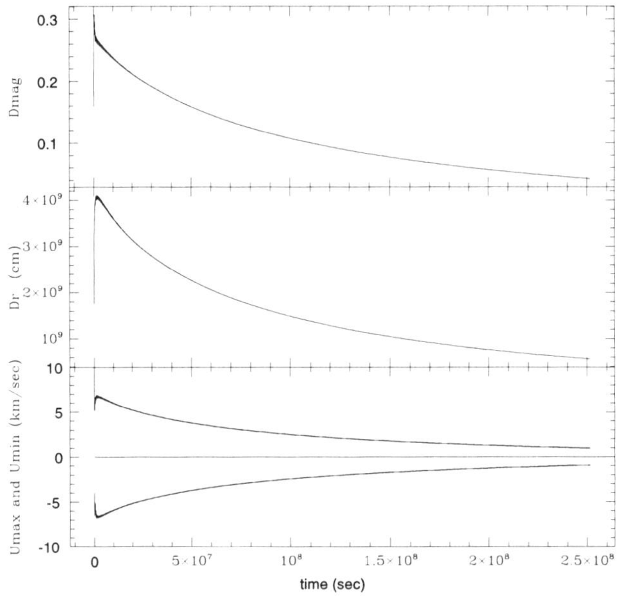

FIGURE 3. Pulsational amplitudes versus the integration time.

order modes have died out. Due to the very small growth rates obtained from the linear analysis ($\eta \sim 1 \times 10^{-4}$), it is necessary to follow the evolution of the linear pulsation for at least 15000 periods.

In Fig. 3 we show the results for $M = 3.5\, M_{\odot}$. We notice that the amplitude variations of luminosity (top), radius (middle) and radial velocity (bottom) tend to decrease with time, but still after almost 15000 periods we cannot assess whether pulsation will reach a stable limit cycle at the level observed in HR 5999 or it will be quenched.

It is worth noting that the same nonlinear approach has also be applied to other models which satisfy the linear period condition but not the evolutionary constraints and in most of these cases we found that pulsation is quenched in few thousands of periods.

5. Conclusions

We have studied the stability properties of PMS stars of intermediate mass for a wide range of luminosity and effective temperature. The period of the pulsation is successfully reproduced both in the fundamental and in the first overtone modes. This restricts the range of mass, luminosity and effective temperature to a narrow region of the parameter space. We have also presented preliminary results for a full amplitude pulsator.

REFERENCES

Baade, D. & Stahl, O. 1989, A&A, 209, 255

Bono, G., Caputo, F., Castellani, V., & Marconi, M. 1996, A&AS, accepted

bono, G. & Stellingwerf, R. F. 1994, ApJS, 93, 233

Kurtz, D. W. & Marang, F. 1995, MNRAS, 276, 191

Molster, L. & Thé, P. S. 1994, in The Nature and Evolutionary Status of Herbig Ae/Be Stars, PASP Conf. Series, ed. J. P. E. van den Heuvel

Palla, F. & Stahler, S. W. 1990, ApJ, 360, L47

Palla, F. & Stahler, S. W. 1993, ApJ, 418, 414
Prez, M. R., Grady, C. A., & The, P. S. 1993, A&A, 274, 381
Prez, M. R., Webb, J. R., & The, P. S. 1992, A&A, 257, 209
Praderie, F., Catala, C., Czarny, J., The, P. S., & Tjin a Djie, H. R. E. 1991, A&AS, 89, 91
Stellingwerf, R. F. 1975, ApJ, 195, 441
Tjin a Djie, H. R. E. 1989, A&AS, 78, 1
Zinnecker, H. & Preibisch, Th. 1994, A&A, 292, 152

New Light on RV Tau Variables

By STEPHEN RUSSELL

Dublin Institute for Advanced Studies

Recent work has shown that RV Tau variables, and stars like them, have s-process under-abundances similar to those found in extreme halo red giants. Yet these variables have much higher metallicities than extreme halo stars. The question is, are they indeed extreme halo stars, but with hydrogen deficient atmospheres, or are they relatively metal rich stars, displaying the effects of over-ionization from shock waves in their atmospheres? These possibilities have been put to the test by comparing the abundances of RV Tau variables, with their companion red giants in globular clusters. Preliminary results are reported in this paper.

1. Background

The region of the HR diagram above the horizontal branch, and to the left of the Asymptotic Giant Branch (AGB), is thinly populated by low mass variable stars, of several different classes. The most luminous stars include the RV Tau variables. These are thought to be stars at the termination of their evolution along the AGB.

In their study of RV Tau variable abundances, Luck and Bond (1989) suggested two possibilities that might explain the systematic under-abundances of s-process (and r-process) elements in RV Tau variables, compared with normal giants of the same metallicity:

(*a*) the stars reflect their original s-process deficiencies relative to Fe, but severe mass-loss has reduced the hydrogen abundance in the atmosphere, thereby raising the Fe/H ratios to the observed levels.

(*b*) the stars have normal s-process abundances, but because of their low 2nd ionization potentials, these elements are selectively over-ionized by Lyman-continuum photons. The key to this possibility is the observed systematic deficiency in Sc in these stars; the only light metal that shares a sufficiently low 2nd ionization potential.

These two alternatives could be tested by observing RV Tau variables in globular clusters, and comparing them with their red giant companions. If a hydrogen deficiency is the cause of the s-process deficiencies, RV Tau variables should show higher [Fe/H] ratios than the normal red giants, but the same [s- process/Fe] ratios. If over-ionization is the cause, RV Tau variables should show the same [Fe/H] ratios as the red giants, but lower [s-process/Fe] ratios.

2. Observations

Observations were made in July 1993, with the Utrecht Echelle Spectrograph (UES) on the 4.3m William Herschel Telescope (WHT), at the Observatorio Roque de los Muchachos, La Palma. The resolution was around 50,000 with the Tek CCD, the wavelength coverage mostly complete from 4650 – 7160 Å (falling off towards the red end), and the signal to noise ratio (S/N) in the range 60 to 110. The stars discussed in this paper are listed in Table 1. ¿From the apparent brightness of M2-V11, it would seem that the phase was close to minimum.

The spectra were reduced using standard IRAF software. Only line strengths of less than 150m Å were used in the analysis, except for two lines of BaII in M2-V11, which were, still, less than 200m Å.

Star	Class	V	$B-V$	$V-R$	[Fe/H]	Ref
M2-V11	RV Tau	12.5-14.0	0.56	...	-1.6†	Demers 1969
M2-HII-89	Red Giant	13.20	1.44	...	-1.7	Suntzeff *et al.* 1988
BD+11 2998	field star	9.27	0.67	0.606	-1.38	Kraft *et al.* 1992

† metallicity of cluster from Zinn & West (1984)

TABLE 1. Photometric parameters of stars in this study

Star	Class	$T_{\rm eff}$	log g	ξ_T	[Fe/H]
M2-V11	RV Tau	5750	0.5	3.0	-1.78
M2-HII-89	Red Giant	4000	0.0	1.7	-1.55
BD+11 2998	field star	5400	2.0	2.0	-1.16

TABLE 2. Physical Parameters of the Stars

3. Abundance Analysis

The atmospheric parameters are shown in Table 2. The temperatures were determined by requiring that the iron abundances determined from FeI, were independent of excitation potential. This was accurate to within ±250K. The gravities were determined by requiring the abundances of iron derived from FeI and FeII lines, should match. The accuracy of log g was $\pm 0.5 cms^{-2}$. The microturbulent velocity ξ_T, was determined by requiring the iron abundances derived from FeI to be independent of equivalent width. This was accurate to $\pm 0.5 kms^{-1}$. The abundances were calculated using the WIDTH6 modeling code, with recent hydrostatic, plane-parallel model atmospheres, assuming LTE.

4. Results

The results are detailed in Table 3, where the solar abundances, except for Fe, come from Anders and Grevesse (1989). The Fe abundance is derived from Johansson *et al.*, (1994). These are shown graphically in Figure 1, which shows the relative abundances of the RV Tau variable M2-V11 and the red-giant M2-HII-89, where:

$$[M/Fe]^* = log(N/\mathrm{Fe})_{\mathrm{M2-VII}} - log(N/\mathrm{Fe})_{\mathrm{M2-HII-89}}$$

The filled circles relate to elements for which there are at least two spectral lines measured from each star. The open circles, without error bars, are for those elements with only one spectral line in at least one of the stars.

The errors in the figure illustrate only the line to line variations. Errors due to uncertainties in the physical parameters are of the order of $\pm 0.2dex$ for all elements in M2-V11, and for all n-capture elements in M2- HII-89. Higher errors due to uncertainties in physical parameters, up to $\pm 0.5dex$, are observed in several of the lighter elements of M2-HII-89.

4.1. *M2-VII*

This star shows over-abundant [M/Fe] ratios, relative to the sun, in the odd-elements Na and Al, the α-elements Si, Ca, and Ti, and in the key light metal Sc.

The s-process elements measured in this star, Y, Zr and Ba, are not under-abundant in a systematic manner. The element Zr is the only one with an apparent under-abundance, but this is only represented by one weak line. Barium appears over abundant, while Y is

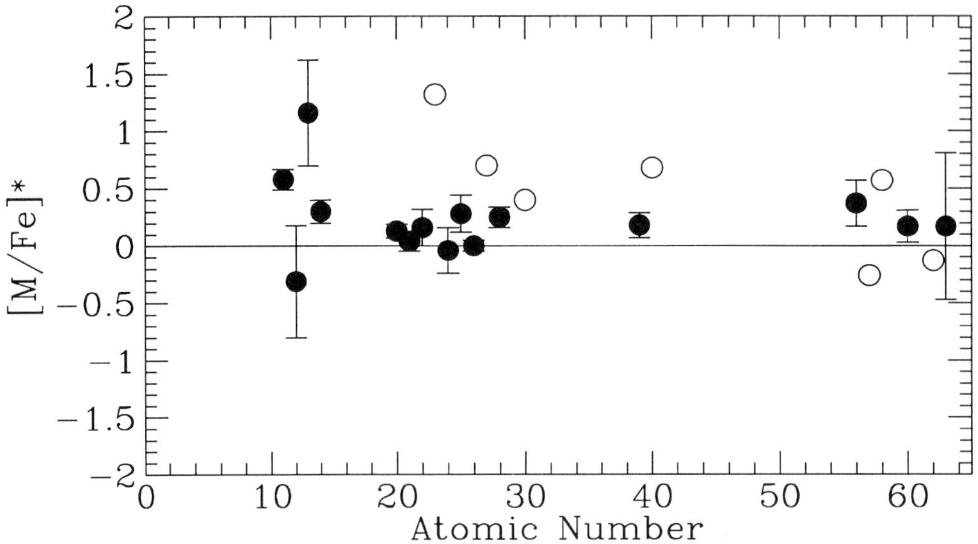

FIGURE 1. Relative metal to iron abundances between the RV Tau star and the red giant

Element	*A*	Solar	M2-V11	sigma	*n*	M2-HII-89	sigma	*n*
C	6	8.56	...	...	...	8.13	0.17	2
Na	11	6.33	4.88	0.03	2	4.53	0.08	2
Mg	12	7.58	5.95	0.48	3	6.49	0.12	4
Al	13	6.47	5.77	0.39	2	4.84	0.25	2
Si	14	7.55	6.17	0.10	7	6.10	0.03	12
S	16	7.21	...	...	...	5.19	...	1
Ca	20	6.36	4.83	0.04	14	4.93	0.05	14
Sc	21	3.10	1.41	0.05	10	1.60	0.07	11
Ti	22	4.99	3.52	0.12	19	3.59	0.10	20
V	23	4.00	3.34	...	1	2.25	0.09	5
Cr	24	5.67	3.77	0.13	11	4.04	0.15	13
Mn	25	5.39	3.47	0.16	3	3.42	0.04	8
Fe	26	7.51	5.73	0.04	54	5.96	0.03	53
Co	27	4.92	3.95	...	1	3.68	0.11	5
Ni	28	6.25	4.65	0.07	11	4.63	0.06	16
Cu	29	4.21	...	...	...	2.46	0.08	2
Zn	30	4.60	3.21	0.16	2	3.04	...	1
Y	39	2.24	0.39	0.09	7	0.44	0.07	9
Zr	40	2.60	0.15		1	-0.66	0.50	2
Ba	56	2.13	0.82	0.20	3	0.68	0.02	3
La	57	1.22	-1.16	...	1	-0.67	0.13	3
Ce	58	1.55	-0.15	...	1	-0.65	0.10	4
Pr	59	0.71	...	...	...	-0.75	0.17	2
Nd	60	1.50	-0.10	0.12	6	-0.04	0.07	10
Sm	62	1.00	-1.27	...	1	-0.91	0.16	3
Eu	63	0.51	-0.78	0.47	2	-0.72	0.43	2

TABLE 3. Measured Abundances

in solar ratio to Fe. The abundance of Ba relies on two strong lines (> 150m Å), and one medium strength line. The two strong lines are very sensitive to errors in the physical parameters. If these two lines are excluded, the remaining line (80m Å) gives a relative abundance of +0.07 dex.

Similarly, all the elements heavier than Ba, the r- process elements (assuming halo-type nucleosynthesis), as a group, are in solar ratio to Fe.

4.2. *M2-HII-89*

This star has apparently high C abundance, possibly indicating significant dredge up of nuclear processed material. The odd-elements Na and Mg are in solar ratio to iron, while most other elements follow a similar distribution to M2-V11.

4.3. *M2-V11 Relative to M2-HII-89*

The odd-elements appear significantly enhanced in M2-V11. Most of the other elements are in very similar ratio to iron as in M2-HII-89. If anything, the s-process elements are actually enhanced in M2-V11, though not significantly. The r-process elements are very similar, and the abundances of Sc are the same. The Fe abundances are the same, within the errors, and agree with values quoted in the literature (see Table 1).

5. Conclusions

What ever the process is that is responsible for producing s-process to Fe under-abundances for stars in the field with $[Fe/H] > -1.6$, it is not observed to be in operation for the stars observed here, in M2 ($[Fe/H] \sim -1.7$). If mass loss were in operation, the [Fe/H] abundance ratio would appear higher in the RV Tau variable, than the red giant. In fact it is a little lower in abundance. If over-ionization were operating, s-process elements would be less in the RV Tau variable. If anything, these elements are higher in abundance. Certainly the element Sc shows no sign of depletion.

Abundances of RV Tau variables collected from the literature by Luck & Bond (1989) (their Fig. 1a), similarly, show no sign of s-process depletion at such low metallicities. It is not obvious why over-ionization would cease at low metallicities. However, it is conceivable that reduced opacities in the atmospheres of such stars may contribute to reduced mass-loss in the lower metallicity stars. Although this sheds little direct light on the solution to the puzzle of why RV Tau variables show such low s-process elemental abundances, if confirmed with the other stars still to be analyzed from this project, it will indicate a definite limit to the metallicity above which the process is effective.

REFERENCES

Anders, E., & Grevesse, N. 1989, Geochim. Cosmochim. Acta, 53, 197

Demers, S. 1969, AJ, 74, 925

Johansson, S. *et al.* 1994, ApJ, 429, 419

Kraft, R. P., Sneden, C., Langer, G. E., & Prosser, C. F. 1992, AJ, 104, 645

Luck, R. E., & Bond, H. E. 1989, ApJ, 342, 476

Suntzeff, N. B., Kraft, R. P., & Kinman, T. D. 1988, AJ, 95, 91

Zinn, R. & West, M. J. 1983, ApJS, 55, 45

Hot Bottom Burning in Asymptotic Giant Branch Stars

By JOHN C. LATTANZIO[1], CHERYL A. FROST[1], ROBERT C. CANNON[2], AND PETER R. WOOD[3].

[1]Department of Mathematics, Monash University, Clayton, Victoria 3168, AUSTRALIA

[2]Department of Physiology and Pharmacology, School of Biological Sciences, University of Southampton, Bassett Crescent East, Southampton, SO16 7PX, UNITED KINGDOM

[3] Mount Stromlo and Siding Springs Observatories, Private Bag, Weston Creek PO, ACT 2611, AUSTRALIA

We review the current state of knowledge about the phenomenon of Hot Bottom Burning as seen in Asymptotic Giant Branch stars. This is illustrated with some results from new $6M_{\odot}$ stellar models.

1. Introduction and Motivation

Hot Bottom Burning (hereafter "HBB") is the colourful name given to the circumstance where the temperature at the bottom of a star's convective envelope is sufficiently high for non-negligible nucleosynthesis to take place. It is perhaps better to think of this as the bottom of the convective envelope penetrating into the top of the hydrogen-burning shell, although these definitions are, of course, equivalent (due to the monotonicity of the temperature profile). It is also sometimes known as "convective envelope burning" or simply "envelope burning".

The first calculations of HBB were performed by Scalo, Despain & Ulrich (1975), inspired by the existence of Li-rich stars in the Galaxy. These calculations extended calculations of Sackmann, Smith & Despain (1974) who also used a diffusion equation for the mixing, but had assumed that during the thermal pulse the intershell convective zone actually joined the convective envelope, producing one large convective zone, extending from the surface into the helium shell. Such was, in fact, the result of one evolutionary calculation, by Schwarzschild & Härm (1967), but this was later shown to be due to the neglect of radiation pressure. No subsequent calculation has ever shown contact between these two convective zones (although this does not rule out such an event in real stars, of course!)

Strong observational impetus for the study of HBB was given by Wood, Bessell & Fox (1983). They investigated the brightest red-giants in the Magellanic Clouds. These were known to have $n(\mathrm{C})/n(\mathrm{O}) < 1$ and hence were not Carbon stars (N-type). Instead it was thought that they may be core helium-burning super-giants. Wood *et al.* showed that these stars contained enhanced absorption bands of ZrO. Zr is an *s*-process element which is produced in thermal pulses on the AGB (*e.g.* Lattanzio *et al.* 1996). Hence these stars were identified as AGB stars which had suffered repeated third dredge-up events without becoming Carbon stars. Because dredge-up of carbon always accompanies the dredge-up of *s*-process elements the explanation proposed by Wood *et al.* was that HBB had burned the carbon into nitrogen via CN cycling.

These same stars were also investigated by Smith & Lambert (1989; see also Smith & Lambert 1990 and Smith *et al.* 1995) who discovered that they were super-rich in ^{7}Li. This was interpreted by these authors as further evidence for HBB, because it is believed that ^{7}Li can be produced by the Cameron-Fowler Beryllium Transport mechanism, to be discussed below. Soon after, theorists independently discovered that the more massive of the intermediate-mass AGB stars developed thin, but hot, zones at the bottom of their convective envelopes (Blöcker & Schönberner 1991, Lattanzio 1992). Typically, the temperatures at the bottom of the convective envelope reached $T_{bce} \simeq 80 \times 10^6$K but only in a region of thickness $\simeq 10^{-4} M_\odot$.

2. The Core-Mass *vs.* Luminosity Relation

Early in the study of AGB stars it was discovered by Paczyński (1970) that there was a linear relationship between the surface luminosity of a star (just prior to the thermal pulse) and the mass of the hydrogen-exhausted core at the same stage. Since then many authors (*e.g.* Wood & Zarro, 1981; Lattanzio, 1986; Boothroyd & Sackmann, 1988) have improved this relation to allow for more recent input physics, as well as the variation due to chemical composition. Havazelet & Barkat (1979) emphasised the importance of radiation pressure (and its variation with core-mass) to the core-mass *vs.* luminosity relation, and Kippenhahn (1981) reconciled apparently different relations by taking this into account. A linear relation is found when there is non-negligible radiation pressure, which means core-masses exceeding about $0.5 M_\odot$. Tuchman, Glasner & Barkat (1983) found that a necessary condition for the luminosity to depend solely on the core-mass was the existence of a thin (in mass) "transition zone"where the thermodynamic variables changed very quickly. They noted that in the case of HBB a core-mass *vs.* luminosity relation may not hold. Scalo *et al.* (1975) seemed also to be aware of this complication. Indeed, Blöcker & Schönberner (1991) found a $7 M_\odot$model where HBB existed and where the luminosity variation was some ten times steeper than the Paczyński (1970) relation. This was soon confirmed by Lattanzio (1992) for a $5 M_\odot$model. Indeed, HBB had been found by Iben (1975) in a $7 M_\odot$model, and this was presumably the reason for a weak mass-dependence in his core-mass *vs.* luminosity relation quoted in Iben (1977). These developments soon led to a systematic study of where HBB could be expected.

3. Nucleosynthesis of ^{7}Li

A method for the production of ^{7}Li was outlined by Cameron (1955) and Cameron & Fowler (1971) and is known as the "Beryllium Transport Mechanism". The idea here is to combine mixing with the usual pp-chain reactions. We begin with some ^{3}He left over from the early main sequence evolution of the star. When the bottom of the convective envelope reaches into the hydrogen shell, the high temperatures found there are able to produce ^{7}Be by ^{3}He$(\alpha,\gamma)^7$Be. Now, if the ^{7}Be is left *in situ*, then it burns by proton capture into ^{8}B which decays into two ^{4}He nuclei, completing the ppIII chain. If, however, convection can mix the ^{7}Be away from the hot region, then electron capture can occur, to produce ^{7}Li. This is the basic idea of the Cameron and Fowler mechanism. Note that it simply involves normal hydrogen burning reactions, but relies on convection to transport the ^{7}Be to cooler regions, thereby enabling the production of ^{7}Li. Of course, on subsequent passes through the bottom of the envelope there is some destruction of the ^{7}Li by ^{7}Li$(p,\alpha)^4$He which completes the ppII chain. So there is always a competition between the production of ^{7}Li (from the seed ^{3}He, until it is all used up) and the destruction of ^{7}Li from proton capture. Initially, production wins, but once the ^{3}He abundance decreases sufficiently then the proton captures dominate and the ^{7}Li content begins to decrease again. Hence the Li-rich phase is only temporary.

The first systematic calculations of ^{7}Li-production by HBB were carried out by Sackmann & Boothroyd (1992). The key ingredient is an algorithm for time-dependent mixing because it is essential that the timescale for mixing be much shorter than the electron capture lifetime for ^{7}Be. This is indeed true for AGB stars in the mass range ~ 4–$7M_\odot$, depending on the composition (primarily the metallicity). The abundance of ^{7}Li very quickly (within about six pulses) grows to $\epsilon(^7\text{Li}) \simeq 4$ from a typical red-giant envelope value of $\simeq -1$ to 1 (recall that $\epsilon(\text{X}) \equiv \log_{10}[n(\text{X})/n(\text{H})] + 12$). Conversely, if the usual assumption of complete instantaneous mixing is made, then the ^{7}Li abundance decreases to values more like $\epsilon(^7\text{Li}) \simeq -5$. This is the ingredient lacking in the attempt by Iben (1973) to produce Li-rich giants. Sackmann & Boothroyd (1992) used a diffusion formulation to model the time-dependence of mixing, and this appears to work well. An alternative has been used by Lattanzio *et al.* (1996, 1997), based on the Thorne-Żytkow calculations of Cannon (1993), which allowed for upward and downward moving streams with some horizontal diffusion between the streams. This also gives results very similar to those obtained by Sackmann & Boothroyd (1992).

Sackmann & Boothroyd (1992) showed that ^{7}Li was produced as soon as T_{bce} exceeded 50×10^6K, and produced abundances in the range $\epsilon(^7\text{Li}) \simeq 4.5$ in stars with $M_{bol} \simeq -6$ to -7. This is in excellent agreement with the observations of Smith & Lambert (1989, 1990) as well as Smith *et al.* (1995), which found Magellanic Cloud AGB stars of M_{bol} between -6 and -7 to have $\epsilon(^7\text{Li})$ in the range 2.2 to 3.8. Note that there are also galactic giants with even higher lithium values, such as WX Cyg with $\epsilon(^7\text{Li}) = 4.7 \pm 0.5$ (Abia *et al.*, 1991).

4. Nucleosynthesis of CNO Elements

One of the main effects of HBB is to take the ^{12}C which is dredged to the surface and process it into ^{14}N via the CN cycle. This will necessarily involve the production of ^{13}C also, and the ^{12}C destruction prevents the star from becoming a Carbon star.

Boothroyd, Sackmann & Ahern (1993) have made extensive calculations of the effects of HBB on the carbon isotopes. Significant destruction of ^{12}C together with production of ^{13}C and ^{14}N required temperatures of at least 80×10^6K. This was found for masses greater than about 4 or $5M_\odot$, depending on the metallicity (decreasing as Z increased). From their model results, these authors expect a maximum luminosity of $M_{bol} = -6.4$ for Carbon stars, as higher luminosities will result in HBB processing the ^{12}C into ^{13}C and ^{14}N. Indeed, despite the shallowness of the burning layer, the mixing timescale is such that the entire envelope is processed through the burning region many times during the interpulse phase and ^{12}C comes into equilibrium with ^{13}C, reaching $^{12}\text{C}/^{13}\text{C} \simeq 3$ (by number) for stars with $M_{bol} \gtrsim -6.3$. Thus there is a narrow region (in luminosity) where the stars may be rich in ^{13}C prior to further processing of ^{12}C (and ^{13}C) into ^{14}N.

This is shown in Figure 1, which shows the surface abundances of ^{12}C, ^{13}C and ^{14}N in a $6M_\odot$model with $Z = 0.02$, taken from Lattanzio *et al.* (1997). One can clearly see the increase in ^{13}C as HBB begins. Soon after, ^{12}C and ^{13}C come into equilibrium. Following this one can clearly see a continued rise in ^{14}N content. Figure 2 shows the ratio of $^{12}\text{C}/^{13}\text{C}$ and $^{12}\text{C}/^{16}\text{O}$ during this phase. Note that the star shows a decrease in the C/O ratio due to the destruction of the ^{12}C which is periodically added via dredge-up.

As pointed out by Boothroyd *et al.* (1993), higher temperatures are required for CN cycling than for the production of ^{7}Li. Hence this would appear to explain the existence of some Carbon stars with high $\epsilon(^7\text{Li})$; these stars have begun HBB at temperatures in excess of 50×10^6K, which produces ^{7}Li, but have not (yet) reached the more extreme values ($\sim 80 \times 10^6$K) required for carbon destruction. However, this picture may be complicated by ^{19}F (see below).

The situation with oxygen isotopes is a little more complicated. The Third dredge-up

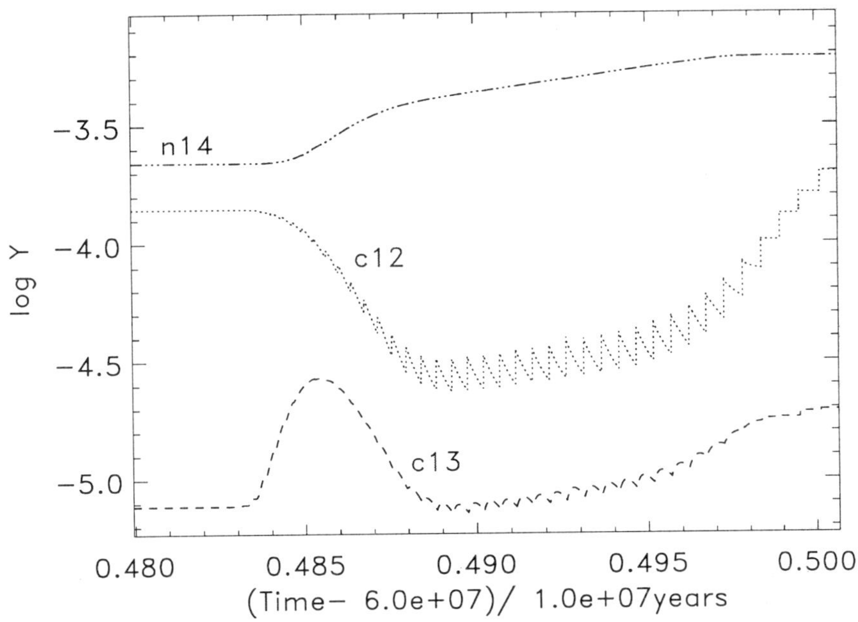

FIGURE 1. Surface abundances during the AGB evolution for a $6M_\odot$model with $Z = 0.02$. Note that increased mass loss, near the end of the evolution, turns off HBB and hence the CN cycling stops during the last few pulses.

appears to have a negligible effect on the oxygen isotopes. HBB initially destroys any ^{18}O which is present, and then follows this by producing some ^{17}O, as shown by Boothroyd, Sackmann & Wasserburg (1995). The complication comes from the fact that many AGB stars do not seem to fit this pattern. This is despite the fact that extremely accurate meteoritic grain analysis (*e.g.* Nittler *et al.* 1994) reveals isotopic ratios which do fit the results of standard first and second dredge-up models quite well (see Boothroyd, Sackmann & Wasserburg, 1994). It now appears that to match the observations of S and C stars we must invoke some form of "deep-mixing" during the AGB evolution (possibly in addition to the first ascent of the giant branch). We cannot discuss this fascinating topic here, but refer the interested reader to Wasserburg, Boothroyd & Sackmann (1995), Charbonnel (1995), Sackmann & Boothroyd (1997) and Boothroyd & Sackmann (1997).

5. Nucleosynthesis of ^{19}F

It was shown by Jorrisen, Smith & Lambert (1992) that there was a correlation between the C/O ratio and the $[^{19}F/^{16}O]$ value in galactic red-giants of various spectral types. Since the C/O ratio is a direct result of dredge-up (at least for masses where HBB is not initiated) this implies that shell-flashes are also a source of ^{19}F enhancement. Some of these enhancements are quite large, reaching values of $[^{19}F/^{16}O]$ as high as 1.5. In an accompanying paper, Forestini *et al.* (1992) outlined possible production mechanisms for this ^{19}F. These ideas were further refined by Mowlavi, Jorissen & Arnould (1996).

Before describing the suspected mechanism, it is necessary to review (briefly) the role played by ^{13}C. This isotope is produced (albeit in small quantities) by CNO cycling, and is believed to be the neutron source acting in most AGB stars via the $^{13}C(\alpha,n)^{16}O$ reaction. It appears, however, that the small amount of ^{13}C produced by hydrogen

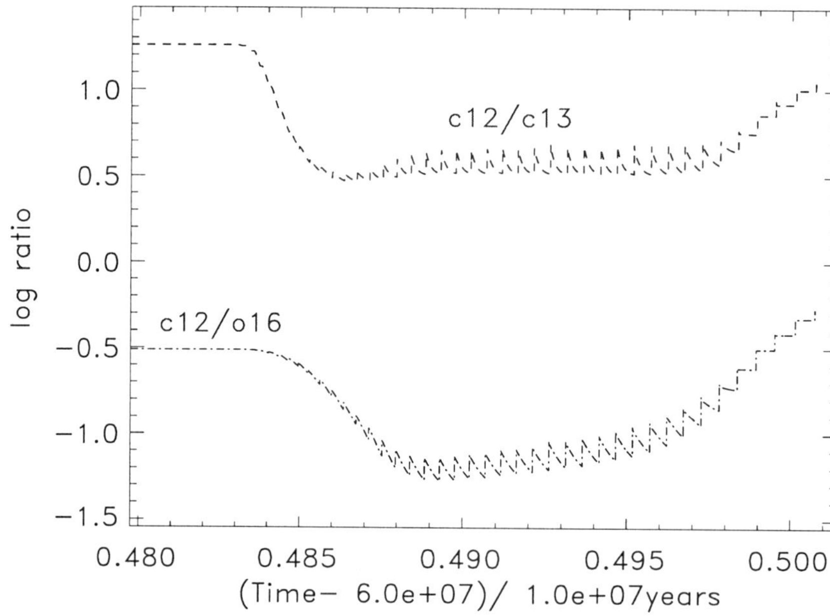

FIGURE 2. Surface abundance ratios during the AGB evolution for a $6M_{\odot} Z = 0.02$ model.

burning is insufficient to produce the large enhancements of *s*-process elements seen in many AGB stars. Hence some extra ^{13}C seems to be required. Iben & Renzini (1982a,b) suggested that this was due to some semiconvection at the bottom of the convective envelope during the dredge-up phase. For further details see Frost & Lattanzio (1995). In this scenario some hydrogen is mixed into the intershell zone, where ^{12}C burns into ^{13}C during the interpulse phase. When this ^{13}C pocket is engulfed by the convective shell at the next pulse, the ^{13}C suffers alpha captures to produce neutrons, as outlined above. But recently Straniero *et al.* (1995a,b) have shown that the ^{13}C burns *between* pulses, during the interpulse phase. Hence the neutrons are released, and the *s*-processing occurs, *in situ* in radiative zones. This has been verified by other authors also, such as Lattanzio *et al.* (1996, 1997) and Mowlavi, Jorissen & Arnould (1996).

The idea behind the production of ^{19}F, is shown in Figure 3. We begin interior to the hydrogen shell, in the region consisting mostly of ^{4}He and the ^{14}N ashes from CNO cycling. ^{14}N captures an alpha particle to produce ^{18}F which then beta decays to ^{18}O. Assuming a source of protons (see below!) ^{18}O then produces ^{15}N via $^{18}O(p,\alpha)^{15}N$. This ^{15}N can then capture another ^{4}He nucleus to produce ^{19}F. But we need a proton source for this chain of reactions to take place. There are two likely sources of protons. ^{13}C left over from CNO cycling suffers (α, n) reactions to produce ^{16}O and neutrons. These neutrons are used to produce the enhancements in *s*-process elements seen in these stars, but they are also available for ^{14}N and ^{26}Al to capture. Each undergoes (n,p) reactions and hence produces protons (together with ^{14}C and ^{26}Mg, respectively). The details of this process depend primarily on the temperature of the intershell region, both during a pulse (when it is convective) and during the interpulse phase. The temperature, of course, also depends on the metallicity of the star as well as the core-mass, and details are provided in Mowlavi, Jorissen & Arnould (1996). An important point for our purposes is that once ^{19}F is dredged to the surface, it is destroyed by any HBB present via $^{19}F(p,\alpha)^{16}O$. Thus stars with enhanced ^{19}F are presumably those of lower masses, where HBB does not take

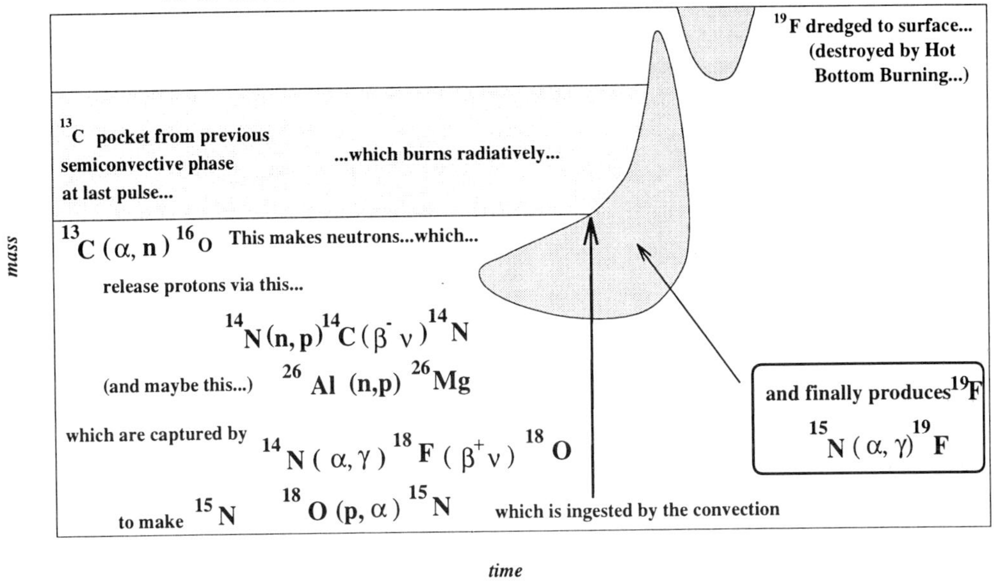

FIGURE 3. A schematic showing how ^{19}F is believed to be produced in AGB star interiors.

place. This is consistent with the correlation between C/O and [^{19}F/^{16}O] and means that we should not find stars with both enhanced ^{7}Li (indicating the presence of HBB) and enhanced ^{19}F, which is destroyed by HBB. In this context the Carbon star WZ Cas may represent a problem, as emphasised by Mowlavi, Jorissen & Arnould (1996).

Finally, we note that to produce the largest observed enhancements of ^{19}F seems to require a substantial source of ^{13}C, just as is needed for the *s*-process abundances. The exact mechanism responsible for producing this ^{13}C remains unknown.

6. Nucleosynthesis of ^{26}Al

The last decade has seen the development of a new technique for the determination of stellar abundances. Meteorities have long been identified as a source of information on the chemical composition of the early solar-system. The development of refined isolation techniques and the use of the ion-microprobe, has enabled the determination of isotope ratios for *individual grains* from *individual stars*. Reviews of this field can be found in Anders & Zinner (1993), Ott (1993), and the many references therein. Further, these isotopic ratios are determined with an accuracy previously never even dreamed of in stellar spectroscopy. Of course, the disadvantage is that we do not know *a priori* which stars produced these grains.

This exciting field has generated many challenges and problems, and a few successes, for stellar models. Primarily, it has led to a new generation of quantitative predictions from stellar models, with much larger nuclear reaction networks than are required purely for evolutionary calculations. In this paper, we want to discuss only the situation with Al. This is found in many grains, such as SiC (believed to form in the carbon-rich atmospheres of Carbon stars), graphite (a very complex and heterogeneous sample of grains) and corundum (Al_2O_3, from both first giant-branch stars and AGB stars). Often we see enhancements of ^{26}Mg, which are taken as evidence for the *in situ* decay of ^{26}Al (lifetime of 720,000y). From this we infer ratios of ^{26}Al/^{27}Al of up to 10^{-2} for corundum grains, and up to 0.1 for the graphite grains. (There are even some SiC grains with ratios heading toward unity! See Zinner (1995) for details.) How can we explain these ratios?

Forestini *et al.* (1992) investigated the production of ^{26}Al in a $3M_{\odot}$stellar model. Although ^{26}Al was indeed produced by proton captures within the hydrogen shell, the maximum abundances reached by this method was relatively low, being $\sim 1.5 \times 10^{-3}$. Of course, with a smaller envelope mass the dilution which occurs with the third dredge-up would be less, and higher concentrations may occur. But it appears that the more natural place to look for higher isotopic ratios (at least within AGB stars) is HBB.

Observations of IRC+10216 by Guélin *et al.* (1995) inferred a value of $^{26}\mathrm{Al}/^{27}\mathrm{Al} =$ 0.04. These authors also calculated HBB models for solar compositions of various masses, and obtained $^{26}\mathrm{Al}/^{27}\mathrm{Al}$ values between 0.01 and 0.08, in excellent agreement with the data. In $6M_{\odot}$models calculated by the authors we have obtained ratios up to 0.6 for metallicity $Z = 0.004$, 0.25 for $Z = 0.008$ and 0.03 for solar composition. Of course, there are many uncertainties in these numbers, not least of which is the effect of dredge-up (Frost & Lattanzio, 1996) and mass-loss (which finally terminates both HBB and dredge-up), together with some quite large uncertainties in the reaction rates, as discussed by Arnould & Mowlavi (1995). Nevertheless, it appears that these models can explain the observed isotopic ratios, at least qualitatively, and we have reason for optimism that a quantitative understanding will soon be reached.

7. Conclusions

There has been an exciting growth in quantitative nucleosynthesis calculations in recent years, inspired by new data (such as pre-solar grain analysis) as well as theoretical devlopments like HBB. At the moment, we appear to have a reasonably good qualitative understanding of the majority of stars, but problems appear when quantitative predictions are tried. The main problem seems to be in determining the edges of convective boundaries, as outlined by Frost & Lattanzio (1996). Other sources of mixing, such as shear instabilities in rotating stars, are also likely to complicate the picture, but may provide a mechanism for understanding the richness and variety of the compositions seen in the photospheres of AGB stars.

REFERENCES

Abia, *et al.*, 1991, ApJL, 245, L1.

Anders, E., & Zinner, E., 1993, Meteoritics, 28, 490.

Arnould, M., & Mowlavi, N., 1995, "AGB Stars: What Should Be Done?" in Proceedings of the 32nd Liege Astrophysical Colloquium, "Stellar Evolution: What Should Be Done?", Liege, July 3-5, 1995, Eds A. Noels *et al.*œ¡, p17.

Blöcker, T., & Schönberner, D., 1991, A&A, 244L, 43B.

Boothroyd, A. I. & Sackmann, I.-J., 1988, ApJ, 328, 641.

Boothroyd, A. I. & Sackmann, I.-J., 1997, ApJ, in press.

Boothroyd, A. I., Sackmann, I.-J., & Ahern, S. C., 1993, ApJ, 416, 762.

Boothroyd, A. I., Sackmann, I.-J., & Wasserburg, G. J., 1995, ApJL, 430, L77.

Boothroyd, A. I., Sackmann, I.-J., & Wasserburg, G. J., 1995, ApJL, 442, 21.

Cameron, A. G. W., 1955, ApJ, 212,144.

Cameron, A. G. W., & Fowler, W. A., 1971, ApJ, 164, 111.

Cannon, R. C., 1993, MNRAS, 263, 817.

Charbonnel, C., 1995, ApJL, 453, L41.

Forestini, M., Paulus, G., & Arnould, M., 1991, A&A, 252, 597.

Forestini, M., Goriely, S., Jorissen, A. & Arnould, M., 1992, A&A, 261, 157.

Frost, C. A., & Lattanzio, J. C., 1995, "AGB Stars: What Should Be Done?" in Proceedings of the 32nd Liege Astrophysical Colloquium, "Stellar Evolution: What Should Be Done?", Liege, July 3-5, 1995, Eds A. Noels *et al.*., p307.

Frost, C. A., & Lattanzio, J. C., ApJ, 473, 383.

Guélin, M., Forestini, M., Valiron, P., Ziurys, L. M., Andreson, M. A., Cernicharo, J., & Kahane, C., 1995, A&A, 297, 183.

Havazelet, D., & Barkat, Z., 1979, ApJ,233,589.

Iben, I., Jr., 1973, ApJ, 185, 209.

Iben, I., Jr., 1975, ApJ, 196, 252.

Iben, I., Jr., 1977, ApJ, 217, 788.

Iben, I., Jr., & Renzini, A., 1982a, ApjL, 259, L791.

Iben, I., Jr., & Renzini, A., 1982a, ApjL, 263, L231.

Jorissen, A., Smith, V. V., & Lambert, D. L., A&A, 261, 164.

Kippenhahn, R., 1981, A&A, 102, 293.

Lattanzio, J. C., 1986, ApJ, 311, 708.

Lattanzio, J. C., 1992, Proc. Astron. Soc. Aust., 10, 120L.

Lattanzio, J. C., Frost, C. A., Cannon, R. C., & Wood, P. R., 1997, Mem. S. A. It., in press.

Lattanzio, J. C., Frost, C. A., Cannon, R. C., & Wood, P. R., 1997, in Proceedings of IAU Symposium 177, "The Carbon Star Phenomenon", Ed. R. F. Wing (in press).

Mowlavi, N., Jorissen, A., & Arnould, M., 1996, A&A, 311,803.

Nittler, L. R., Alexander, C. M. O'D., Gao, X., Walker, R., & Zinner, E. K., 1994, Nature, 370, 443.

Ott, U., 1993, Nature, 364, 25.

Paczyński, B., 1970, Acta Astron., 20, 47.

Sackmann, I.-J., Smith, R. L., & Despain, K. H., 1974, ApJ, 187, 555.

Sackmann, I.-J., & Boothroyd, A. I., 1992, ApJL, 392, L71.

Sackmann, I.-J., & Boothroyd, A. I., 1997, ApJ, in press.

Scalo, J. M., Despain, K. H., & Ulrich, R. K., 1975, ApJ, 196, 805.

Schwarzschild, M., & Härm, R., 1967, ApJ, 150,961.

Smith, V. V., & Lambert, D. L., 1989, ApJ, 345, 375.

Smith, V. V., & Lambert, D. L., 1990, ApJL, 361, L69.

Smith, V. V., Lambert, D. L., Plez, B., & Lubowich, D. A., 1995, ApJ, 441, 735.

Straniero, O., *et al.*, 1995a, in *Nuclei in the Cosmos III*, Eds M. Busso, R. Gallino & C. M. Raiteri, (AIP Press: New York), 407.

Straniero, O., *et al.*, 1995b, ApJL, 440, L85.

Tuchman, Y., Glasner, A., & Barkat, Z., 1983, ApJ, 268, 356.

Wasserburg, G. J., Boothroyd, A. I., & Sackmann, I.-J., 1995, ApJL, 447, L37.

Wood, P. R., & Zarro, D., 1981, ApJ, 247, 247.

Wood, P. R., Bessell, M. S. & Fox, M. W., 1983, ApJ, 272, 99.

Zinner, E.K., 1995, in *Nuclei in the Cosmos III*, Eds M. Busso, R. Gallino & C. M. Raiteri, (AIP Press: New York), 567.

Carbon Burning in SAGB Stars

By ENRIQUE GARCIA-BERRO[1], CLAUDIO RITOSSA[1,2] AND ICKO IBEN, JR.[3]

[1]Departament de Física Aplicada, Universitat Politécnica de Catalunya, Jordi Girona Salgado s/n, Módulo B4 – Campus Nord, 08034 Barcelona, Spain garcia@etseccpb.upc.es

[2]Dipartimento di Astronomia, Università di Bologna, CP 596, I-40100 Bologna, Italy, ritossa@astbo3.bo.astro.it

[3]Astronomy and Physics Departments, University of Illinois, 1002 W. Green St., Urbana, IL 61801 icko@astro.uiuc.edu

The evolution of asymptotic giant branch (AGB) stars has been studied mostly for stars less massive than $\sim 8\ M_\odot$ which develop electron degenerate carbon-oxygen (CO) cores. Stars in the mass range 9–12 $M_\odot$ ignite carbon in a partially degenerate carbon-oxygen core and form electron degenerate oxygen-neon (ONe) cores. For the lower range of masses (up to $\sim 10.5\ M_\odot$) carbon ignites off-center due to a combination of degeneracy and neutrino losses, whereas, for the upper range of masses, carbon ignites at the center. After burning carbon, hydrogen is reignited and these stars become thermally pulsing "super" asymptotic giant branch (SAGB) stars (that is, AGB stars with ONe cores). The primary motivation for constructing thermally pulsing SAGB models is that no such models existed in the literature, probably due to the intrinsically time-consuming and difficult task of following the carbon-burning flame on its way to the center of the star. In this paper, we determine the physical properties of the flame that propagates inward and compare our results for a 9 $M_\odot$ star with the theoretical predictions of a simplified model which calculates the speed from local physical conditions only.

1. Introduction

The range of stellar masses over which, after exhaustion of central helium, stars ignite carbon under conditions of partial degeneracy and ultimately become SAGB stars with ONe cores of masses between ~ 1.10 and $\sim 1.37\ M_\odot$ is very narrow (Nomoto 1984, 1987). Apart from Nomoto's work, this range of masses has been much neglected since the pioneering works of Miyaji et al. (1980) and Woosley et al. (1980), who found that for stars in the mass range $8M_\odot \lesssim M \lesssim 10M_\odot$ the ONe core formed during carbon shell burning would capture electrons on ^{24}Mg and ^{24}Na first, and then on ^{20}Ne and ^{20}F, which finally trigger the explosive ignition of neon and oxygen at very high densities ($\sim 2\times10^{10}$ g cm^{-3}). At these very high densities, fast electron captures on the incinerated material would bring the Chandrasekhar mass below the actual mass of the ONe core and induce gravitational collapse to neutron star dimensions. Although there is a general agreement that electron-capture induced collapse could be successful, a major drawback of this scenario is that no detailed pre-collapse models existed in the literature. For instance, after a second dredge-up episode (when the base of the hydrogen-rich convective envelope penetrates inward into helium- and carbon-rich material), hydrogen is reignited at the base of the convective envelope and, later, helium shell flash thermal pulses begin. In several studies, the thermal instability causing the pulses was suppressed in order to follow the growth of the helium-exhausted core to larger masses and eventually to a mass close to the Chandrasekhar limit (Nomoto 1984, 1987), and little attention was paid to the properties of the thermal pulses themselves. In other studies (Sugimoto & Nomoto 1975, Iben 1977), the suppression mechanism was invoked at different stages of the evolution and a couple of thermal pulses were studied at most. At the first thermal pulse, the surface luminosity is given, to a reasonably good approximation, by the Paczyński-Uus (Paczyński 1970, Uus 1970) core mass-luminosity relation: $L/L_\odot \sim 6 \times 10^4 (M_H/M_\odot - 0.5)$. For a core of mass $M_H \sim 1.2\ M_\odot$, one obtains $L \sim 4.2 \times 10^4$ $L_\odot$, which corresponds to a bolometric magnitude $M_{bol} \sim -6.8$. Consequently, thermally pulsing SAGB stars populate the brightest section of the AGB band. The long period variables (LPVs) which are not carbon stars show excesses of ZrO (Wood et al. 1992) and are Li super-rich (Smith et al. 1995). Since Zr is an s-process element, these LPVs must be thermally pulsing AGB or SAGB stars and, therefore, some of the brightest LPVs could be observational counterparts of models with ONe cores. Finally, in many nova outbursts, excesses of neon are found and in a few cases the observed abundances of neon are as high as 2-3 times the solar abundance of the CNO elements (Livio & Truran 1994). Therefore it is rather unlikely that the observed neon is in the form of ^{22}Ne produced by helium burning on the ashes of the CNO cycle. In such cases, a possible explanation of these overabundances of neon could be that the underlying matter from a white dwarf produced from a SAGB progenitor is involved.

Probably one of the reasons for the absence of thermally pulsing SAGB models in the literature is that the conversion of a CO core into an ONe core involves following the evolution of an unstable nuclear burning flame as it propagates first towards the center of the star and then outwards. The calculation is a very delicate and time-consuming task since the propagation of the flame front is not steady, but is interrupted by a series of rather violent shell flashes whose characteristics are highly dependent on the actual zoning in the model. The flashes are violent because the nuclear reaction rates are very temperature sensitive and the material is partially degenerate. During each thermonuclear runaway, a convective shell develops above the outer edge of the ignited region. During the first flash the convective shell brings fresh carbon to the region where nuclear reactions are important. Consequently, a good fraction of the CO core is incorporated into the convective shell. The increase in temperature in the convective region translates into an increase in neutrino loss and gravothermal energy release rates. Eventually, when the carbon abundance decreases below a threshold value, the carbon-burning rate drops to the extent that the convective shell disappears. During the subsequent flashes the convective shell mixes carbon-depleted material, and, depending on the mass of the model, during the second or third flash a competition between the natural tendency of the convective shell to grow outwards and the tendency of the flame to move inwards ensues, forcing carbon burning to evolve into a steady flame propagating inwards. Because of the near discontinuity in physical variables at the flame front, following the inward motion of the flame requires very good spatial resolution and short time-steps.

We have studied the evolution of stars with masses 9, 10, 10.5 and 11 $M_\odot$ (García-Berro & Iben 1994, Ritossa et al. 1996, García-Berro et al. 1996, Ritossa et al. 1997, Iben et al. 1997), which covers almost the whole range of masses of interest. Although the general trends are similar for all the cases studied, there are significant differences among them. Carbon is ignited at the center in the 11 $M_\odot$ model, whereas, for stars with masses below 11 $M_\odot$, carbon is ignited off-center. The second dredge-up episode occurs during the first carbon shell flash for the 9 $M_\odot$ model and at the end of the carbon burning phase for the 10 $M_\odot$ model. For the 10.5 and 11 $M_\odot$ models the dredge-up episode is replaced by a "dredge-out" episode brought about when a convective shell driven by nuclear burning and gravothermal energy release merges with the convective envelope. Finally, models of masses up to $\sim$ 11 $M_\odot$ reignite hydrogen and experience thermal pulses. A real counterpart eventually loses its hydrogen-rich envelope to form a planetary nebula before electron captures begin in the ONe core. The 11 $M_\odot$ model is a limiting case, with electron captures occurring at the same time thermal pulses occur.

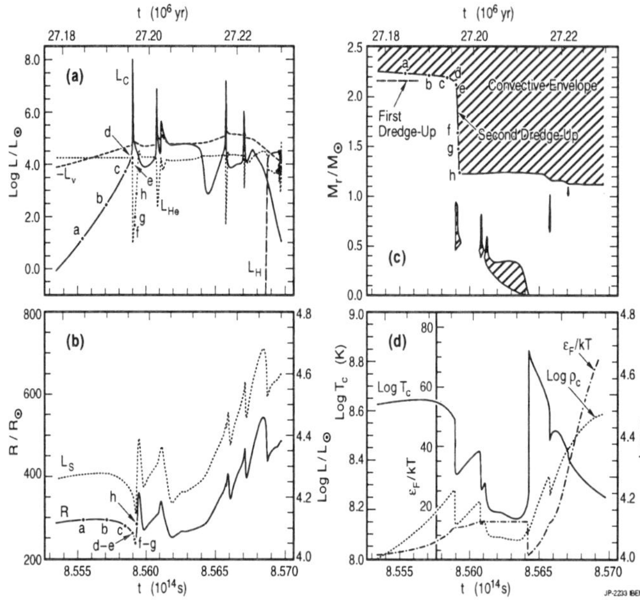

FIGURE 1. Global and internal characteristics of the 9 $M_\odot$ model during the carbon-burning phase. In the upper left-hand panel (a) are shown the carbon-burning luminosity L_C, the helium-burning luminosity L_{He}, the hydrogen-burning luminosity L_H and the neutrino loss rate $-L_\nu$). Stellar radius R and luminosity L_S are shown in the lower left-hand panel (b) and the location of the boundaries of convective layers are shown in the upper right-hand panel (c). Central temperature T_c, density ρ_c, and the electron Fermi energy ϵ_F divided by kT are shown in the lower right-hand panel (d).

Stars with masses larger thant 11 $M_\odot$ undergo electron captures before thermal pulses begin and will probably collapse to neutron star dimensions. In this paper, we describe the evolution of a 9 $M_\odot$ model of population I composition during the core carbon-burning phase. The entire evolution of the 9 $M_\odot$ model, from the zero age main sequence to the thermally pulsing "super" asymptotic giant branch stage, is described in García-Berro et al. (1996). After carbon-burning is completed, the model has an electron-degenerate core composed of an inner oxygen-neon part of mass $\sim$ 1.066 $M_\odot$ and an outer carbon-oxygen layer of mass $\sim$ 0.052 $M_\odot$, and is experiencing thermal pulses driven by helium-burning thermonuclear flashes.

2. Energetics and propagation of the carbon-burning front

The temporal behavior of the global energetics during the carbon-burning phase is shown in Figure 1a (carbon-burning luminosity L_C, helium-burning luminosity L_{He}, and rate of neutrino losses $-L_\nu$). The temporal behavior of global characteristics is shown in Figure 1b (surface luminosity L_S and stellar radius R). The boundaries of internal convective regions are shown in Figure 1c, and the time evolution of the center of the star is shown in Figure 1d (central temperature T_C, density ρ_C, and degree of electron degeneracy, as measured by the electron Fermi energy ϵ_F divided by kT).

Carbon is ignited off center, and a series of carbon flashes accompanied by a convective shell occurs, the first flash being the most powerful one ($L_C \sim 10^8\ L_\odot$). Each carbon flash is "felt" all the way to the surface, as is demonstrated by correlations between variations in stellar radius and luminosity (Fig. 1b) and variations in the nuclear-burning luminosities (Fig. 1a). This behavior is in striking contrast with that in stars of mass

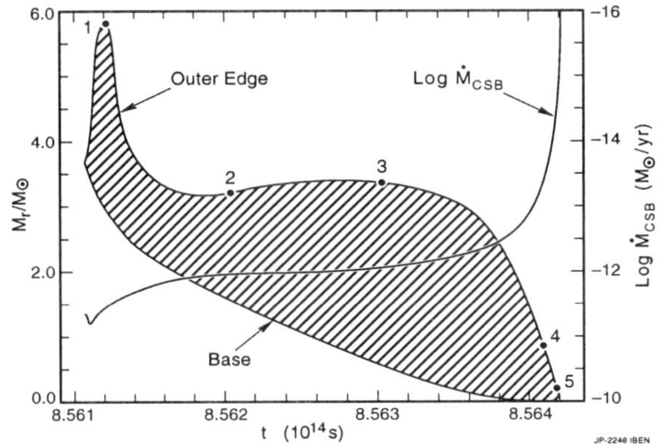

FIGURE 2. The borders of the convective shell and the speed $\dot{M}_{\rm CSB}$ with which the base of the convective shell (the flame front) moves inward during the third episode of carbon burning with an associated convective shell. The numbers show the times associated with models 1-5 for which additional properties are given in the text.

$\gtrsim 11\ M_\odot$ in which all of the energy produced by carbon burning is totally absorbed in the core, isolating the evolution of the core from the rest of the star. A very interesting feature is the strong correlation between the stellar radius and luminosity (Fig. 1b) which is due to the facts that (1) the convective envelope is basically transparent to radiation passing through it and (2) the opacity in the radiative atmosphere above the convective envelope is dominated by H^- aborption. These two reasons lead to $\log R \propto \log L$ (see Ritossa et al. 1996). The correlations between variations in the density and temperature at the center (Fig. 1d) show that each flash is also "felt" at the center, where matter cools and expands in response to each flash.

The 9 $M_\odot$ model experiences the second dredge-up phenomenon during the first and most powerful carbon shell flash, in contrast with more massive models. The reason for this difference is threefold. First, degeneracy is larger in the 9 $M_\odot$ model; second, neutrino losses are larger as well and, consequently, carbon is ignited closer to the helium-carbon interface; and, finally, the carbon flash is much more powerful (see García-Berro et al. 1996 and Ritossa et al. 1996 for detailed discussions of the dredge-up phenomenon).

The convective shell associated with the first carbon shell flash continues for ~ 1000 years (Fig. 1a). At the end of this time, carbon has been partially depleted over most of the region once encompassed by the convective shell. For the next ~ 4000 years, carbon burns slowly inward in mass without the formation of an associated convective shell. A second flash creates a new convective shell which this time mixes carbon-depleted material from the previous convective shell. During the third episode of carbon burning with an associated convective shell, the base of the convective shell continues to move inward in mass until it reaches the center. The borders of the convective shell and the rate $\dot{M}_{\rm CSB}$ at which the base of the shell moves inward in mass are shown in Figure 2.

The structure of the burning zone as it progresses toward the center during the third convective shell episode is worth examining closely. For that purpose we describe properties of models at the five times indicated in Figure 2. We use the words "flame front" (Timmes, Woosley, & Taam 1994) to mean the base of the convective shell and divide the burning zone, or flame, into two parts: a precursor flame (in the radiative region ahead of the front) and the main body of the flame (behind the front) which we define to be coincident with the convective shell. In Figure 3 are shown the structure and dynamics in the neighborhood of the flame front in model 2 (see Fig. 2), which is the most typical model during this phase of the evolution. Panel (a) gives the density ρ and temperature

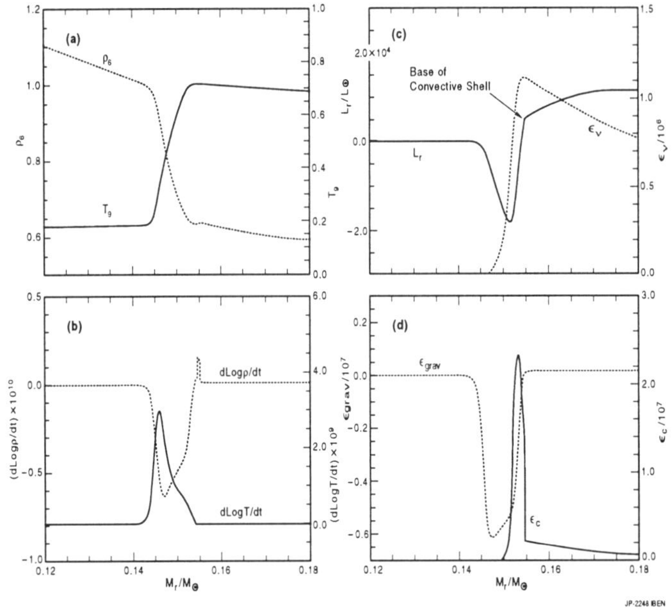

FIGURE 3. Density and temperature profiles (panel a), dynamical quantities (panel b), luminosity profile and neutrino loss rate (panel c), and gravothermal and nuclear energy generation rates (panel d) in the vicinity of the flame front for model 2.

$t\,(10^{14}$ s)	$L_{\rm C}^{\rm RZ}$	$-L_{\nu}^{\rm RZ}$	$L_{\rm grav}^{\rm RZ}$	$L_{\rm C}^{\rm CS}$	$-L_{\nu}^{\rm CS}$	$L_{\rm grav}^{\rm CS}$	$L_{\rm tot}^{\rm CS}$	$L_{\rm CO}$
8.5612	1.24/5	4.25/3	−1.17/5	1.14/5	1.10/5	−2.82/3	2.85/2	2.16/2
8.5620	2.57/4	1.67/3	−2.32/4	2.84/4	4.04/4	1.24/4	2.90/2	2.15/2
8.5630	1.75/4	5.03/2	−1.71/4	4.06/4	5.09/4	1.10/4	6.80/2	2.24/2
8.5641	4.83/2	2.73/0	−4.85/2	8.46/3	2.57/4	1.73/4	1.07/2	2.50/2
8.5642	6.97/0	0.04/0	−7.34/0	1.58/3	6.93/3	5.37/3	2.20/1	2.55/2

TABLE 1. Energetics in the Carbon-Burning Interior as the flame front approaches the center.

T profiles, panel (b) gives the dynamic characteristics $d\log\rho/dt$ and $d\log T/dt$, panel (c) gives the luminosity profile and the neutrino loss rate $-\epsilon_{\nu}$, and panel (d) gives the gravothermal energy-generation rate $\epsilon_{\rm grav}$ and the carbon-burning energy-generation rate $\epsilon_{\rm C}$.

Panels (a) and (b) show that matter ahead of the flame front is being heated and expanded. Panels (c) and (d) show that most of the carbon-burning energy released by the precursor flame is used up in heating and expansion, with very little lost to neutrinos. Of this energy, about two thirds goes into heat and one third goes into the work of expansion. The integrals under the ϵ curves in the radiative zone (RZ) ahead of the flame front (ff) are given (in units of $L_{\odot}$) by the entries in columns 2-4 in the second row of Table 1. The same quantities are tabulated for the rest of the models in their corresponding rows.

The integrals under the ϵ curves within the convective shell (CS) are given by the entries in columns 5–7 in the second row of Table 1, and the sum of the contributions from the convective shell is given in column 8. It is evident that the energies released by carbon burning in the precursor flame and in the the convective shell are comparable. It is also evident that gravothermal energy release contributes significantly to the near

$t\,(10^{14}$ s)	$M_{\rm ff}\,(M_\odot)$	$\Delta M_{\rm pf}\,(M_\odot)$	$\Delta M_{\rm pf}/M_{\rm ff}$	$\Delta r_{\rm pf}$ (km)	$v/v_{\rm cond}$	$L_{\rm C}^{\rm CS}/L_{\rm flame}$
8.5612	2.936×10^{-1}	2.312×10^{-3}	0.008	36	0.48	0.53
8.5620	1.541×10^{-1}	2.603×10^{-3}	0.017	28	1.10	0.48
8.5630	5.428×10^{-2}	7.754×10^{-4}	0.014	20	1.06	0.51
8.5641	1.057×10^{-4}	7.047×10^{-6}	0.067	5	1.42	4.38
8.5642	3.046×10^{-7}	7.615×10^{-8}	0.250	3	1.48	31.0

TABLE 2. Characteristics of the flame front.

balance between energy gains and losses in the convective shell. In fact, the total rate at which nuclear and gravothermal energy is liberated within the convective shell when added to the luminosity at the base of the flame front closely balances (within ~ 3 %) the rate of neutrino losses in the convective shell. In the region beyond the outer edge of the convective shell, neutrino energy losses and gravothermal energy release continue to be important but, once again, there is a near balance (within $\sim$ 10%) between them. In this region, the gravothermal energy release is divided almost equally between cooling and the work of compression. Finally, the last entry in Table 1 gives the luminosity emerging from the helium-exhausted core. Note that this luminosity is very modest when compared with the energy production and loss rates within the core.

A measure of the width of the precursor flame (pf) is given by the distance between the base of the convective shell and the minimum in the luminosity profile in the radiative region ahead of the front (see Figs. 3c and 3d). Table 2 displays the position of the flame front (second entry), the width of the precursor flame (third entry), the ratio $\Delta M_{\rm pf}/M_{\rm ff}$ (fourth entry), and the radial thickness of the precursor flame (fifth entry). In evolving from model 2 to model 3, the distance (in mass) of the flame front to the center and the width of the precursor flame decrease by a factor of about 3, whereas its radial thickness decreases by a factor of roughly 1.5. However, the rate of energy release by carbon burning remains essentially constant (see Table 1). To maintain adequate resolution of the front (about 200 mass shells) and to achieve convergence, the average time step between models for model 3 and also the mass between grid points have been decreased by a factor of about 3. The narrowing of the precursor flame continues as the front moves closer to the center, but, fortunately, the ratio between $\Delta M_{\rm pf}$ and $M_{\rm ff}$ increases as the center is approached and the progression of the flame front towards the center can be followed in a finite time.

It is interesting to compare the speed with which the flame front in our model moves inward to the speed predicted by a theoretical model of flame-front propagation which uses a balanced power condition between *nuclear energy release and neutrino energy losses*, and which finally depends only on local properties of the flame front (Timmes & Woosley 1992, Timmes et al. 1994). The motion of the flame front is calculated on the assumption that the temperature and density ahead and behind of the flame front are fixed. The effect of gravity is neglected and a steady state flame speed is obtained as a function of the input temperatures and densities and of the carbon abundance behind the front. The calculated speeds are reasonably well fit by the expression (Timmes & Woosley 1992, Timmes et al. 1994)

$$v_{\rm cond} \simeq \left(\frac{c\epsilon_{\rm nuc}}{\kappa\rho E}\right)^{1/2} \tag{2.1}$$

where c is the speed of sound, $\epsilon_{\rm nuc}$ is the rate of nuclear energy generation, κ is the opacity, ρ is the density, and E is the energy density, all calculated just behind the flame

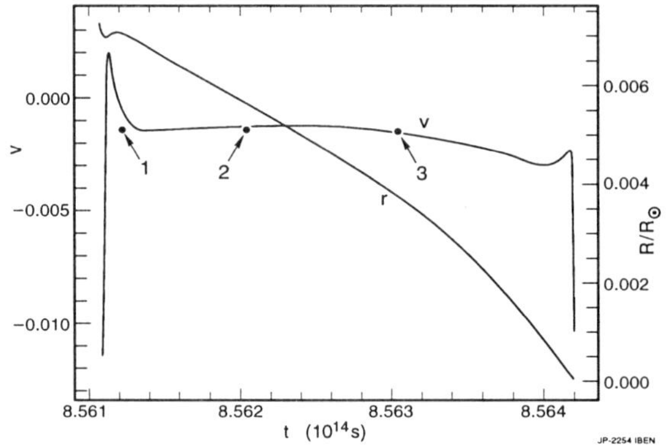

FIGURE 4. The radius and velocity of the base of the convective shell during the third convective episode of carbon burning

front. Timmes et al. (1994) also suggest that L_C^{CS}, the rate of nuclear-energy production in the convective shell, should be given approximately by the luminosity

$$L_{\text{flame}} = \dot{M}_{\text{ff}} Q \left(X_{12}^{+} - X_{12}^{-} \right), \tag{2.2}$$

where $Q \sim 4 \times 10^{17}$ erg g^{-1} is the energy liberated in burning one gram of carbon, X_{12}^{+} and X_{12}^{-} are the abundances by mass of carbon just ahead of the front and just behind the front, respectively, and $\dot{M}_{\text{ff}}$ is the rate at which the flame front propagates inward in mass (see Fig. 2).

The radius r and velocity v of the flame front in our model are shown in Figure 4 as functions of time. Using the densities and temperatures of the flame front in our models 1–3, interpolation in the Tables of Timmes et al. (1994) gives the velocites indicated by the solid circles in Figure 4. The good agreement between the calculated and predicted speeds suggests that, to a very good approximation, the flame speed is, indeed, a function only of local properties and does not depend directly on the gravity or on a detailed balance between the carbon-burning luminosity and the neutrino luminosity. The second to last entry in Table 2 gives the ratio between the actual velocity of the flame front v and the predicted value v_{cond} obtained by applying eq. (2.1). The last entry in Table 2 gives the ratio between the carbon luminosity in the convective shell and the luminosity of the flame obtained by applying eq. (2.2). The ratio $L_C^{CS}/L_{\text{flame}} \sim 0.5$ for models 1 to 3 reflects the fact that approximately half of the energy produced by carbon burning is released in the precursor flame. For models 4 and 5, the results show that the simple theoretical approximation is inapplicable when the flame is very close to the center; eq. (2.2) severely underestimates the luminosity produced by carbon burning.

Of course, the physics of flame burning in a real star are much more complicated than has been assumed in our simple model and there is no guarantee that we have described the burning adequately, no matter how well we have resolved the flame front. Recent studies of the three dimensional nature of flame fronts in Type Ia supernova models emphasize the role of turbulence in corrugating the flame front and accelerating it beyond the laminar speed (e.g., Niemeyer & Hillebrandt 1995a,b; Bravo & García 1995).

3. Conclusions

The evolution of intermediate mass model stars has been computed from the zero age main sequence to the thermally pulsing SAGB phase for the whole range of masses of interest. These models ignite carbon in a partially degenerate (CO) core and ultimately produce degenerate ONe cores. The ignition of carbon has been found to be off-center for the low mass range and at the center for the upper interval of masses. The complications of converting a CO core into an ONe core derive from the fact that these stars develop almost discontinuous nuclear burning flame fronts that propagate inward in mass. We have found that high spatial resolution is needed in order to avoid sharp discontinuities in the structure variables. Resolutions as high as a few tens of meters are needed in order to follow the propagation of the flame front. We find that almost half of the energy of carbon burning is released in a (small) radiative region ahead of the flame front (the precursor flame). As the flame progresses towards the center the front narrows but, eventually, its progression can be followed in a finite time because its distance from the center decreases faster than the thickness of the precursor flame. The computed velocities of the flame front are in very good agreement with the predictions of a simplified model of flame front propagation although the central hypothesis of the model (the balanced power condition between nuclear energy release and neutrino losses) is never fulfilled. Instead, a near balance between nuclear energy release plus gravothermal energy release, on the one hand, and neutrino losses, on the other, prevails.

Acknowledgements This work has been partially supported by the DGICYT (Spain) grant PB94-0827-C02-02 and the NSF (USA) grant AST94-17156. One of us (C.R.) thanks the E.C. for a grant from the "Human Capital and Mobility" Programme ER-BCHGECT920009 to do research at the Universitat Politécnica de Catalunya. It is a pleasure to thank the Centre de Supercomputació de Catalunya (C^4 Consortium) for supplying computer time.

REFERENCES

Bravo, E., García, D., 1995, ApJ, 450, L17

García-Berro, E., & Iben, I. Jr. 1994, ApJ, 434, 306

García-Berro, E., Ritossa, C., & Iben I. Jr. 1996, ApJ, submitted

Iben, I. Jr. 1977, ApJ, 217, 788

Iben, I. Jr., Ritossa, C., & García-Berro, E. 1997, ApJ, in progress

Livio, M., & Truran, J. W. 1994, ApJ, 425, 797

Miyaji, S., Nomoto, K., Yokoi, K., & Sugimoto, D. 1980, PASJ, 32, 303

Niemeyer, J. C., & Hillebrandt, W. 1995a, ApJ, 452, 769

Niemeyer, J. C., & Hillebrandt, W. 1995b, ApJ, 452, 779

Nomoto, K. 1984, ApJ, 277, 791

Nomoto, K. 1987, ApJ, 322, 206

Paczyński, B. 1970, Acta Astron., 20, 47

Ritossa, C., García-Berro, E., & Iben, I. Jr. 1996, ApJ, 460, 489

Ritossa, C., García-Berro, E., & Iben, I. Jr. 1997, in progress

Sugimoto, D., & Nomoto, K. 1974, in Late Stages of Stellar Evolution, ed. R.J. Taylor (Dordrect: Reidel), 105

Timmes, F. X., & Woosley, S. E. 1992, ApJ, 396, 649

Timmes, F. X., Woosley, S. E., & Taam, R. E. 1994, ApJ, 420, 348

Uus, U. 1970, Nauch. Inf., 17, 48

Wood, P.R., Whiteoak, J.B., Hughes, S.M., Bessell, M.S., Gardner, F.F., & Highland, A.R. 1992, ApJ, 397, 552

Woosley, S.E., Weaver, T.A., & Taam, R.E. 1980, in Proc. Texas Workshop on Type I Supernovae, ed. J.C. Wheeler (Austin: Univ. Texas Press), 96

Through the Upper AGB Towards a Planetary: A Hydrodynamical Simulation †

By D. Schönberner, M. Steffen, J. Stahlberg, K. Kifonidis and T. Blöcker

Astrophysikalisches Institut Potsdam, Potsdam, Germany

We present a first exploratory investigation of the dynamical evolution of a dusty stellar wind envelope along the upper AGB and its transformation into a planetary nebula. We find the existence of AGB stars with detached shells to be a natural consequence of the mass loss variations during a thermal pulse. It is also demonstrated that due to the large dynamical effects caused by the ionizing radiation field and the fast wind of the central star, it is impossible to deduce the AGB mass loss history from the planetary's density and velocity distribution. The structure of the halo, however, is still determined by the AGB mass loss history. The rapid decline of mass loss expected in the aftermath of thermal pulses leads to extended shells of low densities and explains halos with sharp boundaries. The density structure and velocity field of our model planetary closely resemble those of observational counterparts. The possibility of a direct comparison between the hydrodynamical properties of the model and its synthetic emission line spectrum allows a physically consistent interpretation of observed Doppler-split emission lines.

1. Introduction

Recent attempts to model the development of planetary nebulae with evolving central stars by solving the coupled equations of radiation hydrodynamics have been remarkably successful, but so far all of these studies start with initial structures which are exclusively based on guesses. This fact is, of course, just the reflection of our ignorance of the dynamical behavior of dusty wind envelopes around AGB stars. The knowledge about the density, velocity, chemical and ionization structures of stellar wind envelopes when the star leaves the AGB appears to us of paramount importance for the understanding of the evolution of proto- and well-developed planetaries as well as for the interpretation of observations. Important questions that have to be investigated are

(*a*) the short and long term behavior of AGB envelopes subject to stellar mass loss variations on time scales smaller than the flight time of a gas parcel,

(*b*) the final structures after the end of the AGB evolution when the heavy mass loss ceases and the remnant starts to increase in effective temperature,

(*c*) whether a well-developed planetary-nebula shell still contains information about the mass loss variations during the previous AGB evolution.

It is the purpose of this contribution to address these questions and to give first answers. We present here the very first attempt to compute the dynamical response of a stellar wind envelope for the last several 100 000 years of the AGB evolution and the following 5 000 years of post-AGB evolution, based on stellar evolution calculations with an adopted mass loss rate that varies as a function of the stellar parameters. It should be emphasized that although the mass loss law has not been derived by first principles (for details see Blöcker, 1995), the stellar evolution calculations and the computations of the response of the envelope presented here are mutually consistent once the physical details of the mass loss have been specified (see below).

† Part of this work was supported by DARA grant WE2 50 OR 9411.

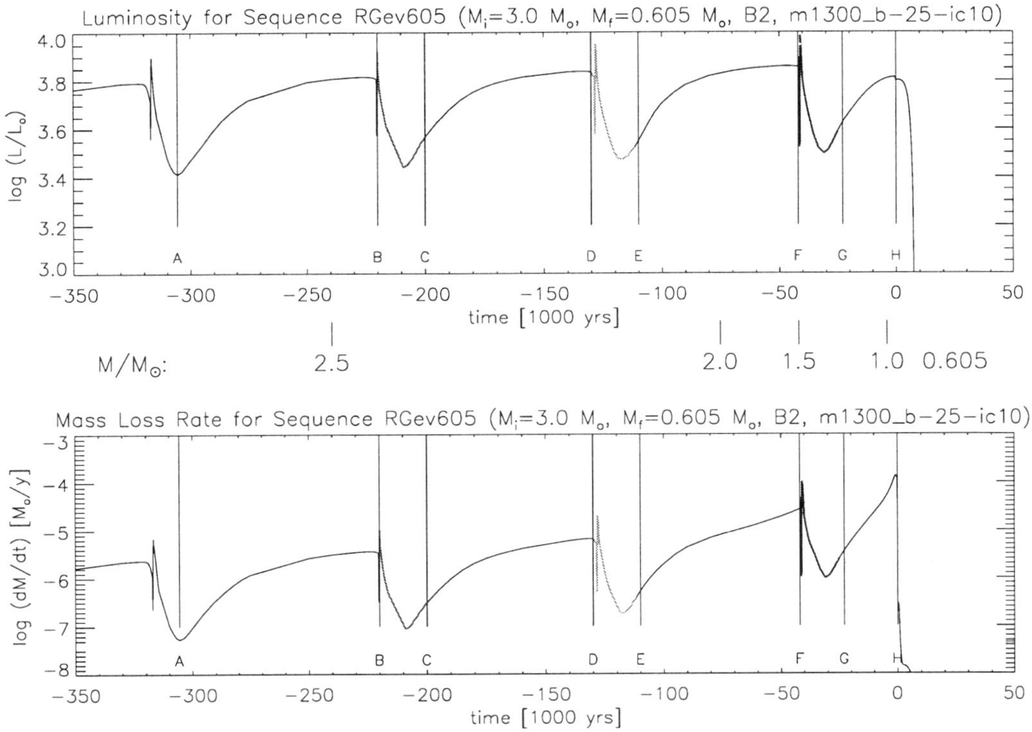

FIGURE 1. Luminosity vs. time (top) and mass loss rate vs. time (bottom) for the last four thermal-pulse cycles of a $3\,M_\odot$ model sequence after Blöcker (1995). Time zero marks the beginning of the model's post-AGB evolution. The current stellar mass is indicated between the two panels. Labels A to H are referred to in the text.

2. Evolution towards the tip of the AGB

Fig. 1 illustrates the temporal evolution of luminosity and mass loss rate for the last about 350 000 years on the AGB and the following few thousand years of the post-AGB evolution of a solar composition stellar model computed by Blöcker (1995). Mass loss reduces the initial mass of $3\,M_\odot$ to a final mass of $0.605\,M_\odot$. The stellar outflow is assumed to be spherically symmetric, and the equations of hydrodynamics are solved for the gas and the dust component, coupled by momentum exchange due to dust-gas collisions. The code is a modification of that developed by Yorke & Krügel (1977) and makes use of the following simplifications:

(*a*) Radiation transfer is considered only for the dust component, i.e. exchange of photons between dust grains and the gas is neglected.

(*b*) The dust temperature is computed from radiative equilibrium, and the gas component (neutral hydrogen) is assumed to have the same (local) temperature. This is justified since we are not interested in computing line emission and since the influence of the gas-pressure gradient on the outflow velocity turned out to be quite small.

(*c*) The dust component consists of single-sized grains, either based on oxygen or carbon chemistry, adopting a fixed dust-to-gas ratio at the dust condensation point.

A more detailed description of this fully implicit radiation hydrodynamics code, including comparisons with solutions for stationary outflows given in the literature, will be published in the near future.

The new approach applied here is to use time-dependent values of stellar mass, luminosity, effective temperature *and* the resulting variable mass loss (as shown in Fig. 1) at the inner boundary of our numerical grid, with a constant flow velocity equaling the local sound speed ($\approx$ 3 km/s). The radiation pressure on the grains and the momentum transfer to the gas leads to an acceleration of the material with typical final outflow velocities around 10 to 15 km/s, in agreement with observations. We started the computations at point (A) in Fig. 1 and continued through the three consecutive thermal pulses until the end of the AGB phase is reached (H) or even a few hundred years beyond, i.e. somewhat into the post-AGB phase.

The pronounced variations of the mass loss rate during the thermal pulses (sections B-C, D-E and F-G in Fig. 1) lead to large deviations of density and velocity from the cases of stationary outflows. Correspondingly, the emerging spectral energy distribution (SED) changes with time. As a consequence of the long lasting mass-loss minima in the aftermath of a pulse, the regions close to the star temporarily become devoid of hotter dust: the envelope is said to be detached and radiates mainly in the far infrared beyond 25 μm, corresponding to thermal emission of dust cooler than about 100 K.

This is illustrated in Fig. 2 for a carbon-rich model, showing the mass-loss variations during the first thermal-pulse cycle (upper panel) and the corresponding path of this object (star plus dusty envelope) in the IRAS two-color diagram (middle panel). Observationally, several objects with well-detached shells are known, and we selected one post-flash hydrodynamical model from our sequence which comes close to the properties of S Scuti (cf. bottom panel of Fig. 2). Corresponding calculations with oxygen-rich shells are given by Steffen et al. (this volume).

Several remarks are in order. (i) In addition to the loops, our dynamical envelopes produce some intrinsic scatter in the two-color diagram, solely due to the deviations from stationary conditions. (ii) As the consequence of a thermal pulse, objects on the upper AGB can develop *detached dust shells* and spend about 10% of their time away from the main two-color relation of dusty AGB stars. We note that our computed SEDs are apparently in good agreement with corresponding observations. (iii) However, one has to be aware that the measured far infrared fluxes may be disturbed by cirrus contamination in some cases. At the same time, they may be (seriously) underestimated if point source fluxes are quoted for objects that are actually extended. In contrast to Ivezić & Elitzur (1995), we believe that modeling dusty envelopes around AGB stars as stationary outflows is not justified *a priori* and may lead to erroneous interpretations!

3. Evolution off the AGB

Observationally, the end of the AGB evolution is characterized by an optically thin dust shell with a somewhat hotter stellar remnant shining through. Obviously, the mass-loss rate must have dropped by orders of magnitudes on a very short time scale, the reason for which is yet unknown. In the mass-loss modeling by Blöcker (1995), the rate is coupled to the period of the fundamental pulsational mode, forcing the transition from the high AGB rate to the much lower Reimers' rate (Reimers, 1977) to occur between periods of 100 and 50 days. It happens that this procedure leads to a mass-loss reduction of about two orders of magnitude within 100 years (cf. Fig. 3).

We extended our AGB gasdynamical simulations somewhat into the post-AGB regime, using the mass-loss prescription shown in the upper panel of Fig. 3, and indeed found a rapid detachment and thinning of the dust shell since any newly formed hot dust is rapidly pushed away from the star and gives no detectable signature. This is illustrated by the sequence of spectral energy distributions in the middle panel of Fig. 3 which covers

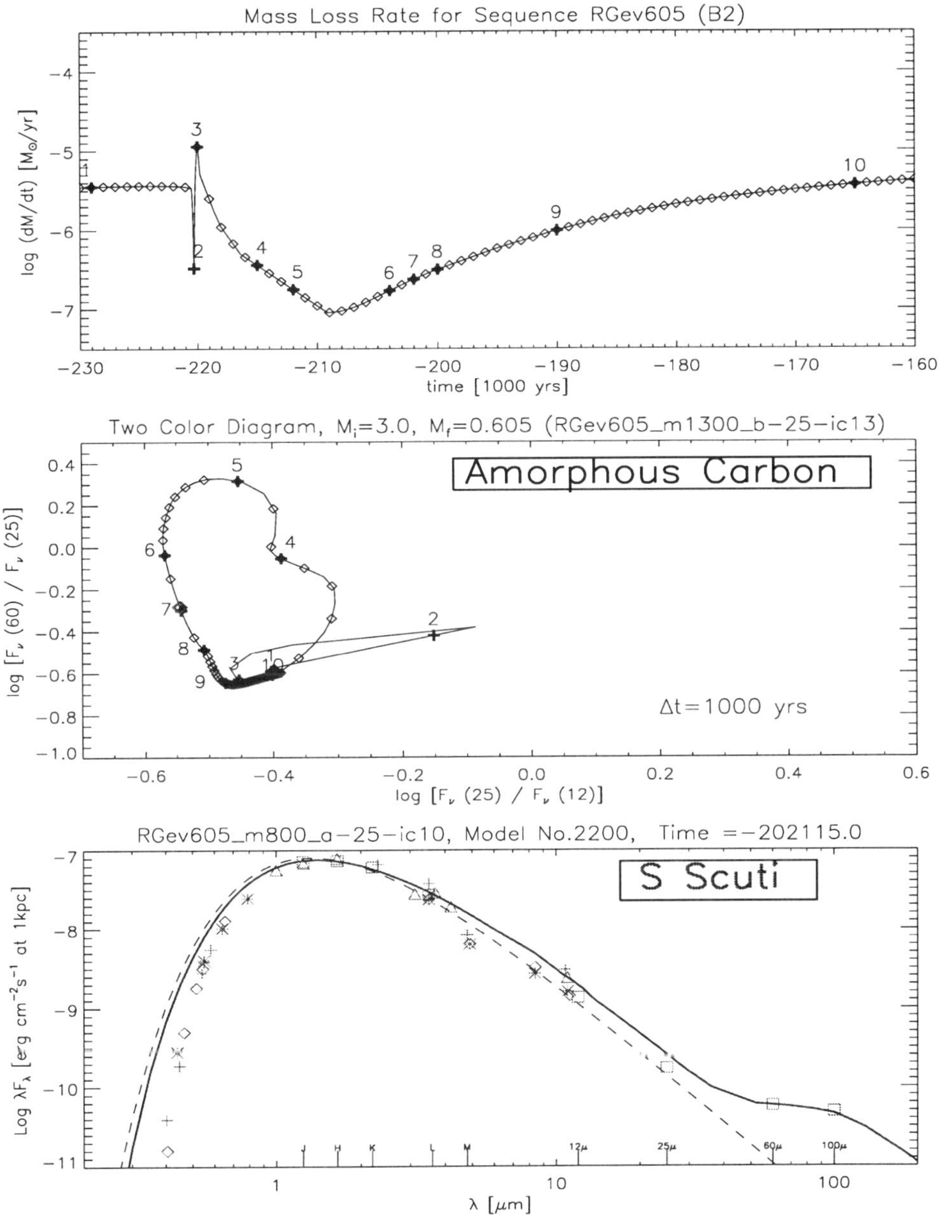

FIGURE 2. **Top:** Temporal change of the mass-loss rate during the first computed thermal-pulse cycle. **Middle:** Corresponding color evolution of a model carbon star, with numbered + signs indicating the same times as in the top panel. Diamonds outline the time evolution in steps of $\Delta t = 1000$ years. The grains consist of amorphous carbon and have a radius of 0.05 μm. The adopted dust-to-gas ratio is $1.5 \cdot 10^{-3}$. **Bottom:** Observed spectral energy distribution of S Scuti (symbols; data from Groenewegen & de Jong, 1994) compared with our theoretical spectrum (solid line) at the time labeled (7). The spectrum of the unobscured central star (dashed line) is that of a black body with $T_{\rm eff} = 2760$ K at this moment. The infrared and IRAS pass bands are indicated on the wavelength scale. Note that we have made no attempt to fit to this particular object!

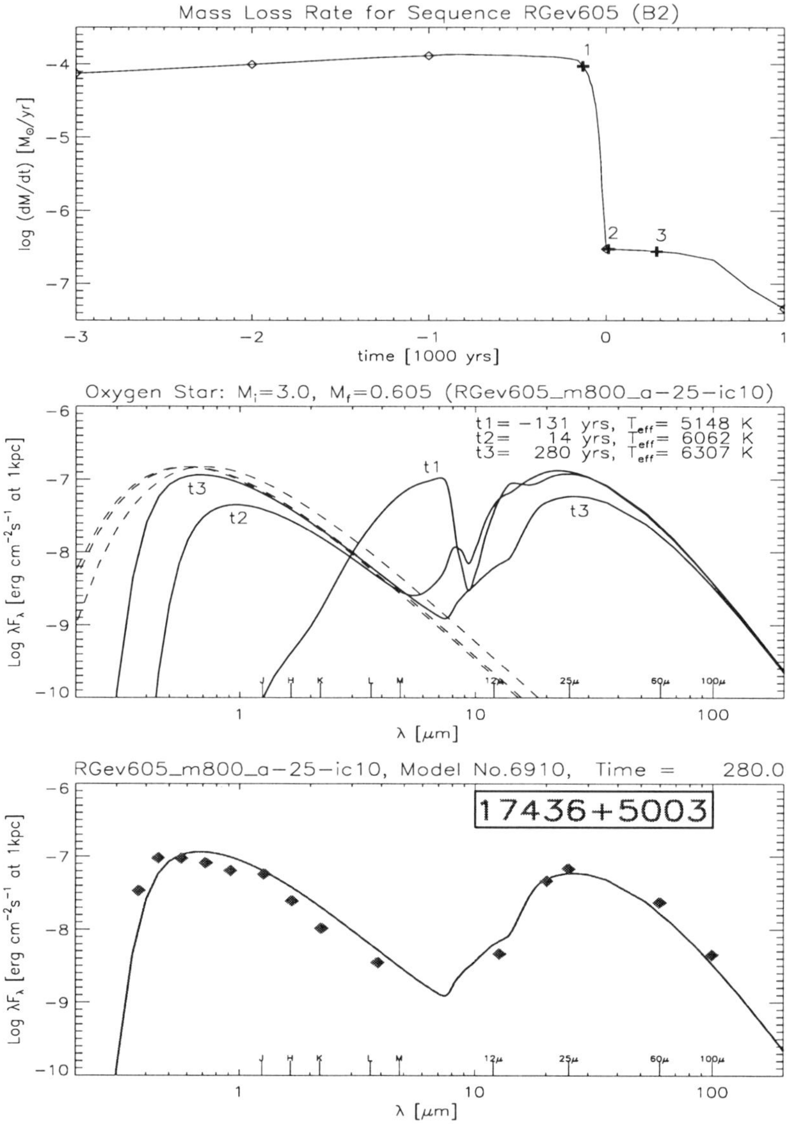

FIGURE 3. **Top:** Adopted mass-loss rate during the transition from the AGB towards higher effective temperatures for an evolutionary track ending up with a remnant of 0.605 $M_{\odot}$(Blöcker, 1995). **Middle:** Spectral energy distributions for models with silicate grains ("Astronomical Silicates") at 3 selected times as indicated in the upper panel. The dashed lines are the corresponding intrinsic spectra of the central star. **Bottom:** Spectral energy distribution of the model at time *t*3 (solid line) compared to observations of *IRAS* 17436 +5003 = HD 161 796 (diamonds; data from Hrivnak *et al.*, 1989). Our central star ($T_{\rm eff} \approx$ 6300 K) seems slightly cooler than the observed object ($T_{\rm eff} \approx$ 7000 K).

a time interval of only about 400 years. For this simulation we adopted an oxygen-based grain type ("Astronomical Silicates"), and the gradual disappearance of the strong silicate absorption feature with increasing shell detachment is clearly seen. At the same time, the totally obscured AGB remnant becomes visible. The agreement of one of our models with the well-known post-AGB object *IRAS* 17436 +5003 = HD 161 796 is encouraging (lower panel). Note that we have not attempted to fit this particular object!

4. Evolution towards a planetary nebula

In the previous section it was explained that Blöcker (1995) chose the beginning of the post-AGB evolution when the remnant's fundamental period reached the value of 50 days, which corresponds to about $T_{\rm eff} = 6000$ K for $M_f = 0.605\, M_\odot$. The mass loss is set to the Reimers prescription (Reimers, 1977) which is then kept until the remnant becomes hot enough for the theory of radiation driven winds Pauldrach *et al.* (1988) to be applicable. A more detailed description of the physical model used for the post-AGB evolution and how mass-loss rate and wind speed vary in the course of evolution can be found in Marten & Schönberner (1991).

The kinematical structure of a model with oxygen-based grains ("Astronomical Silicates"), as obtained from the time-dependent calculations at $t = 0$, is shown in the upper panel of Fig. 4. The density structure is clearly different from the usual assumption of a $\rho \propto r^{-2}$ law: This model reflects the last 100 000 years of AGB evolution. The large density dip near $r = 10^{18}$ cm is caused by the last thermal pulse about 30 000 years ago (cf. Fig. 1), while the rapid density increase towards the inner parts of the shell ($\rho \propto r^{-3}$) is due to the recent increase of mass-loss rate. Further inwards the density increase flattens somewhat ($\rho \propto r^{-1}$). The outflow velocity is rather constant, except for a slight decrease caused by the luminosity dip associated with the last pulse.

This model is then used as input for a new, explicit radiation hydrodynamics code based on an second-order Godunov-type advection scheme (Walder, priv. comm.) in order to follow the evolution into the domain of planetaries. The radiation part of this code considers time-dependent ionization, recombination, heating and cooling for six elements (H, He, C, N, O, Ne) with all of their ionization stages (Marten & Szczerba, 1996).

The lower panel of Fig. 4 gives an example of a well-developed nebular structure of intermediate size after 4603 years of post-AGB evolution. The ionization front has overtaken the relatively fast density wave (B) which had originally been created by ionization heating and is now snowplowing through the halo, the original but now ionized AGB wind. The planetary proper (between C and B) is said to be "density bounded". The fast but tenuous stellar wind terminates through a strong inner shock (near $r = 3 \cdot 10^{16}$ cm), and the high pressure of this shocked wind material compresses the low-velocity gas into a high-density somewhat accelerated shell leading the contact discontinuity (C). The outer shell has much lower density but a positive velocity gradient up to a maximum velocity of 29 km/s. Thus, the combination of ionization and wind interaction has quite naturally created a typical double shell structure: a slow, high-density inner shell surrounded by a faster, more extended low-density outer shell (e.g. Chu *et al.*, 1987).

It is clear from Fig. 4 that the kinematical structure of the planetary proper has no resemblance whatsoever with the mass loss history that originally led to its existence. In particular, the double-shell structure has absolutely *nothing* to do with recurrent helium shell flashes, as has been proposed by Tuchman & Barkat (1980). Rather, this structure is a consequence of photoionization and wind interaction and develops also for simpler initial models, e.g. the "superwind" model of Marten & Schönberner (1991) or even a r^{-2} density structure, and also for other central-star properties.

The situation is different for the halo: The density dip caused by the last thermal pulse on the AGB is still existent, and also the velocity field has not changed much. The surfaces brightness of the model's halo shows a rather sharp outer boundary close to $r = 10^{18}$ cm, comparable to the halo of the double-shell planetary NGC 6826 (Plait & Soker, 1990). Thus, the investigation of the halo's properties seems much more promising for the determination the mass-loss history close to the end of the AGB evolution, although difficult to perform.

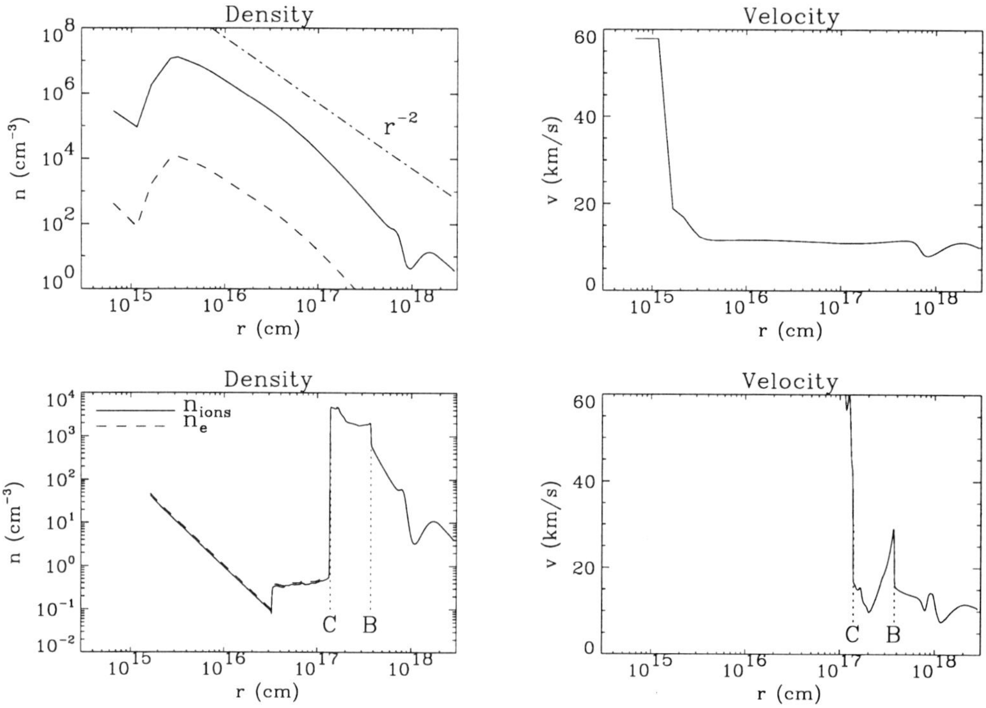

FIGURE 4. **Top:** Kinematical structures of the initial model ($t = 0$) for the post-AGB evolution, with central star paramenters $L = 6310\,L_\odot$, $T_{\rm eff} = 6053$ K, and R $= 72.5\,R_\odot$ as given by Blöcker (1995). **Bottom:** After 4603 years, $L = 5620\,L_\odot$, $T_{\rm eff} = 89\,850$ K, and R $= 0.3\,R_\odot$. The stellar mass-loss rate has declined to $\dot{M} = 1.2 \cdot 10^{-8}\,M_\odot/{\rm yr}$, and the wind speed has accelerated to 2800 km/s. The planetary proper is the region between the contact discontinuity (C) and the outer shock (B).

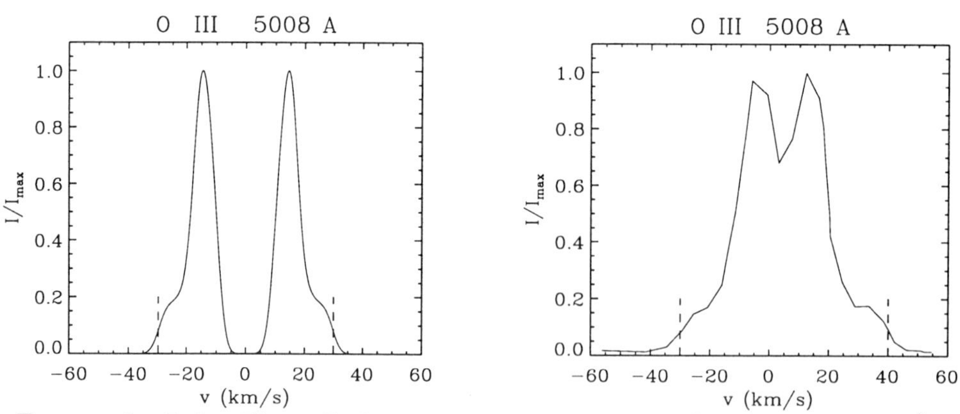

FIGURE 5. **Left:** Thermally broadened line profile of the [OIII] line at 5008 Å measured through the center of our planetary model shown in Fig. 4. **Right:** Corresponding profile measured for NGC 6826, a well-known double shell planetary (Chu *et al.*, 1984, 1987). In both panels, vertical dashed lines indicate the expansion velocities of the planetary's outer shell (see text for details).

Interesting insights into the interpretation of the internal kinematics of planetaries can be gained by computing the detailed Doppler-split emission-line profiles from dynamical models. An example is given in Fig. 5 (left panel), which is based on the nebular model shown in Fig. 4. Obviously, the Doppler splitting at highest intensity reflects the rather uniform expansion ($\approx$ 15 km/s) of the inner, dense shell leading the contact discontinuity (C). The emission in the fainter wings (or shoulders) originates from the less dense but more rapidly expanding outer shell. In order to determine the fastest moving matter in our model (29 km/s), one has to measure at half-maximum intensity of the secondary emission, as indicated by the vertical dashed lines. Finally, note that the "true" expansion velocity of our model planetary, as given by the temporal displacement of the outer rim (B), is presently about 40 km/s. Since this is not a material velocity, it is not accessible to spectroscopic measurements!

Application of these results to the observed profile of NGC 6826 leads to the following conclusions: the inner, bright shell expands with 8 km/s, which is close to the planetary's expansion velocity usually referred to in the literature. However, the secondary wing emissions from the outer shell indicate a much larger expansion velocity, viz. 35 km/s. A similar conclusion has been drawn by Chu *et al.* (1984). But obviously considering the local maxima of the wing emissions, they derived an expansion velocity of only 27 km/s.

It appears to us that the numerical modeling of hydrodynamical structures of planetaries is of paramount importance for any interpretation of observed emission-line profiles in terms of the internal kinematics in planetaries. From our simulations we conclude:

(*a*) In double-shell planetaries, the velocity as measured from the fainter wings (or shoulders) of the line profiles is related to the expansion velocity of the outer shell defining the boundary of the observed planetary proper. This is the velocity to be used for a first order estimate of the planetary's kinematical age.

(*b*) In contrast, the expansion velocity indicated by the splitting of the main line component refers only to the bright inner shell. Being compressed by the bubble's thermal pressure, its expansion velocity depends on the central star's wind properties and is *by no means* appropriate to estimate the planetary's age.

(*c*) The difference between both velocities, and hence the kinematical ages, can be quite large: in our model it is a factor of two, for the real planetary NGC 6826 it is even more than a factor of four!

REFERENCES

Blöcker, T.: 1995, A&A, **297**, 727

Chu, Y-H., Kwitter, K.B., Kaler, J.B. and Jacoby, G.H.: 1984, PASP, **96**, 598

Chu, Y.H., Jacoby, G.H. and Arendt, R.: 1987, ApJS, **64**, 529

Groenewegen, M. & de Jong, T.: 1994, A&A, **282**, 115

Hrivnak, B.J., Kwok, S., Volk, K.: 1989, ApJ **346**, 265

Ivezić, Ž. & Elitzur, M.: 1996, ApJ, **445**, 415

Marten, H. & Schönberner, D.: 1991, A&A, **248**, 590

Marten, H. & Szczerba, R.: 1996, A&A, in press

Pauldrach, A., Puls, J., Kudritzki, R.P. and Méndez, R.: 1988, A&A, **207**, 123

Plait, P. & Soker, N.: 1990, AJ, **99**, 1883

Reimers, D.: 1977, in: *Problems in Stellar Atmospheres and Envelopes*, Baschek B., Kegel W.H., Traving G. (eds.), Springer, p. 229

Tuchman, Y. & Barkat, Z.: 1980, ApJ, **242**, 199

Yorke, H.W. & Krügel, E.: 1977, A&A, **54**, 183

Time-dependent Hydrodynamical Models of Circumstellar Dust Shells around Carbon- and Oxygen-rich AGB Stars †

By M. Steffen[1,2], R. Szczerba[3], A. Men'shchikov[1,4] and D. Schönberner[1]

[1]Astrophysikalisches Institut Potsdam, Potsdam, Germany

[2]Institut für Astronomie und Astrophysik der Universität, Kiel, Germany

[3]Nicolaus Copernicus Astronomical Center, Torun, Poland

[4]Max-Planck-Gesellschaft, AG "Dust in Star Forming Regions", Jena, Germany

We have employed time-dependent two-component hydrodynamics/radiative transfer calculations to investigate the structure, dynamics and emergent spectral energy distribution of dusty circumstellar shells around Carbon and Oxygen stars in the final stages of their AGB evolution. In these models, the wind is driven by radiation pressure on dust grains and subsequent momentum transfer to the gas molecules by collisions. Detailed stellar evolution calculations with a prescribed mass loss rate that varies as a function of the fundamental stellar parameters provide $L_*(t)$, $T_*(t)$, $M_*(t)$, *and* $\dot{M}(t)$ to be used as a time-dependent inner boundary condition for the numerical solution of the coupled equations of hydrodynamics and frequency-dependent radiative transfer which describe a consistent physical model of the evolution of the circumstellar dust/gas shell.

Using our hydrodynamics code to simulate the dynamical response of the circumstellar wind shell to the evolutionary changes of the stellar parameters, we find that the large temporal variations of L and $\dot{M}$ associated with the final thermal pulses near the end of the AGB evolution lead to characteristic, time-dependent signatures in the density structure and in the infrared emission of the circumstellar dust. We present the resulting "loops" in the IRAS two-color-diagram, based on the evolutionary track of a star with an initial mass $M_i = 3.0\ M_\odot$ and a final mass $M_f = 0.605\ M_\odot$, making different assumptions concerning the composition of the dust grains (amorphous carbon or astronomical silicates).

Our models explain the existence of C and O-stars with 60 and 100 μm excess as a natural consequence of the sharp decrease of the mass loss rate following a thermal pulse, leading to the development of a *detached dust shell*. As an illustration, we present a series of synthetic spectra and corresponding 100 μm images showing the time-evolution of the circumstellar dust emission during a thermal pulse, both for a carbon-rich and an oxygen-rich dust shell.

---

1. Introduction

Before turning into planetaries, low to intermediate mass stars (1-8 solar masses) are found to evolve along the Asymptotic Giant Branch (AGB). In this phase, mass loss dominates over nuclear burning and quickly terminates the AGB evolution by reducing the mass of the stellar envelope below some critical limit. Recent observational investigations indicate that mass loss is temporarily interrupted (e.g. Willems & de Jong, 1988; Zijlstra *et al.*, 1992; Olofsson *et al.*, 1996). However, the exact evolutionary behavior of mass loss remains unknown. Most probably, the mass loss interruptions are caused by the rapid luminosity variations occurring when AGB stars undergo thermal pulses.

From the theoretical side, it is presently not possible to derive mass-loss rates along the AGB from first principles. Existing evolutionary calculations resort to semi-empirical

† This work was supported by DARA under grant WE2 50 OR 9411.

mass-loss prescriptions (Vassiliadis & Wood, 1993; Blöcker, 1995). Such modeling leads to pulse-related mass loss variations that occur on time scales which are short compared to the flight time through the envelope. Hence, the situation is far from steady state and the application of time-dependent hydrodynamics/radiative transfer calculations is essential to understand the structure and formation of detached shells.

In the following we present first results of our dynamical model calculations, simulating the physical behavior and observational properties of a dusty stellar wind envelope over the last 325 000 years of AGB evolution, based on stellar evolution calculations by Blöcker (1995) with an adopted mass loss rate that varies as a function of the stellar parameters.

2. Example Results

The temporal evolution of *luminosity* and *mass loss rate* during the last 350 000 years on the AGB of the solar composition stellar model adopted as a test case in this work is illustrated in Fig. 1 of the Schönberner *et al.* paper (this volume). During the final few thermal pulses mass loss reduces the initial mass of $3\,M_\odot$ to a final mass of $0.605\,M_\odot$.

The time-dependent values of stellar mass, luminosity, effective temperature *and* mass loss rate are now used as inner boundary conditions for the hydrodynamics/radiative transfer modeling of the circumstellar gas/dust shell. Assuming $v_{\rm gas} = v_{\rm dust} = v_{\rm sound} \approx$ 3 km/s to hold at the dust formation point (the innermost radius of the model), we find that the pronounced variations of the mass loss rate during the thermal pulses lead to large deviations of the density structure and the velocity field from the conditions of stationary outflows. The emergent spectral energy distribution, and hence the colors of the object, change considerably with time. This is illustrated in Fig. 1, where the path of our model AGB star in the IRAS two-color-diagram is shown for both carbon-rich and oxygen-rich dust composition. In both cases, the thermal pulses strongly modulate the infrared emission of the dust shell, causing these objects to run through conspicuous loops in the two-color-diagram, thereby spending about 10% of their time away from the main color-color relation of dusty AGB stars.

As a consequence of the long lasting mass-loss minima in the aftermath of a pulse, the regions close to the star temporarily become devoid of hotter dust: the envelope is said to be detached and radiates mainly in the far infrared beyond 25 μm, corresponding to thermal emission of dust cooler than about 100 K. This is illustrated in Figs. 2 and 3, showing the temporal evolution of the spectral energy distribution (top panels) as the star undergoes a thermal pulse, both for carbon-rich and for oxygen rich dust. For each of the spectra, the corresponding phase of the thermal pulse cycle may be determined by noting the time given in the figure and referring to the top panel of Fig. 2 of the Schönberner *et al.* paper (this volume). In the lower panels of Figs. 2 and 3 we show the corresponding synthetic images of the model as seen in the infrared at $\lambda\,100\,\mu$m.

REFERENCES

Blöcker, T.: 1995, A&A, **297**, 727

Laor, A. & Draine, B.T.: 1993, ApJ, **402**, 441

Olofsson, H., Bergman, P., Eriksson, K., Gustafsson, B.: 1996, A&A, **311**, 587

Rouleau, F. & Martin, PG.: 1991, ApJ, **377**, 526

Vassiliadis, E. & Wood, P.R.: 1993, ApJ, **413**, 641

Willems, F.J. & de Jong, T.: 1988, A&A, **196**, 173

Zijlstra, A., Loup, C., Waters, L.B.F.M. and de Jong, T.: 1992, A&A, **265**, L5

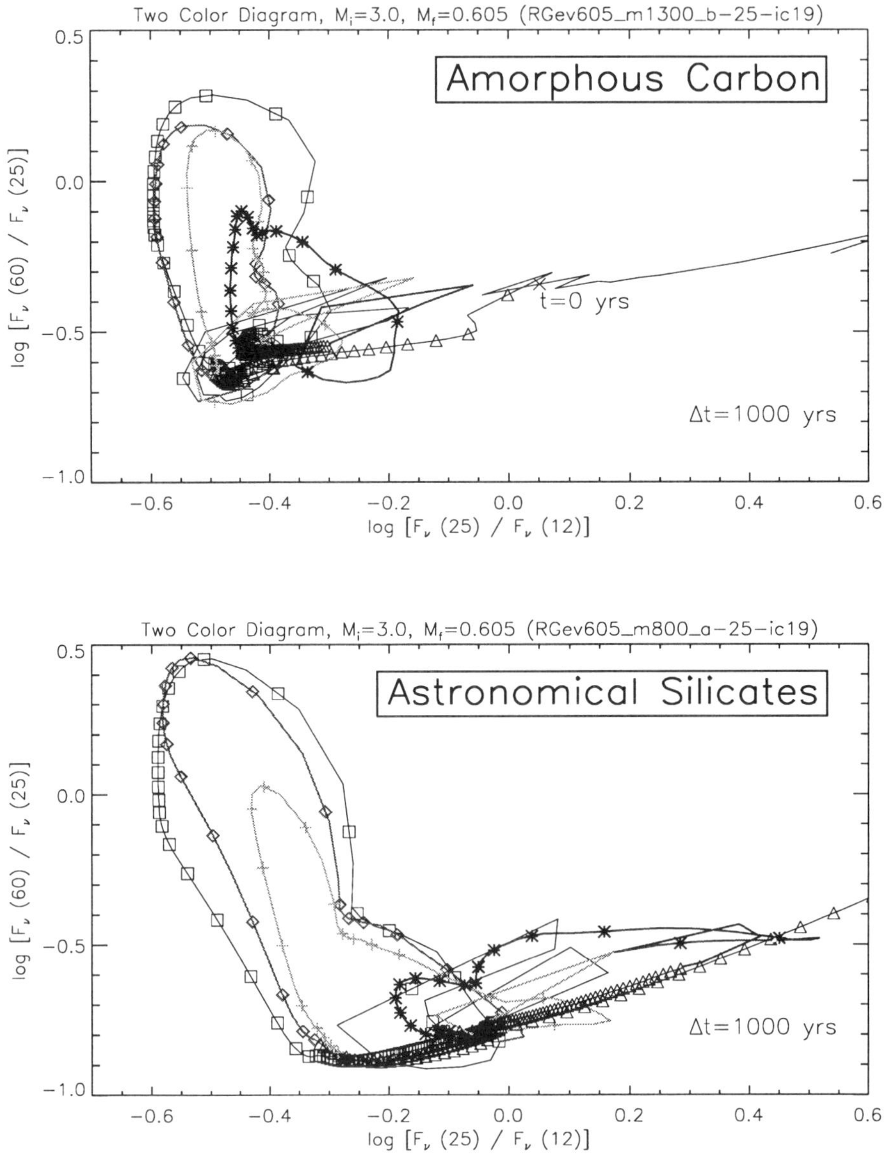

FIGURE 1. **TOP:** Computed evolution of a model *Carbon Star* in the IRAS (60/25) versus (25/12) two-color-diagram, based on a stellar evolution sequence by Blöcker (1995) with $M_i = 3.0\,M_\odot$, $M_f = 0.605\,M_\odot$. The hydrodynamical simulation covers four thermal pulses during the final 325 000 years of AGB evolution. Each of the thermal pulses produces a corresponding loop in the two-color-diagram, indicated by (1) squares, (2) diamonds, (3) plus-signs, and (4) asterisks plotted in equidistant time intervals of 1000 years. The dust grains consist of amorphous carbon and have a radius of 0.05 μm, with optical properties taken from Rouleau & Martin (1991). The adopted dust-to-gas ratio is $1.5 \cdot 10^{-3}$. **BOTTOM:** Computed evolution of a model *Oxygen Star* in the IRAS (60/25) versus (25/12) two-color-diagram, based on the same stellar evolution sequence and covering the same time interval, but now assuming dust grains to consist of astronomical silicates. The grain radius is 0.05 μm, with optical properties adopted from Laor & Draine (1993). The dust-to-gas ratio has been increased to $5 \cdot 10^{-3}$.

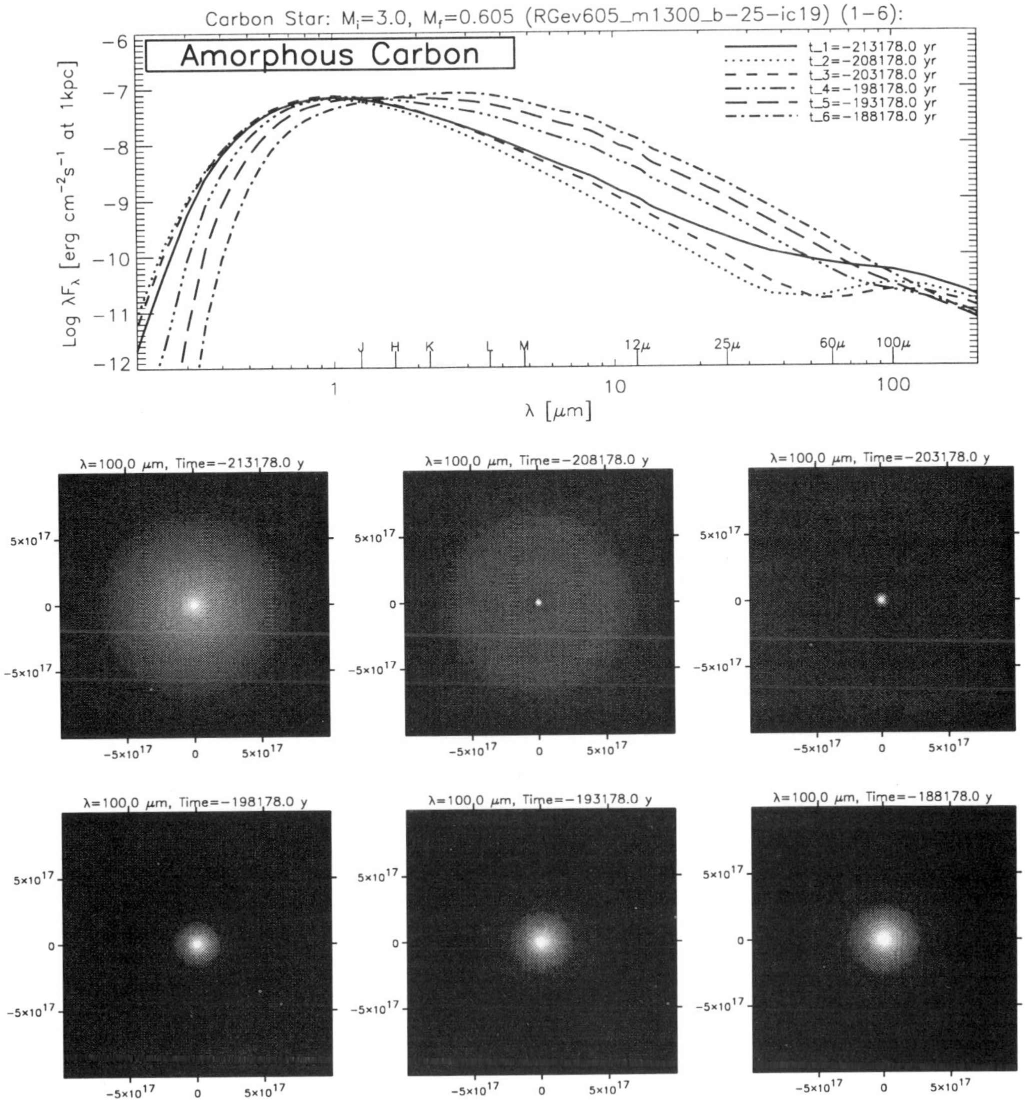

FIGURE 2. **Top:** Time sequence of spectral energy distributions computed from our carbon star model with a dust shell composed of grains of "Amorphous Carbon", demonstrating the time-variability of the stellar spectrum and the circumstellar infrared dust emission during a thermal pulse. The time interval between the spectra is 5000 yrs, with t_2 roughly corresponding to the time of minimum mass loss rate (cf. Schönberner et al. (this volume), top panel of Fig. 2). Note the pronounced excess emission at $\lambda\,60$ and $100\,\mu$m at times t_2 and t_3. **Bottom:** Synthetic images showing the 2D surface brightness distribution (intensity) of the circumstellar shell at $\lambda\,100\,\mu$m. From top left to bottom right, the images are taken at the same times t_1 to t_6 as the spectra displayed above. The intensity scale is identical for all six frames. Due to the low dust-to-gas ratio ($\delta = 1.5 \cdot 10^{-3}$) and relatively high dust outflow velocity ($v_d \approx 15$ to 20 km/s) a detached dust shell is barely discernible in this case.

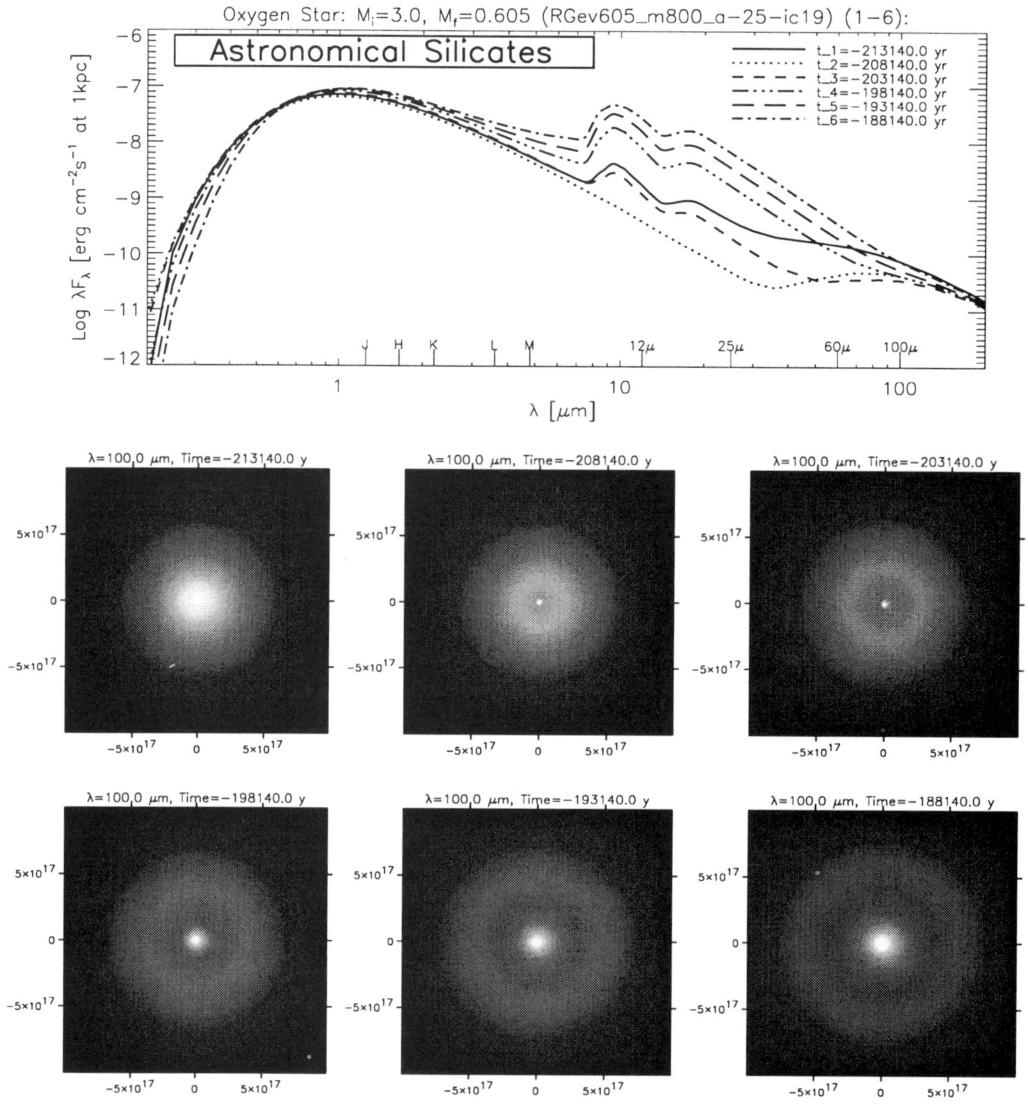

FIGURE 3. **Top:** Time sequence of spectral energy distributions computed from our oxygen star model with a dust shell composed of grains of "Astronomical Silicates", demonstrating the time-variability of the stellar spectrum and the circumstellar infrared dust emission during a thermal pulse. The spectra are shown for the same selected times as those in Fig. 2. Again we find pronounced excess emission at $\lambda\,60$ and $100\,\mu$m at times t_2 and t_3. Note that during the mass loss minimum (t_2) the silicate features at $\lambda\,9.7$ and $18\,\mu$m are completely absent due to the lack of "hot" dust in the inner parts of the shell. **Bottom:** Synthetic images showing the 2D surface bightness distribution (intensity) of the circumstellar shell at $\lambda\,100\,\mu$m. From top left to bottom right, the images are taken at the same times t_1 to t_6 as the the spectra displayed above. The intensity scale is identical for all six frames. Due to the higher dust-to-gas ratio ($\delta = 5\cdot 10^{-3}$) and the low dust outflow velocity ($v_d \leq 10$ km/s) the detached dust shell is clearly visible in this case, even at time t_6 when the IR-spectrum no longer shows any signature of a detached shell.

Chemical Abundances for a Planetary Nebulae Sample Located Outside the Solar Circle

By R. D. D. COSTA[1,2], C. CHIAPPINI[1,3], W. J. MACIEL[1] AND J. A. DE FREITAS PACHECO[2]

[1]Instituto Astronômico e Geofísico - USP, C.P. 9638, 01065-970 São Paulo - SP, Brazil

[2]Observatoire de la Côte d'Azur, B.P. 4229, 06304 Nice Cedex 4, France

[3] SISSA - ISAS, Via Beirut 2-4, 34013 Trieste, Italy

The Galactic disk shows a radial gradient of chemical abundances, with larger values in the inner region and a decreasing towards the edge. In order to check the behavior of this gradient at galactocentric distances larger than that of the Sun, we derived chemical abundances for a sample of planetary nebulae located outside the solar circle. Our results show a decreasing of the galactocentric gradient for distances larger than 9 kpc, and different behaviors for type-I and type-II nebulae.

1. Introduction

Recently Kaufer et al. (1994) found a flat gradient of chemical abundances for galactocentric distances larger than 6 kpc, based on abundances of B-type main sequence stars. As suggested by the authors the apparent controversial results from gaseous nebulae and B stars could be solved if the data from H II regions and PNe are interpreted as showing a gradient only in the inner Galaxy ($R < R_{\odot}$) and a flat profile at larger distances. An equivalent result was recently obtained from galactic chemical evolution models (Chiappini et al. 1996). Other observational works show contradictory results: Henry & Howard (1995), studying three nearby galaxies, find a constant galactic gradient; on the other hand Esteban & Peimbert (1995) describe a flattened gradient in the Galaxy to large galactocentric distances.

In order to investigate the behavior of the galactic radial gradients of abundances at galactocentric distances larger than that of the Sun, we performed spectrophotometric observations of a planetary nebulae (PNe) sample, with all the objects placed outside the solar circle. Physical parameters, kinematic properties and chemical abundances relative to hydrogen of He, O, S, Ar, Ne for the nebulae were derived, and the results are discussed in view of the present controversy.

2. Observations and analysis

Observations were carried out with a Cassegrain Boller & Chivens spectrograph attached to the 1.60 m telescope of the National Laboratory for Astrophysics (LNA), in Brasópolis, Brazil, using a 300 lines/mm grating, which allows a reciprocal dispersion of 4.4 Å/pixel, with resolution of about 9 Å FWHM. This grating allows a spectral coverage of about 5000 Å. Spectrophotometric standard stars were observed each night in order to secure flux calibration.

Reductions were made using the IRAF package, including dark current, bias and flat-field corrections, extraction of the spectra, wavelength calibration, and flux calibration.

Name	RA	Dec.	D	R	Z	V_{LSR}	ΔV	F(Hβ)	V_*
IC 2003	03 56 22	33 52 30	4.68	12.0	-1.2	-22.3	6.7	-11.19	15.0
IC 2165	06 21 43	-12 59 10	2.03	9.2	-0.4	35.2	7.4	-10.90	17.9
J 320	05 05 34	10 42 21	6.06	13.3	-1.8	-37.9	-56.7	-11.39	14.4
J 900	06 25 57	17 47 27	2.76	10.3	0.1	33.8	17.6	-11.32	17.8
K 1-7	05 31 48	06 56 09	5.76	13.0	1.4			-12.13	
K 3-66	04 36 37	33 39 30	6.68	14.1	-1.1	-75.5	-49.8	-12.24	15.4
M 1-8	06 53 34	03 08 24	3.39	10.7	0.1	35.8	-0.7	-12.37	21.4
M 1-9	07 05 19	02 46 59	4.88	12.0	0.4	120.5	71.1	-11.66	15.6
M 1-11	07 11 17	-19 51 03	2.12	9.0	-0.2	10.2	-21.3	-11.84	13.9
M 1-13	07 21 15	-18 08 36	5.32	11.6	-0.2	27.3	-42.5	-11.85	
M 1-14	07 27 56	-20 13 23	3.75	10.2	-0.1	112.7	59.9	-11.56	14.2
M 1-16	07 37 19	-09 38 48	5.45	12.0	0.5	31.9	-35.6	-12.02	16.9
M 1-17	07 40 22	-11 32 30	7.36	13.6	0.7	82.8	-2.4	-12.12	
MaC 2-1	05 03 42	-06 09 34						-12.46	16.7

TABLE 1. Positions and kinematic data for the sample

Name	Type	E(B-V)	He/H	ε(O)	ε(N)	ε(S)	ε(Ar)	ε(Ne)	$\log(\frac{N}{O})$
IC 2003	IIb	0.54	0.118	8.72	7.76	7.02	6.35	7.82	-0.966
IC 2165	IIb	0.60	0.108	8.37	7.40	6.70	5.91	7.59	-0.966
J 320	IIb	0.69	0.137	8.31	6.86	7.04	5.98	7.76	-1.452
J 900	IIb	0.66	0.106	8.66	7.66	6.73	6.05	7.93	-1.003
K 1-7	IIa	1.38	0.134	8.77	8.00	7.10	6.55	8.16	-0.772
K 3-66	IIa	0.62	0.099	7.89	8.16	5.91	5.90		0.266
M 1-8	I	1.01	0.167	8.70	8.58	6.98	6.64	7.91	-0.127
M 1-9	III	0.59	0.094	8.29	7.39	6.41	5.97	8.13	-0.892
M 1-11	IIa	0.52	0.094	7.88	8.65	5.82	5.28		0.771
M 1-13	I	0.86	0.156	9.07	8.47	6.91	7.19	8.51	-0.603
M 1-14	IIb	0.54	0.099	8.40	7.42	6.41	6.32	8.26	-0.983
M 1-16	I	0.75	0.154	8.67	8.53	6.32	6.51	8.08	-0.147
M 1-17	I	0.55	0.138	8.78	8.12	7.11	6.41	8.05	-0.661
MaC 2-1	IIb	1.22	0.111	8.44	7.08	6.99	5.53	7.60	-1.361

TABLE 2. Chemical abundances and classification

Atmospheric extinction was corrected using mean coefficients for the site. The color excess E(B-V) was estimated for each object from the Hα/Hβ ratio, as described by Costa et al. (1996).

Table 1 gives the position and kinematic data for our sample, including distances from the Sun (D), distances to the galactic center projected on the galactic plane (R), and height from the plane (Z), both in kpc, velocities with respect to the Local Standard of Rest (V_{LSR}) and peculiar velocities ($\Delta\ V$), both in km/s. Magnitudes of the central stars and $\log(F_{H\beta})$ are given in the same table. Kinematic parameters were derived following the scheme described by Maciel & Dutra (1992).

3. Analysis and discussion

Electron densities and temperatures necessary to derive ionic abundances were estimated from line ratios. Densities were estimated from the [SII] λ6716/λ6730 ratio and electron temperatures from [NII] λ5754/λ6584 and [OIII] λ4363/λ5007 ratios. When

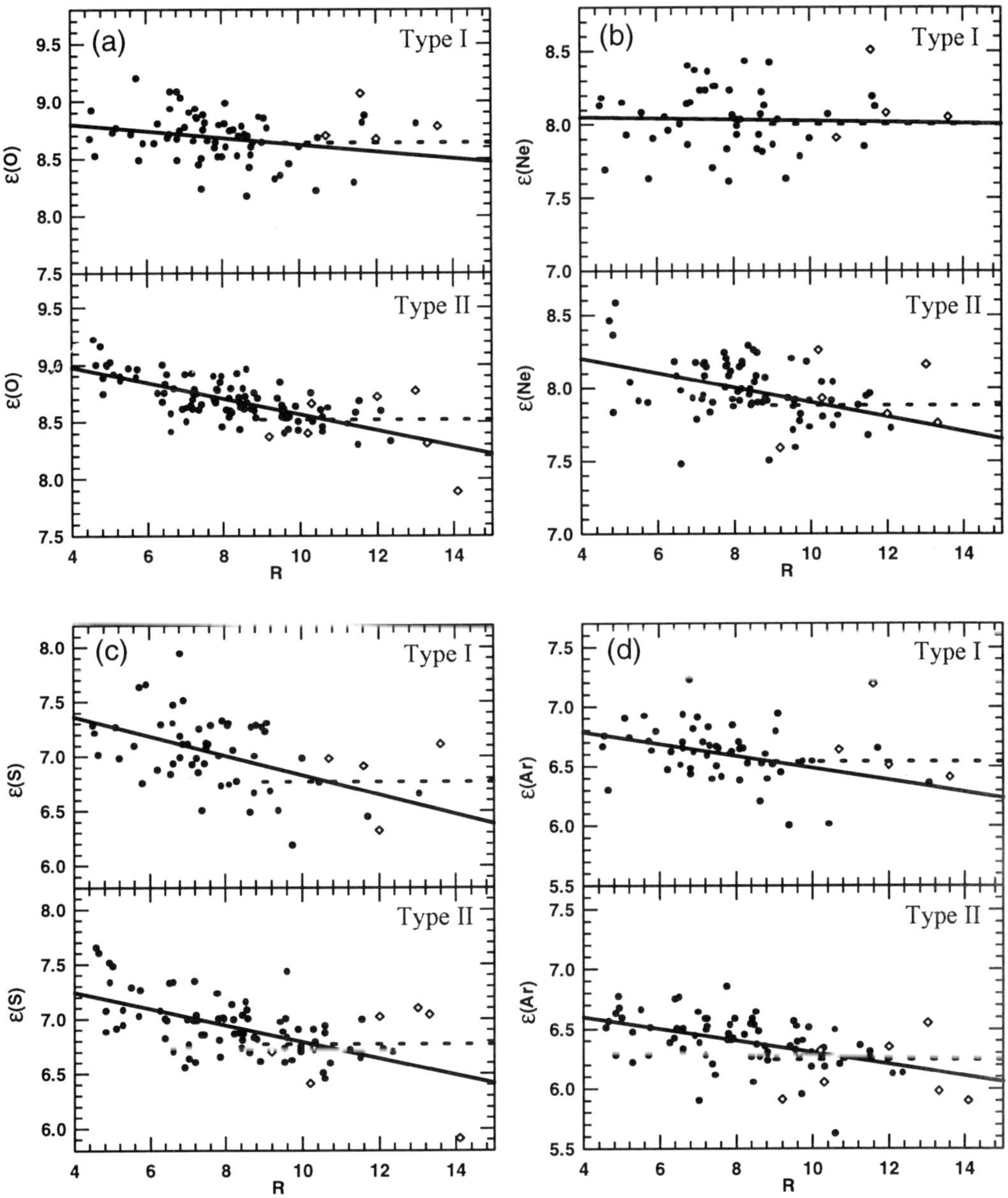

FIGURE 1. Radial gradients for the nebulae. Dots and full lines represent the data from Maciel and Köppen (1994), open diamonds are our new data and dashed lines represent a constant fit of this enlarged sample for R > 9 kpc.

these two temperatures showed differences up to 20 %, a mean value was used, otherwise, [NII] temperatures were used to calculate abundances of lower ionization potential ions like O^+, N^+, S^+, and [OIII] temperatures for higher potential ions like O^{+2}, S^{+2}, Ar^{+2}, Ne^{+2}. Ionic abundances were calculated solving the statistical equilibrium equations for a three-level atom model, including radiative and collisional transitions. Chemical abundances were then derived using ionization correction factors; explicit expressions for them are given, for instance, by Freitas Pacheco et al. (1993). Results for the nebulae are shown in Table 2, where abundances are expressed as $\varepsilon(X)=\log(X/H)+12$. Classification of the nebulae within the Peimbert's scheme and their color excesses are shown in the same table.

Together with Table 2, Figure 1 summarizes our results. It was adapted from Maciel and Köppen (1994, their figs. 1-4) by including our results in their sample. Fig. 1a shows the radial distribution of the oxygen abundance for type-I and type-II PNe; figs. 1b, 1c, 1d show the same respectively for neon, sulphur and argon abundances. Examining the original data (black dots), one can see that all the gradients derived for each chemical element and PNe type, represented by full lines, are based essentially on nebulae closer than 10 kpc from the galactic center. When we include our points (open diamonds) and take all the points farther than 9 kpc from the galactic center, type-I and type-II nebulae behave differently. For type-I, this enlarged subsample is totally compatible with a constant value, within the dispersion of the data. On the other hand, for type-II a residual gradient can still be viewed. In each plot, the dashed line represent a constant mean value displayed for comparison with the gradient previously derived.

As type-I PNe are originated from younger, more massive progenitors ($M > 2.5\ M_\odot$), and type-II from older progenitors, these results are probably related to the chemical evolution of the galactic disk, and should be examined in view of the results from chemical evolution models. As demonstrated by Götz & Köppen (1992), radial gas flows and infall rates with radial dependence can combine to produce abundance gradients, so, time variation of these parameters could produce such differences.

Our data, being derived from PNe, are a complement to previous results like those of Kilian-Montenbruck et al. (1994) and Kaufer et al. (1994), both derived from B-type stars. As pointed out by these last authors, flatten distributions of chemical abundances at large galactocentric distances are not in conflict with other evidences of radial abundance gradients in the Galaxy, but indicate a steeper gradient in the inner part of the disk, and a flatter distribution in the outer regions. In view of these results, new observational data from PNe and other objects towards the galactic anticenter, as well as improved modelling, are clearly needed to better describe this problem.

REFERENCES

Chiappini, C., Matteucci, F., & Gratton, R. 1996, Astrophys. J., , submitted

Costa, R.D.D., Chiappini, C., Maciel, W.J. & Freitas Pacheco, J.A. 1996, Astron. & Astrophys. Suppl. Ser., 116, 249

Esteban, C. & Peimbert, M. 1993, Rev. Mex. A&A Ser. Conf., 3, 133

Freitas Pacheco, J.A., Barbuy, B., Costa, R.D.D. & Idiart, T.P. 1993, Astron. & Astrophys., 271, 429

Götz, M. & Köppen, J. 1992, Astron. & Astrophys., 262, 455

Henry, R.B.C. & Howard, J.W. 1995, Astrophys. J., 438, 170

Kaufer, A., Szeifert, T., Krenzin, R., Bashek, B. & Wolf, B. 1994, Astron. & Astrophys., 289, 740

Maciel, W.J. & Köppen, J. 1994, Astron. & Astrophys., 282, 436

Maciel, W.J. & Dutra, C.M. 1992, Astron. & Astrophys., 262, 271

Víchez, J.M. 1995, in The formation of the Milky Way, ed. E.J. Alfaro & A.J. Delgado , (Cambridge U.P.) p. 163

On the Fate of Born-again Red Giants

By K. WERNER[1,2], S. DREIZLER[3], U. HEBER[4], AND T. RAUCH[1,2,3]

[1]Institut für Astronomie und Astrophysik, Universität Tübingen, Germany

[2]Lehrstuhl Astrophysik, Universität Potsdam, Germany

[3]Institut für Astronomie und Astrophysik, Universität Kiel, Germany

[4]Dr. Remeis-Sternwarte, Bamberg, Germany

Post-AGB stars which are burning hydrogen in a shell and hot white dwarfs may suffer a final helium shell flash. Subsequently the stellar envelope re-expands and the star becomes a "born-again" red giant. What is the consequence for the chemistry in the envelope? In particular, what surface abundances can be expected and how are they affected by mass loss during the second descent from the AGB? Iben & MacDonald (1995) provided the only appropriate theoretical calculation available up to now for this scenario. We present here the results of spectroscopic analyses from hot H-deficient stars, which we believe to have undergone a late helium flash. The surface abundance pattern strongly varies from star to star. An explanation of this variety demands extensive parameter studies in evolutionary calculations.

1. Introduction

The existence of hydrogen-deficient post-AGB stars is still a challenge to stellar evolution theory. (For an extensive discussion see the proc. of the 2nd coll. on H-deficient stars, Jeffery & Heber 1996.) How do R CrB stars, Wolf-Rayet type central stars and non-DA white dwarfs come into being? It has long been realised that the composition of the intershell region closely mimics that of the H-deficient stars and it has been speculated that these post-AGB stars have lost their H-rich envelope. However, it is not easy for the star to get rid of the residual envelope, because the main effect of mass loss during the post-AGB phase is to accelerate the evolutionary rate, i.e. for vanishing envelope mass the star would already be a white dwarf (see e.g. Renzini 1981). A way out of this dilemma could be a final helium shell flash. It has been shown by Fujimoto (1977) and Schönberner (1979) that the intershell convection initiated by the flash may reach the residual envelope. Protons are engulfed and rapidly burned via $^{12}C(p,\gamma)^{13}N(e^{+}\nu)^{13}C(\alpha,n)^{16}O$. But still, strong mass loss must have operated in the case of the Wolf-Rayet central stars and the so-called PG 1159 stars, because the atmospheres of most stars from these groups are dominated by carbon and helium and high admixtures of oxygen. The most extreme case is that of an essentially naked C/O stellar core (see below).

Computing stellar models undergoing a late thermal pulse with subsequent burning of hydrogen is technically extremely difficult. A first attempt to clarify the basic evolutionary features was made by Iben et al. (1983) and Iben (1984), though circumventing the difficulties in an artificial way by neglecting the release of nuclear energy. Nevertheless it could be shown that the star swells briefly to red giant dimensions and it was reasoned that a second super-wind phase thins the H-envelope such that a moderate PN wind is sufficient to remove the H-rich layer. It was not until recently that a single detailed calculation has been presented which includes the effects of time-dependent mixing (Iben & MacDonald 1995). We are now beginning to be able to make quantitative comparisons of theoretically predicted and observed element abundances.

On the empirical side enormous progress has been made in the recent years to perform quantitative spectral analyses of hottest ($T_{\rm eff} > 30\,000$ K) H-deficient post-AGB stars.

This is due to both, improved observational material available from large telescopes (and space-based observatories), and new numerical techniques to cope with the non-LTE problem in model atmosphere calculations. The progress in numerical techniques was triggered by the discovery of a new group of very hot H-deficient stars, named PG 1159 stars after the prototype. In the following we will discuss the results, deliberately excluding the Wolf-Rayet central stars for sake of brevity. We mention, however, that first model atmosphere analysis for these objects became available very recently (Hamann 1996) and that the surface abundances strongly suggest an evolutionary link between them and the more advanced PG 1159 stars.

The central stars of the famous PNe Abell 30 and Abell 78 are today regarded as transition objects between the WR- and PG 1159 groups but they exhibit interesting chemistry peculiarities. It was the discovery of H-deficient zones in the inner parts of these PNe which motivated the first calculations by Iben et al. (1983).

2. Element abundances in hot hydrogen-deficient post-AGB stars

As mentioned, we want to discuss the hottest H-deficient post-AGB stars, excluding the WRs. We are then left with the group of PG 1159 stars which in a broad sense range from (almost) helium-pure to helium-free objects. In between these extremes we find a variety of compositions with He, C, and O as the main constituents. The situation becomes even more complicated because a few PG1159s have a significant amount of hydrogen, whereas most of them seem to be hydrogen-free. The group of objects we are talking about here is almost completely analyzed (see Werner et al. 1996 for the latest review and detailed references) and their positions in the HRD are shown in Figure 1.

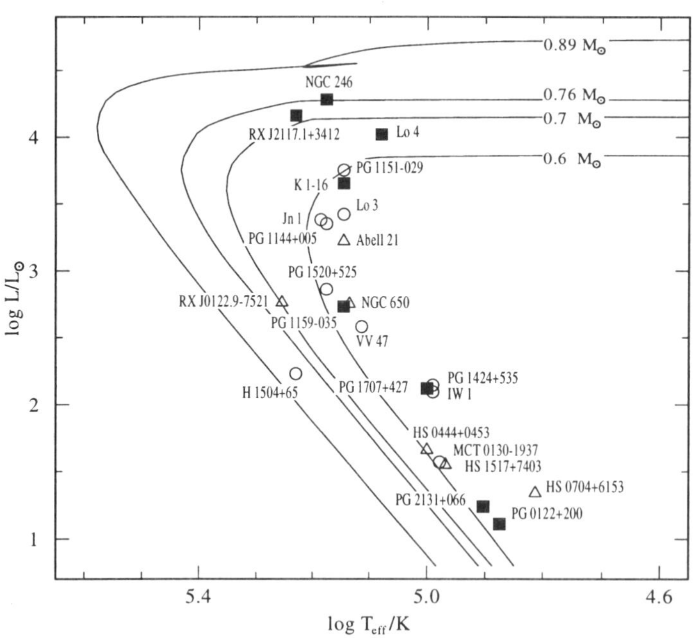

FIGURE 1. Location of the PG 1159 stars in the HRD. Symbols denote pulsators □, non-pulsators ∘, and stars with unknown photometric behavior △). Evolutionary tracks for helium-burners are from Wood & Faulkner (1986).

Let us describe the spectra and chemical abundances of some representative objects in detail. Blue spectra of these selected stars are displayed in Fig. 2 and their deduced atmospheric parameters are summarized in Table 1.

As a reference we use the central star of NGC 7293 (top spectrum in Fig. 2), a hot hydrogen-rich object (Napiwotzki 1995) whose spectrum is dominated by the Balmer lines and a weaker He II 4686Å line. Below that is the spectrum of RX J2117+3412 which is a central star with abundances which are typical for PG 1159 stars: C, He, and O are dominant, in this order. Neon was detected recently and its abundance is as high as 2%, more than 20 times the solar value, and it is indicative for deep intershell matter. The Balmer lines are absent; residual depressions in the spectrum stem from He II lines. The He II 4686Å line has an emission reversal due to the very high $T_{\rm eff}$. Just blueward of this line there is a complex of C IV lines, one component has an emission reversal, the other a pronounced absorption core. This is typical for a low gravity (i.e. high luminosity) star. The carbon lines and He II 4686Å merge to form a broad absorption through which is characteristic for PG 1159 stars. The next spectrum is from HS 2324+3944. Besides the typical PG 1159 trough it has prominent Balmer lines and for this the star is called a hybrid-PG 1159. The atmosphere is C, He, and hydrogen-rich. This clearly shows the important role of processes, which must have mixed up matter processed by hydrogen and helium burning with the original H-rich envelope.

Then follows the central star of K1-27. It has a pure He II absorption line spectrum and is by definition no PG 1159 star (lack of C IV). Two weak emissions are seen instead (near 4604/4620Å and 4944Å) which are due to trace nitrogen. This abundance pattern is typical for intershell matter which was processed by the CNO cycle, and one can interpret the star as an object which for some reason experienced a less extreme mass loss than a PG 1159 star. Then we display the spectrum of the PG 1159 prototype. It has not the distinctly sharp C IV absorption component like RX J2117+3412 which points at a high gravity (i.e. low luminosity). The spectrum below that is from PG 1144+005 and it is obviously very similar to the prototype, but with an important exception: We detect N V 4601/4612Å! Nitrogen is not the element one expects to see if the simple picture from a PG 1159 star as a peeled AGB giant would be correct. The high C and O abundances are met only at deep intershell layers where helium burning took place and which immediately destroys nitrogen. Sofar, this was the only positive detection of nitrogen in a PG 1159 star, but recently the detection of photospheric N in three more cases was announced based on HST UV spectra (Werner et al. 1997). Hence PG 1144+005 plays no longer an exceptional role. Nitrogen was detected long before in the WR–PG 1159 transition stars A 30 and A 78 (see Tab. 1) so that the presence of N now appears to be quite a common phenomenon. Note that on the other hand the oxygen abundance in PG 1144+005 is remarkably low.

One of the most curious objects is H 1504+65. The missing He II 4686Å emission core is a consequence of lacking helium (the broad absorption is due to lines from highly excited O VI), i.e. the star ist H and He-deficient. The atmosphere is a pure carbon/oxygen mixture and the natural interpretation is that we see here the outcome of an extreme case of mass loss that uncovered the C/O core of the AGB giant.

The next spectrum is from a PG 1159 star which is much cooler ($T_{\rm eff}$ = 100 kK) than the prototype (140 kK). As a consequence the emission reversals in the absorption trough have disappeared. If this star cools further down and contracts, then element diffusion will eventually set in and remove the heavy elements (C and O) from the atmosphere. It will appear as a hot He-rich white dwarf (spectral type DO). As long as the WD is hot enough, radiation pressure can keep trace metals in the atmosphere. Hence the pre-history of the chemical surface evolution is wiped out in this stage. We display KPD 0005+5106, the hottest known DO, and we see a He II absorption line spectrum

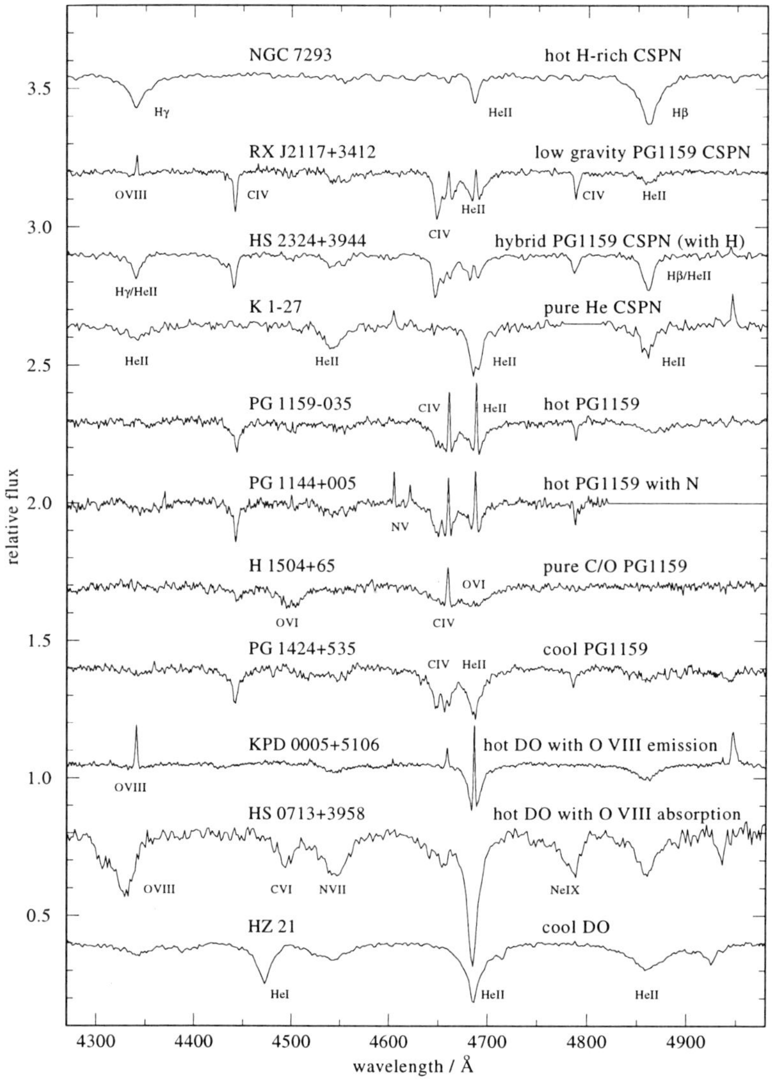

FIGURE 2. Blue spectra of hot hydrogen-deficient post-AGB stars compared to the hydrogen-rich central star NGC 7293 (top). This sample illustrates the variety of abundance patterns and displays signatures of mass loss (O VIII lines) from the hottest helium-rich white dwarfs.

with a 4686Å emission core. What is most interesting in this spectrum and why it is interesting in our context is the appearance of ultrahigh-ionization emission lines, e.g. O VIII 4340Å and 6068Å. Extremely high, non-photospheric temperatures (10^6 K) are required to produce such lines, and we have speculated that their origin is from shock fronts in a wind from the WD. Support for this idea came from the (X-ray) detection of a corona around KPD 0005+5106, which is the first such detection for a WD at all (Fleming et al. 1993).

Hence mass loss may still play an important role even in the WD stage. In fact, we witness a really drastic example in the next spectrum, which is from another DO, HS 0713+3958, and this is among the weirdest discoveries in WD research during the last years. It exhibits strong absorption (!) lines from ultrahigh-ionized metals (C V/C VI, N VI/N VII, O VII/O VIII, and even Ne IX/Ne X)! We have interpreted these asymmetric

TABLE 1. Atmospheric parameters of the stars displayed in Figure 1 and of Abell 78. We list $T_{\rm eff}$ in kK, $\log g$ in cgs units, abundances in % mass fraction.

star	$T_{\rm eff}$	$\log g$	H	He	C	N	O	Ne	remark, spectral type
NGC 7293	107	7.0	83	17					H dominated CSPN
RX J2117+3412	160	6.1		32	50		16	2	PG 1159 CSPN
HS 2324+3944	130	6.2	21	41	37	<1	1		PG 1159 with H
K 1-27	100	6.5	<8	92	< 1	2	<2		He dominated CSPN
PG 1159-035	140	7.0	<8	30	46	<.3	16		PG 1159 prototype
PG 1144+005	150	6.5		39	58	1.5	1.6		PG 1159 with N
H 1504+65	170	8.0			50		50		pure C/O atmosphere
PG 1424+535	100	7.0	<8	30	46	<.3	16		PG 1159 cool
KPD 0005+5106	120	7.0	<3						DO WD, trace CNO
HS 0713+3958	90?	7?							DO, ultra-hot wind
HZ 21	53	7.8		100					cool DO white dwarf
Abell 78	115	5.5		33	50	2	15		WR-PG 1159

lines as a super-hot and dense wind from the white dwarf. It is important, that four other hot DO white dwarfs with similar spectral features were detected so that a significantly high fraction (about 30%) of all known hot DOs shows this phenomenon. Unexpected mass loss events might play an important role also during previous evolution stages of H-deficient pre-white dwarfs. Such an event was serendipitously observed in the PG 1159-type central star of Lo 4, where on a time scale of weeks a strongly enhanced mass loss rate turned the spectrum of the star into a WR-type spectrum and back. The cause for such an event is unknown. Finally, to complete our survey we display at the bottom of Fig. 2 the spectrum of a cool DO white dwarf (by definition showing He I lines, too).

How does this large variety of abundance patterns compare with theoretical predictions? As already mentioned, there exists only one calculation which includes time-dependent mixing and which may be expected to yield reasonable surface abundances. Iben & MacDonald (1995) predict H=3%, He=76%, C=15%, N=5%, O=1%. The hydrogen abundance is below the spectroscopic detection limit, but nitrogen is predicted in a spectroscopically detectable amount. Hence the calculations are of relevance only for the minority of PG 1159 stars in which N was found, e.g. PG 1144+005. The O abundance in this star is also close to the calculation, but the C/He ratio is markedly higher. Another case which comes relatively close to theory is the hybrid-PG 1159 HS 2324+3944. Here He dominates and O is low, both as predicted, yet H is too high. For the majority of PG 1159 stars which have high C and O abundances it is therefore necessary to invoke mass loss to lay bare deeper layers. Hence, many more evolutionary models have to be constructed for a large range in parameter space.

3. Hints from asteroseismology

Several PG 1159 stars are non-radial pulsators so that their interior structure can be investigated by means of asteroseismology. The best studied object is PG 1159-035. The analysis of its pulsational behavior reveals that the thickness of the helium-rich surface layer above the C/O stellar core amounts to about $3\cdot10^{-3}\,M_\odot$ (Kawaler & Bradley 1994). Assuming that the envelope has a homogeneous composition equal to the photosphere, then the He abundance of 33% means that the envelope contains about $10^{-3}M_\odot$ of helium. This value can be explained with the post-AGB evolution of a star which has at most 10^{-2} $M_\odot$ helium when descending from the AGB (Iben & Tutukov 1984) and then undergoes modest mass loss during evolution to the PG 1159 phase.

Another interesting question concerns the total masses of the PG 1159 stars. Is it a high mass which, in contrast to the usual H-rich stars, favors enhanced mass loss and the formation of H-deficient post-AGB stars? Obviously not. The spectroscopic mass determination of the PG 1159 sample gives an average mass of $0.59\,M_\odot$, which is close to the mean mass of the H-rich central stars (Mendez et al. 1988, Napiwotzki 1995) and their successors, the DA white dwarfs (Bragaglia et al. 1995). A recent analysis of DO white dwarfs, the probable PG 1159 successors, arrives at the very same value for the mean mass (Dreizler & Werner 1996). The asteroseismologic mass determinations of the three best studied PG 1159 variables are in accordance with the spectroscopic results (see Werner et al. 1996, and references to asteroseismology works therein).

In summary, the spectroscopic and pulsational analysis of PG 1159 stars has provided us with a wealth of information on their structure. The formation of born-again red giants by a late helium shell flash is predicted from evolutionary theory. If the hot hydrogen-deficient post-AGB stars are indeed the outcome of such a scenario is an open question, until evolutionary calculations are available to explain the observational data. In fact the interpretation of nebular properties of Wolf-Rayet and PG 1159 central stars are difficult to match with this scenario (Gorny & Stasinska 1996) and it is suggested that binary evolution might play an important role (Napiwotzki et al. 1996).

ROSAT and HST data analysis is supported by DARA (grants 50 OR 9409 1, 9602 9).

REFERENCES

Bragaglia A., Renzini A., Bergeron P. 1995, ApJ 443, 735

Dreizler S., Werner K. 1996, A&A 314, 217

Iben I. Jr. 1984, ApJ 277, 333

Iben I. Jr., Tutukov A.V. 1984, ApJ 282, 615

Iben I. Jr., MacDonald J. 1995, in White Dwarfs, ed. D.Koester, K.Werner, LNP 443, 48

Iben I. Jr., Kaler, J.B., Truran, J.W., Renzini, A. 1983, ApJ 264, 605

Jeffery C.S., Heber U. 1996, 2nd colloquium on H-deficient stars, ASP Conf. Ser. 96

Fleming T.A., Werner K., Barstow M.A. 1993, ApJ 416, L79

Fujimoto M.Y. 1977, Pub. Astr. Soc. Japan 29, 331

Gorny S.K., Stasinska G. 1996, in H-def. stars, ed. C.S.Jeffery, U.Heber, ASP Conf. Ser. 96, 209

Hamann W.-R. 1996, in H-deficient stars, ed. C.S.Jeffery, U.Heber, ASP Conf. Ser. 96, 127

Kawaler S.D., Bradley P.A. 1994, ApJ 427, 415

Mendez R.H., Kudritzki R.P., Herrero A., Husfeld D., Groth H.G. 1988, A&A 190, 113

Napiwotzki R. 1995, in White Dwarfs, ed. D.Koester, K.Werner, LNP 443, 176

Napiwotzki R., Haas S., Schönberner D. 1996, in H-deficient stars, ed. C.S.Jeffery, U.Heber, ASP Conf. Ser. 96, 213

Renzini A. 1981, in Effects of mass loss on stellar evolution, ed. C.Chiosi, R.Stalio, Reidel, p. 319

Schönberner D. 1979, A&A 103, 119

Werner K., Dreizler S., Heber U., Rauch T. 1996, in H-deficient stars, ed. C.S.Jeffery, U.Heber, ASP Conf. Ser. 96, 267

Werner K., Dreizler S., Heber U., Rauch T. 1997, in White Dwarfs, ed. J.Isern, M.Hernanz, E.Garcia-Berro, Kluwer, in press

Wood P.R., Faulkner D.J. 1986, ApJ 307, 659

Hyades Cluster White Dwarfs and the Initial-Final Mass Relation

By VOLKER WEIDEMANN

Inst.Astronomie u.Astrophysik, University, 24098 Kiel, Germany

A recalculation of masses for Hyades white dwarfs based on individual distances, recent gravity determinations and redshift data, using thick-layer mass-radius relations not only improves the agreement between $M(R)$ $M(g)$ and $M_{\rm red}$ individually but also removes the discrepancies between the average values discussed by the author (1993, A&A, 275, 158). For the first time the M-R relation for thick-layer models is observationally confirmed. The individual masses cover a larger range which implies differential mass loss for the progenitors. The average initial-final mass relation runs steeper, more like the parameterized form suggested by Iben and Renzini (1983) or by Iben, Tutukov and Yungelson (1996).

1. Introduction

White dwarfs constitute the main final stage of stellar evolution: less than a few percent end up in other ways as neutron stars or black holes. This is due to the fact that stars undergo heavy mass loss during their evolution especially on the AGB when the interior is a degenerate carbon-oxygen core below the Chandrasekhar limiting mass. Depending on details of single or binary evolution the initial masses up to which stars enter the AGB and thus produce white dwarfs are between 8 and 10 solar masses and thus constitute the vast majority of all stars born on the main sequence. The Hyades were crucial in reaching this conclusion since already in 1965 Auer & Woolf and Eggen & Greenstein noticed that the white dwarfs in this cluster must have been produced from parent stars with masses larger than the turn-off mass, above 2 solar masses, which implied heavy mass loss.

In the following years the Hyades white dwarfs found more interest: Van Altena (1965) searched for and discussed membership, Koester & Weidemann (1973) used the cluster white dwarfs in order to improve the distance determination by a fit to the narrow field DA sequence in the color-magnitude diagram, and Weidemann (1977) investigated the question of mass loss toward the white dwarf stage with special emphasis on the Hyades white dwarfs. He found that for simple models of cluster evolution there should be many more white dwarfs and suggested that the older white dwarfs have been lost from the cluster. Are they members of the Hyades supercluster or moving group – the concept of which was the topic of Olin Eggen's research for many years?

Eggen (1985) had listed 56 Hyades supercluster white dwarfs – enough to explain the deficit found by Weidemann (1977). Since Icko Iben had spent some time in Chile with Eggen, using the supercluster concept to investigate giants and to compare with theoretical predictions (Eggen & Iben 1991), it was natural that we re-studied the Hyades case when he came to Kiel afterwards in 1990. We determined as well the cluster and white dwarf parameters as studied the question of evaporation and the evidence for the supercluster - see Weidemann/Jordan/Iben/Casertano 1992 (WJIC). Whereas the dynamical problem was further investigated in Illinois by Casertano et al. (1993) with the conclusion that local dark matter was necessary in order to confine evaporated stars to the neighborhood as a supercluster, I myself re-studied the parameter determination of the Hyades white dwarfs with improved spectroscopic data for the individual stars and checked with them the theoretically predicted mass-radius relation for white dwarfs

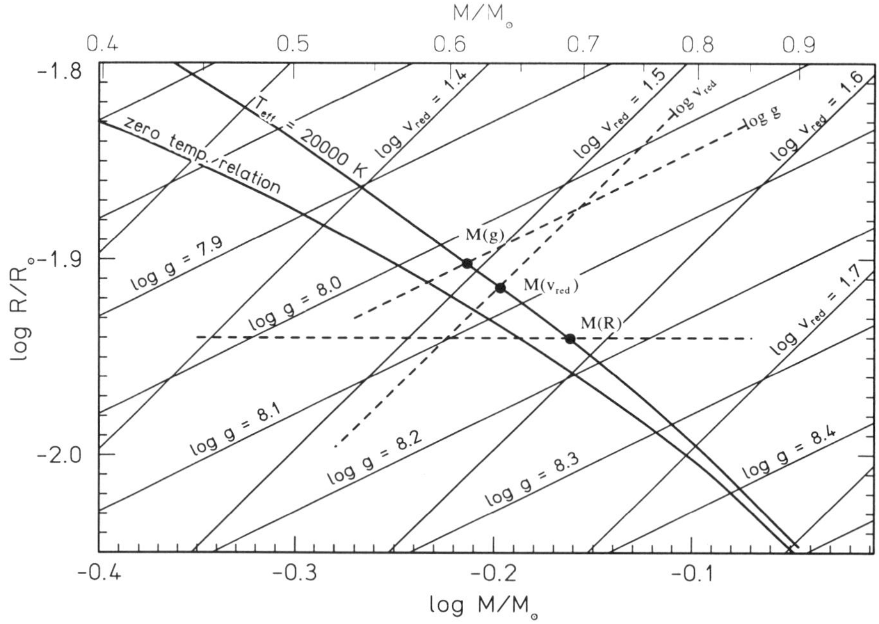

FIGURE 1. Mass determination for white dwarfs, using surface gravity, or redshift or radius and a mass-radius relation for a given effective temperature in a logarithmic mass-radius plane.

(Weidemann 1993a). In another paper I applied these results to the distance problem of the Hyades as it turned out that masses determined from radii (and distances) and masses determined from surface gravity and/or redshifts were discrepant (Weidemann 1993b). I emphasised that the individual masses depend furthermore strongly on the assumed thickness of the outer hydrogen-layer which had been taken to be negligible thin in the investigations of Bergeron et al. (1992) (BSL). Thin layers have also been used in the recent comparative study for DA white dwarfs with redshifts of Bergeron et al. (1995) (BLF). Since at the European White Dwarf Workshop in Kiel 1994 Clemens convincingly showed from pulsation analysis that the hydrogen layers of variable DA stars had to be thick (i.e. of the order of 10^{-4} stellar masses) – see also Bradley (1995) – as predicted by evolutionary calculations for stars leaving the AGB (Schönberner 1979, 1983, Iben 1984) I decided to restudy the Hyades white dwarfs under this assumption for this Meeting. I had hoped to use also the HIPPARCOS results for the Hyades distance determination however due to several delays these were not yet available.

2. White dwarf mass determinations

There are three different possibilities to obtain white dwarf masses - see Weidemann (1990) for a review. In short: if the surface gravity g and the effective temperature T_{eff} has been determined by spectroscopic analysis, and a mass-radius relation is assumed to hold, it is simple to derive $M(g)$– see Fig. 1. If the distance is known, one obtains the radius from T_{eff} and the observed flux, again with the M-R relation one obtains $M(R)$. If the radial velocity is known, the gravitational redshift can be isolated and, since it is proportional to M/R, can be used to obtain M_{red}. In the case of the Hyades, all three possibilities exist. The spectroscopic g-determination has in recent years been

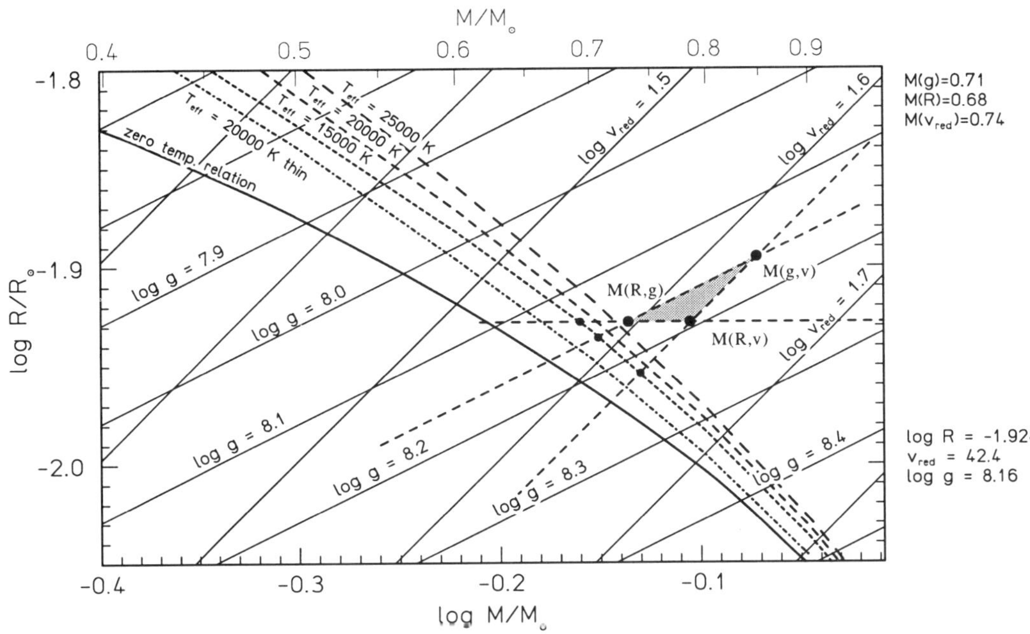

FIGURE 2. Mass and radius determinations without use of mass-radius relations (triangle) and with a thick-layer M-R relation.

improved, due to the analysis of line profiles in CCD spectra by BSL and BLF, however the derived mass $M(g)$ depends on which mass-radius is assumed to hold. The traditional zero-temperature relation of Hamada and Salpeter (1961) has been replaced by finite-temperature relations which for a given mass predict larger radii by an amount which increases with the temperature, but also depends on the thickness of the outer non-degenerate helium- and hydrogen layers. These relations have been calculated by Wood (1990) (mainly thin layers of helium and hydrogen) and recently completed for thick layers (Wood 1995). The derived masses for a given surface gravity, radius or redshift are correspondingly increased, as is best demonstrated in Fig. 1 and 2 in which the logarithmic scales for mass and radius allow the lines of constant surface gravity or redshift to be straight.

We shall use Fig. 2 for the mass determinations of the individual Hyades white dwarfs. In each case one obtains from g, $v_{\rm red}$ and R a triangle the corners of which allow mass determinations without recourse to an M-R relation. The size of the triangle is a measure of agreement of the different methods, and the location of the triangles allows the M-R relations to be checked. The corners of the triangles give $M(g,R)$, $M(g,v)$ and $M(v,R)$, the exact values of which can of course be obtained by calculation from the definition of g and $v_{\rm red}$: $GM(g,R) = gR^2$, $GM(g,v) = v^2c^2/g$ and $GM(v,R) = vRc$, where G is the gravitational constant and c the speed of light. $R(g,v)$ is thus given by vc/g. H.Schmidt (1995) gives the logarithmic form as $\log R = \log v - \log g - .366$ and $\log M = 2\log v - \log g - 5.169$ in solar and c.g.s.units.

2.1. *Example of a Hyades white dwarf*

We give a demonstrative example of the method for the Hyades DA white dwarf HZ 4, (EG 26), WD 0352+096. For the effective temperature and the surface gravity we accept the most recent values obtained by BLF, 14770 K and $\log g = 8.16$. The redshift is taken

TABLE 1. Masses of Hyades DA white dwarfs

WD	$M(g)$	M_{red}	$M(R)$	$M(g,R)$	$M(g,v)$	$M(v,R)$	M_f
EG 26	0.71	0.74	0.68	0.73	0.84	0.79	0.71
EG 29	0.79	0.80	0.82	0.725	0.80	0.77	0.79
EG 36	0.67	0.69	0.71	0.625	0.73	0.675	0.675
EG 37	0.70	0.69	0.655	0.775	0.66	0.72	0.70
EG 39	0.645	0.67	0.645	0.64	0.735	0.685	0.645
EG 42	0.68	0.66	0.685	0.68	0.59	0.63	0.68

from Reid (1996) who re-determined it by Hα observations with the HIRES spectrograph at the Keck telescope. He gives $v_{\text{red}} = 42.4 \pm 1.2\,\text{km/s}$, considerably larger than the 33.3 km/s derived in the earlier determination by Wegner et al. (1989). We thus obtain for $M(g)$ 0.71, M_{red} 0.74 and $M(R)$ 0.68 $M_\odot$, whereas we find for $M(g,R)$ 0.73, $M(g,v)$ 0.84 and $M(v,R)$ 0.79 $M_\odot$. Fig. 2 demonstrates that the triangle is small, but located far above the M-R relation. However, if a smaller redshift value, 38 km/s, $\log v = 1.58$, intermediate between the old and new determination, had been taken the triangle would shrink to nearly zero and the M-R relation would be covered.

But uncertainty arises also in the R determination. The individual distance is obtained from the kinematics by the moving cluster method. It depends on the measured proper motion, the convergence point, the angular distance to the latter and the group velocity. In our example we have used the proper motion as given by Eggen (1985), 0.150 arcsec/yr, convergence point by Schwan (1991), $\alpha = 6^{\text{h}}30\overset{\text{m}}{.}7$, $\delta = +5°59'$, and group velocity by Detweiler al. (1984), 46.6 km/s. It is hoped that the HIPPARCOS results will yield more reliable cluster parameters, which enable not only improved distance determinations but also better calculations of individual radial velocities and thus of the gravitational part of the measured redshift. Unfortunately the proper motion accuracy cannot be improved since the Hyades white dwarfs are too faint for HIPPARCOS observations.

3. The Hyades white dwarfs

In the same way we re-determined the masses of the other Hyades DA white dwarfs LB227 (EG29), WD 0406+169, VR7 (EG36), WD 0421+162, VR16 (EG37), WD 0425+168, HZ7 (EG39), WD 0431+125, and HZ14 (EG42), WD 0438+108. The results are given in Table 1. The last column gives the average final white dwarf mass M_f.

Compared to the earlier determinations it is striking that the derived masses are considerably larger. The average mass has been increased from $0.628M_\odot$ in the WJIC study (1992) to $0.682M_\odot$ found here. This needs some comments. First, we excluded in WJIC EG29, since $M(g)$ and $M(R)$ were too discrepant (see the detailed discussion there). In the meantime the recent spectroscopic analysis by BLF increased $\log g$ from 8.04 (Koester et al. 1979) (KSW) to 8.30 corresponding to $M(g) = 0.79M_\odot$ and yielded agreement with $M(R)$. With other words: the small radius is real and not caused by a normal $0.6M_\odot$ white dwarf at a larger distance! The recent determination of the redshift by Reid (1996) confirmed the old value of 49 km/s, although with a larger uncertainty of ± 6 km, corresponding to $M_{\text{red}} = 0.80 \pm 0.06M_\odot$ and therefore agrees. If EG29 is thus indeed a Hyad its higher mass needs a special explanation which will be discussed below in connection with the initial/final mass relation. Second, the increase of the average masses of $M(g)$ has a secular aspect due to the fact that one went from zero-temperature

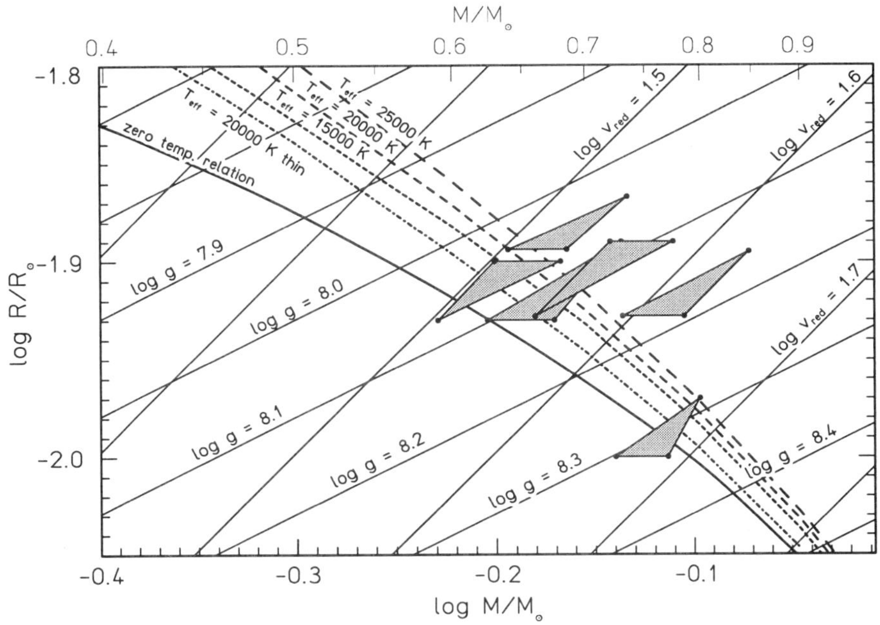

FIGURE 3. Location of Hyades white dwarfs in the M-R plane. Triangles from left to right are for EG42, 39, 36, 37, 26 and 29.

configurations, as used in KSW 1979 and WK 1984, to thin-layer models , as used by BSL 1992, WJIC 1992, and BLF 1995, to thick-layer models, as used by Herwig (1996) and here. The first step raised the average mass from 0.59 to 0.62 $M_\odot$, a second was due to the fact that the g values of BLF 1995 were systematically higher and the sample included the higher mass objects EG26 and EG29, leading to $0.66M_\odot$ (without EG29) or $0.68M_\odot$ (with EG29), whereas the last step to the thick layer models, first done by Herwig (1996) (without EG29), yielded $0.684M_\odot$, practically the same value which we obtained: 0.68 without and $0.70M_\odot$ with EG29. Remarkable however is the fact that the average mass derived from redshift determinations was much less affected: it climbed only from 0.607 $M_\odot$ (WJIC 1992) to $0.68M_\odot$ (without the high mass objects EG26 and EG29) (Reid 1996).This is due to the fact that the lines of constant redshift run much steeper than the lines of constant gravity (see Fig.1) The discrepancy which I had pointed to in my A&A paper (1993b, especially Fig.1) has thus disappeared, as I predicted for thick-layer models. This has also been noticed by Reid (1996) in his recent redshift study. In those days I was reluctant to accept higher Hyades white dwarf masses since they implied a steeper initial/final mass relation which I felt would not be compatible with the observed narrow DA mass distribution. We shall discuss this point below. Does our result imply that thick-layer models have become even more trustworthy? Do they shed new light on the empirical M-R relation?

4. Hyades white dwarfs and mass-radius relation

My investigation (1993a, Fig.3) had demonstrated that the individual points for $M(g,R)$ $M(g,v)$ and $M(v,R)$ cover a wide range across the theoretical M-R relation. Although some evidence in favor of non-zero temperature models was apparent, the data were not sufficient to constrain the thickness of outer layers. A similar conclusion was reached in the detailed study of Schmidt (1996) which covered all white dwarfs for which distances, radii, surface gravity and redshifts were available. The sub-ensemble of the Hyades which he kindly provided to me showed the same large scatter, partly due to the fact that he had used group parallaxes from the literature. Since I have used individual distances and the revised data for surface gravity and redshift the scatter is now considerably reduced.

If one plots the data from Table 1 in Fig.3 it appears that the triangles are of similar small size and, with exception of EG26, cover the theoretical M-R relations. If one forms the average for the four brightest and hottest stars, EG 36, 37, 39 and 42 , one finds an average radius of $\log R = -1.903$ and an average mass of $0.678 M_\odot$ – a value which falls exactly on the M-R relation for thick-layer models of 24000 K. The average temperature of the four stars is 22000 K. We may thus conclude that for the first time it has been possible to show that the Hyades white dwarfs fulfill the theoretically predicted M-R relation, and that thick-layer models are clearly superior to thin-layer models.

Looking at Fig.2 or 3 it is interesting to see how the mass value $M(R)$ would increase for a fixed radius, say $logR = -1.90$: from $0.57 M_\odot$ for a zero-temperature configuration to $0.61 M_\odot$ for thin-layer models and to $0.65 M_\odot$ for thick-layer models (of 20000 K). The radius determination is therefore important. A few remarks must be made in this respect. The uncertainty of the effective temperature and the fluxes (V magnitudes) is probably small, however the distances depend – aside of the cluster parameters – on the proper motion. Here one relies either on catalogue values (McCook/Sion) like Schmidt (1996), or on the list of Eggen (1985) or takes the new determinations from Reid (1992, 1993) as was done by Reid (1996). I calculated radii from all three sources and found mostly agreement. My finally accepted radii values are for EG26, EG29, EG37 and EG39 those of Reid (1996), his table 5, for EG36 I use instead of Reid's -1.946 $logR = -1.930$, in view of Herwig (1996) (corrected for distance change) -1.931 and Schmidt (1995) (also corrected for distance change) -1.925 . For EG42 I used -1.90 instead of Reid's -1.91, again as corrected value from Schmidt (1995) -1.899 exactly like Herwig (1996). The differences may be due to Reid's use of bolometric corrections from Bergeron models whereas Schmidt (1996) uses V fluxes from Koester models. Anyway: it is apparent that the uncertainties in $\log R$ (with the individual distances calculated with the present cluster parameters) are about $\pm$ 0.01 dex. (Possible changes in effective temperatures not included, since they are here all equal, taken from BLF.) As far as future improvements are concerned one should also keep in mind that individual distances determined from proper motions and radial velocities (and thus the redshifts) depend on the assumption of a strictly parallel Hyades stream, deviations of which might cause additional changes of the derived masses and radii.

5. Hyades white dwarfs and the initial-final mass relation

A semi-empirical method to determine initial masses for white dwarfs in open clusters has been developed and applied in several studies during the last 15 years (see Weidemann & Koester 1983, Weidemann 1987). Koester & Reimers observed at the ESO 3.6m telescope, analysed the spectra and published 8 papers since 1981 (see quotations in Koester & Reimers 1996). Herwig (1994, 1996) has redetermined initial masses using improved theoretical information on stellar evolution and white dwarf cooling. He also investigated the influence of core overshooting on the main sequence lifetime which is critical for the determination of the initial mass (cluster age minus white dwarf cooling time equals the progenitor lifetime). For the Hyades, Praesepe and the Pleiad a similar study was made by Reid (1996). Koester & Reimers (1996) confirmed an upper limiting mass for white dwarf (vs.supernova) production around $8M_\odot$ based mainly on new data for NGC 2516, and Reid (1996) concluded from his data on the three clusters that the initial-final mass relation runs straight through the increased Hyades masses, steeper and not bended,(and first flatter) as was suggested in earlier studies (Weidemann & Koester, 1983, Weidemann 1987, 1990). He finds the linear relation suggested in parameterized form by Iben and Renzini (1983) $\eta = 2$ – (not 3 as misprinted below Reid's Fig.5) – to come closest to the new data. On the other hand Herwig (1996), also considering the new Hyades and Pleiades data, suggested a relation which first runs flatter (from $M_f = 0.55$ to 0.6 $M_\odot$for M_i from 1 to $2M_\odot$, then steeper through the Hyades up to $M_f = 0.7$ at $M_i = 3M_\odot$, then bends slowly over to the Pleiad ($M_f = 1$ at $M_i = 5.5M_\odot$) to the limit ($M_f = 1.1$ at $M_i = 8M_\odot$). Indeed this is what one expects for theoretical reasons as the core mass at the first thermal pulse shows a similar shape as function of the initial mass (see Weidemann 1993c and Blöcker 1995). A straight relation all the way from $0.55M_\odot$ at $M_i = 1M_\odot$ upwards through the Hyades mean, $M_f = 0.7$ at $M_i = 2.5M_\odot$ would probably not reproduce the observed mass distribution of white dwarfs with its sharp peak at $0.6M_\odot$ (see Bragaglia et al. 1995) although this has to be checked again by synthetic calculations with galactic evolution models, as has been done by Yuan (1992). However the Hyades results demonstrate another fact: the probability of differential mass loss - well known from the existence of the horizontal branch for stars which left the first giant stage (Iben & Rood 1970) and certainly also expected on the AGB (see Weidemann 1977, 1981). Reid (1996) states that although the progenitor masses for all of the Hyades and Praesepe white dwarfs are nearly the same the remnant masses range from at least 0.6 to $1.0M_\odot$. He concludes that the mass loss is essentially a stochastic phenomenon. I think this conclusion is too strong: for EG 36,37,39 and 42 are the masses nearly equal (Table 1) with progenitor masses within $0.1M_\odot$ (2.5 to $2.6M_\odot$for WJIC, and 2.4 to 2.6 $M_\odot$ for Reid), however the masses for EG26 and especially EG29 are considerably higher. But these stars, especially EG29, may be merger products - a phenomenon often predicted by Iben and Tutukov in their long row of papers on binary star evolution (see Iben et al. (1996) for quotations). On the other hand Yuan (1992) has shown that differential mass loss for single star evolution cannot be more than about ± 0.1 $M_\odot$ since otherwise the resulting calculated mass distribution would be too wide compared with observation. Evidently these questions need further study.

In closing I would like to present an initial-final mass relation which was proposed this year (Iben et al. 1996) by Icko in his ingenious way: he interpolates three points for (M_f, M_i) : $(0.55, 1.0), (0.65, 2.3)$ (from WJIC) and (1.2, 8) with a power-law $M_f = aM^b$ in each interval, yielding $0.55M^{0.20}$ up to the Hyades progenitor range and $0.43M^{0.41}$ up to $M_i = 8M_\odot$. It practically fits the new Hyades results with (0.68 at $2.5M_\odot$) and it is thus very usable for simple calculations of galactic evolution and population synthesis.

I am very much indebted to Stefan Jordan and Holger Schmidt for help at the preparation of the postscript and the drawings.

REFERENCES

Auer, C.H., Woolf, N.J., 1965, ApJ 142, 182

Bergeron,P., Saffer, R.A., Liebert, J. 1992, ApJ 394, 228 (BSL)

Bergeron, P., Liebert, J., Fulbright, M.S. 1995, ApJ 444, 810 (BLF)

Blöcker, T. 1995, A&A 297, 727

Bradley, P.A. 1995, in White Dwarfs, Proc. 9.Europa.Workshop, Kiel, Germany 1994. Lecture Notes in Physics vol. 443, ed. D.Koester & K.Werner, (Berlin: Springer) p.284

Bragaglia, A., Renzini, A., Bergeron, P. 1995, ApJ 443, 735

Casertano, S., Iben Jr., I. 1993, ApJ 410, 90

Detweiler, H.L., Yoss, K.M., Radick, R.R., Becker S.A. 1984, Astron.J. 89, 1038

Eggen, O.J. 1985, Publ.Astron.Soc.Pacific 97, 1029

Eggen, O., Greenstein, J.L. 1965, ApJ 141, 83

Eggen, O.J., Iben Jr., I. 1991, Astron.J. 101, 1377

Hamada T., Salpeter, E.E. 1961, ApJ 134, 683

Herwig, F. 1996, in Proc. 32.Liège Intern. Astrophys. Coll.1995, p.441

Iben Jr., I 1984, ApJ 277, 333

Iben Jr., I., Rood, R.T. 1970, ApJ 161, 587

Iben Jr., I., Renzini, A. 1983, ARA&A 21, 271

Iben Jr., I., Tutukov, A.V., Yungelson, L.R. 1996, ApJ 456, 750

Koester, D., Weidemann, V. 1973, A&A 25, 437

Koester, D., Schulz, H., Weidemann, V. 1979, A&A 76, 262 (KSW)

Koester, D., Reimers, D. 1996, A&A 313, 810

Reid, I.N. 1992, MNRAS 257, 257

Reid, I,N. 1993, MNRAS 265, 785

Reid, I.N. 1996, Astron.J. 111, 2000

Schmidt, H. 1995, Dipl.Thesis Univ.Kiel

Schmidt, H. 1996, A&A 311, 852

Schönberner, D. 1979, A&A 79, 108

Schönberner, D. 1983, ApJ 272, 708

Schwan, H. 1991, A&A 243, 386

Van Altena, W.F. 1969, Astron.J. 74, 2

Wegner, G., Reid, I.N. & McMahan, R.K. 1989, in White Dwarfs, IAU Coll. 114, Dartmouth College 1988, Lecture Notes in Physics vol. 328, ed. G.Wegner (Berlin: Springer), p.378

Weidemann, V. 1977, A&A 59, 411

Weidemann. V. 1981, in Effects of Mass Loss on Stellar Evolution eds. C.Chiosi, R.Stalio (Dordrecht: Reidel) , p.339

Weidemann, V. 1987, A&A 188, 74

Weidemann, V. 1990, Ann.Rev.A&A 28, 74

Weidemann, V. 1993a, in White Dwarfs: Advances in Observation and Theory ed. M.A. Barstow, NATO ASI Series, vol.C 403, (Dordrecht: Kluwer), p.1

Weidemann, V. 1993b, A&A 275, 158

Weidemann, V. 1993c, in Mass loss on the AGB and beyond, Proc. ESO Conference and Workshop La Serena 1992, No.46, ed. H.E. Schwarz, p.55

Weidemann, V., Koester, D. 1983, A&A 121, 77

Weidemann, V., Koester, D. 1984, A&A 132, 195 (WK)

Weidemann, V., Jordan, S., Iben Jr., I. & Casertano, S. 1992, Astron.J. 104, 1976 (WJIC)

Yuan, J.W. 1992, A&A 261, 10

Wood, M.A. 1990, Ph.D. thesis, Univ.Texas at Austin

Wood, M.A. 1995, in White Dwarfs, Proc. 9.Europ.Workshop, Kiel, Germany 1994. Lecture Notes in Physics vol. 443, ed. D.Koester & K.Werner, (Berlin: Springer) p.41

The Role of Mixing and Accretion on the Chemical Evolution of Cool White Dwarf Atmospheres

By MARÍA TERESA RUIZ[1], P. BERGERON[2] AND S. K. LEGGETT[3]

[1]Departamento de Astronomía, Universidad de Chile, Casilla 36-D, Santiago, Chile; mtruiz@das.uchile.cl

[2]Département de Physique, Université de Montréal, C.P. 6128, Succ. Centre-Ville, Montréal, Québec, Canada, H3C 3J7; bergeron@astro.umontreal.ca

[3]NASA Infrared Telescope Facility, P.O. Box 4729, Hilo, HI 96720; skl@galileo.ifa.hawaii.edu

Photometric (B, V, R, I, J, H, K) and spectroscopic observations of 110 cool white dwarf stars have been analyzed with state-of-the-art model atmosphere calculations appropriate for cool white dwarfs ($T_{\rm eff} < 8000$ K) with pure H, pure He, as well as mixed H/He compositions. The simultaneous analysis of the photometric and spectroscopic observations provides a detailed picture of the chemical composition of cool white dwarfs. Our results reveal an inhomogeneous temperature distribution of H-rich and He-rich white dwarfs, and in particular the presence of a non-DA gap in the range $5000 \lesssim T_{\rm eff} \lesssim 6000$ K. We present possible new physical mechanisms to explain the temperature distribution of H-rich and He-rich white dwarfs. We propose a mechanism by which H is accreted onto the surface of He-rich white dwarfs while remaining spectroscopically invisible.

1. Introduction

The spectral evolution of white dwarfs has been a particularly active field of research in the last decade. Complete samples drawn from colorimetric surveys such as the Palomar-Green survey have provided the groundwork for statistical studies of white dwarfs with different spectral types. Fontaine & Wesemael (1987) have discussed evidence that various mechanisms are competing with gravitational settling in altering the atmospheric composition of white dwarfs as they evolve along the cooling sequence.

In contrast with their hotter siblings, the atmospheric composition of cool white dwarfs cannot be simply interpreted from the spectral type alone. For instance, below $T_{\rm eff} \sim 11,000$ K He becomes spectroscopically invisible, and could thus be present in large amounts in cool DA stars (Bergeron et al. 1990b, 1991). The problem becomes critical at lower temperatures ($T_{\rm eff} \lesssim 5000$ K) where H lines become spectroscopically invisible as well. In this range of temperature, the atmospheric composition can only be inferred from a detailed analysis of the energy distribution. Furthermore, the transparency of cool, He-rich atmospheres increases the likelihood of detection of trace elements such as carbon or heavier elements. As a result, the spectra of cool white dwarfs are both more diversified and more complex than those of hotter degenerates (e.g., Wesemael et al. 1993), and the interpretation of these additional features in terms of mechanisms which compete against gravitational settling becomes equally more complex. The most important of these mechanisms — namely convective mixing, convective dredge-up from the core, and accretion from the interstellar medium — are still not fully understood despite significant progress in the last few years.

Our analysis relies on new model atmospheres by Bergeron et al. (1995b), as well as

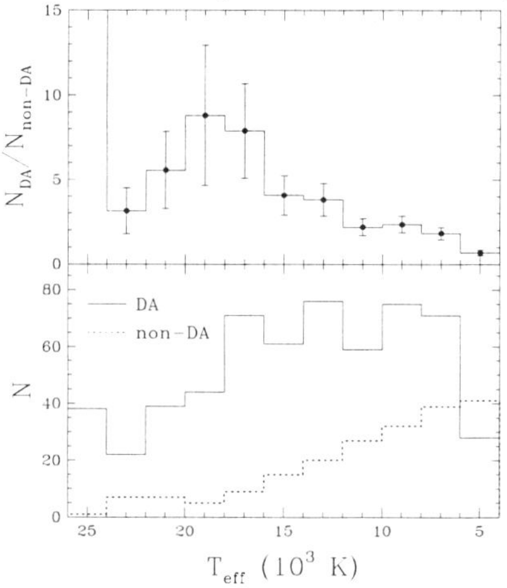

FIGURE 1. Bottom : number of DA and non-DA stars in the McCook & Sion catalog as a function of $T_{\rm eff}$. Top : DA to non-DA ratio as a function of $T_{\rm eff}$ for the same data set; error bars represent the Poisson statistics in each bin.

optical BVRI and infrared JHK photometry, combined with optical spectroscopy (mostly at Hα). Most objects were selected from the catalog of McCook & Sion (1987). Cool white dwarf of all spectral types (DA, DQ, DZ, and DC stars) were included in our sample in order to test the validity of our model atmospheres and obtain a clear picture of the chemical evolution of cool white dwarfs. A detailed account of our observations and analysis will be published elsewhere.

2. Convective Mixing and the DA to non-DA Ratio

Below $T_{\rm eff} \sim 12,000$ K, convective mixing of the thin superficial H atmosphere with the more massive underlying He envelope is believed to be able to turn a H-rich star into a He-rich star, provided the mass of the H layer is small enough (Koester 1976, Vauclair & Reisse 1977, D'Antona & Mazzitelli 1979). Since the bottom of the H convection zone deepens as the white dwarf cools off (see, e.g., Tassoul, Fontaine, & Winget 1990), white dwarfs with thicker H atmospheres will mix at lower temperatures. However, because the H convection zone reaches a maximum depth of $M_{\rm H}/M_\star \sim 10^{-6}$ around $T_{\rm eff} \sim 6000$ K, DA white dwarfs with thicker H layers will never mix. Evidence for convective mixing in cool white dwarfs has been discussed by Sion (1984) who showed that the ratio of DA to non-DA stars drops from 6 : 1 in the range $10.5 \leq M_V \leq 11.5$ to a value of 2 : 1 for $11.5 \leq M_V \leq 13.5$ (see also Greenstein 1986 for a similar result).

Using the photometric calibrations of Bergeron et al. (1995c), we have reevaluated the above statistics. From the photometry published in the catalog of McCook & Sion (1987) we estimate temperatures for each object assuming a mass of 0.6 $M_\odot$ for each star, with pure H and pure He compositions for the DA stars and non-DA stars, respectively. Results are shown in Fig 1 . At $T_{\rm eff} \sim 12,000$ K, the DA to non-DA ratio in Fig 1 drops by 50% remaining constant at 2 : 1 in the temperature range where convective mixing is believed to occur, i.e. $12,000 \gtrsim T_{\rm eff} \gtrsim 6000$ K. The last bin where the ratio further drops below 1 : 1 corresponds to the temperature range where H lines disappear. However, many of these cool stars revealed the presence of Hα when observed at sufficiently high signal-to-noise ratio (see also Greenstein 1986). Therefore, it is entirely plausible that the ratio remains constant at 2:1 even lower temperatures, with no strong evidence for convective mixing below 12,000 K.

3. Accretion from the Interstellar Medium

Elements heavier than carbon will diffuse downward at the bottom of the He convection zone with time scales much shorter than the evolutionary time scales (see Dupuis et al. 1992 and references therein). Therefore cool white dwarfs should have metal-depleted atmospheres. Yet magnesium, calcium, and iron lines are often observed in the spectra of cool, He-rich white dwarfs — the DZ stars. The only viable explanation for the presence of these metals is provided by the accretion-diffusion model in which a balance is achieved between the gravitational settling of metals at the base of the He convection zone and the accretion of these elements at the surface of the star.

The almost complete absence of cool white dwarf spectra showing simultaneously H and metallic features remains the biggest puzzle in our understanding of the accretion mechanism. Unlike heavier elements, H accreted from the interstellar medium remains homogeneously diluted within the He convection zone, and the H-to-He abundance ratio can only increase with time. This implies that no cool non-DA stars should exist. Michaud & Fontaine (1979) and Wesemael & Truran (1982) have proposed two different screening mechanisms to selectively accrete heavy elements but not H. The first mechanism uses the electric field of the He plasma to prevent protons from accreting at the surface of the white dwarfs, while the second is the so-called propeller mechanism, discussed by Illarionov & Sunyaev (1975), where protons are expelled from a combination of magnetic field and rotation. Both models, however, require that H be ionized near the stellar surface. Although there are probably enough ionizing photons in relatively hot white dwarfs to ionize H ($T_{\rm eff} \gtrsim 8500$ K, Wesemael & Truran 1982), these screening mechanisms are expected to become less efficient in cooler stars which produce a negligible amount of ultraviolet flux.

4. Results and Discussion

4.1. *Observational evidence for chemical evolution*

The stellar masses of all white dwarfs in our sample having well determined trigonometric parallaxes, are displayed in Fig 2 as a function of effective temperature; stars with no trigonometric parallax measurements are shown at the bottom of the plot with $M <$ 0. Note that some of the coolest non-DA stars in the top panel actually have H-rich compositions in the bottom panel.

The results presented in the bottom panel of Fig 2 , contain crucial information about the chemical evolution of cool white dwarfs. Certainly the most striking feature of the temperature distribution of H-rich and He-rich atmospheres is the existence of a temperature range, $6100 \gtrsim T_{\rm eff} \gtrsim 5100$ K, in which the number of He-rich stars is greatly reduced — a non-DA gap. Excluding the C_2H stars, which have mixed compositions, we find only three He-rich objects in this temperature range, all three exhibiting some peculiarity, with energy distributions that do not fit well a model atmosphere with pure He.

Since our current sample is strongly biased against DA stars (by selection), further observations of an unbiased sample can only increase the statistical significance of this non-DA gap. Above and below the gap, He-rich stars are found in large numbers, however, and it is clear that some physical process is at work to alter the chemical composition of these stars as they cool off. In particular, we are forced to conclude that most, if not all, He-rich white dwarfs above the non-DA gap will turn into H-rich stars as they reach $T_{\rm eff} \sim 6000$ K (and become He-rich again at $T_{\rm eff} < 5000$ K). This result is totally unexpected since it is generally believed that convective mixing will transform

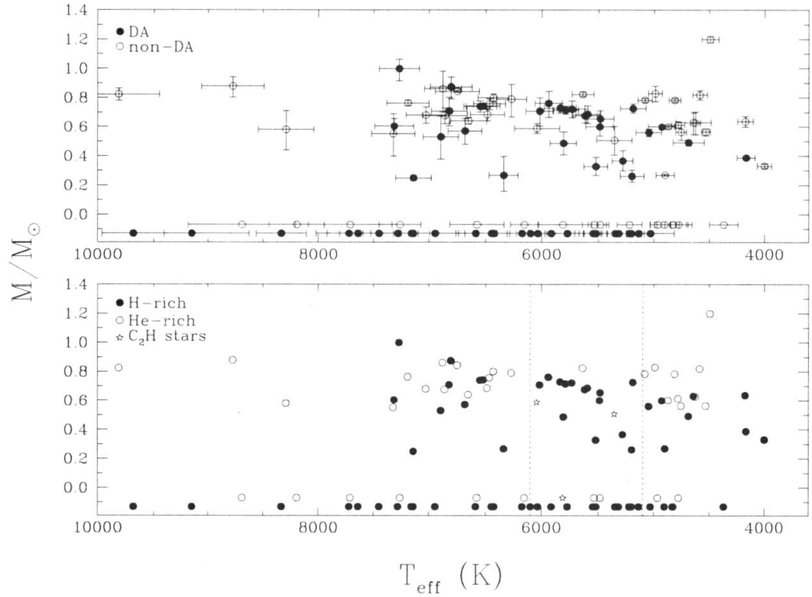

FIGURE 2. Masses of white dwarfs as a function of $T_{\rm eff}$; stars with no trigonomertic parallax measurements are shown at the bottom of each panel with M < 0. In the top panel, stars are divided in DA and non-DAs, individual uncertainties are shown for M and $T_{\rm eff}$. In the bottom panel, different symbols are used to distinguish the main atmospheric constituent of each object, dotted lines mark the non-DA gap discussed in the text.

H-rich atmospheres into He-rich atmospheres. The only mechanism that could produce the He-rich to H-rich transition is accretion of H from the interstellar medium. However, such a transition would occur gradually over time, and one would expect to observe a broad spectrum of He-to-H abundance ratios. Yet the results of our analysis have clearly shown that such objects are extremely rare, and that the energy distributions of all cool white dwarfs are consistent with pure H or pure He models, with the exception of the C_2H stars. Thus a simple accretion model cannot account for the observations.

There is a peculiar class of non-DA white dwarfs, in the temperature range 6000 – 8500 K, whose energy distributions are better reproduced by pure H models, the most extreme cases, lie very close to the blue edge of the non-DA gap. This result strongly suggests that these objects may be about to turn into H-rich atmospheres, yet their spectra are featureless. Clearly, these results cannot be easily explained within our current physical understanding of mixing and accretion mechanisms, and more exotic models need to be invoked to account for the observations.

4.2. *New Physical Mechanisms*

The gas pressure of He-dominated atmospheres is about four order of magnitudes larger than that of H atmospheres as a result of the lower opacity of He (Bergeron et al. 1995b). Under such extreme physical conditions, the atomic levels of H, when present in small amounts in a pure He atmosphere, are strongly perturbed by the interactions with He — an excluded volume effect in this case, and one needs to consider in detail the occupation probability of the H atomic levels (Hummer & Mihalas 1988). In Figure 3, we show the occupation probability of the first three atomic levels of H as well as that of H^- as a function of $T_{\rm eff}$ at the photosphere of pure He atmosphere models (i.e., H is considered a trace element). These results show in particular that the $n = 2$

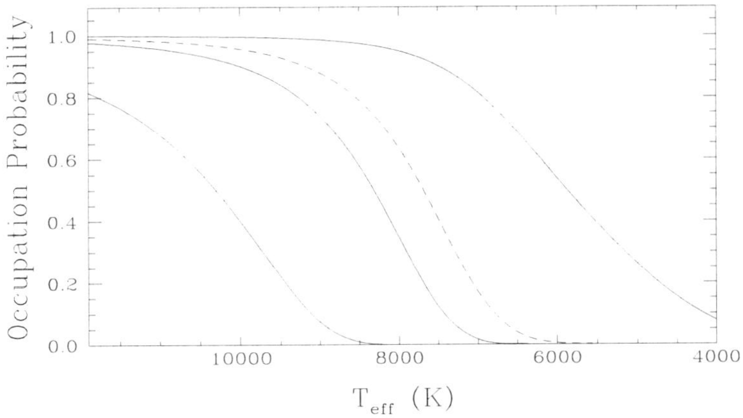

FIGURE 3. Occupation probability of trace H calculated at the photosphere of pure He models as a function of $T_{\rm eff}$. The solid lines correspond, from right to left, to the n=1, 2, and 3 atomic levels while the dashed line represents the occupation probability of H^- .

to 3 transition (Hα) is never allowed in extreme He-rich atmospheres in this range of $T_{\rm eff}$. Such extremes are certainly found in typical cool DBA stars with H abundances of $N(\mathrm{H})/N(\mathrm{He}) \sim 10^{-5}$. We thus conclude from these considerations that line quenching provides a natural explanation for the disappearance of the DBA phenomenon at low effective temperatures ($T_{\rm eff} < 8000$ K).

On the other hand, the occupation probability of the bound level of H^- which represents the dominant H opacity source in cool white dwarf atmospheres, drops rapidly below $T_{\rm eff} \sim 8000$ K and so the bound-free opacity of H^- is quenched in such atmospheres and the contribution of trace amounts of H to the total opacity remains negligible. Consequently, since H makes an insignificant contribution to the total opacity, the atmospheric pressure remains unaffected ($P \propto \kappa^{-1}$). Hence even larger amounts of H can be added to the atmosphere without changing its properties.

The situation is fundamentally different when we consider the inverse process of increasing the He abundance in a H-rich atmosphere. In such atmospheres, the atmospheric pressure is already low as a result of the large H opacity. Hydrogen is so opaque at these temperatures that the atmospheric pressure will remain relatively low — with respect to pure He atmospheres — even at very large He abundances, and the bound level of H^- will easily survive under these conditions. It is thus possible to envision two cool white dwarf atmospheres with otherwise identical effective temperatures, surface gravities, and H-to-He abundance ratios, say $N(\mathrm{He})/N(\mathrm{H}) \sim 1$, to have different atmospheric pressure conditions. The key factor between the two is the total opacity. Therefore, the He-to-H abundance ratio is not necessarily sufficient to determine the physical conditions of the atmosphere — it is also important to specify how these conditions were achieved. If one starts with a H-rich atmosphere, the opacity will remain high up to large He abundances, while if one starts with a He-rich atmosphere, the opacity will remain low because H^- is quenched.

It thus seems entirely possible through a simple quenching effect to "hide" large amounts of H in a cool white dwarf atmosphere. Exactly how much H can be hidden remains to be determined through more detailed calculations of H in dense plasmas. An immediate application of this mechanism is to provide a natural explanation for the long-standing problem of the accretion of H . Hydrogen accreted onto a He-rich white dwarf atmosphere remains invisible because of the low occupation probability of H^-. The obvious appeal of this mechanism is that it relies upon well understood atomic physics principles. Furthermore, it operates even better at low effective temperatures where most previous screening mechanisms become inefficient. With this simple quenching model, such screening mechanisms are no longer required.

If for some reason the atmospheric pressure is reduced and the bound level of H^- becomes available, the H accreted on a cool ($T_{eff} \lesssim 8000$ K) He-rich atmosphere white dwarf may suddenly become visible. This process can suddenly turn what appears to be a non-DA star into a DA star. We suggest that the sudden non-DA to DA transition which occurs near $T_{eff} \sim 6000$ K is the result of such a transition.

The energy distribution of non-DA stars immediately preceding the transition would appear somewhat peculiar. They would resemble H-rich atmospheres with the exception that the importance of the bump at 1.6 μm caused by the H^- bound-free edge would be greatly reduced, and Hα could not be seen because the $n = 3$ level is quenched. Such peculiar objects do exist in our sample. We found three examples of this peculiar class of white dwarfs whose energy distributions are H-like, except for the absence of both the characteristic 1.6 μm bump and Hα. More importantly, all of these objects are found above and close to the blue edge of the non-DA gap. We thus suggest that white dwarfs with such peculiar energy distributions have atmospheres filled with H whose H^- bound-level is in the process of recombining. Hα will reappear only when the atmospheric pressure drops below the threshold level at which the $n = 3$ level of H becomes available.

This accretion/quenching scenario requires initial white dwarf atmospheres which are sufficiently He-rich to provide the high atmospheric pressures required to quench at least the $n = 3$ level (i.e. a DC star), and to further quench the bound-level of H^- below $T_{eff} \sim 8000$ K. Cooling DB stars certainly meet this criterion, as well as cooling DBA stars with their typical abundances of $N(\mathrm{H})/N(\mathrm{He}) \sim 10^{-5}$. But cooling DB and DBA stars account for only 20% of all white dwarfs at higher temperatures and so the observed 2 : 1 DA to non-DA ratio below $T_{eff} \sim 12,000$ K (see Fig. 1) requires an additional mechanism to feed the non-DA channel. Convective mixing is the only mechanism capable of turning DA stars into non-DA stars in this temperature range. We thus conclude from these considerations that convective mixing must occur in cool DA white dwarfs to account for the increase of non-DA stars below 12,000 K.

The mixing temperature is a function of the H layer mass. We discuss this point more quantitatively by using new envelope models generously calculated by G. Fontaine for the purpose of our analysis. These calculations are similar to those of Tassoul et al. (1990) with the exception that the Los Alamos opacities and the equation of state of Fontaine, Graboske, & Van Horn (1977) have been replaced by the OPAL opacities and the equation of state of Saumon, Chabrier, & Van Horn (1995) for H and He. The location of the bottom of the H convection zone is shown as a function of T_{eff} in Figure 4 for the 0.4, 0.6, and 0.8 $M_\odot$ models; the additional calculations shown as a dashed line are discussed below. These new results are comparable to those of Tassoul et al. (1990). In particular, the bottom of the H convection zone of the 0.6 $M_\odot$ model reaches a maximum depth of $q(\mathrm{H}) \sim 10^{-6}$ at $T_{eff} \sim 6000$ K. Also, the maximum depth reached by the H convection zone depends sensitively on the total mass, while the effective temperature at which this phenomenon occurs is only weakly mass dependent.

If the H layer mass of a DA star is small enough, the bottom of the H convection zone can penetrate the deeper and more massive underlying He envelope. As shown in Figure 4, the mixing temperature depends on the thickness of the H layer as well as the stellar mass of the white dwarf. It is generally believed that when convective mixing occurs, H-rich atmospheres (DA) are turned into He-dominated atmospheres (non-DA). Our earlier conclusion that convective mixing occurs as early as $T_{eff} \sim 12,000$ K implies that the H layer mass in some DA stars is smaller than $q(\mathrm{H}) = 10^{-10}$. Furthermore,

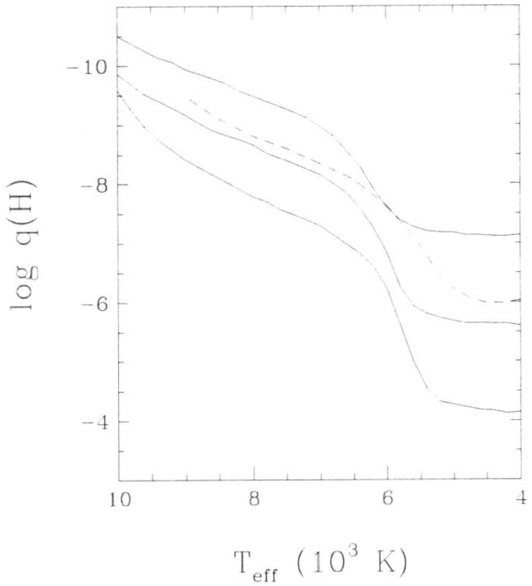

FIGURE 4. Location of the bottom of the H convection zone in terms of the mass fraction q(H) as a function of $T_{\rm eff}$ in models with (from bottom to top) 0.4, 0.6, and 0.8 $M_\odot$. The dashed curve corresponds to a 0.6 $M_\odot$ model with detailed model atmospheres as surface boundary conditions.

since DA stars are most likely characterized by a spectrum of H layer masses — up to $q(\rm H) \sim 10^{-4}$ according to stellar evolution models — we should expect the DA to non-DA ratio to decrease with decreasing $T_{\rm eff}$, as DA white dwarfs with thicker H layers are gradually turned into non-DA stars through the 6000–12,000 K temperature range ($10^{-10} \lesssim q(\rm H) \lesssim 10^{-6}$ for a 0.6 $M_\odot$ white dwarf). Even though a detailed account of the number of DA and non-DA stars in this temperature range would require a more thorough analysis of white dwarfs above $T_{\rm eff} \sim 7500$ K, such as that performed here for the cooler white dwarfs, the results shown in Figure 1 indicate that this ratio is relatively constant at 2 : 1 between 6000 and 12,000 K. Nevertheless, if our description of the convective mixing process is accurate, an additional mechanism needs to be invoked to replenish the DA population below $T_{\rm eff} \sim 12,000$ K, and to compensate the effects of convective mixing on the observed DA to non-DA ratio. The non-DA to DA phase transition proposed above is certainly the most appealing possibility. If both the phase transition and the convective mixing mechanisms operate simultaneously below 12,000 K, we should be able to find these peculiar non-DA white dwarfs (i.e. those with energy distributions that are better fit with H-rich models) even at high effective temperatures. However, the small total number of hotter white dwarfs in our sample precludes us from performing a more detailed statistical analysis. Thus, the phase transition may occur at all temperatures, but the process is simply accelerated below 6000 K as new (hydrogen) opacity sources become available.

The picture we have outlined so far is obviously not complete since He-rich white dwarfs reappear in fairly large numbers below $T_{\rm eff} \sim 5000$ K, and some of them are even present within the non-DA gap. According to the results shown in Figure 4, standard envelope models predict that convective mixing can no longer occur below $T_{\rm eff} \sim 5500 - 6000$ K. This corresponds to the temperature at which the bottom of the H convection zone reaches its maximum depth. However, these envelope models rely upon very approximate gray atmospheres as initial boundary conditions, and it has been suggested that the use of more detailed boundary conditions may affect the envelope structure, and in particular the location of the bottom of the H convection zone. In order to explore this, a new set of envelope models has been kindly calculated by G. Fontaine using our detailed model atmospheres as a boundary condition for the envelope calculation. The results for a 0.6 $M_\odot$ white dwarf are reported in Fig 4. The use of detailed atmospheres has affected not only the location of the bottom of the H convection zone, but also the effective temperature at which the maximum depth is reached. Convective mixing in 0.6 $M_\odot$ white dwarfs can now proceed at a much cooler effective temperature of $T_{\rm eff} \sim 5000$ K, which corresponds precisely to the temperature at which He-rich white dwarfs reappear in large numbers in our sample. These improved envelope models also account for the presence of some non-DA stars in the non-DA gap.

The existence of this cooler threshold is thus certainly supported by our analysis. The number of He-rich white dwarfs below 5000 K is much larger than that between 5000 and 6000 K, however. This result can be explained if several DA stars in the non-DA gap have H layer masses close to $q(\rm H) \sim 10^{-6}$ while only a few have thinner layers. It can also be noticed from Fig 2 that below 6000 K, He-rich white dwarfs have, on average, larger masses than their cool H-rich counterparts. The results of Fig 4 show that in less massive white dwarfs the H convection zone reaches deeper into the atmosphere and so all low-mass white dwarfs may have preferentially thicker H envelopes. It may become too speculative to argue whether these cool He-rich white dwarfs will turn into DA stars at even cooler temperatures. We just note for the time being that if H is pressure ionized, as suggested in Fig 3, H may not be able to contribute to the total opacity anymore.

This work was supported in part by FONDECYT grant 1950588, by a "CATEDRA PRESIDENCIAL", by the NSERC Canada, by the Fund FCAR (Québec), by the NSF grant 93-15372, and by a Chrétien International Research Grant.

REFERENCES

Bergeron, P., Saumon, D., & Wesemael, F. 1995b, ApJ, 443, 764

Bergeron, P., Wesemael, F., & Beauchamp, A. 1995c, PASP, 107, 1047

Bergeron, P., Wesemael, F., & Fontaine, G. 1991, ApJ, 367, 253

Bergeron, P., Wesemael, F., Fontaine, G., & Liebert, J. 1990b, ApJ, 351, L21

D'Antona, F., & Mazzitelli, I. 1979, A&A, 74, 161

Dupuis, J., Fontaine, G., Pelletier, C., & Wesemael, F. 1992, ApJS, 82, 505

Fontaine, G., Graboske, H. C., Jr. & Van Horn, H. M. 1977, ApJS, 35, 293

Fontaine, G., & Wesemael, F. 1987, in IAU Colloquium 95, The Second Conference on Faint Blue Stars, eds. A. G. D. Philip, D. S. Hayes & J. Liebert (Schenectady: L. Davis Press), 319

Greenstein, J.L. 1986, ApJ, 304, 334

Hummer, D. G., & Mihalas, D. 1988, ApJ, 331, 794

Illarionov, A., & Sunyaev, R. 1975, A&A, 39, 185

Koester, D. 1976, A&A, 52, 415

McCook, G. P., & Sion, E. M. 1987, ApJS, 65, 603

Michaud, G., & Fontaine, G. 1979, ApJ, 229, 694

Saumon, D., Chabrier, G., & Van Horn, H. M. 1995, ApJS, 99, 713

Sion, E. M. 1984, ApJ, 282, 612

Tassoul, M., Fontaine, G., & Winget, D. E. 1990, ApJS, 72, 335

Vauclair, G., & Reisse, C. 1977, A&A, 61, 415

Wesemael, F., Greenstein, J. L., Liebert, J., Lamontagne, R., Fontaine, G., Bergeron, P., & Glaspey, J. W. 1993, PASP, 105, 761

Wesemael, F., & Truran, J. 1982, ApJ, 260, 807

Neutrino Energy Loss in Stellar Interiors

By NAOKI ITOH
AND AKINORI NISHIKAWA

Department of Physics, Sophia University, 7-1 Kioi-cho, Chiyoda-ku, Tokyo, 102, Japan

The results of the calculations of the neutrino energy loss rates due to pair,photo-, plasma, bremsstrahlung, and recombination processes are summarized in the form of analytic fitting formulae and tables. Cares have been taken in order to serve the convenience of the users of these results. We have tried to make the present paper as self-contained as possible. The present paper is intended to serve as useful physical input data for stellar evolution computations. We have published the numerical data and the FORTRAN codes for the results of the neutrino energy loss rates in the CD-ROM series of the American Astronomical Society. In the CD-ROM we give three options of the FORTRAN codes. The first option comprises the results with analytic fitting formulae only. The second option comprises the results with analytic fitting formulae for the bremsstrahlung and recombination neutrino processes, and the numerical tables for the pair, photo-,and plasma neutrino processes with a standard quadratic interpolation procedure. In this option, the grid of log T for the pair neutrino energy loss rate is 0.2. The third option is a variant of the second option with the replacement of the grid of log T for the pair neutrino energy loss rate to be 0.05.

1. Introduction

In the past decade two of the present authors (N.I. and Y.K.) together with their collaborators have systematically investigated the neutrino energy loss in stellar interiors based on the Weinberg-Salam theory (Itoh & Kohyama 1983; Itoh et al. 1984 a,b,c; Munakata, Kohyama, & Itoh 1985; Kohyama, Itoh, & Munakata 1986; Munakata, Kohyama, & Itoh 1987; Itoh et al. 1989; Itoh et al. 1992; Kohyama et al. 1993; Kohyama et al. 1994). They have dealt with the pair, photo-, plasma, bremsstrahlung, and recombination neutrino processes. Dicus (1973), Dicus et al. (1976), Schinder et al. (1987), Braaten (1991), Braaten & Segel (1993), and Haft, Raffelt, & Weiss (1994) also calculated the neutrino energy loss rates using the Weinberg-Salam theory. The calculations of the neutrino energy loss rates based on the Feynman-Gell-Mann theory were summarized by Beaudet, Petrosian, & Salpeter (1967).

In 1989 Itoh et al. (1989) summarized the calculations of the neutrino energy loss rates due to the pair, photo-, plasma, and bremsstrahlung processes. However, major improvements on the numerical result of the plasma neutrino energy loss rates took place more recently (Braaten 1991; Itoh et al. 1992; Braaten & Segel 1993; Kohyama et al. 1994). The neutrino energy loss rates due to the recombination process have been also accurately calculated very recently (Kohyama et al. 1993). Therefore, it is useful to summarize the results of the neutrino energy loss rates due to the pair, photo-, plasma, bremsstrahlung, and recombination processes and update the results of Itoh et al. (1989). In this paper we take care to serve the convenience of the users of the results of the neutrino energy loss rates. For that purpose we present analytic fitting formulae and numerical tables. We try to make the present paper as self-contained as possible. The present paper is intended to serve as useful physical input data for stellar evolution computations. We have published the numerical data and the FORTRAN codes for the results of the neutrino energy loss rates in the CD-ROM series of the American Astronomical Society.

The bremsstrahlung neutrino process essentially depends on the states of ions. Therefore, different methods must be applied to the calculations of the neutrino energy loss rates corresponding to different regimes of densities and temperatures. Thus analytic fitting formulae are best suited to express the numerical results of the bremsstrahlung neutrino process. The results of the recombination neutrino process can be expressed in a semi-analytical way. Therefore, analytic fitting formulae are also best suited to express the numerical results of this process.

Regarding the expression of the numerical results of pair,photo-, and plasma neutrino energy loss rates, in the accompanying CD-ROM version, we give three options. The first option uses analytical fitting formulae for these three processes. The accuracy of the fitting formulae are generally better than 10% where the respective neutrino process is the most dominant process. We will describe the accuracy and the validity region of the fitting formulae in more detail where we describe the explicit forms of the fitting formulae. The reason why we give this option of the analytical fitting formulae is that this is efficient and time-saving, although the accuracy may not be too high in some density-temperature regions. For those who do not wish their neutrino-loss subroutine become too heavy, we recommend this option. It is our belief that for ordinary stellar evolution computations which do not require higher accuracy than 10% for the neutrino energy loss rates this option should be sufficient.

The second option comprises the numerical tables for the pair, photo-, and plasma neutrino processes with a standard quadratic interpolation procedure. In this option, the grid of log T for the pair neutrino energy loss rate is 0.2. The third option is a variant of the second option with the replacement of the grid of log T for the pair neutrino energy loss rate to be 0.05. Users who wish to require higher accuracies for the neutrino energy loss rates may use these options according to their choice of the burdens of of the neutrino-loss subroutine.

The full details of the calculations have been described in Itoh et al. (1996).

2. COMPARISON OF VARIOUS NEUTRINO PROCESSES

In Figures 1,2 we show the contributions of the various neutrino processes for the case of $\sin^2\theta_W = 0.2319$ and $n = 2$, considering the three chemical elements of ^{4}He, ^{12}C, and ^{56}Fe, corresponding to the temperatures $T = 10^7, 10^8$K. In Figures 3 we show the most dominant neutrino process for a given density and temperature for the chemical element of ^{12}C, assuming $\sin^2\theta_W = 0.2319$ and $n = 2$.

3. CONCLUDING REMARKS

We have summarized the results of the calculations of the neutrino energy loss rates due to pair, photo-, plasma, bremsstrahlung, and recombination neutrino processes based on the Weinberg-Salam theory. A wide density-temperature regime $1 \leq \rho/\mu_e \leq 10^{14}\text{gcm}^{-3}\text{and}10^7 \leq T \leq 10^{11}$K has been considered. We have published the numerical results of the present paper and the FORTRAN codes for the results of the neutrino energy loss rates in the CD-ROM series of the American Astronomical Society. It is hoped that the present paper would meet the needs of the researchers who wish to use the present results in their studies of stellar structure and stellar evolution.

One of the authors (N.I) wishes to thank Icko Iben on the occasion of his 65th birthday for his advice and encouragement in the last quarter of century that helped to complete the present work.

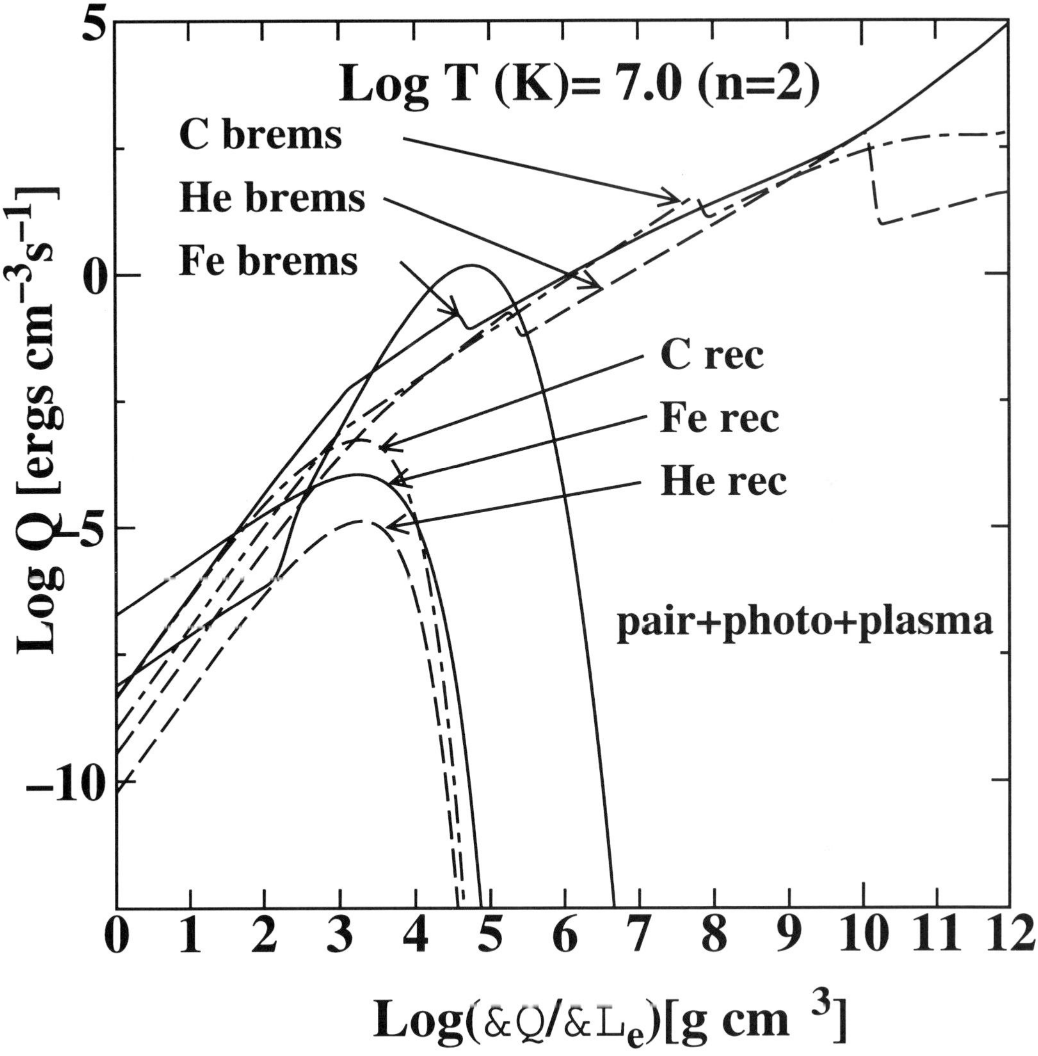

FIGURE 1. Neutrino energy-loss rates resulting from bremsstrahlung and recombination processes for $T = 10^7$K.

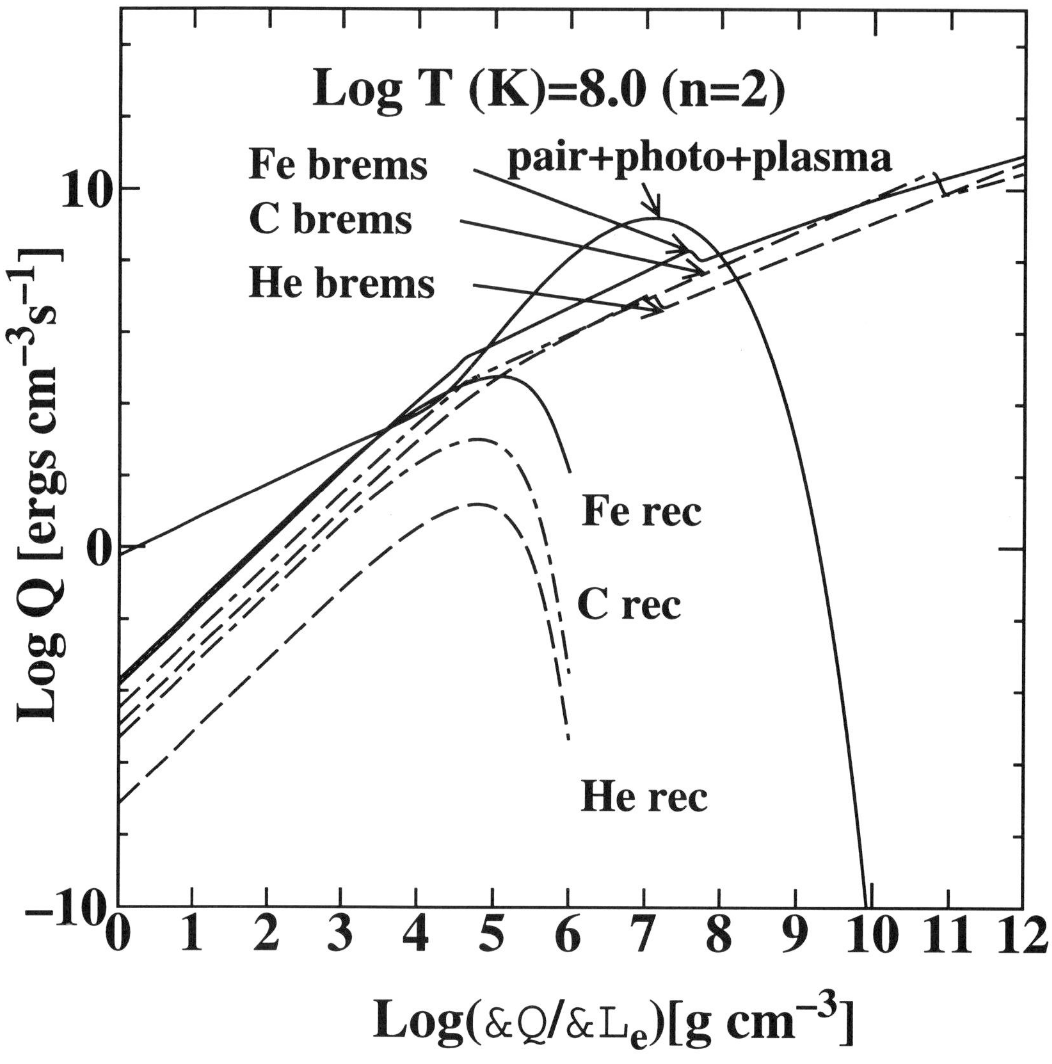

FIGURE 2. Same as Fig.1, for $T = 10^8$K.

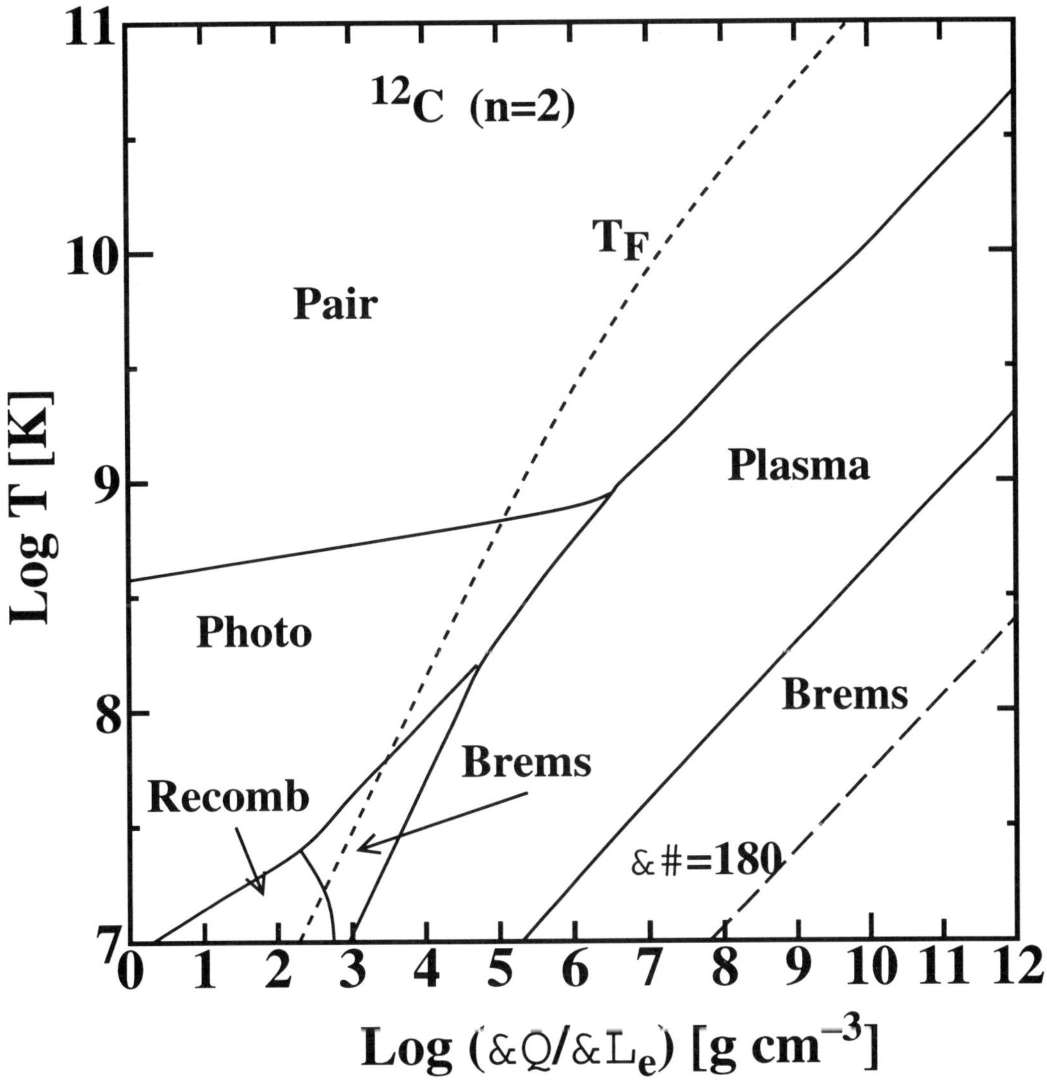

FIGURE 3. Most dominant neutrino process for a given density and temperature in the case of ^{12}C matter. T_F is the electron Fermi temperature. The line marked $\Gamma = 180$ is the melting curve of the ionic Coulomb solid.

REFERENCES

Beaudet, G., Petrosian, V., and Salpeter, E. E. 1967, ApJ, **150**, 979

Braaten, E. 1991, Phys. Rev. Lett., **66**, 1655

Braaten, E., and Segel, D. 1993, Phys. Rev., **D48**, 1478

Dicus, D. A., 1973, Phys. Rev., **D6**, 941

Dicus, D. A., Kolb, E. W., Schramm, D. N., and Tubbs, D. L. 1976, ApJ, **210**, 481

Flowers, E. 1973, ApJ, **180**, 911

Haft, M., Raffelt, G., and Weiss, A. 1994, ApJ, **425**, 222

Itoh, N., Adachi, T., Nakagawa, M., Kohyama, Y., and Munakata, H. 1989, ApJ, **339**, 354; erratum **360**, 741(1990)

Itoh, N., Hayashi, H., Nishikawa, A., & Kohyama, Y. 1996, ApJS, 102, 411

Itoh, N., Hayashi, H., Nishikawa, A., & Kohyama, Y. 1995, AAS CD-ROM series, Volume V

Itoh, N.,and Kohyama, Y. 1983, ApJ, **275**, 858

Itoh, N., Kohyama, Y., Matsumoto, N., and Seki, M. 1984a, ApJ, **280**, 787; erratum **404**, 418(1993)

Itoh, N., Kohyama, Y., Matsumoto, N., and Seki, M. 1984b, ApJ, **285**, 304; erratum **322**, 584(1987)

Itoh, N., Kojo, K., and Nakagawa, M., 1990, APJS, **74**, 291

Itoh, N., Matsumoto, N., Seki, M., and Kohyama, Y. 1984c, ApJ, **279**, 413

Itoh, N., Mutoh, H., Hikita, A., and Kohyama, Y. 1992, ApJ, **395**, 622; erratum **404**, 418(1993)

Jancovici, B. 1962, Nuov. Cimento, **25**, 428

Kohyama, Y., Itoh, N., and Munakata, H. 1986, ApJ, **310**, 815

Kohyama, Y., Itoh, N., Obama, A., and Hayashi, H. 1994, ApJ, **431**, 761

Kohyama, Y., Itoh, N., Obama, A., and Mutoh, H. 1993, ApJ, **415**, 267

Munakata, H., Kohyama, Y., and Itoh, N. 1985, ApJ, **296**, 197; erratum **304**, 580(1986)

Munakata, H., Kohyama, Y., and Itoh, N. 1987, ApJ, **316**, 708

Ogata, S., and Ichimaru, S. 1987, Phys. Rev., **A36**, 5451

Pethick, C. J., and Thorsson, V. 1994, Phys. Rev. Lett., **72**, 1964

Pinaev, V. S. 1964, Soviet Phys. - JETP, **18**, 377

Salam, A. 1968, in Elementary Particle Physics, ed. N. Svartholm (Stockholm: Almqvist and Wiksells), p. 367

Schinder, P. J., Schramm, D. N., Wiita, P. J., Margolis, S. H., & Tubbs, D. L. 1987, ApJ, **313**, 531

Weinberg, S. 1967, Phys. Rev. Lett., **19**, 1264.

Binary Stars Below the Turnoff in Globular Cluster Color–Magnitude Diagrams†

By ADRIENNE M. COOL

Dept. of Physics and Astronomy, San Francisco State U., San Francisco, CA 94132, USA

Color–magnitude diagrams of stars below the main-sequence turnoff in globular clusters are rich with information about binary stars. The resolving power of the Hubble Space Telescope (HST) significantly increases the ease with which binaries can be located in globulars using CMDs, and enables searches to be extended into the dense, central regions of clusters where binaries tend to congregate. Recent results of HST studies of NGC 6397 are used for illustration.

1. Introduction

Globular clusters have long served as crucibles for stellar evolution theory. Color–magnitude diagrams (CMDs) provide critical measurements against which predictions concerning the evolution of single stars are tested. By contrast, as recently as 1980, it seemed that globulars had little to contribute to efforts to understand binary star evolution, as they appeared to be essentially devoid of binaries (Trimble 1980). Since then, improved techniques have uncovered significant numbers of binaries in numerous clusters (see Pryor *et al.* 1996 and Mateo 1996 for recent reviews). It has also become apparent that the fates of cluster binaries—and in some cases even the origins—are intimately linked to the unusual environments in which they find themselves. In turn, the fate of clusters is linked to that of the binaries. Thus the study of binaries in globular clusters provides insight into the evolution of both member and host.

Here we discuss the use of CMDs for identifying cluster binaries below the main-sequence turnoff. In particular, we focus on the utility of HST in searching for semi-detached and detached binaries below the main-sequence turnoff. (See reviews above for discussions of contact binaries, including blue stragglers.) Many binary stars occupy distinctive locations in CMDs and can be recognized as binaries, or binary candidates, on that basis alone (see illustration in Fig. 1). This method complements those that rely on photometric or radial-velocity variability, as it is sensitive to a different (but overlapping) subset of the underlying binary population. The method has the advantage of requiring only a single epoch of observation, although, as will be seen, it often does not select uniquely for binaries. HST studies of the nearby cluster NGC 6397 will be used by way of illustration, and some recent results concerning binaries in this cluster will also be presented.

2. Semi-detached binaries

Semi-detached binaries ought be the easiest of binaries to locate in globular cluster CMDs. When material spills onto a compact star from a main-sequence companion, the resulting accretion disk outshines the stellar components, and has distinctive spectral signatures. Neutron-star accretors are a particularly extreme example, as their usually high X-ray luminosities enable them to be readily detected in the handful of globulars

† Based on observations with the NASA/ESA *Hubble Space Telescope*, obtained at the Space Telescope Science Institute, which is operated by AURA, Inc., under NASA contract NAS5-26555.

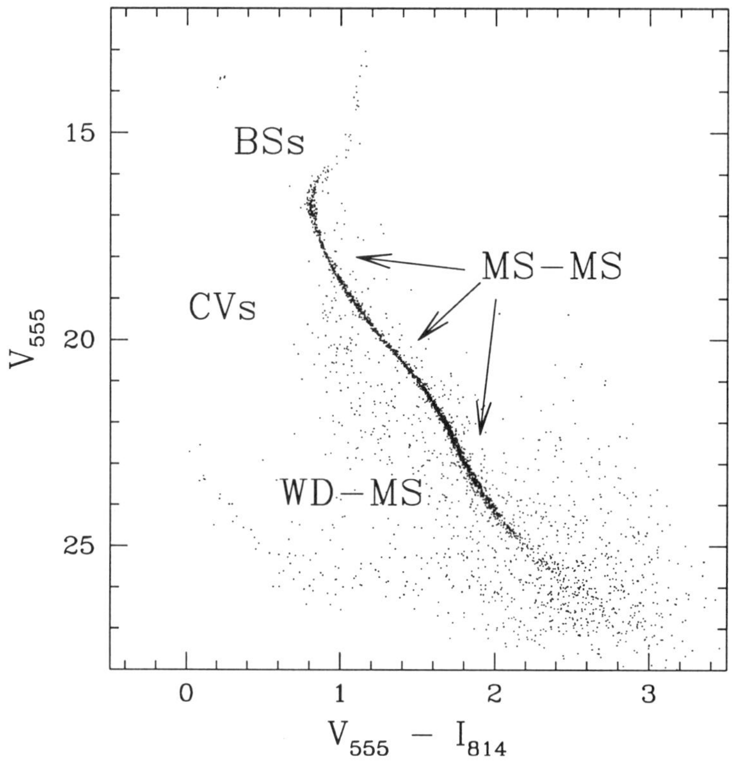

FIGURE 1. CMD for stars in the globular cluster NGC 6397 obtained from HST/WFPC2 images. This diagram combines measurements made in two fields, one ~5′ from the cluster center (stars below the turnoff) and one near the cluster center (stars near and above the turnoff). The large majority of stars on either side of the main sequence are field stars, owing to the cluster's low Galactic latitude. Regions that may contain various types of cluster binaries are indicated (MS = main sequence; WD = white dwarf; BS = blue straggler; CV = cataclysmic variable).

that contain them (see, *e.g.*, Verbunt *et al.* 1995). X-ray imaging has also revealed a larger population of much fainter sources in globulars, many of which are likely to be accreting white dwarfs, or cataclysmic variables (CVs) (Hertz & Grindlay 1983). At longer wavelengths, the hot accretion disk generally places compact binaries blueward of the main sequence in optical/UV CMDs, and its usually prominent emission lines can be targeted in narrow-band imaging searches.

In practice, it has required the resolving power of HST to uncover more than a very small number of these compact binaries in the optical/UV, as they appear to be confined largely to cluster cores. Ground-based searches, which were effectively limited to the envelopes of clusters, yielded few convincing candidates (Shara *et al.* 1988, Cool 1993). But Hα imaging studies with HST of the central regions of two globulars have both been fruitful, yielding a total of five CV candidates (Cool *et al.* 1995, Bailyn *et al.* 1997), three of which have since been confirmed spectroscopically with HST/FOS (Grindlay *et al.* 1995). Objects with the right sorts of absolute magnitudes to be CVs have also been identified in the central regions of other clusters studied in the UV with HST (Shara *et*

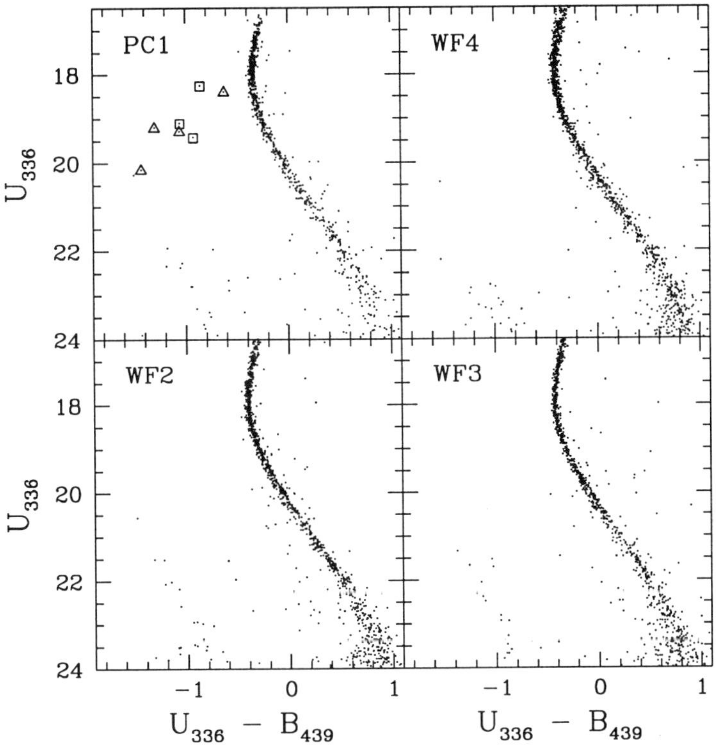

FIGURE 2. CMDs for stars near the center of NGC 6397 from HST/WFPC2 data. The PC1 chip (upper left) was approximately centered on the cluster core. The triangles and squares mark flickering and non-flickering UV-bright objects, respectively (see text), both of which are centrally concentrated. Fainter UV-bright objects are cluster white dwarfs. The bend in the upper main sequence is a bandpass effect, not the turnoff, which is closer to $U_{336} \sim 16$.

al. 1995, Sosin & Cool 1995). But it remains to be seen how many of these objects are compact binaries, and initial indications are that some may not be (Shara *et al.* 1995).

That UV excess is not a unique indicator of binarity is borne out by recent HST/WFPC2 UBVI observations of NGC 6397 (Cool *et al.* 1997). Preliminary CMDs are shown in Fig. 2. Seven UV-bright stars are clearly visible in the PC1 chip, which was roughly centered on the cluster core. Three of these stars (the brightest three triangles) are the spectroscopically confirmed CVs. Four other stars (two of which were previously known from pre-repair HST/FOC UV images) occupy a similar region in the CMD. But are these four objects CVs, like the other three?

The answer appears to be no, except in one case. Fig. 3 compares two CMDs of stars in the PC1 chip. While the seven stars stand out clearly from the main sequence in the $U - B$ color, in the redder $V - I$ color they separate into two groups. The three known CVs and one of the other stars lie on or near the main sequence in $V - I$, while the other three remain separated from the main sequence by a magnitude or more. A similar dichotomy is found in the light curves for these stars. The subset that is on or near the main sequence in $V - I$ shows the flickering characteristic of CVs, while there is no sign of photometric variability in the other three (see sample light curves in Fig. 4).

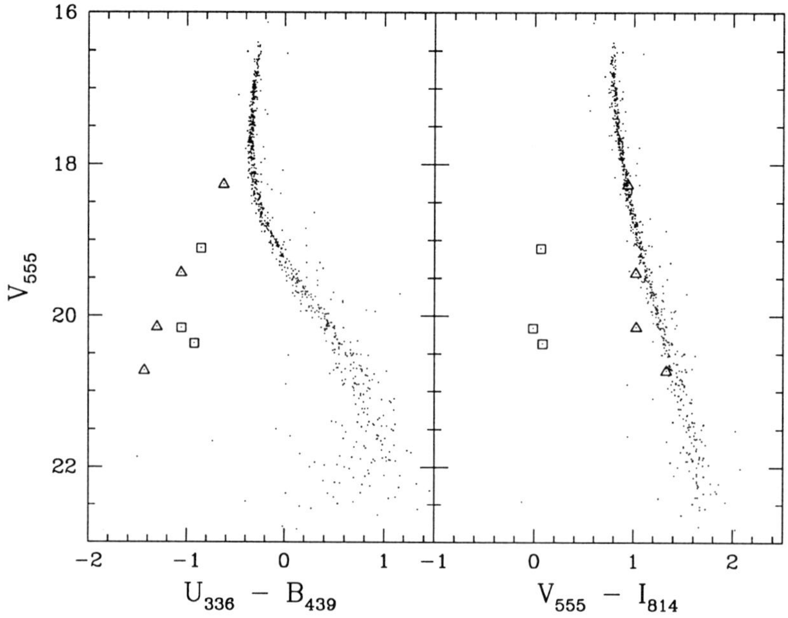

FIGURE 3. CMD of stars in the same field as in Fig. 2 (PC1 chip only), but with different combinations of filters on the x and y axes. Symbols are as in Fig. 2.

It thus appears that NGC 6397 contains two classes of faint, UV-bright stars. The flickerers show every indication of being CVs. These stars have rather ordinary $V - I$ colors, but significant UV excesses. A similar pattern is seen for the CV candidates in NGC 6752 (Bailyn *et al.* 1997). This may the result of relatively massive secondaries that should be favored in the stellar interactions that form these compact systems, and which would dominate the light at the longer wavelengths. The non-flickerers, by contrast, appear simply to be very blue (as opposed to having UV excesses), and given their lack of photometric variability are likely to be isolated stars. However, they too appear linked in some way to stellar interactions, as they are to be found only in the densest part of the cluster (see Fig. 2). It remains to be seen whether a similar dichotomy exists among the faint UV-bright stars identified in other clusters. In any case, both types of stars appear to represent important links in the interplay between stellar and cluster evolution.

3. Detached binary stars

In the absence of accretion disks, binary stars can still in principle be identified on the basis of their positions in CMDs. Detached MS–WD pairs will appear between the WD and main sequences, while MS–MS pairs will appear above and to the right of the main sequence defined by single H-burning stars (see Fig. 1). The so-called "second sequence" of equal-mass binaries in clusters has been discussed extensively by Romani & Weinberg (1991), and candidate binaries have been found in the CMDs of several clusters (see summary in Hut *et al.* 1992). Since the binary's position in the CMD depends only on its total magnitude and color, the method is potentially sensitive to a wide range of

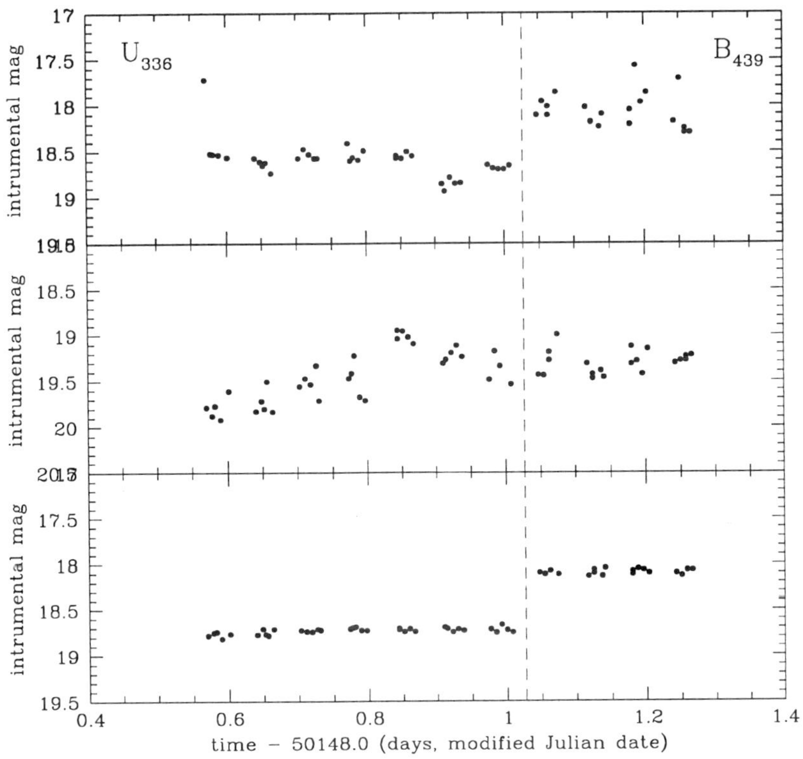

FIGURE 4. Light curves for three of the UV-bright stars shown in Figs. 2 and 3, first in the HST U filter, then in the B filter. The gaps are due to earth occultation; very high points are due to cosmic ray hits. A previously known CV, a new CV candidate (based on this light curve), and a non-flickering UV-bright star are shown in the top, middle, and bottom panels, respectively.

binary periods; and unlike variability searches, it is essentially independent of inclination angle. The main limitation is in the range of mass ratios that result in CMD positions sufficiently well separated from those of single stars. Thus, photometric accuracy is key.

Fig. 5 is an expanded version of Fig. 1 that illustrates the anticipated positions for MS–MS binaries with mass ratios of 1, 0.5, and 0.25. With a MS width of a few hundredths of a magnitude, binaries with mass ratios in the range 0.5–1.0 have colors that distinguish them quite effectively over much of the length of the MS, although the separation is greatest in regions where the main sequence slope is shallowest. Locations are also shown for three hypothetical sequences of detached MS–WD binaries. The three dotted curves below and to the left of the MS represent the positions that would be occupied by WDs with apparent magnitudes $V_{555} = 23$, 24, and 25, if paired with MS companions with apparent magnitudes from the turnoff down to $V_{555} = 26.5$. The colors of these binaries are distinguishable from those of isolated WDs for even the faintest MS companions, and for all MS companions that are $\lesssim 2$ magnitudes brighter than the WD.

It is evident from Fig. 5 that in the case of NGC 6397, detached binaries cannot be identified uniquely on the basis of their positions in the CMD. The cluster's low Galactic latitude ($b = -12\,^\circ$), together with its Galactic longitude of $l = 338\,^\circ$, result in significant

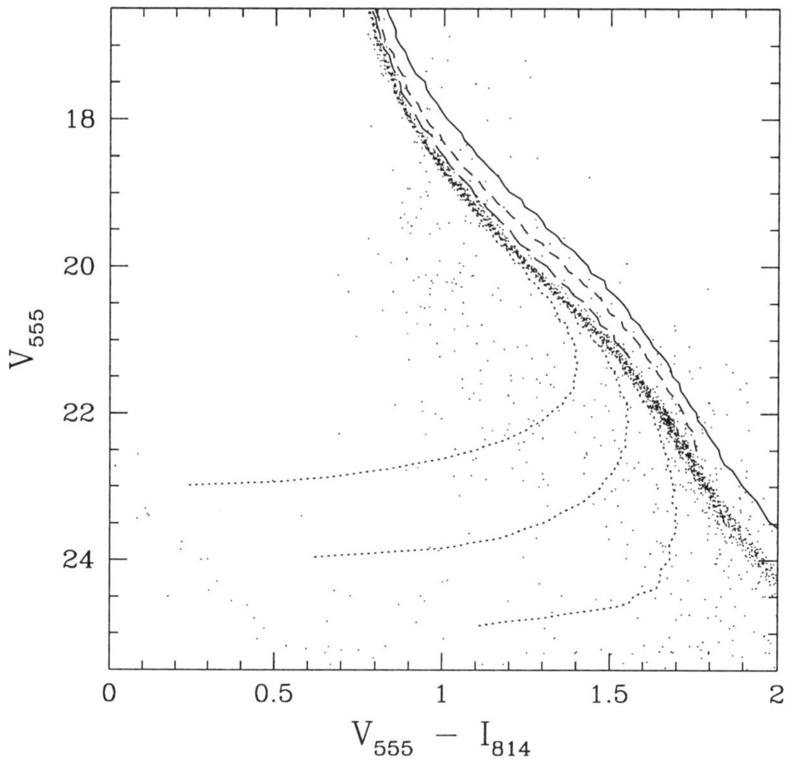

FIGURE 5. An expanded view of part of the CMD shown in Fig. 1, with curves overlaid that represent locations anticipated for detached binary stars. The solid, short-dash, and long-dash curves represent MS–MS binaries with mass ratios of 1, 0.5, and 0.25, respectively. The three dotted curves below the main sequence represent WD–MS binary sequences in which a WD of magnitude $V_{555} = 23$, 24, and 25 (top to bottom) is joined by a binary companion from the MS.

numbers of background and foreground stars occupying the same regions of the diagram as those occupied by the binaries. For this field in the envelope of NGC 6397, one can still place a useful upper limit to the MS–MS binary fraction, however. By comparing the numbers of stars in strips on either side of the MS, we estimate an upper limit of ~2% to the fraction of binaries with mass ratios in the range 0.5–1.0. This is quite a low fraction, despite its encompassing a relatively small range of mass ratios. However, given mass-segregation effects, this value may not be representative of the cluster as a whole.

The severity of field-star contamination can be reduced (though not eliminated) by observing regions closer to the cluster center, where the ratio of cluster to field stars is higher (see, *e.g.*, Fig. 2; analyses of detached binaries in these fields are in progress and will be reported elsewhere). Fortunately, for higher-latitude clusters, the contamination can be negligible, and binary fractions can be estimated directly from the numbers of stars seen to the upper right of the MS. Rubenstein & Bailyn (1997) have recently found MS–MS binary fraction in excess of 15% in the central regions of NGC 6752.

For WD–MS detached binaries, the situation is more difficult. The number of observable WDs is already small, and the number of WD–MS binaries is clearly going to be smaller. In Fig. 5, for example, where about 20 WDs are visible, a hypothetical 10% binary fraction would imply that of order 2 potentially observable WD–MS binaries would be present in the same field. Given such small numbers, even a few field stars can confuse matters. One possible solution is multicolor observations, in which field stars and cluster binaries may segregate into different parts of a 3-color diagram (see, *e.g.*, Richer *et al.* 1996). Alternatively (or in addition), cluster membership may be determined through proper-motion observations. For clusters with relatively large proper motions, it should be possible with the WFPC2 to detect the motion of individual cluster stars against the background of field stars in a matter of a few years.

4. Conclusion

Binary stars can occupy distinctive regions in globular cluster CMDs, depending on the nature of the binary components and the degree of contact between them. The resolving power of HST makes it possible to use CMDs to assess the binary populations below the turnoff even in and near dense cluster cores. Improved statistics concerning these classes of binaries will complement and help to cross-check studies that use other techniques to pinpoint cluster binaries. As binary statistics improve, it will begin to be possible to compare their populations in various evolutionary stages, which should help unravel the complex interplay between binary-star and cluster evolution.

I thank J. Anderson, C. Bailyn, J. Grindlay, I. King, and C. Sosin for discussions. This work was supported by NASA grant NAG5-1607 and by grant GO-5929 from STScI.

REFERENCES

Bailyn, C. *et al.* 1997, ApJ, 473, L31

Cool, A., *et al.* 1997, in preparation

Cool, A., Grindlay, J., Cohn, H., Lugger, P., & Slavin, S. 1995, ApJ, 439, 695

Cool, A. 1993, Ph. D. thesis, Harvard U.

Grindlay, J., Cool, A., Callanan, P., Bailyn, C., Cohn, H., & Lugger, P. 1995, ApJ, 455, L47

Hertz, P., & Grindlay, J., 1983, ApJ, 275, 105

Hut, P. *et al.* 1992, PASP, 104, 981

Mateo, M. 1996, in *The Origins, Evolution, and Destinies of Binary Stars in Clusters*, ASP Conf. Series, Vol. 90, eds. E. Milone and J.-C. Mermilliod, p. 21

Pryor, C., *et al.* 1996, in *Dynamical Evolution of Star Clusters – Confrontation of Theory and Observations*, Proc. of IAU Symp. No. 174, eds. P. Hut and J. Makino, p. 193

Richer, H., *et al.* 1996, in *The Origins, Evolution, and Destinies of Binary Stars in Clusters*, ASP Conf. Series, Vol. 90, eds. E. F. Milone and J.-C. Mermilliod, p. 65

Romani, R., & Weinberg, M. 1991, ApJ, 372, 487

Rubenstein, E., & Bailyn, C. 1997, ApJ, 474, 701

Shara, M., Drissen, L., Bergeron, L., & Paresce, F. 1995, ApJ, 441, 617

Shara, M., Kaluzny, J., Potter, M., & Moffat, A. 1988, ApJ, 328, 594

Sosin, C., & Cool, A. 1995, ApJ, 452, L29

Trimble, V. 1980, IAU Symp. 85, ed. J. Hesser, p. 259

Verbunt, F., Bunk, W., Hasinger, G., & Johnston, H. 1995, A&A, 300, 732

Evolution of Novae, He Novae and the Final He Shell Flashes

By MARIKO KATO

Department of Astronomy, Keio University, Hiyoshi, Kouhoku-ku, Yokohama, 223, Japan

This paper reports that one of the most important long-standing questions of nova field, what causes the speed class of novae, has been essentially solved. The light curves and the evolutional time scale of various type of phenomena, i.e., classical nova, slow nova, symbiotic nova, He shell flash, last He shell flash and the evolution of planetary nebula nuclei, are explained by the single scheme of optically thick wind theory with different parameters such as the white dwarf mass and the chemical composition of the envelope. Light curve analysis for individual novae has accurately determined the white dwarf mass which enable extensive study of novae and binaries. In classical novae, very fast light curves correspond to very massive white dwarfs with heavy elements enrichment envelope, and the slower light curves less massive white dwarfs (e.g., $\sim 1.0\, M_\odot$ for the moderately fast classical nova, nova Cygni 1978). Recurrent novae, slow novae, and symbiotic novae are understand in the various mass of the white dwarf with essentially solar composition. Very fast light curves of recurrent novae correspond to the extreme massive white dwarf ($\sim 1.38 M_\odot$), and the slow outbursts in symbiotic nova the less massive white dwarfs ($\sim 0.6 M_\odot$). The theoretical light curves and the evolutional time scale for He novae, the final helium shell flashes, and planetary nebular nuclei are also obtained. The evolutional time-scale of planetary nebular nuclei is much shortened due to the effects of the new opacity. This optically thick wind changes main frame of studies of binary evolution which contains an accreting white dwarf: the loss of the mass and the angular momentum due to the wind essentially determine the binary evolution.

1. Introduction

There were two long-standing problems in the theoretical study of nova filed; one is what determines the nova speed class, and the another one is the super-Eddington luminosity at the optical peak of nova outbursts.

Recently, the opacity table is revised (Rogers & Iglesias 1992, Iglesias & Rogers 1993, Iglesias & Rogers 1996), in which the opacity shows a large enhancement at the temperature of $\log T$ (K) ~ 5.2. This large peak in the opacity causes strong winds and then we can get a drastically shorter evolutional time-scales in calculations of nova outbursts.

With the OPAL opacity, Kato (1994) and Kato & Hachisu (1994) have succeeded in reproducing nova light curves by quasi-evolutional method based on the optically thick wind theory. Their theoretical light curves are reliable for qualitative comparison with observational data to derive the binary parameters such as the white dwarf mass. From the light curve fittings in various type of novae Kato (1995a, 1995b, 1996, 1997a) showed that the difference of the speed class is caused by the binary parameters such as the white dwarf mass. Prialnik & Kovetz (1995) have calculated time-dependently many nova outbursts with comprehensive sets of the white dwarf mass, the mass accretion rate and the thermal condition of the white dwarf. They showed that these difference in parameters causes the variety of nova duration. These two complementary works have shown that the variety of nova outbursts is understood by a single scheme based on the thermonuclear runaway model with different white dwarf conditions. The present paper concentrates how the optically thick wind has been succeeded in explaining nova speed class.

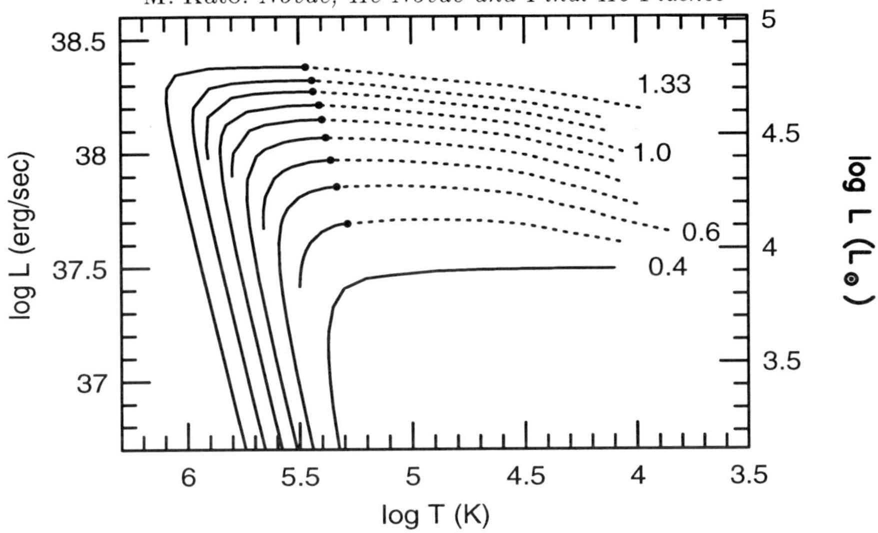

FIGURE 1. The evolutional tracks of the decay phase of classical novae in the H-R diagram. The white dwarf mass is attached. The optically thick wind occurs in the dashed region and stops at the small circles near $\log T = 5.3 - 5.5$. Some of the curves ended at the point that hydrogen nuclear burning extinguishes. The chemical composition of the envelope is assumed to be uniform with $X = 0.35$, $Y = 0.33$, O $= 0.2$, C $= 0.1$ and $Z = 0.02$

2. Optically Thick Winds for Degenerate Core

Optically thick wind is a continuum-radiation wind in which the acceleration occurs deep inside to the photosphere (Friedjung 1966). The structure of the mass losing envelope is obtained by solving the equations of motion, continuity, radiative transfer in diffusion approximation, and energy conservation with steady-state approximation. This steady-state approximation is quite good because the envelope reaches a thermal equilibrium in the decay phase of novae and also in the central star of the planetary nebulae. The equations and the numerical method are summarized in detail by Kato & Hachisu (1994). The wind mass loss rate is uniquely determined as an eigenvalue and the surface temperature is accurately determined in this method. From these two values we can calculate reliable light curve of novae. This method is only the one that can reproduce reliable nova light curves. In the present paper, the OPAL opacity (Iglesias & Rogers 1993) is used for nova and He nova envelopes, and the updated OPAL opacity (Iglesias & Rogers 1996) is used for solar composition envelope.

The evolution of the decay phase of nova, He nova, last He shell flash and the evolution of the central star of planetary nebula nuclei can be followed by a sequence of the static and the steady wind solutions. Because in many cases of such phenomena the luminosity is very close to the Eddington luminosity, the large enhancement of the new opacity causes considerable change of the internal structure of the envelope at temperature $\log T \leq 5.3$, i.e., the lower side of the opacity peak and often causes strong acceleration of the winds. Therefore we get a very rapid evolution of nova outburst. Even when the acceleration is not enough to cause the wind such as in the planetary nebula nuclei, the change of the envelope structure is substantial to shorten its evolution time.

Figure 1 shows the region where the optically thick wind occurs in the H-R diagram for a cold white dwarf with C-O rich envelope. The optically thick wind occurs in most of the white dwarf masses ($M_{\rm WD} > 0.4\,M_\odot$) for a wide range of the surface temperature. The wind mass loss rate strongly depends on the surface temperature and weakly depends on the white dwarf mass, i.e., $\sim 10^{-4}\,M_\odot$ yr^{-1} at $\log T \sim 4.0$ and quickly decreases with

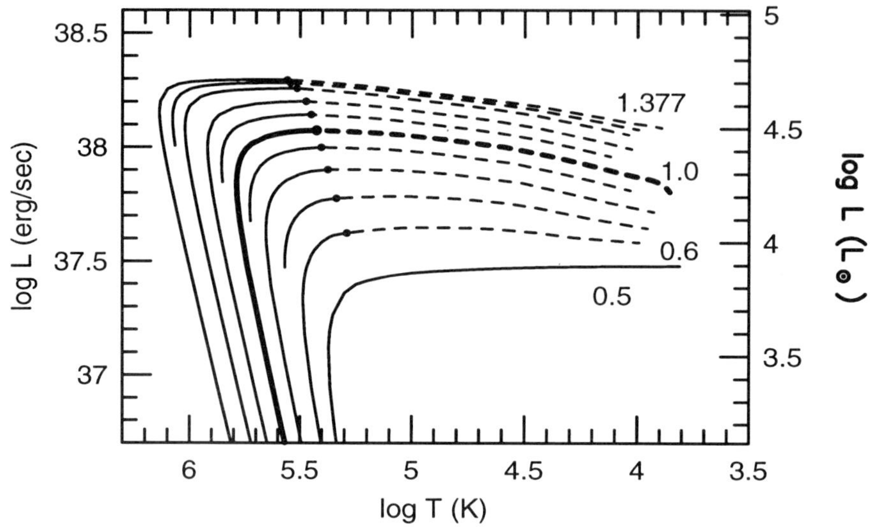

FIGURE 2. The evolutional tracks of the decay phase of recurrent novae, slow novae and symbiotic novae in the H-R diagram. The optically thick wind occurs in the dashed region. The chemical composition of the envelope is assumed to be $X = 0.7$ and $Z = 0.02$.

the increase of the temperature to several $\times 10^{-7} M_\odot$ yr^{-1} near the point where the wind mass loss stops (for the details, see Kato & Hachisu (1994)).

These tracks in the H-R diagram represent the decay phase of classical novae. After the peak of the visual luminosity, the star reaches a thermal equilibrium with very extended envelope. The strong wind mass loss reduces the envelope mass and the star moves leftward along the horizontal track in figure 1. The mass loss rate is large when the envelope is very extended, and quickly decreases as the star moves leftward. The optically thick wind stops at the point marked by the small circles. After that the star continues to move leftward with the decrease of the envelope mass due to hydrogen burning.

Figure 2 shows the similar H-R diagram but for the cold white dwarf with hydrogen-rich envelope. The optically thick wind also occurs in the wide parameter range ($M_{\rm WD} > 0.5\, M_\odot$). These tracks can be used to follow the recurrent novae, slow novae and symbiotic novae as described below.

The optically thick wind also widely occurs in the evolutional course of planetary nebulae nuclei (Kato 1997a, 1997b), the last He shell flash and the He nova (Kato 1996).

3. What Determines the Speed Class of Novae

Kato (1994) showed that the theoretical light curves obtained from the optically thick wind method have enough quality for quantitative comparison with observational data of nova Cygni 1978. From the quantitative fitting of the observational and the theoretical light curves, we can determine the white dwarf mass, the distance to the star and the chemical composition of the envelope, if we have enough observational information. Light curve analysis for many nova outbursts has clarified that the wide range of nova duration corresponds to the range of variation of the white dwarf mass and the chemical composition of the envelope. This section reviews light curve fittings for various objects and show how the optically thick wind theory solved the question of cause of the nova speed class.

	Core mass ($M_\odot$)	total duration (yr)	Wind phase (yr)
classical nova (fig.1)	0.5	28	7.4
	0.6	12.8	3.9
	1.0	1.3	0.61
	1.33	0.13	0.11
solar composition (fig.2)	0.6	3600	no wind
	0.7	700	no wind
	1.0	28	7.7
	1.3	1.3	0.63
	1.377	0.17	0.13
last He shell flash	0.6	3900	no wind
	1.05	75	47
	1.3	11	8.4
He nova	1.3	9	6.6
	1.377	2.5	2.2

TABLE 1. Evolutional time-scale of various nova outbursts

3.1. *Classical Novae*

The evolutional time-scale and the theoretical light curves are obtained from the sequence of steady-state and the static solutions (Kato & Hachisu; 1994). Figure 3 shows the light curve fitting for the moderately fast classical nova, nova Cygni 1978. As the star moves leftward in the H-R diagram, the mean photon energy emitted from the photosphere increases, and the UV flux has a maximum value after the optical peak.

The evolutional time scale strongly depends on the white dwarf mass. The massive white dwarf has an less massive envelope whereas the wind mass loss rate weakly depends on the white dwarf mass, therefore the massive star shows a rapid evolution. The chemical composition of the mass-losing envelope is assumed to the one close to the observed value. From this figure we can see that the white dwarf mass is about $1.0\,M_\odot$. The dependence of the light curves on the chemical composition is very weak for CO/Ne rich envelopes if CO/Ne content is greater than 20 percent). Considering other effects such as the companion motion and the extended photosphere, the accuracy of the light curve fitting is $\Delta M_{\rm WD} = \pm 0.05\,M_\odot$ in this case. Kato (1995a) also shows the light curve fitting of classical nova, Nova Mus 1983 in which the observational light curves of visual, UV and X-ray are consistent with the theoretical curves for less massive white dwarf of 0.5–$0.6\,M_\odot$.

In this way the evolutional time scale strongly depends on the white dwarf mass and the chemical composition. Table 1 summarizes the typical time-scales of the decay phase of various type of nova/degenerate core. The total duration is the time for a star to move from $\log T_{\rm eff} = 4.0$ to the extinguish point of hydrogen burning which is well known as the X-ray turn off time.

We can see that the duration time in table 1 covers the variation of the evolutional time-scale of classical novae. The short time-scale in fast classical nova corresponds to a massive white dwarf and the long time-scale a less massive white dwarf. We can see that the white dwarf mass in nova system is not necessarily as massive ($> 1.3\,M_\odot$) as regarded before in order to get a short decay time as observed. With the OPAL opacity which causes massive winds, we get very rapid evolution in theoretical nova outbursts and then we need not always assume a very massive white dwarf in nova system.

Table 1 shows the decay time-scales for various type of nova outburst. We can see that

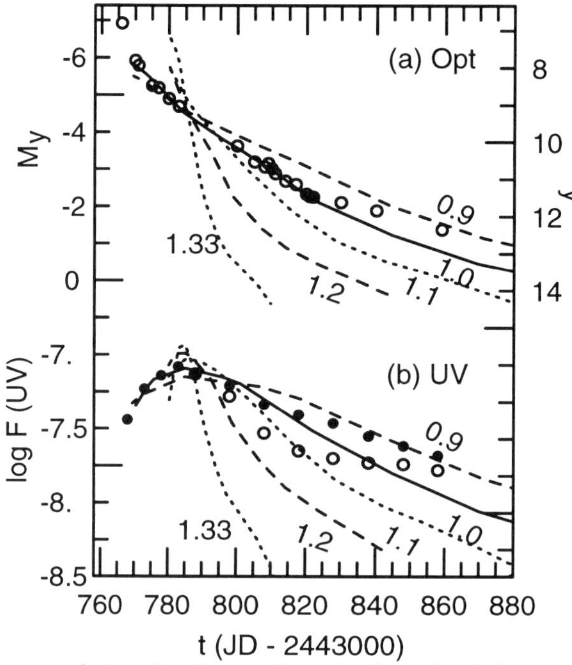

FIGURE 3. Light curve fitting for the moderately fast classical nova V1668 Cyg (nova Cyg 1978). Theoretical curves are denoted by thick curves, which are shown by the white dwarf masses. (a) Optical light curves. The left ordinate shows the absolute y-magnitude for the theoretical curves and the right ordinate the apparent y-magnitude for observed data. Data are taken from Gallagher et al. (1980). (b) UV light curves. Open circles denote the UV flux (1140 Å- 3290 Å) and filled circles give the summation of the UV and IR fluxes (> 12000 Å) in units of erg cm^{-2} sec^{-1} (Stickland et al. 1981). In theoretical UV flux, $F = L_{UV}/4\pi D^2$, the distance to the star $D = 2.88$ kpc is assumed.

the entire region of the white dwarf mass distribution, i.e., 0.5–1.33 $M_\odot$, corresponds to the distribution of classical nova duration.

3.2. *Recurrent Novae*

Recurrent novae show very rapid decay time-scales from a few tens of days to several months. The light curve fitting of the recurrent nova, U Sco, is shown in figure 4. The observational data by Sekiguchi et al. (1988) show a good agreement with the theoretical light curve for the white dwarf of mass of 1.377 $M_\odot$. Other two recurrent novae, RS Oph and T Pyx, show a slower decline rate that corresponds to relatively less massive white dwarf (> 1.0 $M_\odot$).

Kato (1995a) has pointed out that the light curves of four recurrent novae, U Sco, T CrB, V394 CrA and V745 Sco show very similar light curves. It is very interesting because their light curves are all consistent with the ~ 1.377 $M_\odot$ white dwarf which is almost the upper limit of an accreting white dwarf (Nomoto, Thielemann & Yokoi: 1984), i.e., just before the collapse to trigger a type Ia supernova explosion.

In this way, the light curve fitting arises an interesting question for the origin and the evolution of such extremely massive white dwarfs and the relation with type Ia supernova progenitor. It is especially of interest if these massive white dwarfs will continue to glow and make collapse to have a type Ia supernova explosion. In order to estimate whether the white dwarf mass are increasing or decreasing, we need to accurately determine envelope

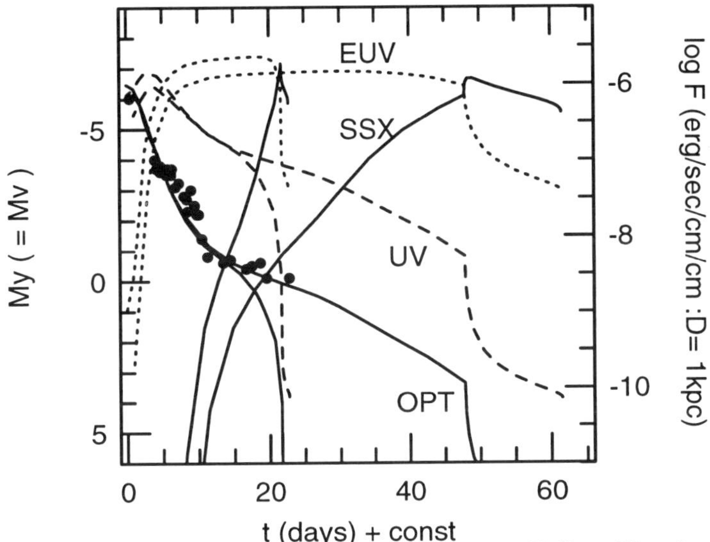

FIGURE 4. Theoretical light curve fitting of recurrent nova U Sco. The observational data by Sekiguchi et al. (1988) are plotted by dots ($t = 4$ corresponds to May 16, 1987 and $M_y = 6$ correspond 20.8 mag). The dot at $t = 0.43$ is the maximum value extrapolated from previous outbursts. The two envelope models for $1.377\,M_\odot$ are shown with different chemical composition; $X = 0.7$ and $Z = 0.02$ (curves ended at $t = 61.5$ when the hydrogen nuclear burning extinguishes) and $X = 0.1$ and $Z = 0.02$ (curves ended at $t = 22.8$). In the model with $X = 0.7$ the photospheric temperature is $\log T = 3.89$ at $t = 1$ and in the model with $X = 0.1$, $\log T = 4.07$ at $t = 0$. Optically thick wind ceases at $t = 46.8$ ($X = 0.7$) and at $t = 21.7$ ($X = 0.1$).

parameters. As shown in figure 4, multiwavelength observation such as in extreme UV or supersoft X-ray are helpful to determine such parameters.

3.3. *Slow Novae and Symbiotic Novae*

Slow nova shows no large heavy element enrichment. Kato (1995a) showed that the slow decline of the light curve and the expansion velocity of slow nova RR Pic are consistent with those of the theoretical curve for less massive white dwarf of mass 0.8–0.9 $M_\odot$.

Symbiotic novae show very slow light curves. For example, AG Peg moved from $\log T_{\rm ph} = 4.0$ to 5.0 in 70 years (Kenyon et al.: 1993). Such a long time scale is consistent with the optically thick wind model of $\sim 0.6\,M_\odot$. The wind mass loss rate increases in time in AG Peg (Kenyon et al.: 1993, Nussbaumer et al.: 1995). The expansion velocity of the wind are very small when the photospheric temperature is low and increases in time. These properties are consistent with those in optically thick wind models (Kato 1997a).

3.4. *Evolution of Planetary Nebula Nuclei and Final He Shell Flashes*

The strong peak in OPAL opacity shortens the evolutional time of the central star of planetary nebulae. Kato (1996,1997b) has recalculated the evolutional track of planetary nebula nuclei for various core mass and showed that the optically thick wind occurs for massive core ($\geq 0.8\,M_\odot$). The evolutional time from the AGB stage to the white dwarf branch is much shortened compared with previous calculations with the old opacity. For example in a massive core of $1.0\,M_\odot$, the optically thick wind occurs that makes the evolutional time scale as short as 28 year. In less massive core of $0.6\,M_\odot$ no optically thick wind occurs, but the evolutional time scale becomes 2100 years from $\log T = 3.8$ to 4.9 which is about one half of the results by Blöcker and Schönberner (1990).

The final He shell flash is a nova-lick event that occurs after a star evolves to white dwarf region (Iben et al. 1983). The theoretical time-scale calculated by optically thick wind method is shown in table 1.

3.5. *What Determines the Nova Speed Class*

In this way, the light curve fitting enables us to determine the white dwarf mass, chemical composition and the distance to the star. The light curve analysis of novae clarified this by quantitative fittings of theoretical curves with observational data. The upper and the lower limits of the white dwarf mass, i.e., $0.5\,M_\odot$ to $1.38\,M_\odot$ seem to correspond to the fastest and the slowest limits of decline rate of novae.

The light curve analysis provides a new method of mass determination of the white dwarfs. It will be very useful in nova systems because the binary parameters are often poorly known.

4. Optically Thick Wind in Accreting WDs and its Effects on Binary Evolution

The optically thick wind occurs not only in nova outbursts and the planetary nebula nuclei, but also in accreting white dwarfs with steady hydrogen burning.

Many scenarios of binary evolution and progenitor models of type Ia supernovae (Munari & Renzini 1992, Iben & Tutukov 1996) involve an accreting white dwarf with high accretion rate ($> 10^{-7}\,M_\odot\ \mathrm{yr}^{-1}$). With such high mass accretion rates the steady nuclear burning supports the high luminosity and the hydrogen-rich envelope largely expands of which radius strongly depends on the mass accretion rate. In such scenarios the accreting white dwarf is expected to receive all of the accreted material and will grow to the Chandrasekhar limit to make a supernova explosion.

Such a scenario, however, must be modified in the presence of the optically thick wind that inevitably occurs for white dwarf with extended envelope. Figure 2 shows that the optically thick wind occurs in a very wide region of the H-R diagram that can also be applied to accreting white dwarfs with high extended envelope.

Once the wind occurs, the large mass loss due to the optically thick wind (10^{-4} – $10^{-7}\,M_\odot\ \mathrm{yr}^{-1}$) carries the angular momentum and the mass away from the binary, which seriously affects the binary evolution. Therefore, scenarios of binary evolution should be modified in the presence of the wind. Hachisu, Kato and Nomoto (1996) have shown that this wind mass loss essentially governs the binary evolution and followed the binary evolution until the white dwarf reaches Chandrasekhar mass as a progenitor of type Ia supernova. In this way the effect of the winds on binary evolution has just begun to study, but it will be the new standard of binary study that provide a new channel to type Ia supernovae.

5. Conclusions

One of the most important long-standing questions, what causes the nova speed class, has been essentially solved by the light curve analysis based on the optically thick wind theory by Kato and the evolutional calculation by Prialnik. The another important problem of theoretical reproduction of the super-Eddington luminosity at the optical peak is still unsolved.

As shown in section 3 the variety of the evolutional speed in classical nova is explained as the variety of the white dwarf mass. The light curve analysis shows that the white dwarf in novae is not always massive ($\sim 1.3\,M_\odot$) as ever believed but can be less massive

($\sim 1.0\,M_{\odot}$ in moderately first nova, nova Cyg 1978). The difference of the evolutional speed in recurrent novae, slow novae, and the symbiotic novae is explained as the difference of the white dwarf mass with essentially solar metallicity envelope. The most rapid type of nova corresponds to extremely massive white dwarf and the slowest one the lowest limit of the white dwarf mass. Therefore the speed class of various type of novae has been understand by the single general scheme of the optically thick wind theory for different white dwarf masses with various chemical composition of the envelope.

The evolution of He novae, the final helium shell flashes, and planetary nebular nuclei also can be followed by the optically thick wind theory. The evolutional time-scale of planetary nebular nuclei is much shortened with the use of the OPAL opacity.

Occurrence of the optically thick winds seriously affect the binary evolution that contains an accreting white dwarf with high mass accretion rate. The binary evolution is essentially determined by the mass and the angular momentum loss due to the wind. It provides a new channel of binary evolution that leads to type Ia supernova.

This work has been supported in part by the Grant-in-Aid for Scientific Research (05640314, 08212208) of the Japanese Ministry of Education, Science and Culture.

REFERENCES

Blöcker, T. & Schönberner, D. 1990, AAp, 240, L11

Friedjung, M. 1966, MNRAS, 132, 317

Gallagher, J. S., Kaler, J. B., Olson, E. C., Hartkopf, W. I., & Hunter, D. A. 1980, PASP, 92, 46

Hachisu, I., Kato, M. & Nomoto, K., 1996, ApJ, 470, L97

Iben, I. Jr., Kaler, J.B., Truran, J.W., & Renzini, A., 1983, ApJ 264, 605

Iben, I. Jr & Tutukov, A.V., 1996, ApJS, 105, 145

Iglesias, C.A. & Rogers, F.J., 1993, ApJ., 412, 752

Iglesias, C.A. & Rogers, F.J., 1996, ApJ 464, 943

Kato, M., 1994, AA, 281, L49

Kato, M. 1995a, in Cataclysmic Variables, eds. Bianchini,A., Della Valle, M. & Orio M. (Dordrecht:Kluwer) p.243

Kato, M. 1995b, in Flares and Flashes, eds. Greiner, J., Duerbeck, H. & Gershberg, R.E. (Berlin:Springer) p.237

Kato, M. 1996, in Supersoft X-ray Sources, ed. Greiner, J., (Berlin:Springer) p.15

Kato, M. 1997a, in Proceedings of Physical Processes of Symbiotic Binries , in press

Kato, M. 1997b, ApJS, submitted

Kato, M. & Hachisu, I. 1994, ApJ 437,802

Kenyon, S., Mikolajewska, J., Mikolajewski, M., Polidan, R.S. & Slovak, M.H., 1993, AJ, 106, 1573

Nomoto, K., Thielemann, F. & Yokoi, K. 1984, ApJ, 286, 644

Prialnik, D. & Kovetz, A. 1995, ApJ, 445, 789

Munari, U. & Renzini, A., 1992, apJ, 397, L87

Nussbaumer, H., Schmutz, W & Vogel, M., 1995, AA, 293, L13

Rogers,F.J. & Iglesias, C.A., 1992, ApJS, 79, 507

Sekiguchi, K. Feast, M.W., Whitelock, P.A., Overbeek, M.D., Wargau,W. & Spencer Jones, J., 1988, MNRAS, 234, 281

Stickland, D. J., Penn, C. J., Seaton, M. J., Snijders, M. A. J. & Storey, P. J. 1981, MNRAS, 197, 107

Element Diffusion in Novae

By DINA PRIALNIK[1] AND ATTAY KOVETZ[2]

[1]Department of Geophysics and Planetary Sciences
[2]School of Physics and Astronomy
Tel Aviv University, Ramat Aviv 69978, Israel

Enhanced concentrations of heavy elements are characteristic of all nova ejecta, their source being material dredged up from the progenitor white dwarf's core. The degree of mixing between core and accreted material may be inferred from the observed abundances. We consider diffusion during the accretion phase, followed by convection at the onset of the thermonuclear runaway as the mixing mechanism responsible for the composition of nova ejecta and show that most of the observed abundances and abundance ratios can be reproduced. The white dwarf's composition is generally *not* reflected in the abundances of the ejecta: whereas a large fraction of the carbon is always turned into nitrogen, oxygen is in some cases unaffected and in others almost completely destroyed. A fraction of novae appear to show extreme enhancements of neon and heavier elements, demanding ONeMg-rich white dwarf progenitors. Yet very high neon abundances remain difficult to account for. The question whether diffusion-convection is sufficient or whether another mixing mechanism must also be invoked remains open.

1. Introduction

The abundance anomalies observed in nova ejecta, exhibiting various degrees of enrichment of helium and heavier isotopes support and, at the same time, constrain the standard model for nova outbursts: a thermonuclear runaway (TNR) on the surface of an accreting white dwarf (WD), leading to mass ejection. The main characteristics of the nova ejecta composition derived from observations, as given in Table 1, are:

1. *All* novae show *some* degree of He-CNO enrichment.
2. There is *simultaneous* enrichment of He *and* CNO and/or O-Ne-Mg.
3. *No* apparent correlation is found between the helium mass fraction Y and the heavy element mass fraction Z.
4. As a rule, $C/N < 1$ $[(C/N)_\odot \approx 2.8]$, while $0.1 \lesssim O/N \lesssim 4$ $[(O/N)_\odot \approx 8.7]$.

Although the theoretical prediction of significantly enriched ejecta preceded the actual observation, the enrichment mechanism has eluded workers in this field for quite a long time. In principle, there are three possible sources for the unusual composition: (a) the accreted matter itself is rich in He, CNO and heavier isotopes; (b) heavy elements are synthesized during outburst; (c) core material is dredged-up into the accreted hydrogen-rich matter. The first possibility is highly unlikely, the donor being a K or M main sequence star. In fact, no stars are known with outer layers of the nova ejecta composition. Regarding the second, the energy required to power a nova outburst in all its aspects is readily supplied by burning a tiny fraction of the accreted mass and temperatures are not high enough to produce elements heavier than helium. Hence there remains the possibility of mixing of WD material with the accreted mass. Two mechanisms have been suggested: (1) diffusion of hydrogen into the core, which causes ignition to take place below the core boundary and results in convective mixing of core material lying above the ignition radius into the envelope (Prialnik & Kovetz 1984, Kovetz & Prialnik 1985); (2) turbulent mixing due to shear induced instabilities at the boundary between the WD and the accretion disk (Kippenhahn & Thomas 1978, MacDonald 1983). Quantitative studies have mostly considered the diffusion-convection mechanism. We shall show that this mechanism can by itself account for the bulk of observed abundances.

Nova	X	Y	Z	$\frac{C}{N}$	$\frac{O}{N}$	$\frac{Ne}{Z}$	$\frac{m_{acc}}{m_{env}}$	$\frac{m_{rem}}{m_{env}}$	$\frac{m_{core}}{m_{env}}$
V1370 Aql	0.053	0.088	0.859	0.250	0.364	0.605	0.076	0.474	0.450
T Aur	0.470	0.400	0.130	...	0.646	...	0.671	0.244	0.085
V842 Cen	0.410	0.230	0.360	0.571	0.143	0.003	0.586	0.103	0.311
IV Cep	0.400	0.210	0.390	...	...	...	0.571	0.082	0.347
V693 CrA	0.290	0.320	0.390	0.058	1.500	0.436	0.414	0.334	0.251
V1500 Cyg	0.570	0.270	0.160	1.415	1.220	0.063	0.814	0.050	0.136
V1668 Cyg	0.450	0.230	0.320	0.336	0.929	0.021	0.643	0.073	0.284
HR Del	0.443	0.480	0.077	...	1.741	0.039	0.633	0.328	0.039
DQ Her	0.310	0.310	0.380	0.431	1.538	0.010	0.443	0.300	0.257
V827 Her	0.360	0.290	0.350	0.363	0.067	0.002	0.514	0.225	0.261
CP Lac	0.600	0.260	0.140	...	...	...	0.857	0.023	0.120
GQ Mus	0.270	0.320	0.410	0.084	0.018	0.003	0.386	0.359	0.255
V2214 Oph	0.340	0.260	0.400	...	0.194	0.043	0.486	0.207	0.308
RR Pic	0.527	0.430	0.043	3.793	1.897	...	0.753	0.229	0.018
CP Pup	0.350	0.170	0.480	...	...	...	0.500	0.058	0.442
V977 Sco	0.510	0.390	0.100	...	0.714	0.260	0.728	0.207	0.065
V443 Sct	0.488	0.450	0.062	...	0.132	0.002	0.697	0.272	0.031
PW Vul	0.540	0.280	0.180	0.291	0.345	0.004	0.771	0.078	0.151
QU Vul	0.440	0.460	0.100	0.093	2.533	0.400	0.628	0.316	0.056
QV Vul	0.677	0.270	0.053	...	4.100	0.019	0.966	0	0.034

TABLE 1. Nova abundances (Warner 1995) and derived mass ratios

The source of helium excess may be either a helium-rich outer layer of the WD progenitor — a relic of its early evolutionary stages — or the hydrogen-exhausted envelope remnant of a previous nova outburst. Since helium and heavier elements enrichments are concomitant, the helium layer overlying the WD core cannot be too thick. We shall show that the common helium source for the ejecta is the helium-rich remnant that forms during each nova outburst, to be destroyed and reformed in the next.

2. Constraints on element mixing

At the onset of an outburst, the nova progenitor consists of two parts — core and envelope. The burning front separating them is located at the depth reached during the accretion phase by the downward diffusing hydrogen in the diffusion scenario, or at the bottom of the unstable, turbulent layer, in the shear mixing scenario. The explosive ignition of hydrogen leads to the formation of a convective zone that sweeps the entire envelope. Thus, regardless of the element distribution prior to outburst, the convective envelope's composition becomes well mixed before any matter is ejected. Hence observed abundances do not bear direct witness to the mixing mechanism.

The nova envelope's mass m_{env} consists of three originally different contributions: the accreted mass m_{acc} of solar composition, the remnant mass m_{rem}, which has the metalicity of the envelope formed at the previous outburst and is depleted of hydrogen, and the mass dredged up from the WD core m_{core}, which has no hydrogen or helium:

$$m_{env} = m_{acc} + m_{rem} + m_{core} \tag{2.1}$$

$$[m_{env}X]^{(n+1)} = [m_{acc}X_\odot + m_{rem} \cdot 0 + m_{core} \cdot 0]^{(n)} \tag{2.2}$$

$$[m_{env}Y]^{(n+1)} = [m_{acc}Y_\odot + m_{rem}(X+Y) + m_{core} \cdot 0]^{(n)} \tag{2.3}$$

$$[m_{env}Z]^{(n+1)} = [m_{acc}Z_\odot + m_{rem}Z + m_{core} \cdot 1]^{(n)}, \tag{2.4}$$

where superscripts distinguish between a new outburst $(n+1)$ and its predecessor (n). As evolutionary calculations spanning several consecutive cycles (Prialnik & Kovetz 1995) show that a regular pattern develops and nova characteristics do not change from one cycle to the next, we may drop the superscripts and obtain the abundances in terms of the masses involved (Prialnik 1990, Sparks 1995). We may define two ratios α and β related to the masses and express them in terms of abundances, as follows:

$$\alpha = m_{acc}/m_{env} \qquad \beta = m_{rem}/m_{env}, \tag{2.5}$$

$$1 - \alpha - \beta = m_{core}/m_{env} \tag{2.6}$$

$$X = \alpha X_{\odot} \quad \Longrightarrow \quad \alpha = \frac{X}{X_{\odot}} \tag{2.7}$$

$$Y = \alpha X_{\odot} + \beta(X+Y) \quad \Longrightarrow \quad \beta = \frac{Y/X - Y_{\odot}/X_{\odot}}{1+Y/X}. \tag{2.8}$$

We may now compute the values of α and β for novae with known compositions and deduce the extent of dredge up of core material from equation(2.6). The results are given in Table 1. A similar, but more elaborate, method has been proposed by Fujimoto & Iben (1992), but they include second-order effects which make the mass ratios model-dependent and hence no sharp distinction can be made between observations and theory, and uncertainties associated with each become mingled. As the fraction of core material mixed into the envelope turns out to be at most 0.45 and usually much smaller than that, in excellent agreement with the predictions based on the diffusion-convection mechanism, we may conclude that if turbulent mixing will eventually be found efficient, it will have to be restrained, so as to agree with the constraints imposed by observations.

3. Calculated nova ejecta abundances — CO WDs

In Table 2 we summarize the results obtained by Kovetz & Prialnik (1996) in a systematic study involving calculations of evolutionary sequences of nova outbursts through several cycles. Initial models have a homogeneous composition of C and O in equal mass fractions and masses of $0.65\,M_{\odot}$, $1.00\,M_{\odot}$, $1.25\,M_{\odot}$, and $1.40\,M_{\odot}$; core temperatures are 1×10^7 K ($T_7 = 1$), 3×10^7 K ($T_7 = 3$), and 5×10^7 K ($T_7 = 5$). The range of mass transfer rates extends from $\sim 10^{-8}\,M_{\odot}$ yr^{-1} to $10^{-11}\,M_{\odot}$ yr^{-1}, the limits varying with the WD mass. The accreted material is assumed to have solar composition (according to Anders & Grevesse 1989). The calculated abundances (mass fractions) listed are obtained by integrating over the entire ejected mass; $Z = X_{\rm C} + X_{\rm N} + X_{\rm O}$, since production of heavier elements is found to be negligible. The nova type (speed class 'sc') is indicated in the fourth column: vs – very slow nova; s – slow nova; m – moderate nova; f – fast nova; vf – very fast nova.

We now focus on ranges of variation of properties and correlations between properties, which are far more reliable than specific, individual, values obtained for them, either observationally, or by model calculations. This is particularly true in the case of nova abundances, which are characterized by large uncertainties (sometimes by factors of 2–4 between different determinations, Livio & Truran 1994).

3.1. *Mass fractions*

The general conclusions that emerge from Table 2 is that calculated abundances assuming CO progenitors are very similar to those derived from observations (except for the relatively few novae with high Ne content that demand O-Ne-Mg progenitors): all nova models do indeed show *some* degree of He-CNO enrichment and there is *simultaneous*

M	T_7	$\dot{M}$	sc	m_{ej}	Y	Z	$\frac{C}{N}$	$\frac{O}{N}$	$\frac{O}{CN}$	$[\frac{^{13}C}{^{12}C}]$	$[\frac{^{15}N}{^{14}N}]$	$[\frac{^{17}O}{^{16}O}]$
0.65	1	- 9	s	1.81(-4)	.250	.194	0.207	1.084	.898	1.38	-1.41	2.11
0.65	1	-10	s	2.99(-4)	.244	.191	0.220	1.114	.913	1.64	-1.02	2.06
0.65	3	- 9	vs	1.21(-4)	.296	.110	0.053	0.905	.860	1.44	-1.97	2.17
0.65	3	-10	vs	1.11(-4)	.257	.233	0.130	0.977	.865	1.66	-1.59	1.89
0.65	5	- 9	vs	1.00(-4)	.264	.229	0.146	0.987	.861	1.58	-2.07	1.68
0.65	5	-10	vs	6.72(-5)	.257	.257	0.158	0.991	.856	1.56	-2.08	1.70
0.65	5	-11	vs	5.31(-5)	.251	.323	0.071	0.887	.828	1.48	-2.07	1.84
1.00	1	- 8	s	2.22(-5)	.332	.100	0.064	0.680	.639	1.46	-1.66	2.46
1.00	1	- 9	m	5.28(-5)	.297	.127	0.216	0.765	.629	1.76	0.01	2.56
1.00	1	-10	f	9.73(-5)	.275	.162	0.834	1.097	.598	1.87	1.72	2.91
1.00	3	- 8	s	2.18(-5)	.330	.101	0.062	0.689	.649	1.45	-1.73	2.46
1.00	3	- 9	s	3.10(-5)	.302	.137	0.082	0.777	.718	1.50	-1.56	2.42
1.00	3	-10	s	2.71(-5)	.259	.256	0.159	0.946	.817	1.75	-0.40	2.26
1.00	5	- 8	s	2.28(-5)	.390	.027	0.057	0.251	.238	1.44	-1.52	2.58
1.00	5	- 9	s	1.90(-5)	.299	.180	0.067	0.809	.758	1.46	-1.73	2.34
1.00	5	-10	m	1.42(-5)	.281	.264	0.067	0.832	.780	1.46	-1.86	2.23
1.00	5	-11	s	1.21(-5)	.245	.379	0.070	0.859	.802	1.47	-1.90	2.17
1.25	1	- 8	f	4.24(-6)	.369	.102	0.082	0.306	.283	1.47	-1.38	2.47
1.25	1	- 9	vf	1.07(-5)	.340	.133	0.480	0.101	.068	1.88	1.03	3.17
1.25	1	-10	vf	2.18(-5)	.321	.158	0.604	0.120	.075	1.97	2.05	2.31
1.25	3	- 8	f	4.16(-6)	.367	.104	0.081	0.332	.307	1.48	-1.26	2.47
1.25	3	- 9	f	6.00(-6)	.341	.135	0.130	0.303	.268	1.61	-0.29	2.60
1.25	3	-10	f	5.58(-6)	.304	.258	0.301	0.717	.551	1.80	0.45	2.63
1.25	5	- 8	m	3.99(-6)	.366	.107	0.079	0.367	.340	1.47	-1.33	2.47
1.25	5	- 9	f	3.69(-6)	.342	.157	0.085	0.522	.481	1.48	-1.36	2.47
1.25	5	-10	f	2.79(-6)	.307	.273	0.093	0.731	.668	1.48	-1.68	2.40
1.25	5	-11	f	2.41(-6)	.276	.412	0.122	0.843	.752	1.57	-0.62	2.30
1.40	1	- 9	vf	4.89(-7)	.480	.147	0.501	0.002	.002	1.93	1.05	2.20
1.40	1	-10	vf	6.89(-7)	.406	.231	1.227	0.002	.001	1.90	1.54	2.27
1.40	3	- 9	vf	2.93(-7)	.479	.153	0.290	0.002	.002	1.72	0.44	2.20
1.40	3	-10	vf	2.69(-7)	.464	.232	0.422	0.002	.002	1.84	0.90	2.59
1.40	5	- 9	vf	2.18(-7)	.495	.154	0.227	0.002	.002	1.64	0.02	2.23
1.40	5	-10	vf	1.66(-7)	.467	.285	0.299	0.061	.047	1.74	0.36	2.68
1.40	5	-11	vf	1.21(-7)	.421	.465	0.347	0.403	.299	1.72	0.24	2.57

TABLE 2. Results of evolutionary calculations

enrichment of He *and* CNO isotopes; there is no correlation between Y and Z; $C/N < 1$, with one exception, and the range of variation is very similar to that inferred from observations; O/N ranges from 0.002 to 1.114, but very high O/N ratios, as are sometimes observed, are *not* obtained; O/(C+N) may be very low, indicating *depletion* of oxygen.

3.2. *Isotopic ratios*

Novae have been repeatedly invoked as sources of the rarer C, N and O isotopes, although their contribution to the dominant isotopes appears to be negligible (e.g. Audouze & Tinsley 1976, Truran 1982, Peimbert & Sarmiento 1984, Sarmiento & Peimbert 1985). Sneden & Lambert (1975) set upper limits — $^{12}C/^{13}C \geq 1.5$ and $^{14}N/^{15}N \geq 2$ — based on the analysis of Nova DQ Her's spectrum. Isotopic ratios obtained from model calculations (relative to solar values) are listed in the last three columns of Table 2. The interesting conclusion is that ^{13}C and ^{17}O are overabundant by large factors in all cases, while ^{15}N is overabundant in some cases and underabundant in others. We suspect that the *average* $^{14}N/^{15}N$ production ratio for novae — allowing for the relative relevance of different models — will turn out to be closer to the solar (interstellar medium) value of ~ 300 than previously thought. Consequently, novae should be reconsidered potential nitrogen sources in galaxies, as suggested by Williams (1982), since the obstacle of ^{15}N production in too large amounts may not exist.

4. The WD composition

As the WD is the source of the peculiar abundances of nova ejecta, its own composition and, in particular that of the outer layers, is of crucial importance. An extensive study of the evolution of close binary systems leading to the formation of WDs was performed by Iben & Tutukov in 1985 and its results regarding WD masses and abundance distributions has been invaluable to nova theory. According to Iben & Tutukov, WDs originating from stars of initial masses $\lesssim 10\,M_\odot$ have a hydrogen-rich outer layer of at most $10^{-3}\,M_\odot$and below it a $\sim 0.01 M_\odot$ thick shell, over which Y decreases from 0.9 to 0.1. Such a shell could provide CNO and He enrichments simultaneously, but only for a limited time, since only in a small fraction of this shell does Y exceed the observed values. Below this shell is the CO core with C and O *in equal mass fractions* (as is usually assumed in nova model calculations) and a few percent neon. Stars in a very narrow initial mass range around $10\,M_\odot$were found to produce O-Ne WDs. The formation of such an WD was recently studied in great detail by Rittosa, Garcìa-Berro & Iben (1996).

4.1. *Erosion of the He-rich outer layer*

The implicit assumption of the dredge-up model for the heavy element enrichment of nova ejecta is that the initial helium layer of the WD is gradually eroded during the first outbursts of any given nova system. To test this assumption, the crucial question one has to ask is: does a WD with a thick outer helium rich layer lose or gain mass by nova outbursts? In the former case the outer layer would be gradually reduced and eventually obliterated; in the latter, it would be augmented by each outburst, growing beyond the reach of any mixing mechanism. This question has been addressed by Iben *et al.*(1991, 1992) and by Prialnik & Livio (1995), by performing calculations (including diffusion) in which accretion took place onto a WD with a helium surface layer. In spite of quantitative differences, both studies agree that if the WD mass is not too low nor the accretion rate too high, the helium layer which exists at the surface of newly formed WDs will indeed be eroded in nova outbursts.

We note that, due to selection effects, the average WD mass in observed novae is well above the average single WD mass (e.g. , Ritter *et al.*1991). Furthermore, many of the systems which do have low-mass WDs, will have low secular mean accretion rates, thus allowing for an erosion of the helium layer to occur. Finally, being characterized by relatively large envelope masses, low-mass WDs are more susceptible to mass ejection resulting from the deposition of drag energy by the secondary star (e.g. Shankar, Livio & Truran 1991). Thus, if the effects of drag energy deposition are included in evolutionary calculations, they might result in an erosion of the helium layer even for low mass WDs accreting at relatively high rates.

4.2. *Variable C/O ratio*

Assuming a nova progenitor composition of C and O in equal mass fractions is justified by the calculations of Iben & Tutukov (1985), and, indeed, almost the entire range of observed abundances is reproduced by this single WD composition. Nevertheless, it is interesting to consider the two extreme (academic) cases of pure C and pure O WDs. First, the WD models of Iben & Tutukov have a thin outer layer where $X_C/X_O > 1$. Before this relatively thin layer is eroded by repeated nova outbursts, the material dredged up from the WD core will be richer in C than in O. Secondly, a question that still remains to be answered concerns the very high O/N in the ejecta that are sometimes observed, which may be solved by dredged up material having an extremely high O concentration. Furthermore, we should ask whether nova progenitors may be inferred to be half C half O WDs from the agreement between CO models and observations. A small sample of multicycle evolutionary calculations for pure C and pure O progenitors was performed by Kovetz & Prialnik (1996). The general trend is for the accreted (and ejected) masses, as well as peak temperatures attained at outburst, to increase with increasing oxygen content. *Hot* oxygen-rich WD progenitors are bound to lead to substantially more massive ejected shells than WDs poorer in oxygen (regardless of WD mass or accretion rate). At the same time, the heavy element content of the ejecta would be lower. Surprisingly, perhaps, the O/(C+N) ratio obtained for the ejecta does not necessarily increase with increasing oxygen content in the WD progenitor. When the material dredged up from the core is devoid of carbon, triggering of the CNO cycle hydrogen burning occurs at higher temperatures. At very high temperatures, however, oxygen is destroyed. This is the reason for the low oxygen content of the ejecta of massive WD progenitors (see also Starrfield, Sparks & Truran 1985, Kovetz & Prialnik 1994). In order to produce exceptionally high O/N [or O/(C+N)] in the ejecta, a low-mass, slowly accreting, oxygen-rich WD progenitor is required. In conclusion, a great deal of caution must be exercised when guessing the WD composition from observed abundances of nova ejecta.

4.3. *The Ne problem*

According to their neon content, observed novae may be roughly divided into two categories: those with a modest enrichment in neon ($\sim 1\%$), in spite of a considerable enrichment in CNO nuclei, and those which appear to show a very high neon enhancement (0.1 or more, by mass), even when allowance is made for the uncertainties in abundance determinations (see seventh column of Table 1). While the former have C-O WD progenitors, the latter are commonly presumed to arise from O-Ne-Mg WDs.

4.3.1. *Ne content of CO WDs*

In spite of the high temperatures attained at outburst, the only abundant species in the ejecta of CO nova models are the isotopes of C, N and O. Neon production is entirely insignificant. Hence, it appears that the only way to obtain any Ne in the ejecta would be to have it present in the WD composition. Indeed, Iben and Tutukov find that CO WDs formed in close binary systems contain a few percent of Ne beside the major C and O constituents, precisely the amount indicated by observations.

4.3.2. *O-Ne-Mg WD progenitors*

Nova evolutionary calculations on O-Ne-Mg WDs, assuming *artificial* mixing between core and accreted material show that Ne enriched ejecta can be obtained (Politano *et al.*1995, Coc *et al.*1995). Including a mixing mechanism — such as diffusion-convection — should not change the general conclusions of these studies. However, there remain several problems that still require solutions.

1. As Ritossa *et al.*(1996) point out, a O-Ne-Mg WD has a 0.01 $M_\odot$outer layer composed primarily of O an C, with ^{25}Mg as the dominant trace element, but essentially no Ne.

Neon will thus appear in nova ejecta only after about 1000 outbursts, when this layer will have been eroded.
2. Observations show that even when neon is very abundant, heavier elements (Na-Fe) are not, although such elements are present in relatively large proportions in the WD core and, moreover, their abundance is found to be further increased at outburst.
3. Neon is found to be destroyed (to a larger or lesser extent) at outburst (e.g. , Politano *et al.*1995), hence it must be more abundant in the core of the WD progenitor than it is found in nova ejecta.

Thus, although we have made considerable progress in understanding the puzzling composition of nova ejecta, we still have some way to go.

This research was supported by the Israel Science Foundation administered by the Israel Academy of Sciences and Humanities.

REFERENCES

Anders, E. & Grevesse, N.1989,Geochim. Cosmochim. Acta,53,197

Audouze, J. & Tinsley, B. M.1976,Ann. Rev. Astr. Ap.,14,43

Coc, A., Mochkovitch, R., Oberto, Y., Thibaud, J. P. & Vangioni-Flam, E.1995, A&A, 299, 479

Fujimoto, M. Y. & Iben, I. Jr.1992, ApJ, 399, 646

Iben, I. Jr., Fujimoto, M. Y. & MacDonald, J.1991, ApJL, 375, L27

Iben, I. Jr., Fujimoto, M. Y. & MacDonald, J.1992, ApJ, 388, 521

Iben, I. Jr. & Tutukov, A. V.1985,ApJS,58,661

Kippenhahn, R. & Thomas, H.-C.1978,A&A,63,265

Kovetz, A. & Prialnik, D.1985, ApJ, 291, 812

Kovetz, A. & Prialnik, D.1994, ApJ, 424, 319

Kovetz, A. & Prialnik, D.1996, ApJ, in press

Livio, M. & Truran, J. W.1994, ApJ, 425, 797

MacDonald, J.1983, ApJ, 267, 732

Peimbert, M. & Sarmiento, A.1984, Astr. Express, 1, 94

Politano, M., Starrfield, S., Truran, J. W., Weiss, A. & Sparks, W. M.1995, ApJ, 448, 807

Prialnik, D.1990, in Physics of Classical Novae, eds. A. Cassatella & R. Viotti (Berlin: Springer) p.351

Prialnik, D. & Kovetz, A.1984, ApJ, 281, 367

Prialnik, D. & Kovetz, A.1995,ApJ, 445, 789

Prialnik, D. & Livio, M.1995,Publ. Astron. Soc. Pac., 107, 1201

Prialnik, D., Shara, M. M. & Shaviv, G.1979, A& A, 72, 192

Ritter, H., Politano, M., Livio, M. & Webbink, R. F.1991, ApJ, 376, 177

Ritossa, C., Garcìa-Berro, E. & Iben, I. Jr.1996, ApJ, 460, 489

Sarmiento, A. & Peimbert, M.1985,Rev. Mexicana Astron. Astrof.,11,73

Shankar, A., Livio, M. & Truran, J. W.1991, ApJ, 374, 623

Sneden, C. & Lambert, D. L.1975,MNRAS,170,533

Sparks, W. M.1995, in Cataclysmic Variables, eds. A. Bianchini, M. Della Valle & M. Orio (Dordrecht: Kluwer) p.211

Starrfield, S., Sparks, W. M. & Truran, J. W.1985,ApJ,291,136

Truran, J. W.1982, in Essays in Nuclear Astrophysics, eds. C. A. Barnes, D. D. Clayton & D. N. Schramm (Cambridge: Cambridge Univ. Press), p.467

Warner, B.1995, Cataclysmic Variable Stars, (Cambridge: Cambridge Univ. Press) p.278

Williams, R. E.1982,ApJ,261,L77

White Dwarf Heating in Response to Radial and Tangential Accretion in Close Binaries†

By EDWARD M. SION

Department of Astronomy and Astrophysics, Villanova University, Villanova, PA 19085, USA; emsion@ucis.vill.edu

The comprehensive study of the long term evolution of accreting white dwarfs in close binaries by Iben (1982, ApJ, 259, 244) helped to elucidate the physics of recurrent shell flashes, provided the first numerical estimates of compressional heating, and of the long term effect of accretion on the thermal evolution of the underlying degenerate core. Along with other long term white dwarf accretion studies, a theoretical framework has emerged for understanding radial and tangential accretion and for interpreting the quiescent surface temperatures/luminosities of white dwarfs in dwarf novae, symbiotic variables, symbiotic novae and supersoft sources. Far ultraviolet Hubble Space Telescope studies of exposed accreting hot degenerates in dwarf novae during quiescence provide fundamental empirical tests of this framework by yielding rotation rates, chemical abundances of accreted atmospheres, the thermal response of the accreting white dwarf to heating by irradiation, compressional heating and shear mixing, and the first evidence of sustained accretion belts on the white dwarf accreter.

1. Introduction

The underlying accreting degenerate dwarfs in dwarf novae, classical novae, recurrent novae, symbiotic novae and symbiotic variables are the central engines of their observed outbursts, either as potential wells for the release of gravitational potential energy due to accretion or as the sites of explosive thermonuclear shell burning. However, direct spectroscopic knowledge of their physical properties and how accretion and explosive burning affect those properties over time, has remained elusive. The first attempt to derive their temperatures in a systematic way was by Smak (1985) who used emission line intensities assuming an inverse square photoionization flux from the white dwarf/boundary layer. At the same time, the first important individual photospheric analyses based upon *IUE* and optical spectra were presented by numerous authors (*e.g.*, Panek & Holm 1985, Mateo & Szkody 1985; Shafter *et al.* 1985) who determined temperatures of individual objects based upon photospheric analyses. Based upon 12 objects with photospheric analysis, it became clear that the median and mean effective temperatures of these objects was hotter than the average field DA white dwarf (Sion 1991) but the details of the physical interpretation were obscure.

The first rigorous, quantitative exploration of the detailed physics of accreting white dwarfs was the quasi-static evolution study by Iben (1982) which provided a theoretical framework with which to compare their observed quiescent luminosities/effective temperatures. These sequences also provided the first published detailed quantitative evolutionary results elucidating the underlying physics of shell flashes. Conveniently, the range of accretion rates explored by Iben (1982) spanned the observed range now thought to be associated with accreting white dwarfs in supersoft sources, symbiotic

† Based on observations with the NASA/ESA Hubble Space Telescope, obtained at the Space Telescope Science Institute, which is operated by the Association of Universities for Research in Astronomy, Inc., under NASA contract NAS5-26555

variables, symbiotic novae, and cataclysmic variables (*e.g.*, dwarf novae, classical novae, and nova-like variables.

2. Long Term Radial Accretion Studies

The white dwarfs in symbiotic and supersoft binaries should be the hottest in view of their presumably higher accretion rates. However, until the early 1990s, only long term (more than one shell flash) sequences of accreting high mass ($M_{wd} \geq 0.8\,M_\odot$) white dwarfs had been studied (Paczynski & Zytkow 1978, Sion *et al.*1979, Iben 1982, Prialnik 1986, Livio, Prialnik & Regev 1988; Sion & Starrfield 1986 and references therein) or evolution was followed using much colder initial models with lower accretion rates (Shara, Prialnik & Kovetz 1993) than one likely encounters among symbiotic variables. The results of evolutionary quasi-static accretion studies at high white dwarf mass ($0.8\,M_\odot$, $1.0\,M_\odot$, $1.2\,M_\odot$, $1.3\,M_\odot$, $1.4\,M_\odot$) have revealed that for accretion rates $\dot{M} > 2.7 \times 10^{-7}\,M_\odot$, expansion to a red giant structure occurs, for accretion rates in the narrow range 2.7×10^{-7} to $1 \times 10^{-7}\,M_\odot/\mathrm{yr}$, stable hydrogen burning at the same rate as the rate of accretion, occurs, while for $\dot{M} < 1 \times 10^{-7}\,M_\odot/\mathrm{yr}$, repetitive hydrogen shell flashes occur. Subsequent studies (*e.g.*, Prialnik & Kovetz 1995) have tentatively confirmed this behavior for white dwarfs of much lower mass. The large grid of hydrodynamic evolutionary models (followed through many shell flash cycles) by Prialnik and Kovetz (1995) extended down to $0.65M_\odot$. Key questions, which can only be definitively answered with similar long term evolution sequences, remain: Do low mass hot white dwarfs have properties or responses to accretion which could be bear directly on such questions as the nature of the observed short-time scale variability of symbiotics, the evolution of the growing underlying helium ash layer, their luminosities during the plateau phase of a shell flash, and their heat conductivity, thermal timescales and response to accretion over the lifetime of a symbiotic, relative to more massive white dwarfs of the same core temperature?

Regarding the helium ash layer, low mass, very hot white dwarfs accreting hydrogen-rich gas at rates typical of symbiotic variables are susceptible to non-degenerate *helium shell flashes.* In these low mass models compressional heating due to accretion is very important and the hydrogen shell flashes have a profound effect on the deep envelope thermal structure and cannot be ignored in studying the evolutionary behavior of the underlying helium ash layer. Convection above the helium burning shell, driven by the very steep temperature gradients, should mix triple α-processed material into the H-rich layer. Some of the hot white dwarfs in symbiotic variables should therefore exhibit the products of helium burning at the stellar surface. Accordingly, the outflow from the hot component in symbiotic variables containing *low mass hot* degenerates which have suffered He shell flashes, should be spectroscopically similar to the WC subluminous Wolf-Rayet-like central stars of planetary nebulae (Sion & Ready 1992; see also Shara, Prialnik & Kovetz 1993).

As shown recently by Iben & Tutukov (1996), whether the He burns in an accreting white dwarf, will depend on the thermal history of the accreting star, the mass of the white dwarf, M_{wd}, and the accretion rate, $\dot{M}$. It appears from the published studies to date which included the evolution of the helium layer, that the range of helium layer masses which may undergo helium burning is approximately 1×10^{-4} to $1 \times 10^{-1}\,M_\odot$ (Iben & Tutukov 1996).

Both Sion & Ready (1992) and Shara, Prialnik & Kovetz (1993) found sub-outbursts during and just after the long outburst phase of a low mass accreting white dwarf. These small outbursts have timescales of years to tens of years and may be analogous to the mini-pulses (Becker 1996) seen in AGB stars. During such mini-pulses, the luminosity of the accreting white dwarf may vary by an order of magnitude or more thus possibly accounting for some brightness fluctuations in symbiotic behavior (cf., Yungelson, Livio, Tutukov & Kenyon 1995).

It is clear that further long term quasi-static and hydrodynamic evolutionary model sequences of accreting white dwarfs are badly needed. This need is especially acute in the parameter space regime of low white dwarf mass ($M_{wd} < 0.8\,M_{\odot}$), "hot" cores ($T_c > 1 \times 10^7$ K) as well as "cold" cores ($T_c < 10^7$ K), accretion onto C-O cores as well as He cores, with OPAL or TOPBASE opacities and the latest neutrino rates. Studies of model structures resulting from a variety of thermal histories would also be illuminating. Since few previous long term evolutionary model sequences appropriate to symbiotics and supersoft sources have been constructed with element diffusion, kinetic energy of infall of the accreted matter, time variable accretion, mass loss and the burning of the accumulated helium ash layer, there is a clear need for the inclusion of all these effects. It is also essential that authors publish as much model detail as possible to allow other investigators to know exactly what they did and to compare with their results and account for any differences.

2.1. *The Surface Luminosity of a Radially-Accreting White Dwarf*

Iben (1982) showed that the long term, time averaged surface luminosity of a radially-accreting white dwarf should be

$$(L_s) = (L_{g,acc}) + L_{cool} + L_{nuc} + L_{rad} - (L_\nu) \tag{2.1}$$

where L_{nuc} is the luminosity from nuclear burning, L_{cool} is the thermal cooling lumnosity, L_{rad} is the luminosity resulting from the instantaneous re-radiating of the kinetic energy of infall, L_ν is the neutrino cooling luminosity and $L_{g,acc}$ is the long term net release of gravitational energy in response to compressional heating and structure changes due to accretion regardless of the details of the accretion process or how the matter is processed through nuclear burning. This quantity is given approximately by

$$(L_{g,acc}) \simeq ([GM_{wd}\dot{M}/R_{zt}] - [GM_{wd}\dot{M}/R_{wd}]) \tag{2.2}$$

$$\simeq 0.17 G\dot{M}M_{wd}/R_{wd} \tag{2.3}$$

where $\dot{M}$ is the accretion rate, M_{wd} is the white dwarf mass, R_{wd} is the white dwarf radius,R_{zt} the zero temperature (Hamada-Saltpeter) radius, and the factor 0.17 is the long term, time-averaged fraction of the total accretion (gravitational) energy liberated in response to accretion, *i.e.*, compressional heating/structure changes due to the added matter (Iben 1982). This fraction is found to range between 0.17 and 0.25 and could only be derived from actual long term quasi-static evolutionary sequences with radial accretion. If the long term tangential accretion via a disk can be approximated as radial accretion, then the above equation should provide the surface luminosity due to compressional heating and structure changes in response to accretion.

In recent evolutionary sequences with time variable radial accretion followed through many dwarf nova outburst cycles (Sion 1995), it is expected that the white dwarf, if followed long enough, will heat up until an equilibrium state is established between heating and cooling. While such a state is clearly approached in the model sequences reported, some sequences are closer to the equilibrium state than others, due to their longer total evolution times. It is important and useful to estimate analytically, what the expected long term equilibrium end state should be after many dwarf nova cycles.

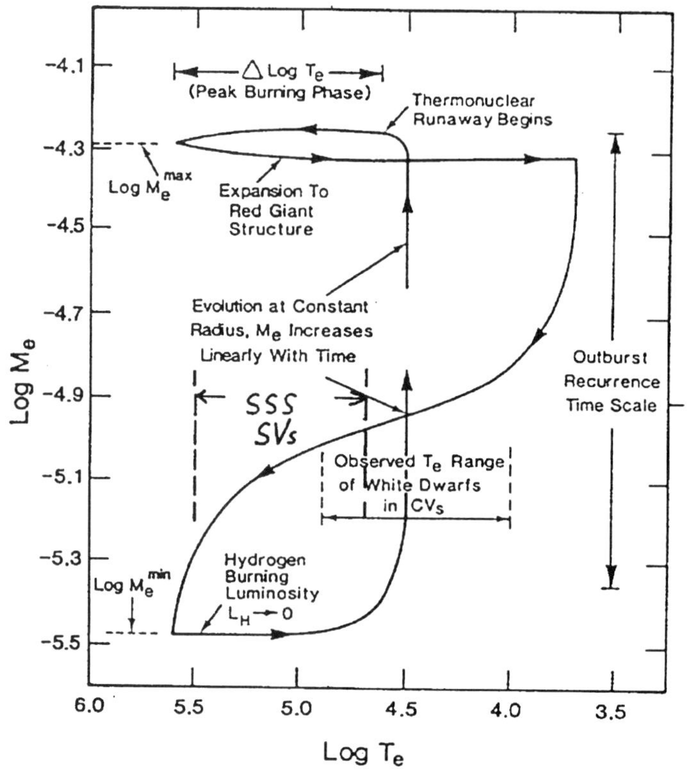

FIGURE 1. Mass M_e in the hydrogen-rich envelope as a function of $T_{\rm eff}$ scaled from Iben (1982) to roughly correspond to a $1\,M_\odot$ white dwarf accreting at a rate $10^{-10} M_\odot$/yr. The schematic evolution from the point where expansion t o a red giant structure occurs down to where the H-envelope mass is reduced to M_e^{min} (by burning and mass loss) is shown. The approximate observed $T_{\rm eff}$ ranges for exposed white dwarfs in CVs and for white dwarfs in supersoft sources (labelled SSS) and symbiotic variables (labelled SVs) are indicated with verical dashed lines. Note that SSS and SVs require evolutionary tracks between flashes, at much higher $T_{\rm eff}$ and, hence, higher time-averaged accretion rates (adapted from Sion 1985).

Balancing on average the rate of energy deposition to the rate at which the energy is radiated away, the equilibrium temperature should be approximately

$$T_{\rm eff} = [\frac{(0.17GM_{wd}\dot{M})}{4\pi R^3 \sigma}]^{0.25} \tag{2.4}$$

Inasmuch as accreting white dwarfs in other close binaries might be expected to yield a similar equilibrium $T_{\rm eff}$ in response to long term average accretion, it is possible to indicate in the envelope mass, Log M_e, versus Log $T_{\rm eff}$ plane the approximate expected locations of the accreting white dwarfs in the various types of accreting systems as shown in Figure 1. (adapted from Sion 1985).

2.2. *The Physics of Compressional Heating: A Comparison with Other Sources of Heat Input*

In order to gain further insight into the physics of compressional heating, it is useful to examine its basic physics and compare its magnitude with other potentially important possible sources of heat input into the white dwarf, arising from time-averaged accretion and dwarf nova outbursts. There are two other sources of heat input with which to compare: (1) accretion luminosity shining inward (associated with the kinetic energy of infall) and; (2) tangential accretion onto an equatorial belt with shear mixing. Characteristic numerical estimates are available based upon recent simulations (e.g., Sparks *et al.* 1993; Pringle 1988) and will be compared with compressional heating.

The evolutionary sequences of white dwarf models undergoing time-variable accretion were generated with a Henyey-type stellar evolution code with an accretion algorithm which treats spherical (radial) accretion. The accreting material is assumed to rain down onto the stellar surface with essentially zero velocity (*i.e.*, it is assumed to "land softly"). This assumption of course completely excludes the effect of the kinetic energy of infall of the accreting matter. In other words, the spherical accretion treatment implicitly assumes that all of the kinetic energy gained by the accreted matter in flowing from the donor star has been dissipated before this matter settles onto the surface of the star (at the same temperature and density as of the matter already at the white dwarf surface).

What happens physically when the new material is added to the star? The white dwarf undergoes pressure changes from model to model at any given M_r which is caused by readjustment of hydrostatic equilibrium. The white dwarf responds with (1) radius changes in the outer layers and (2) compensation for the weight of the newly added matter. However since r changes only very slightly for the largest part of envelope matter in mass fraction space M_r/M, most of the relative pressure change is due to compensation for the additional weight of the accreted gas. The change in total radius is also very small since the amount of mass added during a given dwarf nova outburst is also very small.

Energy conservation for a given mass element may be written:

$$dL_r/dM_r = \epsilon_g \tag{2.5}$$

where:

$$\epsilon_g = -du/dt - PdV/dt \tag{2.6}$$

$$= -du/dt + (P/\rho^2)d\rho/dt \tag{2.7}$$

$$= -C_p dT/dt + (\delta/\rho)dP/dt \tag{2.8}$$

and:

$$\delta = -(d\ln\rho/d\ln T)_p \tag{2.9}$$

Since the compression of the added material is essentially adiabatic (cf., Giannone & Weigert 1967):

$$dU/dt \approx p/\rho^2(d\rho/dt). \tag{2.10}$$

Since $dP/dt > 0$, it follows that $dT/dt > 0$, and the work done against the pressure is absorbed as internal energy (Giannone & Weigert 1967).

When the mass accretion first begins the temperature rises rapidly in the outer layers where the matter of lowest density is appreciably adiabatically compressed by the weight of the newly added mass, while the relative temperature increase becomes smaller with decreasing M_r/M toward the deeper envelope regions. Since the slope of $T(M_r)$ decreases rapidly outward in the non-degenerate layers above the core, $\dot{T}/T$ (a temperature "velocity"), which increases outward, produces a temperature maximum very close to the surface initially. At $M_r/M < 0.9$ the temperature remains essentially constant. In these deeper layers where electron degeneracy dominates, the large electron conductivity rapidly distributes the energy gain by compression over the entire interior thus smoothing out any temperature differences in the deep interior.

Inclusion of the kinetic energy of infall of the accreting matter typically assumes that the radiated energy is equal to one-half of the gravitational potential energy of the accreted mass at the white dwarf surface; namely:

$$L_{acc} \approx 0.5 G M_{wd} \Delta M / R_{wd} \tag{2.11}$$

This accretion luminosity, the accreted rotational kinetic energy rate, shines inward heating the white dwarf to a maximum temperature, T_{max}, in the outburst, down to a depth in the white dwarf atmosphere roughly equal to the depth in the original atmosphere where $T(M_r) = T_{max}$ (cf. Pringle 1988). Hence the higher the surface temperature to which the white dwarf is elevated at the peak of the outburst, the deeper is the radiative heat diffusion and thus the longer the thermal cooling timescale at the cessation of the accretion outburst episode.

The two heat sources are therefore quite distinct; the kinetic energy of infall governs (dominates) the surface luminosity of a white dwarf in a dwarf nova via the necessity of the star to rid itself of a large heat source shining inward (*e.g.*, at $\dot{M} = 1 \times 10^{-9} M_\odot/\mathrm{yr}$, $\dot{E}_k >> L_{wd}$) and is deposited in the the outermost surface layers. Compressional heating, on the other hand, affects the entire stellar structure (to varying degrees, as a function of mass fraction M_r/M). If the kinetic energy of infall were included, an accretion luminosity of 16 $L_\odot$ (corresponding to an accretion rate of $1 \times 10^{-8} M_\odot/\mathrm{yr}$ onto a 0.6 $M_\odot$ white dwarf) would lead to a very high white dwarf surface temperature ($T_{\mathrm{eff}} > 1 \times 10^5$ K) compared to the undisturbed (initial) $T_{\mathrm{eff}} = 27{,}000$ K. Compressional heating, if operative over a long timescale, (and depending upon the accretion rate, white dwarf mass and outburst/quiescence timescale), could become comparable to or exceed the heating effect associated with accretion luminosity shining inward.

If shear mixing of the accreted matter into the white dwarf envelope is treated, then the conversion of rotational kinetic energy into heat provides a luminosity due to shear which emerges as surface luminosity (Sparks, Sion, Starrfield & Austin 1993). For a $0.93 M_\odot$ white dwarf accreting onto an equatorial belt of width $\pm 30°$ at a rate 2×10^{-8} for seven days. The maximum temperature of the white dwarf was 10^5 K. By comparison, the same model with radial (spherical) accretion and accretion luminosity shining inward at the surface, reached a peak temperature of 64,000 K. The higher temperature associated with shear mixing results from the fact that part of the energy continues as kinetic energy into the envelope before it is given up as an extra luminosity source due to shear. The accretion luminosity shining inward by kinetic energy of infall, on the other hand, is partly radiated away as high energy photons right at the surface.

Part of the rotational kinetic energy goes into spinning up the white dwarf. Indeed, Popham and Narayan (1995) have shown that the boundary layer luminosity must also include the fraction of the rotational kinetic energy that goes into spinning up the white dwarf. Hence for a rotational kinetic energy $1/2 \dot{M} \Omega_k (R_{wd})^2 \times R_{wd}^2$ and accretion-imparted angular momentum $\dot{M} \times \Omega_{wd} \times R_{wd}^2$, Popham and Narayan (1995) have shown that the boundary layer luminosity should be:

$$L_{bl} = (G M_{wd} \dot{M} / R_{wd}) \times [1/2 - j \Omega_{wd} / \Omega_k(R_{wd}) + 1/2(\Omega_{wd}^2 / (\Omega_k(R_{wd})^2))] \tag{2.12}$$

where j is the angular momentum accretion rate in units of the Keplarian angular momentum rate at the stellar surface ($j = \dot{J} / (\dot{M} \Omega_k (R_{wd}) R_{wd}^2)$ and Ω_{wd} is the angular velocity of the white dwarf. For a white dwarf not rotating near breakup, $j \simeq 1$. The predicted boundary layer luminosity for a 0.6 $M_\odot$ white dwarf accreting at the rate $10^{-10} M_\odot/\mathrm{yr}$ and rotating at 600 km/s, corresponding to VW Hydri's white dwarf in quiescence, is $2 \times 10^{+32}$ ergs/s. This value is close to the actual surface luminosity of VW Hyi's 22,000 K white dwarf.

3. Observed Properties and Heating of White Dwarfs in CVs Using HST

Far UV flux declines have been observed in several dwarf novae during quiescence. Any UV flux decline, if due to a decrease in flux from the accretion disk during dwarf nova quiescence, would contradict the standard disk instability theory which predicts the disk should brighten through the quiescent interval because the accretion rate is gradually increasing. But these flux declines could be accounted for if they are due to cooling of the white dwarf in response to the accretion event. HST observations of three systems have provided conclusive evidence of white dwarf cooling in response to the accretion event.

In U Geminorum a number of investigations demonstrated that the white dwarf in U Gem dominates the system light during quiescence in the far UV (Panek & Holm 1984; Kiplinger *et al.* 1989; Long *et al.* 1994). The work by Kiplinger *et al.* 1989 presented inconclusive evidence of white dwarf cooling using *IUE* . However, Long, Sion, Szkody and Huang (1994) observed U Gem in quiescence with HST/FOS at 13 days and 70 days after a normal dwarf nova outburst. They found a flux decline of 28% between the two observations and, that the white dwarf had cooled from 39,400 K down to 32,100 K, between the two observations.

For the eclipsing system OY Car, Cheng *et al.* (1994) compared the white dwarf FOS spectra in early quiescence to the the spectra at later times in quiescence. They determined that the temperature of the white dwarf just after the outburst was 19,500 K but had cooled to 17,400 K about 3 months after outburst. They found a cooling of 8000 K in a two week interval.

HST high speed UV photometry of the eclipsing system Z Cha by Wood *et al.* (1993) from outburst far into quiescence revealed no evidence of an extended boundary layer even after outburst. The temperature of the white dwarf was 17,600 K at the decline of outburst and cooled to its normal quiescent temperature of 15,600 K 16 days later.

In WZ Sagittae, the dwarf nova with the longest interoutburst time the temperature of the white dwarf just after the December, 1980 outburst was 30,000 K. Since then the white dwarf surface temperature has been declining and had cooled to 12,500 K in 1989 (LaDous 1990) based upon *IUE* measurements. More recently the white dwarf appears to be somewhat hotter (see below). The cooling measurements of the white dwarf in VW Hydri are discussed in the next section.

4. Tangential Accretion and the First Evidence of Sustained Accretion Belts

Accretion belts should be a natural consequence of tangential accretion onto a white dwarf (Kippenhahn & Thomas 1978) but have never been identified spectroscopically. Among dwarf novae with exposed white dwarfs, the SU UMa-type dwarf nova VW Hydri is nearly ideal for a search because it undergoes both normal outbursts and superoutbursts, as well as clearly revealing its white dwarf during quiescence (Mateo & Szkody 1985; Sion *et al.* 1995a,b), as the dominant far UV flux source. HST spectroscopic determinations of the white dwarf $T_{\rm eff}$, log g, photospheric chemical abundances and axial rotation rate have been reported (Sion *et al.*1995a,b) but the signature and contribution of UV emitting gas in Keplarian motion near or on the white dwarf during quiescence has not yet been characterized. Sion *et al.* (1996) reported an analysis of a second FOS spectrum obtained only 1 day after a normal outburst, comparatively analyzed with an earlier post-superoutburst (hereafter post-SOB) FOS spectrum.

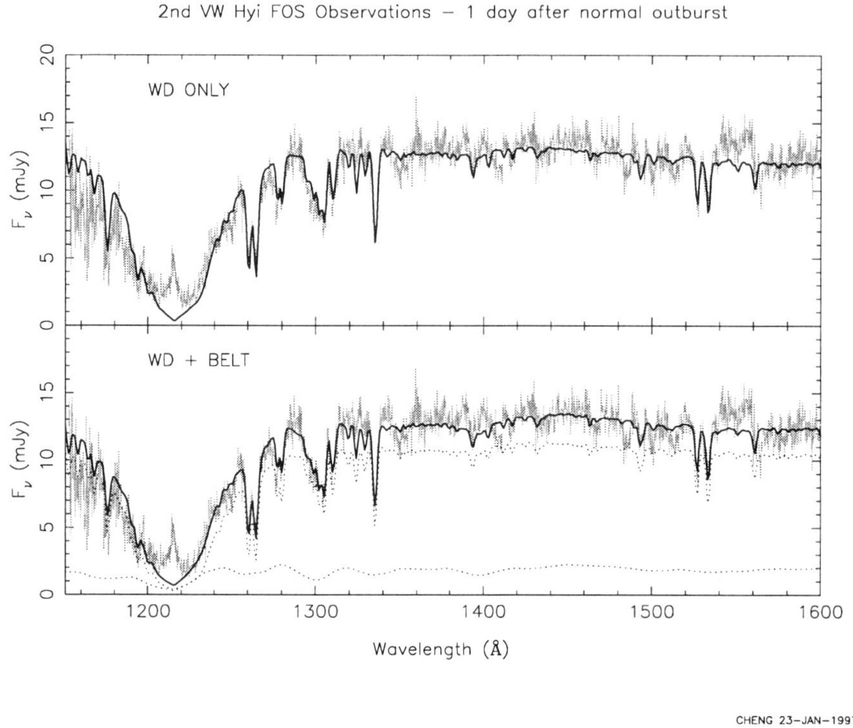

FIGURE 2. (a): The best fit synthetic spectrum of a white dwarf alone, to the FOS spectrum one day post-outburst. ($T_{\rm eff}$ = 23,500 K, log g = 8, $V \sin i$ = 300 km/s, see text for details). (b): The best fit two-temperature component model (white dwarf plus accretion belt) to the new FOS spectrum. The white dwarf has $T_{\rm eff}$ = 22,500, log g = 8.0, V sin i = 300 km/s and contributes 89% of the flux, plus an accretion belt with V = 3350 km/s, log g = 6, emitting area and flux contribution 11% and 16% respectively (see text for details).

Hubble FOS observations of VW Hydri one day after the end of a normal dwarf nova outburst reveal a heated white dwarf with deep broad Lymanα, narrow metallic absorption features and evidence of a hotter Keplerian-broadened component manifested in quiescence as a broad continuum hump (Fig. 2). Any successful composite model in quiescence must consist of a far UV flux-dominant, white dwarf, (replicating the broad Lyman α wings and core, with the high gravity and moderately slow rotation preserving the narrow photospheric metals) while simultaneously providing the curved continuum signature of a Keplerian broadening component. The model atmosphere (TLUSTY; Hubeny 1988), model disk (DISKSYN; Hubeny 1995) and spectrum synthesis (SYNSPEC; Hubeny, Lanz & Jeffrey 1994) codes and details of our χ^2 minimization fitting procedures can be found in Sion *et al.* (1995a,b) and Huang *et al.* (1996a,b) and for the sake of brevity will not be repeated here.

We define an accretion belt to mean a rapidly spinning, narrow belt on the white dwarf surface itself, centered at the white dwarf equator. A given belt is constructed by taking appropriate values of log g and $T_{\rm eff}$ for a stellar atmosphere, rotationally broadening the synthetic spectrum to high velocity and multiplying the fluxes by the fractional area of the belt relative to the white dwarf surface area. We found a large improvement in the χ^2 fits when including such a belt in combination with the white dwarf synthetic fluxes. The parameters of our best reduced χ^2 fits to both FOS spectra using white dwarf plus accretion belt fits are given in table 1. The accretion belt temperatures span a range

	Fos1 (10 days Post-SOB)	Fos2 (1 Day Post-NOB)
C	0.2	0.5
N	3.0	5.0
O	0.5	2.0
Fe	2.0	0.5
Si	0.15	0.2
other Metals	0.15	0.3

TABLE 1. White Dwarf Photospheric Abundances

from 26,000 K to 31,000 K. Hotter belts (e.g. $T_{belt} > 50,000\,$K) of smaller area are ruled out because Si IV becomes too shallow and the absorption wings vanish while species like N V will become detectable but are not present in the data.

Our best reduced χ^2 fit to the data reveals (1) a DAZQ white dwarf with $T_{\rm eff} = 22,500\,$K, log $g = 8$, and photospheric abundances C = 0.5 solar, N = 5.0 solar, O = 0.02 solar, Si = 0.2 solar, Fe = 0.5 solar with all other metals being 0.3 solar and; (2) a rapidly spinning accretion belt with $V_{\rm rot} = 3350$ km/s, $T_{\rm belt} = 26{,}000\,$K, log $g = 6.0$, with a fractional belt area of approximately 11%. Our earlier FOS spectrum obtained 10 days after superoutburst (Fig. 3) reveals a cooler DAZQ white dwarf (20,500 K), lower metal abundances and a smaller fractional area (3%) for the accretion belt. Thus one day after outburst, the white dwarf is 2000 K hotter, the accretion belt fractional area is a factor of three greater, the accretion belt temperature is cooler than at 10 days post-superoutburst and the accreted atmosphere has relatively higher metal abundances. Finally the accretion belt maintained during quiescence may provide a natural explanation for the 14s soft X-ray oscillations, requires a deeper source of heating (compression and shear mixing) and implies a lower limit to the viscous spindown timescale of 10 days.

The Hubble FOS observation was carried out on 1995, January 28 (19:33:01 UT) approximately 1 day after the return to optical quiescence following a normal outburst which began on Day 24 (24 January 1995). Details of the observations are given in Sion *et al.*(1996).

There are several important differences between this new FOS spectrum and the post-SOB spectrum analyzed in Sion *et al.*(1995a) and Huang *et al.* (1996b). In the new spectrum the continuum curvature between 1400Å and 1550Å is flatter and less pronounced while the overall flux level is a factor of 1.3 higher and the Lyman α absorption feature is measurably narrower. The latter two differences almost certainly indicate a higher overall white dwarf surface temperature (one day post-normal outburst) while the flatter continuum is consistent with a hotter white dwarf photosphere and/or a reduced contribution from gas with Keplerian broadening. From HST spectroscopy of VW Hydri in superoutburst (Huang *et al.* 1996a) and *IUE* spectra taken in outburst *e.g.*, Verbunt *et al.* 1987), it is clearly evident that pronounced continuum curvature between 1380Å and 1550Å is a spectroscopic hallmark of disk/ring/belt UV emitting gas in Keplerian motion near the white dwarf surface. This curvature is still present in spectra at quiescence but considerably less pronounced.

Our synthetic spectral fitting results for the physical state and surface chemistry of the white dwarf and belt one day after the end of a normal outburst have provided a crucial comparison with the characteristics of the white dwarf and belt 10 days after the end of a superoutburst, analyzed earlier by Sion *et al.* (1995a) and Huang *et al.* (1996b). Our results reveal: (1) the white dwarf is hotter by 2000 K one day after normal outburst than the corresponding temperature ten days post-superoutburst; (2)

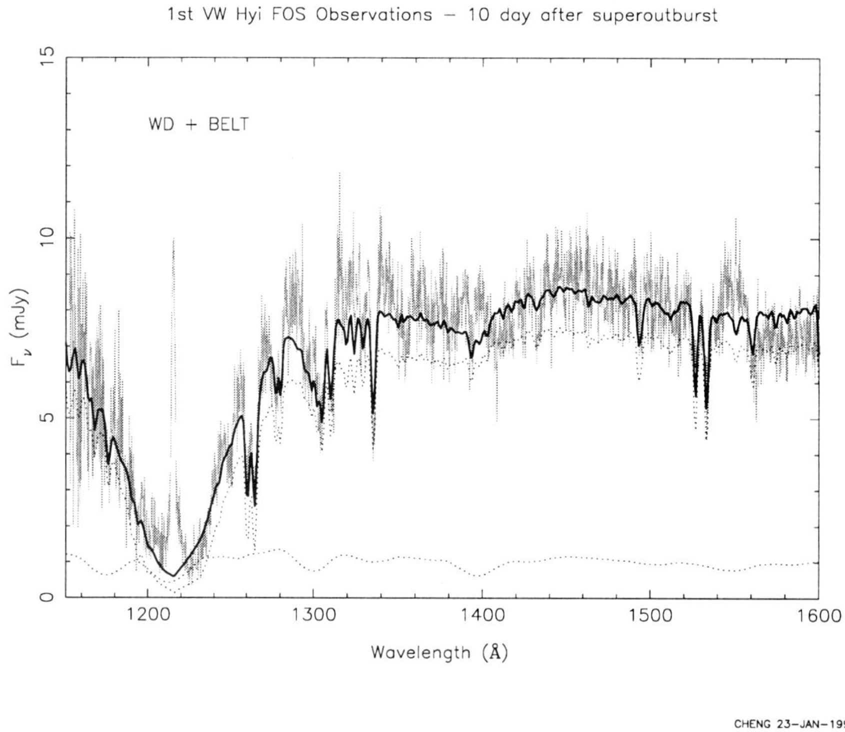

FIGURE 3. The FOS spectrum of VW Hydri obtained 10 days after superoutburst. The best fit parameters differ slightly from Sion *et al.* 1995a due to the inclusion of the H_2 quasi-molecular satellite absorption features and a rotational broadening of 300 km/s.

the chemical abundances of the white dwarf atmosphere one day post-outburst (except for Fe) appear elevated relative to the same elements in the white dwarf atmosphere 10 days post-SOB, *i.e.*, the white dwarf was more metal-enriched one day post-normal outburst than 10 days post-SOB. It is not implausible that only one day after a normal outburst, the spectroscopic signature of the the injection of freshly accreted matter is stronger than it is ten days after a superoutburst because the accreted matter has not had as much time to dilute its chemical abundance due to lateral spreading as well as downward diffusion; (3) The fractional area of the accretion belt we derive at one day post-outburst is larger than the fractional area 10 days post-superoutburst, possibly due to the more recent occurrence of the outburst accretion pulse; (4) the accretion belt temperature 1 day post-normal outburst is cooler (by 5000 K) than the accretion belt at 10 days post-SOB.

Since the white dwarf in VW Hydri sustains a rapidly spinning accretion belt during quiescence, maintaining its temperature and spin due to repeated accretion episodes every 2-3 weeks, a number of implications follow from the existence of the maintained accretion belt: (1) the presence of a belt up to 10 days after superoutburst implies a lower limit to the viscous spindown timescale of 10 days; (2) deeper heating is required than the depth of heating associated with a downward-directed boundary layer radiation field. Since the VW Hyri boundary layer luminosity is very low (indicating a low temperature or non-existent boundary layer), it is very unlikely that irradiation alone heats the belt. Instead, shear mixing or compressional heating must provide the required deep heating. Of these, shear mixing is the more efficient, i.e. same degree of heating for less mass accreted than for pure radial compression. A rough estimate of the expected heating of

	FOS1 (10d post-SOB)		FOS2 (1d post-NOB)	
	WD	BELT	WD	BELT
log g	8.0	6.0	8.0	6.0
V_{rot} (km/s)	300	3350	300	3350
Ly α satellite	yes	yes	yes	yes
T/1000 (K)	20.5	31.0	22.5	26.0
Scale factor($1 \times 10^{-2)}$)	3.402	0.114	3.453	0.423
emitting area (%)	96.8	3.20	89.1	10.9
Flux Contribution (%)	87.5	12.5	84.0	16.0
Distance (pc)	65		62	

TABLE 2. White Dwarf + Accretion Belt Fits

an equatorial surface region due to axisymmetric compression by accreting matter can be made from radial accretion calculations by assuming the amount of compressional heating scales linearly with the surface area of the accretion zone. For parameters appropriate to the superoutburst and white dwarf of VW Hydri (2 week outburst duration, 0.6 $M_{\odot}$ white dwarf, $\dot{M} = 1 \times 10^{-8} M_{\odot}/\mathrm{yr}$, $T_{\mathrm{eff}} = 20,000\,\mathrm{K}$), and an assumed belt area of 10%, the model sequences of Sion (1994; see table 2) predict a belt temperature approximately 10,000 K hotter than the quiescent photosphere, or 30,000 K. This is very close to what we derive for the belt temperature from our synthetic spectral fitting of the post-superoutburst spectrum.

Finally, on theoretical grounds an accretion belt is plausibly expected in VW Hydri. As matter accretes tangentially at the equator and does so every 2-3 weeks, an equilibrium state should be established such that the equatorial region spins rapidly while the higher latitudes and polar regions do not. The controlling parameter for this equilibrium state is the Richardson number, which implies a gradient of differential rotation going both inward into the white dwarf and poleward along the white dwarf surface.

I am deeply grateful to my collaborators on HST projects, Paula Szkody, Min Huang, Fuhua Cheng, Ivan Hubeny, Warren Sparks, and Janet Mattei for their energy, expertise and encouragement. It is a pleasure to thank Beth Jewell for assistance with preparing the manuscript. This work is supported by NASA through grants GO3836.01-91A and GO6084.01-95A from the Space Telescope Science Institute, which is operated by the Association of Universities for Research in Astronomy, Inc., under NASA Contract NAS5-26555.

REFERENCES

Becker, S. 1996, private communication

Cheng, F.H., Marsh, T.R., Horne, K., & Hubeny, I. 1994, in The Evolution of X-Ray Binaries, eds. S.S.Holt & C.S. Day, AIP Conf. Proc. 308, (New York: AIP Press), 197

Giannone & Weigert 1967, Z.Astrophys., 67, 41

Huang, M., Sion, E.M., Hubeny, I., Cheng, F.,& Szkody, P. 1996a, ApJ, 458, 355

Huang, M., Sion, E.M., Hubeny, I., Cheng, F., & Szkody, P. 1996b, AJ, in press

Hubeny, I. 1988, Comput.Phys.Comm, 52, 103

Hubeny, I. 1995, unpublished

Hubeny, I., Lanz, T., & Jeffrey, S. 1994, Newsletter on Analysis of Astronomical Spectra (St. Andrews Univ.), 20, 30

Iben, I. 1982, Ap.J., 259, 244

Iben, I. & Tutukov, A. 1996, ApJS, 105, in press

Kippenhahn, R., & Thomas, H.-C. 1978, A&A, 63, 265

Kiplinger, A., Sion, E.M., & Szkody, P. 1991, ApJ, 366, 569

La Dous, C.1990, private communication

Livio, M., Prialnik, D., & Regev, O. 1989, ApJ, 341, 299

Long, K., Blair, W.P., Bowers, C.W., Davidsen, A.F., Kriss, G.A., Sion, E.M., & Hubeny, I. 1993, ApJ, **405**, 327

Long, K., Sion, E.M., Huang, M., & Szkody P. 1993, ApJL, in press

Mateo, M. & Szkody, P. 1984, AJ, 89, 863

Paczynski, B., & Zytkow, A. 1978, Ap.J., 22, 604

Panek, R., & Holm, A. 1984, ApJ, 277, 700

Popham, B., & Narayan, R. 1995, ApJ, 442, 337

Prialnik, D. 1986, ApJ, 310, 222

Prialnik, D., & Kovetz, A. 1995, ApJ, 445, 789

Prialnik, D., & Livio, M. 1995, PASP, 107, 120

Pringle, J.E. 1988, MNRAS, 230, 587

Schwarzschild, M., & Harm, R. 1965, ApJ, 142, 855

Shafter, A. *et al.* 1985, ApJ, 290, 707

Shara, M.M., Prialnik, D., & Kovetz, A.1993, ApJ, 406, 220

Sion, E.M. 1985, ApJ, 297, 538

Sion, E.M. 1991, AJ, 102, 295

Sion, E.M. 1995, ApJ, 438, 876

Sion, E.M., Acierno, M.J., & Tomczyk, S. 1979, ApJ, 230, 832

Sion, E.M., & Ready, C. 1992, PASP, 104, 87

Sion, E.M. & Starrfield, S.G. 1986, ApJ, 303, 130

Sion, E.M., & Starrfield, S.G. 1994, ApJ, 42, 261

Sion, E.M., Szkody, P.Cheng, F., & Huang, M. 1995a ApJ, 444, L97

Sion, E.M., Huang, M., Szkody, P., & Cheng, F. 1995b ApJ, 445, L31

Sion, E.M., *et al.* 1996, ApJ, 471, L41

Smak, J. 1985, Acta Astr., 34, 161

Sparks, W., Sion, E.M., Starrfield, S., & Austin, S. 1993, in *Cataclysmic Variables and Related Physics*, eds. O. Regev and G. Shaviv, (*Annals of the Israel Physical Society*, **10**, p. 96).

Szkody, P., & Sion, E.M. 1989, in White Dwarfs, ed. G. Wegner, (Berlin: Springer-Verlag).

Verbunt, F. 1987, A&AS, **71**, 330.

Verbunt, F. Hassall, B., Pringle, J., Warner, B., Marang, F. 1987, MNRAS, 225, 113

Wood, J., Robinson, E.L., Bless, R.C., Dolan, J.F., Elliot, J.L., & VanCitters, G.W. 1993, in *Cataclysmic Variables and Related Physics*, eds. O. Regev and G. Shaviv, (*Annals of the Israel Physical Society*, **10**, p. 19).

Youngelson, L., Livio, M., Tutukov, A., & Kenyon, S. 1995, ApJ, 447, 656

An Analysis of Long-term Light Curves of Four Novalike Variables

By
ZDRAVKA KRAICHEVA[1]†, VALLERY STANISHEV[1], IGLIKA SPASSOVSKA[2] AND VASIL POPOV[1]

[1]Institute of Astronomy, Bulgarian Academy of Sciences, 72 Tsarigradsko Shosse Blvd., BG-1784 Sofia, Bulgaria

[2]Korea Astronomy Observatory, Taejon, South Korea

The long-term light curves of the novalikes TT Ari, KR Aur, AM Her and MV Lyr, were compiled and analysed for solar-like cyclical behaviour. The frequency analysis showed long-term cyclical modulations of the brightness of the stars, which can be ascribed to changes of the radii of the late type secondaries in order of $\Delta R/R \simeq 10^{-4}$-$10^{-5}$. Periods of about 250 - 400 days were also suspected for all investigated stars.

1. Introduction

The main peculiarity of the observed long-term light curves of cataclysmic binaries is the existence of continuous cyclic fluctuations of the brightness in timescales from months to years. According to current ideas these light variations are due to variability of accretion disk luminosity, produced by changes in mass transfer from the late type secondary. In the last years (see Bianchini, 1992) the presence of solar type cycles in the secondaries of cataclysmic binaries has been assumed as a possible reason for the variations in the mass transfer rate. Warner (1988) and Bianchini (1988, 1990) showed that these cycles could be detected through investigations of the changes of quiescent luminosity, dwarf nova outburst intervals and orbital periods, as well as the behaviour in the dips observed in novalikes. In the polars, the presence of "swinging dipole" mechanism (Andronov 1987) has been proposed also as a reason for cyclic variations of the mass transfer rate.

The difficulties in the investigations of long-term behaviour of CVs are connected mainly with the fragmentary nature of the observations often extended over only a few decades. Consequently, the number of light curves analysed for long-term cyclicity are rather limited. At present the best sources of data for testing the presence of solar-like cycles in cataclysmic binaries are light curves compiled from plate archives and amateur visual observations. We report here results from analysis of long-term light curves of VY Scl type novalike systems TT Ari, KR Aur, MV Lyr and of the polar AM Her.

2. Data and data analysis

On a basis of our own photographic and photoelectric observations and data from literature we compiled the long-term light curves of the stars (Fig. 1). The main characteristics of the curves are given in Table 1. We analysed the quiescent luminosity (high state) and, separately, dips luminosity including intermediate and low states using the period dispersion minimization method described by Stellingwerf (1978). For studying of high state luminosities all measurments in intermediate and low states and the values which were obvious outliers from the quiescent level have been removed. In the case of

† *E-mail:* gstar@bgearn.acad.bg

Star	Type	t_0 [JD]	Length [yr]	High state P_{cyc},[yr]	Δm,[mag]	ΔR/R	P_{short},[day]	Low state P_{dip},[yr]
TT Ari	VY	2425502	66	6	0.033	1.2×10^{-5}	400?	12.6
				5	0.034	1.2×10^{-5}		
KR Aur	VY	2411335	100	8.16	0.38	1.4×10^{-4}	250-300?	10.1
								3.4
AM Her	AM	2425507	66	3.53	0.28	10^{-4}	300?	5.38
				1.72	0.24	8.9×10^{-5}		1.7
MV Lyr	VY	2430517	50	13.5	0.34	1.2×10^{-4}	321?	
				7.27	0.40	1.4×10^{-4}	340?	

TABLE 1. Long-term cycles in novalikes

dips analysis we excluded magnitudes brighter than mean value for the high state. Wenzel and Fuhrmann (1983), Andronov et al. (1988) and Richman et al. (1994) suspected periodicity on a time scale of ~300-400 days for TT Ari and MV Lyr. From analysis of dips events in the light curves of these stars Bianchini (1990) supposed also the presence of 12.6 yr and 11-12 yr cycles in the light curves of these stars. So, we averaged the data over different time intervals using moving average. The smooting windows were equal to one day, one year and 1000 days.

3. Results and discussion

TT Arietis. The historical light curve of TT Ari is based on 855 mean nightly B magnitudes obtained during 1928-1994. We didn't embrase the time interval of 1979-1985 low state in our study. Analysis of all three average light curves gave 6 yr, as well as 5 yr periodicities in high state luminosity (Fig.1). The sinusoidal fits with these periods and peak to peak amplitude ~0.033 mag equaly well describes the 1000 days mean light curve (Fig.2). Consequently the available data didn't enable us to prefer anyone of them. Figure 2 shows an overall slow decreasing of brightness and superimposed cyclical behaviour. Peak corresponding to 392^d period is seen in θ-statistics of night mean light curve. The dips' analysis confirmed 12.6 yr cycle found by Bianchini (1990). Deviations from the mean value are up to ±3 yr.

KR Aurigae. For investigation of long-term behaviour of KR Aur we used 1295 mean nightly B magnitudes obtained from 1890 to 1994. The light curve (Fig.1) shows that during the first 43 years of this interval the star stayed in high state. Dips' behaviour of KR Aur was begun after year 1933. We analysed these two intervals separately. In the high state luminosity KR Aur showed 8.16 yr cycle (Fig.1) as well as a clearly seen cycle with duration 250-300 days. The high state magnitudes averaged over 1000 days and sinusoidal fit with 8.16 yr period and amplitude 0.38 mag are shown in Fig.2. Frequency analysis of annual mean curve in the second interval gave a mean cycle from 3.4 yr for the dips and 10.1 yr for the dips at low state only. Deviations from these mean values reach 0.85 yr and 2 yr respectively.

AM Herculis. The complete long-term light curve in B color was constructed for the time interval 1928-1994. The data consist of 1117 individual points each representing a 1 night mean. The long-term behaviour of AM Her during 66 years is characterised with a 30 years interval of predominant active state since year 1951 and two intervals, before and after it, of predominant intermediate or low state. All three intervals were analysed separately. The investigation of night mean luminosity in high state revealed variation with periodicities 3.53 yr, 1.72 yr and 300 days. The θ-statistics of the data averaged

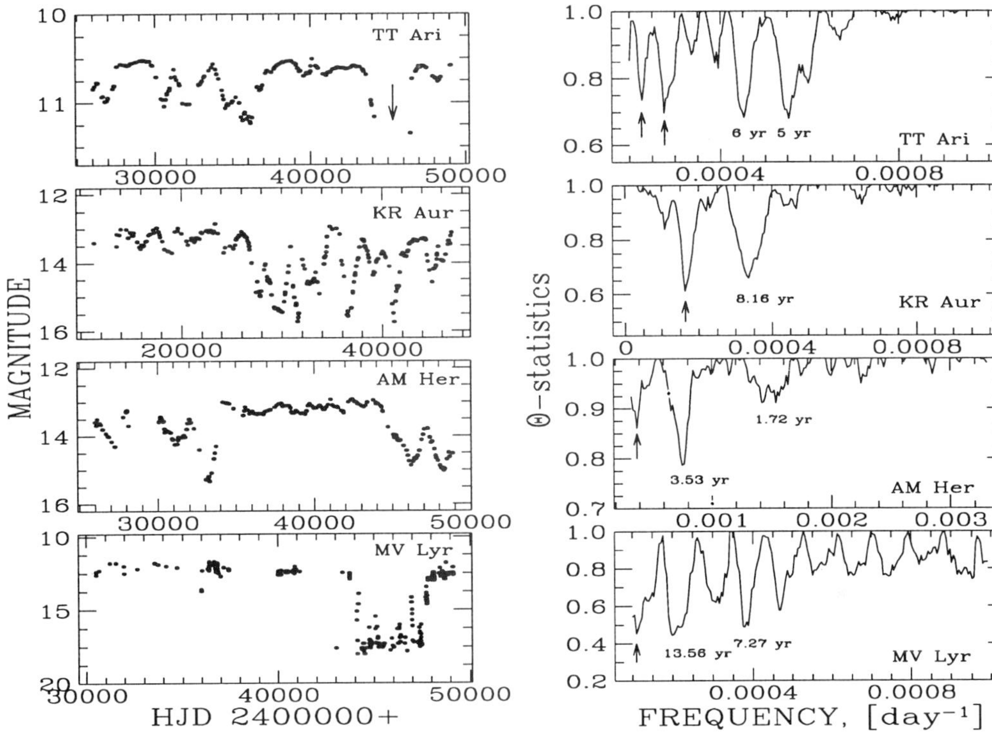

FIGURE 1. *Left panel*: The long-term light curves of TT Ari, KR Aur and AM Her averaged over 1000 days and MV Lyr averaged over a day. The arrow marks TT Ari low state; *Right panel*: Frequency analysis of the data in high state averaged over 1000 days for TT Ari and KR Aur, over a year - AM Her and over a day - MV Lyr. Subharmonics are marked with arrows.

over one year is seen in Fig.1. The best fit to the annual mean magnitudes was obtained using two sinusoids with periods P_1=3.53 yr and P_2=1.72 yr and amplitudes 0.28 mag and 0.24 mag respectively (Fig.2). Frequency analysis of two other parts of the curve shows strong peak corresponding to period of 5.38 yr. In the Figure 2 we presented annual mean B magnitudes and sinusoidal fit with this period. It is seen that star's brightness after 1981 is modulated with shorter, ~1.7 yr, period. The same modulations could be seen in the part of the curve before 1951 which is enough well completed. We must note that these modulations were found in AM Her high state (Fig.1), but the peak corresponding to them is very broad, and we couldn't exactly determine its location.

MV Lyrae. We did not succeed to complete well enough the light curve of MV Lyr since exact values of Wenzel and Fuhrmann(1983) and Andronov et al.(1988) data were not available. Because of this we analysed only high state luminosity and found as most probable 13.56 yr, 7.27 yr, 340 days and 321 days cycles. The best fit to the high state light curve was obtained using two sinusoids with periods P_1 = 13.56 yr and P_2 = 7.27 yr and amplitudes 0.34 and 0.49 mag (Fig.2).

Our results showed long-term cyclycal modulations of the brightness in timescales of years for all four novelikes. Acording frequency analysis cycles exist both in high and low state. Duration of the cycles is in the interval of values for CVs given by Warner (1988) and Bianchini (1992) and for single stars given by Wilson (1978). We estimated the changes of the radii of the secondaries in high state using Eq. 19 in Warner (1988). The results shown in Table 1, coincide with Gilliland's (1982) values for changes of solar

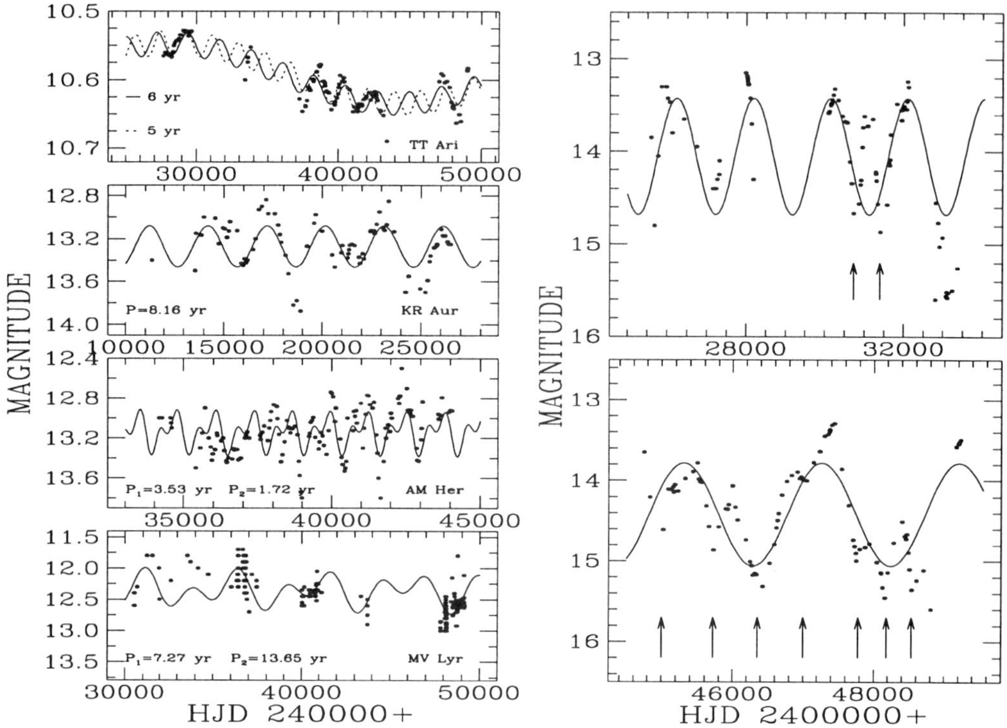

FIGURE 2. *Left panel*: The high state mean light curves as described in Fig.1 (*Right panel*) and the best fit to them; *Right panel*: Annual mean B magnitudes in AM Her dips state (points). Solid line present sinusoidal fit with period 5.38 yr. The arrows mark the minima of 1.7 yr modulations.

radius and Warner's (1988) estimations of changes of the radii of other CVs. We didn't exclude the possibility that high and dips' states are parts of a longer, about a few decades, cycle as can be seen from Fig.1, as well as the change of the duration of the 11 yr-like cycles in case of transition from high to low state.

REFERENCES

Andronov, I.L., 1987, Ap. Sp. Sci., 131, 557

Andronov, I.L., Fuhrmann, B., Wenzel, W., 1988, Astron. Nachr., 309, 39

Bianchini, A., 1988, IBVS, No 3136

Bianchini, A., 1990, AJ, 99, 1941

Bianchini, A., 1992, *ASP Conference Series*, 29, 284

Gilliland, R., 1982, ApJ, 253, 399

Richman, H.R., Applegate, J.H., Patterson, J., 1994, PASP, 106, 1075

Stellingwerf, R.F., 1978, ApJ, 224, 953

Warner, B., 1988, Nature, 336, 129

Wenzel, W., Fuhrmann, B., 1983, MVS, 9, No 6, 175

Wilson, O.C., 1978, ApJ, 205, 612

Binary Degenerate Dwarfs: Theory and Observations

By ALEXANDER TUTUKOV[1,2], LEV YUNGELSON[1] AND ICKO IBEN Jr.[2]

[1]Institute of Astronomy of the Russian Academy of Sciences, 48 Pyatnitskaya Str., 109017 Moscow, Russia

[2]Astronomy Department, University of Illinois, 1002 W. Green Street, Urbana, IL 61801

Systems containing two white dwarfs can descend in less than a Hubble time from wide or close binary stars with initial masses of components exceeding $\sim 0.8\ M_\odot$. The degenerate components may be composed of helium, of mixtures of carbon and oxygen, or of mixtures of oxygen and neon (with traces of sodium and magnesium). They may be detached or semidetached. We present a numerical model of this population in which the total Galactic formation rate of binary degenerate dwarfs is ~ 0.1 yr^{-1} in close binaries and ~ 0.006 yr^{-1} in wide ones. The estimate of the Galactic birthrate of binaries containing a neutron star or a black hole accompanied by a white dwarf is ~ 0.001 yr^{-1}. The main model parameters of close binary degenerates are in reasonable agreement with the parameters of their currently observed counterparts.

Mergers of components of double degenerates under the influence of angular momentum loss via gravitational wave radiation can result in type Ia supernova explosions (~ 0.003 yr^{-1}) or in the formation of semidetached binaries with degenerate helium donors (~ 0.01 yr^{-1}), some of which may perhaps manifest themselves as AM CVn stars, helium nondegenerate subdwarfs (~ 0.014 yr^{-1}), and hydrogen-deficient giants (~ 0.001 yr^{-1}).

---

1. Introduction

Degenerate dwarfs are the end result of the nuclear evolution of stars with initial masses between ~ 0.8 and $\sim 10.5\ M_\odot$. Stars less massive than $\sim 0.8\ M_\odot$ require longer than a Hubble time to evolve off the main sequence and stars more massive than $\sim 10.5\ M_\odot$ evolve into neutron stars. Since a considerable (or even dominant) proportion of stars are components of binary and multiple systems, and since mass loss occurring during the evolution of stars with moderate masses is non-disruptive, it is natural to expect that double degenerates are well represented in the stellar world. However, luminosities of degenerate stars are, as a rule, very low, and related selection effects create enormous difficulties for the observational study of their multiplicity. As a result, only twenty pairs of degenerate dwarfs with measured orbital separations or periods are currently known. They belong to three families: six are wide binary degenerates (Greenstein 1986), eight are detached close binary degenerates (Saffer, Liebert, & Olszewski 1988; Bragaglia *et al.* 1990; Marsh 1995; Marsh *et al.* 1995; Holberg *et al.* 1995), and six are semidetached close binary degenerates (see reviews by Ulla 1994; Warner 1995).

We have applied the scenario code for the population synthesis of binary stars to construct a numerical model of these three families (Tutukov & Yungelson 1992, 1996; Iben, Tutukov, & Yungelson 1997). The model combines an estimate of the birthrate of binaries, which is a function of the masses of primaries M_1, mass ratios of components $q = M_2/M_1$, and semimajor axes of orbits a, with a description of evolutionary transformations which binaries may experience in a Hubble time. It is assumed that all stars are born in binaries† and that the star formation rate has been constant for 15×10^9 yr.

† For different assumptions on the initial incidence of binaries all model results may be rescaled.

The birthrate function has been taken as (e.g., Iben & Tutukov 1984)

$$d^3\nu = 0.2 d\log(a) M_1^{-2.5} dM_1 f(q) dq \ \ \mathrm{yr}^{-1}.$$

Here M_1 and a are in solar units. For close binaries and for wide binaries with $q \leq 0.3$, $f(q) = 1$; for wide binaries with $q > 0.3$, $f(q) \propto q^{-2.5}$. The common envelope parameter $\alpha_{\rm CE}$, which is crucial for the properties of populations arising after mass exchange episodes, has been set equal to 1. For additional details see Yungelson & Tutukov in these proceedings and references therein.

In the model, most newly formed white dwarfs (WDs) descend from initial primaries and have main-sequence companions. The corresponding Galactic birthrates are 0.28 yr^{-1} for close and 0.48 yr^{-1} for wide binaries. Some WDs are single degenerate stars, which owe their origin to the merger of components of initially close binaries (0.16 yr^{-1}) or to the disruption of wide binaries due to mass loss which, terminating the AGB stage, occurs on a time scale that can be much shorter than the orbital period. Binary degenerates arise when the secondary reaches the WD stage. In the present paper we consider certain parameters of the model of the population of binary degenerates and compare them with the properties of the observed pairs of degenerate dwarfs. The formation of close double degenerates (both detached and semidetached) is a result of the action of two mechanisms of angular momentum loss (AML): frictional braking of components inside a common envelope and radiation of gravitational waves. Thus, these systems provide an important means for testing several vital ideas of the current theory of binary stellar evolution.

2. Wide binary degenerate dwarfs

Components of wide binaries do not overflow their Roche lobes during their evolution. Most wide binaries forming degenerate carbon-oxygen WDs survive the mass-loss event which accompanies the formation of the WD. The large orbital eccentricity typical of wide binaries inhibits the disruptive role of mass loss even for binaries with a large mass ratio of components and long orbital periods. The model birthrate of white dwarfs in Galactic wide binaries is $\sim 0.48\ \mathrm{yr}^{-1}$, but the birthrate of WDs with a WD companion in wide systems is only $\sim 0.006\ \mathrm{yr}^{-1}$. Since we assume that secondaries in the initial systems have a distribution over mass similar to the Salpeter power law, most degenerates produced by the model which are in wide binaries have low mass main-sequence companions which simply do not have enough time to evolve to the WD stage in a Hubble time. However, the real distribution of low mass binaries over the mass ratios of their components is still poorly known. This makes the numbers given by the model rather uncertain.

Old white dwarfs in wide binaries are very dim and have a typical age of $\sim 4 \times 10^9$ yr (Greenstein 1986). With this age, the estimated number of wide binary WDs in the Galaxy is $\sim 2.4 \times 10^7$. Taking the volume of the Galaxy equal to 3×10^{11} pc^3, the average distance between them is ~ 25 pc. For six known degenerate dwarfs in wide pairs the average distance from the Sun is 30 pc, close to our estimate. This may mean that Greenstein's (1986) list of wide double degenerates close to the Sun is essentially complete.

The projected component separations in the observed wide binaries with degenerate components range from $1.5 \times 10^3\ R_\odot$ to $5 \times 10^5\ R_\odot$, overlapping almost the whole range of semimajor axes of orbits of wide binaries.

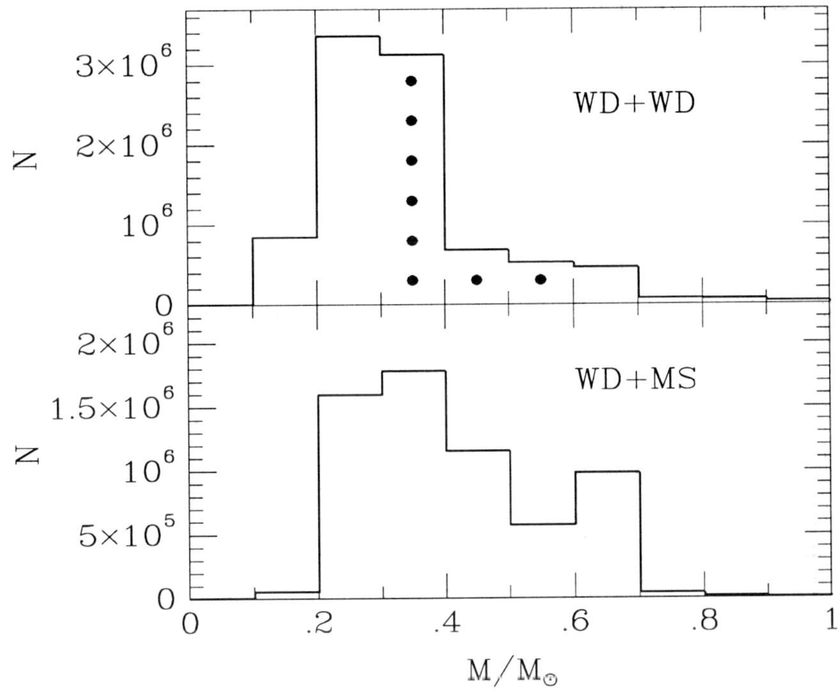

FIGURE 1. Distribution over mass of degenerate primaries in detached close binaries. The upper panel shows the model distribution for double degenerates and the lower panel shows the distribution for degenerate dwarfs in close pairs with low mass ($M \leq 0.3\ M_\odot$) main-sequence stars. Filled circles show masses of primaries of known close degenerate pairs.

3. Detached close binary degenerate dwarfs

Eight degenerate dwarfs with close but detached dim low mass companions are known. The companions are almost certainly also white dwarfs. All observed WDs in these systems (with two possible exceptions) have masses less than 0.4 $M_\odot$ and belong to the low mass tail of the observed mass spectrum of DA white dwarfs which contains about 10 % of all objects (Bragaglia, Renzini, & Bergeron 1995). In fact, five out of the eight known binary WDs were found by Marsh and his collaborators in a study aimed at detecting duplicity among low surface gravity WDs. Since currently no ways are known to form a helium WD except in a close binary, the high duplicity of low mass degenerate dwarfs seems quite reasonable.

White dwarfs in the sample of Marsh and his co-authors are relatively bright ($L \gtrsim 0.01\ L_\odot$), hence, young. Iben & Tutukov (1986) cooling curves give for them ages of $t_c \lesssim 10^8$ yr. We have constructed a model for the population of close binaries with a young ($t_c \leq 10^8$ yr) helium, carbon-oxygen, or oxygen-neon WD as the brighter component (Iben *et al.* 1997). The dimmer companion can be a He, CO, or ONe WD or a low-mass ($M \leq 0.3\ M_\odot$) main-sequence (MS) star†. The total birthrate of such binaries in our Galaxy is ~ 0.17 yr^{-1}. In $\sim 63\%$ of the systems, the dimmer secondary is a WD. In

† The reason for this restriction is that the visual brightness of a 10^8 yr old WD is comparable to that of a 0.3 $M_\odot$ MS star. For the known systems, consideration of (infra)red magnitudes excludes the presence of low mass MS companions more massive than $\sim 0.1\ M_\odot$.

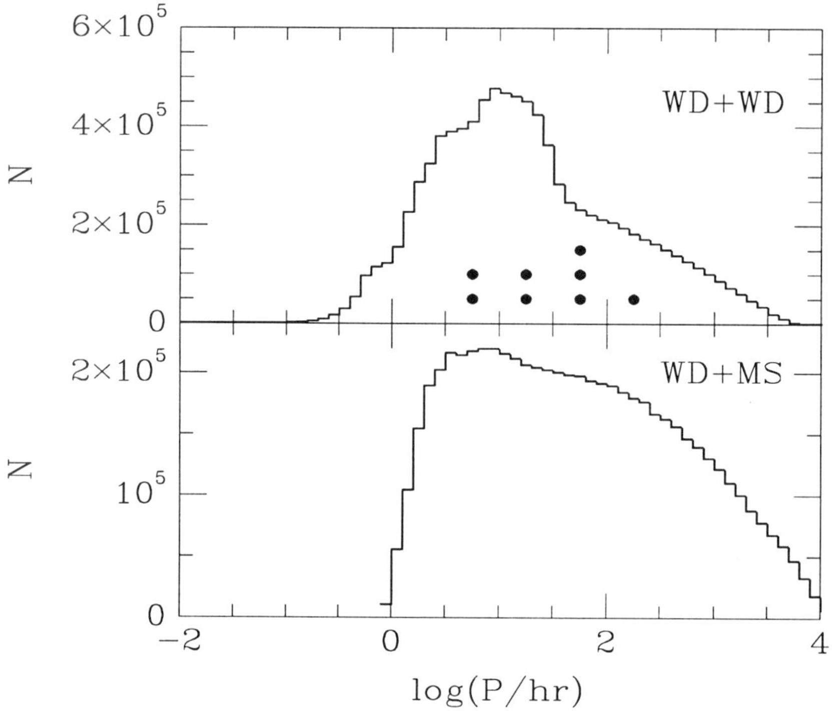

FIGURE 2. The distribution over orbital period of close detached binaries with degenerate dwarfs. The upper panel shows the model for double degenerates and the lower panel shows the model for degenerate dwarfs with low mass ($M \leq 0.3\ M_\odot$) main-sequence companions. Filled circles mark positions of known double degenerates grouped in $\Delta \log(P_{orb}/\mathrm{hr}) = 0.5$ intervals.

$\sim$ 82% of the systems containing two WDs, the brighter (initial secondary) component is a helium WD ($M \lesssim 0.5\ M_\odot$, Fig. 1). This explains, why, in at least 7 out of 8 known close binary degenerates, the brighter component really is a helium WD. The distribution of close double degenerates over their orbital period is shown in Figure 2. Observed close double degenerates fit the model distribution quite well. Periods of 18 precataclysmic binaries (King *et al.* 1994) are also in good agreement with the model period distribution of WD + MS binaries.

The evolution of the short period systems under consideration is determined by AML via gravitational wave radiation. The shortest period systems with MS components evolve into cataclysmic binaries with $P_{\rm orb} \leq 2^{\rm h}\!.0$. The merger of two CO white dwarfs may result in a type Ia supernova (SN Ia) explosion if the total mass of the system exceeds the Chandrasekhar mass ($M_{\rm Ch} \sim 1.4\ M_\odot$). In our model, the occurrence rate of such events is $\sim$ 0.003 yr^{-1}(Tutukov, Yungelson, & Iben 1992), comparable to the estimate of the Galactic SN Ia rate (van den Bergh & Tammann 1991). Mergers of two helium white dwarfs may result in the formation of nondegenerate helium stars, which may, perhaps, be identified with blue helium subdwarfs. The rate of such mergers in the model is $\sim$ 0.014 yr^{-1}(Tutukov & Yungelson 1992). Formation of a subdwarf may be the fate of the double-lined WD 1101+364 ($P_{\rm orb} = 0^{\rm d}\!.145$). The model estimate of the space density of post-merger helium stars is, within a factor of 2 to 3, consistent with the observational estimates (Heber 1986; Downes 1986). For the single-lined WD 2331+290 ($P_{\rm orb} = 0^{\rm d}\!.17$), if the mass of the secondary is sufficiently large, one may expect eventual transformation into a hydrogen-deficient giant (Iben *et al.* 1996)†. Our estimate of the formation rate of such objects via the merger of an old CO WD with a young helium WD is $\sim 0.0013\ \mathrm{yr}^{-1}$.

The average M_V of four of the close detached binary degenerates discovered by Marsh *et al.* (1995) is $\approx 9^{\mathrm{m}}.81$. The age of a 0.3 $M_\odot$ helium WD with such brightness is about 10^8 yr (Iben & Tutukov 1986). Since the model frequency of formation of close double degenerates is $\sim 0.11\ \mathrm{yr}^{-1}$, the number of them in the Galaxy with $M_V \lesssim 9^{\mathrm{m}}.81$ is $\sim 10^7$. The average model distance between them for a Galactic volume $\sim 3 \times 10^{11}\ \mathrm{pc}^3$ is about 70 pc. The average distance to the dwarfs in the Marsh *et al.* sample is about 120 pc. The average distance to the three closest Marsh *et al.* objects is 74 pc. Thus, also in this case, the model produces an estimate of the space density of detached close binary degenerates in the Galaxy which is in reasonable agreement with the observations. Of course, this conclusion is still rather weak because of the small number of known systems.

4. Semidetached binary degenerate dwarfs

These systems have orbital periods among the shortest known. If they evolve into cataclysmic binaries, one may perhaps identify them with "AM CVn-type stars", after the prototype of the group. Orbital periods of six known objects cover the range from 17.2 min (AM CVn) to 46.5 min (GP Com). Degenerate helium donors fill their Roche lobes, and AML via the radiation of gravitational waves is the main driving force for their evolution, supporting continuous mass exchange between components. The model birthrate of these binaries is $\sim$ 0.01 $\mathrm{yr}^{-1}$and their lifetime is $\sim 10^{10}$ yr (Tutukov & Yungelson 1996). Hence, their total number in the Galaxy is $\sim 10^8$. This gives their average Galactic space density as $\sim 3 \times 10^{-4}\ \mathrm{pc}^{-3}$, and the average distance between them $\sim$ 15 pc. The characteristic orbital periods of these binaries are larger than one hour (Fig. 3, lower panel).

Before comparing the model data with the observations, at least one rather evident selection effect preventing easy discovery of semidetached double degenerates has to be taken into account. The main source of radiation in these systems is an accretion disc. As their mass-exchange rate is $\sim 10^{-13}\ \mathrm{M}_\odot\ \mathrm{yr}^{-1}$, typical systems are extremely dim and can be found only in the closest vicinity of the Sun. The number of stars in the model sample with V $< 13^{\mathrm{m}}.0$ is only $\sim$ 10. These 10 are among the most intinsically bright of the entire group, with typically $\dot{M} \gtrsim 10^{-9}\ \mathrm{M}_\odot\ \mathrm{yr}^{-1}$. For such a large $\dot{M}$, helium accretion discs may be stable (Smak 1983), acting against discovery of the systems. Thus, the small model number of potentially "observable" objects (Fig. 3, top panel), at least does not contradict observations (the AM CVn stars are indicated by solid circles). The model distribution of orbital periods of "observable" systems is peaked at much lower periods (typically in the range 0.1 — 0.4 hr) than the bulk of the theoretical population (Fig. 3, lower panel). The observed sample of AM CVn stars evidently represents a short and bright initial stage in the evolution of semidetached double degenerates.

From a purely theoretical point of view, the radii of cooling low mass donors are rather uncertain. In the "standard" model (Fig. 3, solid histograms), we approximated the radii of WDs by the radii of zero-temperature degenerate spheres given by Nauenberg (1972). To check the influence of the donor radius on the results, we computed a test model adopting radii twice as large as in the standard model (Fig. 3, dashed histograms). Orbital period distributions are evidently sensitive to the choice of donor radius. Although

† A redetermination of P_{orb} for WD 0957-666, from $1^{\mathrm{d}}.15$ (Bragaglia *et al.* 1990) to $87^{\mathrm{m}}.5$ (Moran, Marsh, & Bragaglia 1996), makes this system also a promising merger candidate.

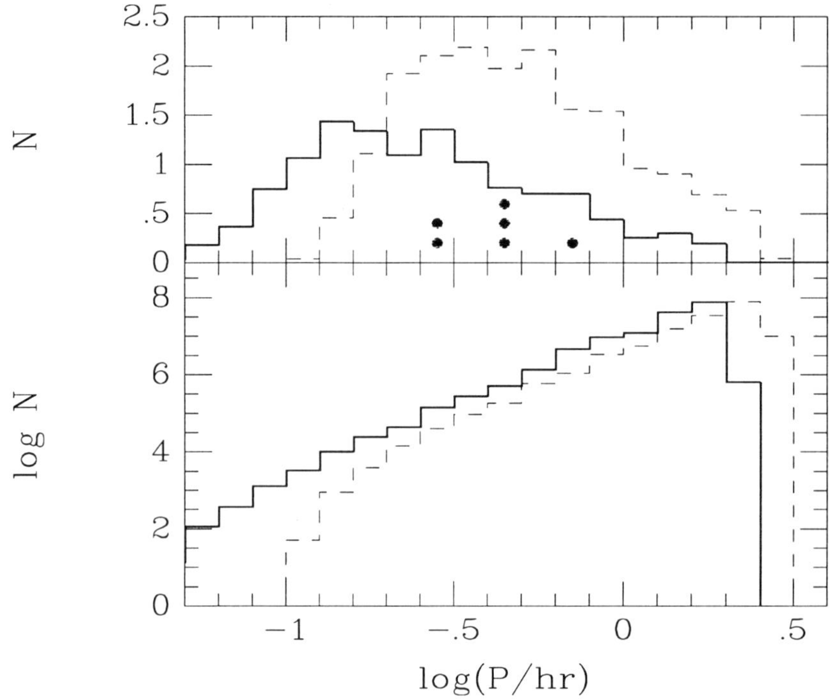

FIGURE 3. Distributions over orbital period of semidetached systems with degenerate helium white dwarf donors. The lower panel gives distributions for the total sample and the upper panel gives those for the sample brighter than V= $13^{m}_{.}0$. The solid histograms give results for the standard model (see the text) and the dashed ones give those for a donor radius twice as large as in the standard model.

the "doubled radius" model histogram fits the observed distribution somewhat better (Fig. 3, upper panel), the standard model histogram does not contradict the observations. The same is true if one applies mass-radius relations for degenerate stars based on different equations of state (Tutukov & Yungelson 1996).

5. Degenerate dwarfs in pairs with neutron stars and black holes

According to the scenario model, a small fraction of degenerate dwarfs are in binaries with a neutron star or black hole companion. The model birthrate of such binaries in our Galaxy is ~ 0.0006 yr^{-1} in wide binaries and ~ 0.0013 yr^{-1} in close ones.

Neutron stars (NSs) and black holes (BHs) in wide binaries are, naturally, remnants of primaries, and CO degenerate companions always arise after the formation of the relativistic component. It is possible that neutron stars existing in wide binaries are "silent": in the absence of an efficient mass exhange between components they do not reach short enough spin periods to produce the radiopulsar stage. This may explain the absence of a well populated family of wide binary radiopulsars (Iben & Tutukov 1996).

Close binaries of the type under consideration can be separated into two families: NS (or BH) + WD binaries (birthrate $\nu \approx 0.0008$ yr^{-1}) with circular orbits and WD + NS (or BH) binaries ($\nu \approx 0.0004$ yr^{-1}) with noncircular orbits. However, a small number of observed objects, combined with an assumed lifetime of $\sim 10^9$ yr (e.g. Bell *et al.* 1995), gives a semiempirical estimate of their Galactic birth rate of only $\sim 10^{-6}$ yr^{-1} (Lorimer 1996). This latter estimate is consistent with the occurrence rate of accretion induced collapses (AIC) in close binaries (Iben, Tutukov, & Yungelson 1995). However, we recognize that, in our model, the neutron star in most NS + WD and WD + NS binaries is formed not via AIC but via the collapse of the core of an initially massive star, as in the case of normal radiopulsars. It is unclear why WD + NS systems are not represented among observed binary radiopulsars in the proportion suggested by the model. A possible solution is that their evolutionary history prevents a long lifetime in the pulsar state. Other solutions may be hidden in the uncertainties of the model. Among these uncertainties are the distribution of initial binary parameters; masses of cores after different stages of component evolution; uncertainty in the treatment of common envelopes. Last but not least, our estimates were obtained assuming zero "kick" velocities, possibly overestimating the birthrate of neutron stars remaining in binaries.

Some close binaries containing degenerate and relativistic stars have, after their formation, short enough orbital periods to be driven into contact in less than a Hubble time by AML via the radiation of gravitational waves. As a result, semidetached low mass X-ray binaries with orbital periods of a few minutes can arise. They are close relatives of AM CVn binaries. Systems MXB 1820-30, 1916-053, and 2259+586 may belong to this family (see Fig. 5 in Iben *et al.* 1995).

6. Conclusion

Numerical models of the population of binary degenerates in our Galaxy help to distinguish four main families of these binaries and to find their basic parameters. The model space densities of wide, close detached, and semidetached binary degenerates are consistent with observations.

Recent studies by Marsh and his collaborators have significantly increased the number of known detached close binary degenerates. Their primary components appear to be, as predicted by the scenario model, mainly helium WDs. Among currently known systems of this type there are two (or even three) binaries able to merge in less than a Hubble time under the influence of gravitational wave radiation. This strongly supports the idea that mergers of close binary WDs produce blue helium subdwarfs, hydrogen-deficient giants, and type Ia supernovae.

There are at least five double degenerates which are so wide that variations of their radial velocities are not recognizible. On the other hand, they are so close that at the typical distance of several tens of pc their components cannot be resolved as components of visual pairs. However, they are still identified as binaries due to their composite DA + DC spectra (Greenstein & Liebert 1990; Bergeron, Greenstein, & Liebert 1990; Bergeron, Ruiz, & Leggett 1993; Liebert *et al.* 1993). It is difficult to compare the data on these white dwarfs with model predictions since we are not well acquainted either with observational selection effects which govern their discovery or with family relationships between white dwarfs of different spectral subtypes.

The problem of the "overproduction" in our model of close binaries containing a neutron star with a white dwarf companion and the problem of the expected observational characteristics of such binaries remain open.

A. T. and L. Y. would like to acknowledge financial support of the Organizing Committee which enabled them to attend the conference. I. I. would like to thank the Organizing

Committee and all participants for making his attendance possible. This study was supported in part by the Russian Foundation for Fundamental Research grant 96-0216531 and in part by the United States National Science Foundation grant AST94-17156.

REFERENCES

Bell, J. F, Kulkarni, S. R., Bailes, M., Leitch, E. M., & Lyne, A. G. 1995, ApJ, 452, L121

Bergeron, P., Greenstein, J. L., & Liebert, J., 1990, ApJ, 361, 190

Bergeron, P., Ruiz, M.-T., & Leggett, S. K. 1993, ApJ, 407, 733

Bragaglia, A., Greggio, L., Renzini, A., & D'Odorico, S. 1990, ApJ, 365, L13

Bragaglia, A., Renzini, A., & Bergeron, P. 1995, ApJ, 443, 735

Downes, R. S. 1986, ApJSS, 61, 569

Greenstein, J. 1986, AJ, 92, 867

Greenstein, J., L., & Liebert, J. 1990, ApJ, 360, 362

Heber, U. 1986, A&A, 155, 33

Holberg, J. B., Saffer, R. A., Tweedy, R. W., & Bastrow, M. A. 1995, ApJ, 452, L133

Iben, I. Jr., & Tutukov, A. V. 1984, ApJ, 282, 615

Iben, I. Jr., & Tutukov, A. V. 1986, ApJ, 311, 742

Iben, I. Jr., & Tutukov, A. V. 1996, ApJ, 456, 738

Iben, I. Jr., Tutukov, A. V., & Yungelson, L. R. 1995, ApJSS, 100, 233

Iben, I. Jr., Tutukov. A. V., & Yungelson, L. R. 1996, ApJ, 456, 750

Iben, I. Jr., Tutukov. A. V., & Yungelson, L. R. 1997, ApJ, January 20 issue

King, A. R., Kolb, U., de Kool, M., & Ritter, H. 1994, MNRAS, 269, 907

Liebert, J., Bergeron, P., Schmidt, G. D., & Saffer, R. A. 1993, ApJ, 418, 426

Lorimer, D. R. 1996, in Compact Stars in Binaries, ed. J. van Paradijs, E. P. J. van den Heuvel, & E. Kuulkers (Dordrecht: Kluwer), p. 235

Marsh, T. R. 1995, MNRAS, 275, L1

Marsh, T. R., Dhillon, V. S., & Duck, S. R. 1995, MNRAS, 275, 828

Moran, C. K. J., Marsh, T. R., & Bragaglia, A. 1996, in progress

Nauenberg, M. 1972, ApJ, 175, 417

Saffer, R. A., Liebert, J., & Olszewski, J. 1988, ApJ, 334, 947

Smak, J., 1983, Acta Astron., 33, 333

Tutukov, A. V., & Yungelson, L. R. 1992, SvA, 69, 526

Tutukov, A. V., & Yungelson, L. R. 1996, MNRAS, 280, 1035 26

Tutukov, A. V., Yungelson, L. R., & Iben, I. Jr. 1992, ApJ, 386, 197

Ulla, A. 1994, SpSciRev, 67, 241

van den Bergh, S., & Tammann, G. 1991, ARA&A, 29, 363

Warner, B. 1995, ApSpSci, 225, 249

Population Synthesis for Close Binary Stars

By LEV YUNGELSON AND ALEXANDER TUTUKOV

Institute of Astronomy of the Russian Academy of Sciences,
48 Pyatnitskaya Str., 109017 Moscow, Russia

We describe the basic concepts of numerical population synthesis for binary stars and provide two examples of its applications. First, we estimate the rate of mergers of neutron star binaries and show that first-generation LIGO interferometers may be able to detect up to ~ 100 events per year. Second, we show that in old populations ($\sim 10^{10}$ yr) the only candidate systems able to support the rate of SNe Ia of $\sim 10^{-3}$ yr^{-1} are merging white dwarfs which ignite carbon directly. In young populations ($\sim 10^{9}$ yr) systems with nondegenerate helium star donors in which carbon ignition may be provoked by He detonation are also able to give a similar contribution to the total rate of SNe Ia.

1. Introduction

Population synthesis is a method for calculation of theoretical statistics of a certain class of objects and for finding relations between the objects of different classes.

The main goals of population synthesis are:

• to understand how very different or even exotic binary stars (e.g., X-ray binaries or close interacting pairs of degenerate stars) descend from unevolved main-sequence binaries which differ only in masses of components and their separations;

• to estimate the incidence of binaries of different classes and occurence rates of certain *events* in them (e.g., explosions of Novae or Supernovae);

• to find the distributions of binaries of different classes over parameters like masses of components, orbital periods, luminosities of components, spatial velocities;

• to understand selection effects which govern the properties of the observed samples of objects;

• to understand the gaps both in theory and observations.

We shall describe the basic concepts of the population synthesis in Sec. 2. Several examples of applications will be provided in Sec. 3.

2. Basic concepts

The basic idea of population synthesis for binary stars is combination of the birthrate of a binary system with the *evolutionary scenario* for this system. In this context, *scenario* is the sequence of transformations of a binary system, which it can experience in Hubble time. The term "scenario" was used for the first time by van den Heuvel & Heise (1972) to describe formation of massive X-ray binary systems.

The birthrate of binaries is usually treated as a function of initial mass of primary component M_{10}, mass ratio of components $q_0 = M_{20}/M_{10}$, and orbital separation A_0 or orbital period P_0. This data comes from the statistical studies of binary stars. In the series of papers of present authors and their collaborators quoted below, the birthrate function, based on study of properties of eclipsing, spectroscopic, and visual binaries, was taken in the form

$$d^3\nu = 0.2 d\log(A_0/R_\odot)(M_{10}/M_\odot)^{-2.5} d(M_{10}/M_\odot) f(q_0) dq_0 \ \ \mathrm{yr}^{-1}.$$

For minimal mass of forming stars 0.1 $M_\odot$, it corresponds to the current rate of star formation in the Galactic disc. It was assumed that all stars are born in binaries with

$10 \le A_0/R_\odot \le 10^6$. Thus, our estimates are upper limits and, for another assumptions on initial incidence of binaries, the results have to be rescaled. Our studies of the distributions of binaries over q_0 suggest $f(q_0) = 1$ for close binaries and for wide binaries with $q_0 \le 0.3$; for the wide ones with $q \ge 0.3$, $f(q_0) \sim q_0^{-2.5}$. Distributions over M_{10}, q_0, and A_0 may be also treated as the parameters of the model.

For construction of evolutionary scenarios all binaries are separated into two classes: *close* and *wide* ones. In close binaries components exchange matter via Roche lobe overflow (RLOF) or via accretion of stellar wind. Interaction of both kinds may be accompanied by mass and angular momentum loss (AML) from the system. A specific way of interaction is evolution of the system inside common envelope (CE), engulfing whole system (Paczyński 1976), which is also accompanied by mass and momentum loss. For particular situations one has also to consider AML via gravitational waves radiation or magnetically coupled stellar wind. The outcome of RLOF depends on the evolutionary stage and inner structure of the donor at the instant of RLOF (particularly, on presence of deep convective envelope) and q. For accretors, one has to consider, e.g., the effects of rejuvenation caused by accretion or the effect of thermonuclear runaways in the accreted matter. These data, as well as initial - final mass relations, lifetimes in particular evolutionary stages etc., are provided by computations of stellar models. In wide binaries components evolve independently, like single stars. The only data necessary for construction of scenarios for wide systems are the initial - final mass relations and the lifetimes in particular stages. Evidently, the sets of models used in different simulations differ, reflecting the state of art in the theory of stellar evolution.

For any initial combination of (M_{10}, M_{20}, A_0), one may follow all evolutionary transformations up to a certain "final" configuration (an example is given in Sect. 3.1). If one knows the birthrate of this "binary" and lifetimes in particular stages, then, for any history of star formation, evolving a "binary" for Hubble time, it is possible to compute its contribution to the past or present population of stars of different types. Integration over whole space of initial parameters or Monte-Carlo simulations for a large sample of initial "binaries" give a complete model of the population of binaries. Applying selection effects, one gets a model of the "observed" population. *It is important to note, that scenarios are not prescribed but generated, based on transformations suggested by the theory of stellar evolution.* Objects of the same class may be formed by several scenarios, hence, one may expect variations inside the class. Scenarios may *predict* existence of certain configurations and thus stimulate both theory and observations.

The (incomplete) list of applications of numerical population synthesis for binary stars includes massive binaries and products of their evolution like Wolf-Rayet stars, high-mass X-ray sources, pulsars, binary neutron stars (or black holes) and spectrum of gravitational waves radiated by them (Dewey & Cordes 1987; Meurs & van den Heuvel 1989; Yungelson & Tutukov 1991; Tutukov & Yungelson 1993a,b, 1994; Romani 1992; Podsiadlowski *et al.* 1992; Iben *et al.* 1995a; Dalton & Sarazin 1995; Portugies Zwart & Verbunt 1995; Lipunov *et al.* 1996 and references therein); Be-binaries (Pols *et al.* 1991); cataclysmic variables (Politano 1988, 1996; de Kool 1992; Kolb 1993; Tutukov & Yungelson 1995, 1996); low-mass hard and supersoft X-ray sources (Iben *et al.* 1995b; Rappaport *et al.* 1994; Yungelson *et al.* 1996); nuclei of planetary nebulae (de Kool 1990; Yungelson *et al.* 1993; Han *et al.* 1995), degenerate dwarfs in binaries (Lipunov & Postnov 1988; Tutukov & Yungelson 1992; de Kool & Ritter 1993; Yungelson *et al.* 1994; Iben *et al.* 1997); statistics of supernovae (Tutukov *et al.* 1992; Tutukov & Yungelson 1994, 1995; Ruiz-Lapuente *et al.* 1995; Canal *et al.* 1996); hydrogen-deficient giants (Iben *et al.* 1996); symbiotic stars (Yungelson *et al.* 1995); dwarf carbon stars (de Kool & Green 1995); an analysis of color-magnitude diagrams of open clusters with ac-

Type of the system	NS+NS	BH+NS	BH+BH
Merger rate in the Galaxy	3×10^{-4}	1×10^{-5}	1×10^{-6}
Merger rate in the Universe	1×10^{6}	6×10^{4}	6×10^{3}
Rate of events from $r \leq 200$ Mpc	100	5	0.5

TABLE 1. The merger rates of NS and BH binaries (yr^{-1})

count of binary evolution (e.g., Pols & Marinus 1994). Recently, population synthesis for binaries was discussed by de Kool (1996) and Lipunov *et al.* (1996).

3. Examples of applications of numerical population synthesis

3.1. *Merger of close binary neutron stars and black holes*

Several gravitational wave observatory projects are currently under development. Coalescing neutron star (NS) or black hole (BH) binaries are, theoretically, the best understood sources of GWR and they are quite promising candidates for detection even by the first-generation detectors (Thorne 1996). They may also emit γ-rays (Paczyński 1986) and eject r-process elements into the interstellar medium (Eichler *et al.* 1989). Detection of mergers would allow accurate determination of the Hubble constant, the masses of merging stars, and the distance to them (Shutz 1986).

Neutron stars in binaries may descend from stars with $11.5 \lesssim M/M_\odot \lesssim 50$, while BH from more massive stars. Let us introduce the following abbreviations: MS - main-sequence star, He - helium star, SN Ib - explosion of type Ib Supernova, MXRB - massive X-ray source, NS - neutron star. Neutron star binaries, which are close enough to merge in Hubble time, may be formed via the following *scenario* (Tutukov & Yungelson 1973): MS + MS $\rightarrow$ RLOF $\rightarrow$ He + MS $\rightarrow$ SNIb $\rightarrow$ NS + MS $\rightarrow$ MXRB $\rightarrow$ RLOF $\rightarrow$ NS + He $\rightarrow$ RLOF $\rightarrow$ SNIb $\rightarrow$ NS + NS. It is crucial that in this scenario three RLOFs accompanied by formation of CEs occur. It is also important that SNe explosions don't disrupt the system. On its way to the NS+NS stage, for a short time ($\sim 10^3 - 10^4$ yr) the system shows up as a massive X-ray source (van den Heuvel & Heise 1972). Before merger, the system may manifest itself as a binary pulsar similar to, e.g., PSR 1913+16. We assign the SNe in the above scenario type Ib. However, if the mass of the RLOF remnant is not high and its stellar wind is weak, presupernovae may partially retain their hydrogen envelopes. Then they would manifest themselves as SNe II, probably, with linear light curves or as type IIb objects. Total model rate of SNe II and SNe Ib is 0.02 yr^{-1}, consistent with the observed combined rate of these SNe (e.g., Van den Bergh & McClure 1994).

Table 1 shows model merger rates ν_m of NS and BH binaries (Tutukov & Yungelson 1993a). The numbers referring to the Universe were obtained, following Phinney (1991), by scaling the Galactic merger rate with the ratio of B-band luminosities of the Universe and the Galaxy. The estimate for $r \leq 200$ Mpc corresponds to the limit of sensitivity of the first-generation LIGO interferometers. Our estimates of ν_m are quite high, like another estimates obtained in population modelling (e.g., Lipunov *et al.* 1994). But the low incidence of binary NS among detected radiopulsars suggests their birthrate is between 2×10^{-7} and 10^{-5} yr^{-1} (Bailes 1996). It is possible that typical coalescence timescale is much shorter than typical pulsar lifetime. Then, most merger candidates will be missed in pulsar surveys (Tutukov & Yungelson 1993a). Another solution may be complete spiral-in of the first-born NS into its companion in the second or third CE

Donor	MS/SG	He star	He-mantle star	He WD	RG	CO WD
Accretion mode	RLOF	RLOF	RLOF	RLOF	Wind	Merger
	(1)	(2)	(3)	(4)	(5)	(6)
	Direct carbon ignition (Chandrasekhar SN)					
Young population	10^{-5}	10^{-4}	10^{-5}	10^{-5}	10^{-6}	10^{-3}
Old population	–	–	–	10^{-5}	10^{-6}	10^{-3}
	Indirect carbon ignition (sub-Chandrasekhar SN)					
Young population	10^{-4}	10^{-3}	–	–	10^{-4}	–

TABLE 2. Occurence rates of SNe Ia in candidate progenitor systems

event (Taam 1996). Also, overproduction of NS+NS binaries may result from neglect, in our model, of kick velocities imparted to NS at birth. The birthrate of binary NS may be reduced $\sim$ 30 times, if kick velocity is at least $\sim$ 500km s^{-1} (Lipunov *et al.* 1996). However, even if we overestimate ν_m by two orders of magnitude, the prognosis for detection of the mergers of NS binaries is still optimistic.

It is worthwhile to notice the following. We estimated ν_m for BH+BH binaries for assumed $M_{BH} = 5\ M_\odot$. The volume of space which produces detectable events scales with total mass of the system as $M_{tot}^{2.5}$ (Thorne 1987). If, e.g., $M_{BH} = 10\ M_\odot$, the rates of detectable NS+NS and BH+BH mergers become comparable.

3.2. *Progenitors of type Ia supernovae*

While it is almost commonly accepted that SNe Ia are thermonuclear explosions of accreting white dwarfs (WDs), the nature of the systems in which SNe Ia occur is still unclear. Therefore, it's worthwhile to estimate expected SNe Ia rates ν_{Ia} from different candidate systems in the populations of different age. The candidate systems include configurations in which "direct" ignition of carbon occurs in the central parts of WD when its mass grows to $M_{Ch} \approx 1.4 M_\odot$ (Chandrasekhar SN, henceforth, C-SN) and configurations with "indirect" ignition of carbon, which is provoked by ignition of He in an accreted layer with $\Delta M \approx 0.2 M_\odot$ (sub-Chandrasekhar SN, henceforth, SC-SN).

Currently, six "explosive" configurations are thought to exist, differing in the nature of donor and accretion mode. They are listed in columns (1) to (6) of Table 2 which gives for every configuration our estimates of its contribution to the current Galactic ν_{Ia}. Only order of magnitude values are given, because they depend within factor $\sim 2-3$ on model parameters, mainly, on $f(q_0)$ and efficiency of energy deposition into CEs. Constant star formation rate through 15×10^9 yr is assumed. We give estimates for young and moderate age populations ($\sim 10^9$ yr) and for an old one ($\sim 10^{10}$ yr). Column (1) refers to accretion of hydrogen from Roche lobe filling main-sequence (MS) or subgiant (SG) companion (Whelan & Iben 1973). These systems, under appropriate conditions, may manifest themselves as cataclysmic variables and/or supersoft X-ray sources. Column (2) refers to the accretion of pure He from low-mass nondegenerate He star in a system evolving under the influence of AML via GWR. Peculiarity of these systems is $\dot{M} \lesssim 3 \times 10^{-8}$ M$_\odot$ yr^{-1} (Savonije *et al.* 1986; Tutukov & Fedorova 1989), just in the range required for SC-SN (Woosley & Weaver 1994). Observed counterparts of candidate systems for SC-SN with He donors are not known yet. They may be hidden among blue helium subdwarfs. In case (3) the donor experiences BB-case of mass exchange and loses mass from a thick He mantle surrounding a CO degenerate core (e.g., Iben & Tutukov 1994). In case (4) Roche lobe is filled by a low-mass degenerate helium dwarf (Tutukov & Yungelson 1979, 1996) and evolution is driven by AML via GWR. In case (5) WD accretes from stellar

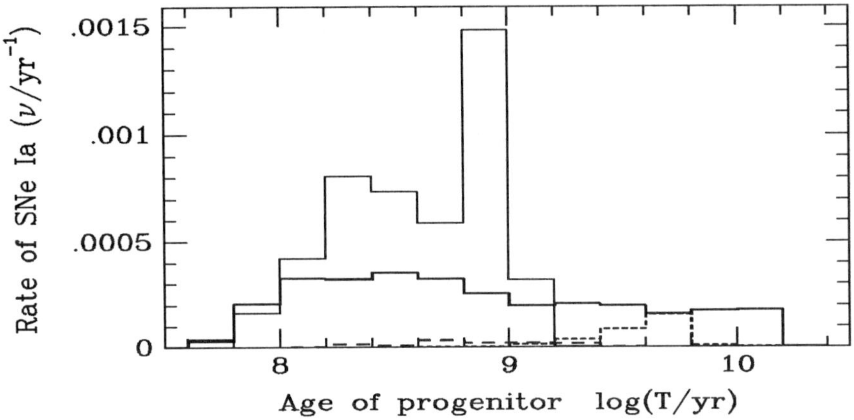

FIGURE 1. Contribution of stars of different age to the present SNe Ia rate. Thick full line - SNe from mergers of WDs; thin full, dotted, and dashed lines - sub-Chandrasekhar SNe in systems with nondegenerate He donors, with subgiant donors, and symbiotic stars, respectively.

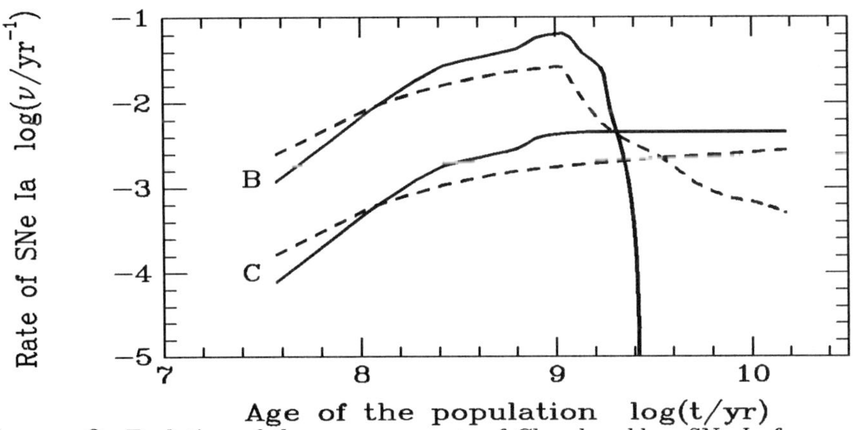

FIGURE 2. Evolution of the occurence rate of Chandrasekhar SNe Ia from mergers of WDs (dashed lines) and sub-Chandrasekhar SNe Ia in systems with nondegenerate He donors (full lines). Case B corresponds to the 10^9 yr long star formation burst, case C - to the continuous star formation for 15×10^9 yr. Total mass of stellar population is the same in both cases.

wind of red (super)giant (RG) companion. The system may manifest itself as a symbiotic star (Tutukov & Yungelson 1976; Yungelson *et al.* 1995). Finally, in case (6) merger of two CO white dwarfs happens. Possible observational features of candidate systems are discussed by Branch *et al.* (1995).

Chandrasekhar SNe Ia are possible, in principle, in all cases (1) through (6). However, when hydrogen is accreted, (cases 1 and 5) efficiency of mass accumulation is limited by Novae explosions, which erode WDs [see Yungelson *et al.* (1995, 1996) where secular evolution of WDs is modelled taking into account relations between masses of accreted and ejected shells after extensive grid of Novae computations from Prialnik & Kovetz (1995)]. When He is accreted from a nondegenerate donor (case 2), helium usually ignites before M_{Ch} is accumulated. Helium-mantle stars (case 3) have $\dot{M} \sim 10^{-6}$ $M_\odot$ yr^{-1}, slightly below the limit of stable He burning. Therefore, one may expect that most of the accreted He is lost in weak helium shell flashes (see Fig. 16 in Iben & Tutukov 1996). The same is true for the systems with He dwarf donors (case 4). This leaves merger of CO WDs (Tutukov & Yungelson 1979; Iben & Tutukov 1984; Webbink 1984) as the most viable candidate for C-SNe Ia. Recent discovery of six close double degenerates (Marsh 1995; Marsh *et al.* 1995; Holberg *et al.* 1995) with properties in close agreement with predicted ones (Iben *et al.* 1997) renders a strong support to the merger scenario. However, calculations of merger process of WDs (e.g., Rasio & Shapiro 1995) still didn't tell us, whether its outcome is really a SN explosion. One may expect (D. Arnett, this conference) that C-SN Ia may happen if the matter is accreted very slowly. In accretion is very rapid, degeneracy may be lifted, nondegenerate CO star may form, which ends its evolution by formation of an ONe WD or core collapse. Systems with MS/SG-donors do not produce C-SN in old populations, because in the latter typical mass of WDs in these systems is low (Tutukov & Yungelson 1995). Systems with nondegenerate He donors and He-mantle donors are always young ($T \lesssim 10^9$ yr).

Massive ($\sim 0.2\ M_\odot$) helium layer which may detonate and provoke ignition of carbon (Livne & Glasner 1990) may be accumulated via stable burning of hydrogen at the surface of accreting dwarf (cases 1 and 5) if $\dot{M} \sim 10^{-7}$ $M_\odot$ yr^{-1}. Supersoft X-ray sources with SG donors (van den Heuvel *et al.* 1992) and symbiotic stars (Munari & Renzini 1992) were suggested as suitable sites. However, our estimates which take into account efficiency of accretion and accumulation of He (Yungelson *et al.* 1995, 1996) show that ν_{Ia} from these systems falls an order of magnitude short of observed rate. The only viable candidates are systems with He star donors, which also have "proper" $\dot{M} \sim 10^{-8}$ $M_\odot$ yr^{-1}.

Two types of candidate systems which may produce desired $\nu_{Ia} \sim 10^{-3}$ yr^{-1} (Van den Bergh & McClure 1994) have progenitors with different typical ages (Fig. 1). Progenitors of WDs in merging pairs have $5 \lesssim M/M_\odot \lesssim 10.5$ with lifetime $T \lesssim 10^8$ yr and their lifetime prior to merger depends mainly on the separation of components. Hence, C-SN may be represented in the populations of any age. Progenitors of He-donors have lifetimes $\sim 10^8$ yr and RLOF has to happen in less then $\sim 10^9$ yr, because otherwise He star becomes a WD before RLOF. The critical mass of He is transferred in $\sim 10^9$ yr. Hence, SC-SN may be represented in relatively young populations *only.*

An important implication is that SC-SN have not to explode in old elliptical galaxies and galactic bulges. Figure 2 shows evolution of the rates of SNe Ia from two types of the progenitors in a model which had star formation rate constant for 15×10^9 yr and in a model in which the same mass of stars was formed in 10^9 yr. The first model is appropriate, e.g., for the Galactic disc, while the second - for elliptical galaxies. In the disc SNe Ia of both kinds may be represented, while in "old" systems SC-SN have to extinguish. If both types of SNe Ia can contribute to ν_{Ia} in approximately equal proportions, one may expect diversity in the properties of observed SNe Ia. Considerable difference in the luminosities of SNe Ia was found by Schaefer (1996), in volume limited samples of SN. Particularly, $\sim 60\%$ of objects are subluminous. This is just the property of models of indirect double-detonation SNe Ia, as compared to "standard" direct detonation models (Höfflich & Khokhlov 1996). For SC-SN one may expect some spread in properties of exploding stars resulting from differences in the masses of critical layers, which are functions of thermal history of the underlying dwarfs, their mass, and $\dot{M}$. Spread in the properties of SNe Ia may be also expected in the C-SN models, due to the difference in the total mass of objects. Mean total mass of merging pairs declines from 2 $M_\odot$ for $T \lesssim 3 \times 10^8$ yr to 1.6 $M_\odot$ for $T \gtrsim 3 \times 10^9$ yr.

Using population synthesis based on the scenario approach to describe stellar evolution proves to be efficient for understanding features of very different objects. It allows us to analyse their genesis and statistical properties. Its important advantage is ability to predict existence of certain stellar configurations and thus provide guidelines for observations. Discovery of several close double degenerates is an excellent example of the latter ability.

We acknowledge support of the Organizing Committee which enabled us to attend the conference and support from RFBR grant 96-0216531.

REFERENCES

Bailes, M. 1996, in Compact Stars in Binaries, ed. J. van Paradijs, E. P. J. van den Heuvel, & E. Kuulkers, (Dordrecht: Kluwer), 213

Branch, D., Livio, M., Yungelson, L. R., Boffi, F. R., & Baron, E. 1995, PASP, 107, 1019

Canal, R., Ruiz-Lapuente, P., & Burkert, A. 1996, ApJ, 456, L101

Dalton, W., & Sarazin, C. L. 1995a, ApJ, 440, 280; 448, 369

Dewey, R., & Cordes, J. 1987, ApJ, 321, 780

Eichler, D., Livio, M., Piran, T., & Schramm, D.N. 1989, Nat, 340, 126

Han, Z., Podsiadlowski, Ph., & Eggleton, P. 1995, MNRAS, 272, 800

Höfflich, P. & Khokhlov, A. 1996, ApJ, 457, 500

Holberg, J. B., Saffer, R. A., Tweedy, R. W., & Bastrow, M. A. 1995, ApJ, 452, L133

Iben, I. Jr., & Tutukov, A. V. 1984, ApJSS, 54, 335

Iben, I. Jr., & Tutukov, A. V. 1994, ApJ, 431, 264

Iben, I. Jr., & Tutukov, A. V. 1996, ApJSS, 105, 145

Iben, I. Jr., Tutukov, A. V., & Yungelson, L. R. 1995a, ApJSS, 100, 217

Iben, I. Jr., Tutukov, A. V., & Yungelson, L. R. 1995b, ApJSS, 100, 233

Iben, I. Jr., Tutukov, A. V., & Yungelson, L. R. 1996, ApJ, 456, 750

Iben, I. Jr., Tutukov, A. V., & Yungelson, L. R. 1997, ApJ, 475, 291

Kolb, U. 1993, A&A, 271, 149

de Kool, M. 1990, ApJ, 358, 189

de Kool, M. 1992, A&A, 261, 188

de Kool, M., & Ritter, H. 1993, A&A, 267, 397

de Kool, M. & Green, P. J. 1995, ApJ, 449, 236

de Kool, M., 1996, in Evolutionary Processes in Binary Stars, ed. R. A. M. J. Wijers, M. B. Davies, & C. A. Tout, (Dordrecht: Kluwer), 365

Lipunov, V. M., & Postnov, K. A. 1988, ApSpSci, 145, 1

Lipunov, V. M., Postnov, K. A., & Prokhorov, M. E. 1994, ApJ, 423, L121

Lipunov, V. M., Postnov, K. A., & Prokhorov, M. E. 1996, ApSpSciRev, 9, 1

Livne, E. & Glasner, A. 1990, ApJ, 361, 244

Marsh, T. R. 1995, MNRAS, 275, L1

Marsh, T. R., Dhillon, V. S., & Duck, S. R. 1995, MNRAS, 275, 828

Meurs, E. J. A., & van den Heuvel, E. P. J. 1989, A&A, 226, 88

Munari, U., & Renzini, A. 1992, ApJ, 397, L87

Paczyński, B., 1976, in Structure and Evolution of Close Binary Systems, ed. P. Eggleton, S. Mitton, & J. Whelan (Dordrecht: Reidel), 75.

Paczyński, B. 1986, ApJ, 308, L43

Phinney, E.S. 1991, ApJ, 380, L17

Podsiadlowski, Ph., Joss, P. C., & Hsu, J. J. L. 1992, ApJ, 391, 246

Politano, M. 1988, PhD Thesis, Univ. Ill. Urbana-Champaign

Politano, M. 1996, ApJ, 465, 338

Pols, O. R., Coté, J., Waters, J. B. F. M., & Heise, J. 1991, A&A, 241, 419

Pols, O. R., & Marinus, M. 1994, A&A, 288, 475

Prialnik, D., & Kovetz, A. 1995, 445, 789

Portugies Zwart, S. F., & Verbunt, F. 1995, A&A, 309, 179

Rappaport, S., Di Stefano, R., & Smith, M. 1994, 426, 692

Rasio, F., & Shapiro, S. L. 1995, 448, 887

Romani, R. W. 1992, ApJ, 399, 621

Ruiz-Lapuente, P., Burkert, A., & Canal, R. 1995, ApJ, 447, L69

Savonije, G. J., de Kool, M., & van den Heuvel, E. P. J. 1986, A&A, 155, 51

Schaefer, B. E. 1996, ApJ, 464, 404

Shutz, B.F. 1986, Nat, 323, 310

Taam, R. E. 1996, in Compact Stars in Binaries, ed. J. van Paradijs, E. P. J. van den Heuvel, & E. Kuulkers, (Dordrecht: Kluwer), 3

Thorne, K. 1987, in 300 Years of Gravity, ed. S. W. Hawking & W. Israel (Cambridge: CUP), 330

Thorne, K. 1996, in Compact Stars in Binaries, ed. J. van Paradijs, E. P. J. van den Heuvel, & E. Kuulkers, (Dordrecht: Kluwer), 153

Tutukov, A. V., & Fedorova, A. V. 1989, SvA, 33, 606

Tutukov, A. V., & Yungelson, L. R. 1973, Nauchn. Informatsii Astron. Sov., 27, 70

Tutukov, A. V., & Yungelson, L. R. 1976, Astrofisika, 12, 521

Tutukov, A. V., & Yungelson, L. R. 1979, Acta Astron., 29, 665

Tutukov, A. V., & Yungelson, L. R. 1992, SvA, 36, 266

Tutukov, A. V., & Yungelson, L. R. 1993a, MNRAS, 260, 675

Tutukov, A. V., & Yungelson, L. R. 1993b, Astron. Zh., 70, 812

Tutukov, A. V., & Yungelson, L. R. 1994, MNRAS, 268, 871

Tutukov, A. V., & Yungelson, L. R. 1995, in Cataclysmic Variables, ed. A. Bianchini, M. Della Valle, & M. Orio (Dordrecht: Kluwer), 495

Tutukov, A. V., & Yungelson, L. R. 1996, MNRAS, 280, 1035

Tutukov, A. V., Yungelson, L. R., & Iben, I. Jr. 1992, ApJ, 386, 197

Van sen Bergh, S., & McClure, R. D. 1994, ApJ, 425, 205

van den Heuvel, E. P. J., & Heise, J. 1972, NatPhysSci., 239, 67

van den Heuvel, E. P. J., Bhattacharya, D., Nomoto, K., & Rappaport, S. A. 1992, A&A, 262, 97

Webbink, R. F. 1984, ApJ, 227, 355

Whelan, J. C., & Iben, I. Jr. 1973, ApJ, 186, 1007

Woosley, S. E., & Weaver, T. A. 1994, ApJ, 423, 371

Yungelson, L. R., & Tutukov, A. V. 1991, in Wolf-Rayet Stars and Interrelations with Other Massive Stars in Galaxies, ed. K van der Hucht, & B. Hidayat (Dordrecht: Kluwer), 459

Yungelson, L. R., Tutukov, A. V., & Livio, M. 1993, ApJ, 418, 794

Yungelson, L. R., Livio, M., Tutukov, A. V., & Saffer, R. 1994, ApJ, 420, 336

Yungelson, L. R., Livio, M., Tutukov, A. V, & Kenyon, S. 1995, ApJ, 447, 656

Yungelson, L. R., Livio, M., Tutukov, A. V., Truran, J. W., & Fedorova, A. V. 1996, ApJ, 466, 890

On The Evolution of Accreting White Dwarfs with Rotation-Induced Mixing

By MASAYUKI Y. FUJIMOTO[1] AND ICKO IBEN, Jr.[2]

[1]Department of Physics, Hokkaido University, Sapporo 060, JAPAN

[2]Departments of Physics and of Astronomy, University of Illinois, Urbana-Champaign, IL 61018, USA

In close interacting binary systems, accreted material carries angular momentum and the resulting differential velocities in the accretor can cause hydrodynamical instabilities which generate turbulent motions, resulting in mixing and heat transfer. Adopting a treatment of rotation-induced mixing proposed by Fujimoto (1993), the evolution of accreting white dwarfs is calculated through the ignition of hydrogen shell flashes. Material mixing takes place to the degree necessary to reproduce the enrichment of heavy elements observed in classical nova ejecta. This indicates that accreted angular momentum plays an important role in the evolution of accreting white dwarfs, and, in addition, gives support to the adopted treatment of rotation-induced mixing. The evolution of merging white dwarfs and the relevance of rotation-induced mixing to the merger model of type Ia supernovae are also discussed within the same framework.

1. Introduction

The evolution of accreting white dwarfs has been a central problem in astrophysics for several decades. There are a wide variety of applications ranging from nova explosions, symbiotic stars, and supersoft X-ray sources to type Ia supernovae. Most studies of the evolution of accreting stars through the resultant thermonuclear flashes have assumed spherical symmetry, with material piling up layer by layer on the surface. It is not obvious that this simplifying assumption is valid since, in close binary systems, material transferred from the companion star carries angular momentum.

Actually, it has been recognized that some of the main features of the classical nova phenomenon cannot be accounted for very well by spherically-symmetric models, even though gross features can possibly be understood in terms of such models. In particular, existing models have failed to explain satisfactorily the overabundances of CNO and other heavier elements observed in ejecta (e.g., see Truran 1990). In the case of X-ray bursts, where the accretor is a neutron star instead of a white dwarf, and the time between outbursts is short enough to allow the monitoring of many outbursts in one observing run, the evidence against spherically symmetric models is more straightforward. Burst properties are observed to vary greatly and rather sporadically without apparent correlation with the accretion rate, even though the accretion rate remains constant within a factor of ~ 2; most bursts recur at much shorter intervals and with much smaller burst energy, by up to more than a factor of ten, than predicted by the spherically-symmetric models (see Fujimoto *et al.* 1987 and references therein).

These observations suggest the presence of mixing mechanism(s) in the radiative region near the surface. In the case of novae, particle diffusion has been proposed as a major mixing mechanism by Prialnik & Kovetz (1984). Numerical models constructed with particle diffusion and convective mixing demonstrate that the dredge-up of white dwarf material indeed takes place and contributes to the enhancement of CNO elements in the nova ejecta (Prialnik & Kovetz 1984, 1995; Kovetz & Prialnik 1985; Iben *et al.* 1992).

Yet, the degree of mixing is not always large enough to explain the observations, in particular for higher accretion rates (Fujimoto & Iben 1992). Moreover, it is difficult to understand the irregular variability of X-ray bursters in terms of particle diffusion alone.

The purpose of the present paper is to explore the possible mixing caused by the accretion of angular momentum. Differential rotation, which is continuously fed by the accretion of angular momentum, gives rise to hydrodynamical instabilities; turbulent motions are generated and lead to the redistribution of angular momentum, and concomitantly, to material mixing and heat transport. In the context of the nova phenomenon, Kippenhahn & Thomas (1978) invoked the Kelvin-Helmholtz instability, and proposed the formation of an accretion belt around the equator of the accretor. MacDonald (1983) subsequently argued that mixing in the horizontal direction is much more efficient than in the vertical direction, and Kutter & Sparks (1989) pointed out that the Kelvin-Helmholtz instability alone is not capable of extracting enough angular momentum from accreted material to cause strong shell flashes.

In studying rotation-induced mixing, the evaluation of the efficiency of turbulent mixing is crucial. Previous explorations have invoked combinations of known kinds of instabilities, including (in analogy with instabilities in plane-parallel flows) the Kelvin-Helmholtz instability and its modification by thermal diffusion (Townsent 1958; Zahn 1974), the thermal instability of axisymmetric modes (Goldreich & Schubert 1967), and even the Eddington-Sweet circulation (e.g., Endal & Sofia 1976; Pinsonneault *et al.* 1989). It is true that a precise theory of the generating mechanisms and properties of turbulence has yet to be established. Even so, the problem can be formulated in a more general and self-consistent way than in most previous work. Here, we adopt a formulation by Fujimoto (1993), which is based on an analysis of the stability of rotating bodies to adiabatic perturbations of nonaxisymmetric modes.

Assumptions and methods of computation are briefly summarized in §2. The results of calculations of the evolution of accreting white dwarfs with rotation-induced mixing are presented in §3 and compared with the results of calculations when only particle diffusion is taken into account. The case of two merging carbon-oxygen (CO) white dwarfs is also examined to test the role of heat transfer accompanied by material mixing, and the relevance to type Ia supernova models is discussed. A summary and discussion are given in §4.

2. Assumptions and method

We compute the evolution of accreting white dwarfs by utilizing the model equations described in Fujimoto (1993) where turbulent mixing is approximated by a diffusion process. The viscosity ν due to turbulent motions is taken to be proportional to the upper bound $\omega_{\rm i,max}$ on the growth rates of the hydrodynamical instabilities of nonaxisymmetric, adiabatic modes, and we set

$$\nu = \alpha\, \omega_{\rm i,max}\, l_{\rm max}^2, \tag{2.1}$$

where $l_{\rm max}$ is a characteristic length which is set equal to the pressure scale height and the radius in the vertical and horizontal directions, respectively, since these are the maximum possible values, and α is a parameter which is presumably much smaller than unity.

We assume that the instabilities of axisymmetric modes dominate in feeding the turbulence, and contributions of other modes of instabilities are ignored. The effect of the Eddington-Sweet circulation is not taken into account either since in our case differential rotation is promoted exclusively by the supply of angular momentum at the surface.

For material mixing in a stable stratification, work must be done against gravity; in the present study, we restrict the energy source to the kinetic energy of differential rotation. We may define the Richardson number Ri as

$$Ri = N^2/|\boldsymbol{D}|^2, \tag{2.2}$$

where N is the Brunt-Väisälä frequency and $\boldsymbol{D}$ is the gradient of the rotational velocity, and denote by the flux-Richardson number Rf the fraction of the available kinetic energy that can be converted into the potential energy of mixed material. With these numbers, the material diffusivity in the vertical direction $K_\rho^{(V)}$ is related to the turbulent viscosity in the vertical direction $\nu^{(V)}$ by

$$K_\rho^{(V)} = \nu^{(V)}(Rf/Ri). \tag{2.3}$$

Extrapolating the results of laboratory experiments, it follows that $Rf \simeq 0.20$ (Fujimoto 1988). For mixing in the horizontal direction, $K_\rho^{(H)} = \nu^{(H)}$.

Although the interaction between convection and rotation has not yet been properly understood, it is expected that convection tends to smear out shears; in the convection zone, therefore, the diffusion coefficients are approximated by

$$\nu = K_\rho = v_{\rm conv}\, l_{\rm mix}, \tag{2.4}$$

where $v_{\rm conv}$ is the average velocity of convective elements calculated from the mixing-length theory with a mixing length $l_{\rm mix}$, which is set equal to the pressure scale height.

The equations of stellar structure and evolution are coupled with the equations for angular momentum and element diffusion and, using the prescriptions developed by Kippenhahn & Thomas (1970), all equations are reduced to a one dimensional form. The boundary condition for angular velocity Ω is replaced by the relationship

$$\dot{M}r^2\Omega + 4\pi r^4 \rho\nu(d\Omega/dr) = \dot{M}R^2\Omega_{\rm K} \tag{2.5}$$

evaluated at the base of the atmosphere, and we have assumed a steady state distribution for Ω above the base of the atmosphere. Here, $\dot{M}$ is the mass accretion rate, R is the surface radius, and $\Omega_{\rm K}$ is the angular velocity of Keplerian rotation at the surface $(= \sqrt{GM_{\rm WD}/R^3})$. Similarly, for the abundance X_i of the $i^{\rm th}$ element,

$$\dot{M}X_i + 4\pi r^2 \rho K_\rho(dX_i/dr) = \dot{M}X_{i,0}, \tag{2.6}$$

where subscript 0 denotes the abundance in the accreted material.

3. Results of computations

3.1. *Models of nova explosions*

We have computed the evolution of an accreting CO white dwarf of mass $1\,M_\odot$ through a hydrogen shell flash for several sets of accretion rates and values of α in eq. (2.1); the initial models are the steady state models of Iben *et al.* (1992), assumed to be rotating uniformly in synchronism with the orbital motion at the rate $\Omega = \sqrt{G\,M_\odot/\,R_\odot{}^3}$. The characteristics of resultant hydrogen shell flashes are summarized in Table 1; included in the table for comparison are results of model computations with particle diffusion (Iben *et al.* 1992) and results for a model without mixing.

Figure 1 shows the changes in the distribution of angular velocity Ω in the white dwarf model when $\dot{M} = 10^{-8}\,M_\odot\,{\rm yr}^{-1}$ and $\alpha = 10^{-3}$. The solid vertical line marks the location of the stellar surface in the initial model. Accreted angular momentum is efficiently transported inwards, and the stellar interior rotates almost uniformly. The degree of differential rotation is measured by the Richardson number defined by eq. (2.2) with $|\boldsymbol{D}| = (d\Omega/d\log r)$; it remains larger than $\sim 10^4$ over most of the interior. During

$\dot{M}(M_\odot\ \mathrm{yr}^{-1})$	α	$M_{\mathrm{acc}}^{a)}(M_\odot)$	$M_{\mathrm{cs}}^{b)}(M_\odot)$	$M_{\mathrm{cs}}/M_{\mathrm{acc}}$	Z_{CNO}	$L_{\mathrm{H}}^{\mathrm{max}}(L_\odot)$
10^{-8}	10^{-3}	4.2×10^{-6}	7.4×10^{-6}	1.8	0.47	2.1×10^{8}
10^{-8}	10^{-5}	3.1×10^{-6}	6.6×10^{-6}	2.1	0.58	1.8×10^{8}
10^{-9}	10^{-5}	6.9×10^{-6}	1.40×10^{-5}	2.0	0.53	3.1×10^{9}
10^{-10}	10^{-5}	1.40×10^{-4}	1.84×10^{-4}	1.3	0.27	3.3×10^{12}
10^{-8}	Iben *et al.*	6.0×10^{-6}	8.4×10^{-6}	1.4	0.30	1.5×10^{8}
10^{-9}	Iben *et al.*	1.33×10^{-5}	1.75×10^{-5}	1.3	0.26	1.5×10^{9}
10^{-10}	Iben *et al.*	1.45×10^{-4}	1.79×10^{-6}	1.2	0.21	3.3×10^{11}
10^{-8}	no-mix.	3.2×10^{-5}	3.2×10^{-5}	1.0	0.02	2.0×10^{8}

a) the amount of accreted mass before the iginition of a shell flash
b) the maximum mass of the convective shell during a flash

TABLE 1. Characteristics of Resultant Hydrogen Shell Flashes

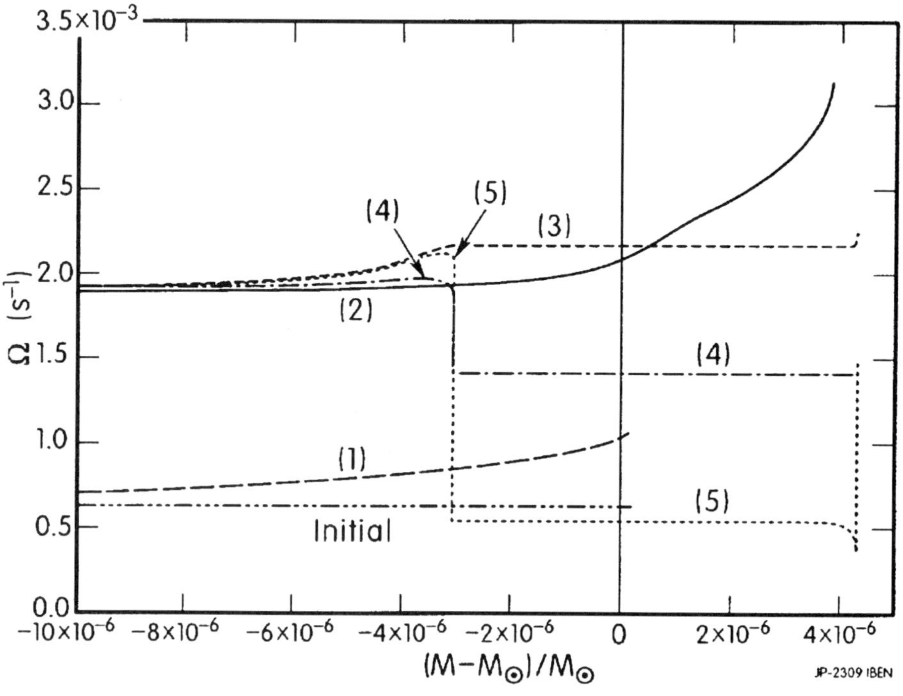

FIGURE 1. Changes in the angular velocity during the evolution of an accreting white dwarf of mass $1\,M_\odot$ through a hydrogen shell flash for $\dot{M} = 10^{-8}\,M_\odot\,\mathrm{yr}^{-1}$ and $\alpha = 10^{-3}$.

the accretion phase (solid line), Ω grows larger rather rapidly in the outer layers as the density there decreases with increasing temperature brought about by accretion-induced heating (see eq. [2.5]); it should be noted, however, that Ω is still much smaller than the Keplerian rotation rate at the surface $\Omega_K \simeq 0.9\,\mathrm{sec}^{-1}$. When flash convection appears, the angular velocity is effectively averaged in the convection zone, and then, it decreases as the convection zone expands.

Composition profiles are shown in Figure 2 at a stage during the accretion phase before the ignition of a shell flash and at a stage after flash convection sets in and becomes fully developed. Despite the efficient transfer of angular momentum, material mixing

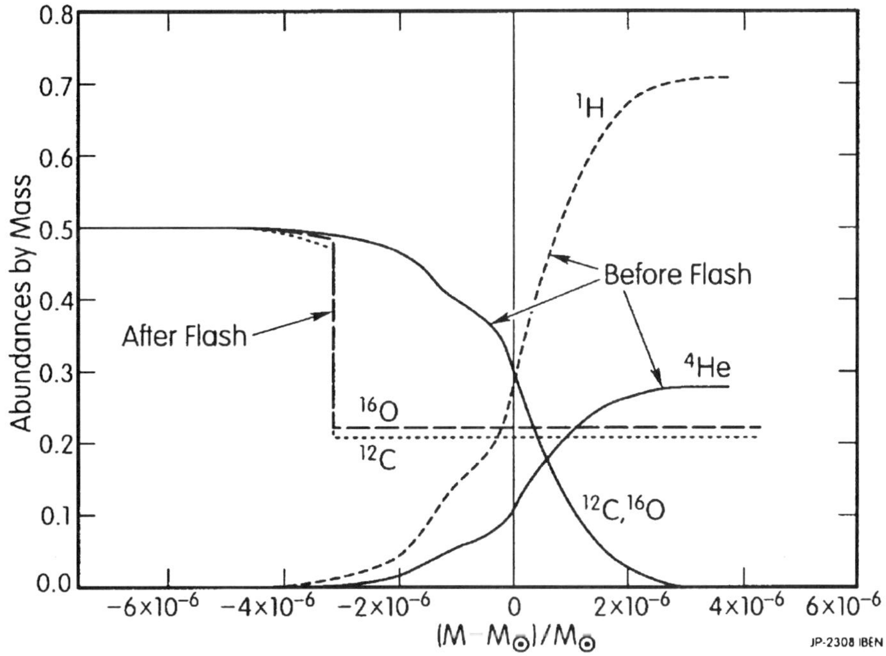

FIGURE 2. Composition profiles before flash convection sets in (solid lines) and after flash convection fully develops (dotted lines) for the same model described in Fig. 1.

remains moderate because of a large Richardson number (see eq. [2.3]); $K_\rho \simeq 10^2$, while $\nu \simeq 2 \times 10^6\,\mathrm{cm\,sec^{-1}}$ near the bottom of the accreted layer. Finally, white dwarf material is dredged up to the surface by flash convection, which penetrates into a layer where the hydrogen concentration (dashed line in Fig. 2) is as small as $X \sim 0.01$, as in models with particle diffusion.

Because of the transport of fuel into layers below the original surface and also because of the heating of these layers, when a shell flash is ignited, the accreted mass is much smaller than in a comparable model without mixing; the difference amounts to a factor of ~ 10 for when $\dot{M} = 10^{-8}\,M_\odot\,\mathrm{yr}^{-1}$.

The resultant abundances by mass of CNO elements Z_{CNO} in the convective zone are given in the sixth column of Table 1. They depend only slightly on α. Smaller α tends to decrease ν, but the increase in the gradient of the rotation velocity competes in such a way as to offset the change in ν. In contrast, K_ρ rather increases owing to the reduction in the Richardson number. With increasing accretion rate, the degree of mixing increases, as is intuitively expected; the dependence is weak, however, for higher accretion rates ($\gtrsim 10^{-9}\,M_\odot\,\mathrm{yr}^{-1}$), since the entropy gradient becomes smaller and the stratification is determined mainly by the difference in the mean molecular weights between the accreted and white dwarf matter.

3.2. *the case of merging CO white dwarfs*

Merging white dwarfs in close binary systems have been proposed as candidates for type Ia supernovae (Iben & Tutukov 1984, Webbink 1984). Rotation-induced mixing or, more specifically, heat transfer accompanied by material mixing may play an important role because of large inflow rates of angular momentum. In order to investigate the effect,

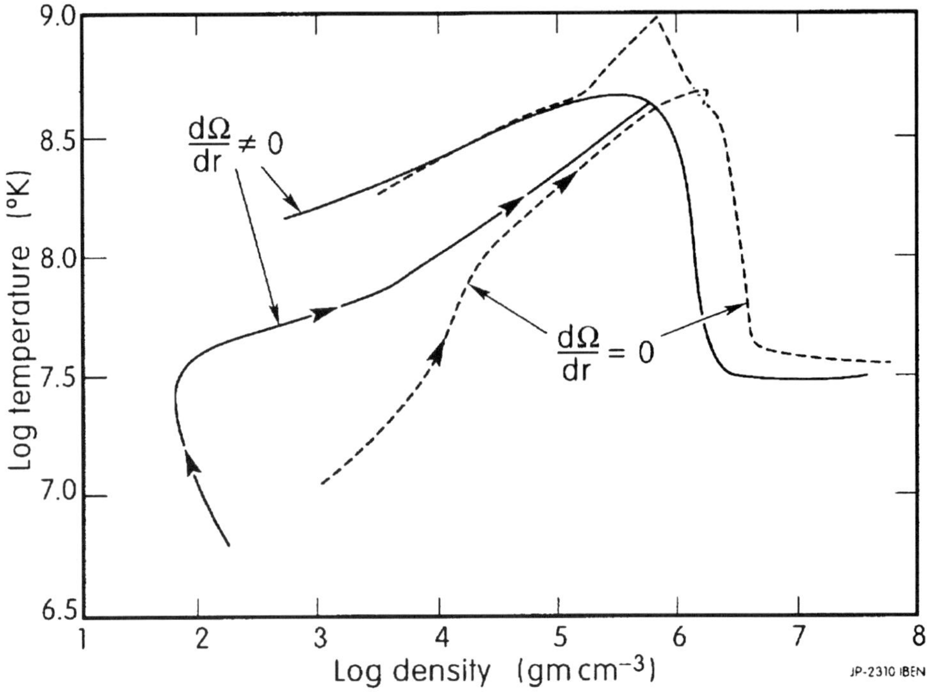

FIGURE 3. Evolutionary loci of a shell near the original surface of a $1\,M_\odot$ white dwarf accreting at rate $10^{-5}\,M_\odot\,\mathrm{yr}^{-1}$ with (solid line with arrows) and without (dashed line with arrows) rotation-induced mixing. Structure curves at the end of the computations are also given.

we have computed with and without rotation-induced mixing the evolution of a CO white dwarf of $1\,M_\odot$ which is accreting CO material. The accretion rate is taken to be $\dot{M} = 10^{-5}\,M_\odot\,\mathrm{yr}^{-1}$, about a half of the rate limited by the Eddington luminosity $(= 4\pi c R_{\mathrm{WD}}/\kappa_{\mathrm{el}} \sim 1.6 \times 10^{-5}\,M_\odot\,\mathrm{yr}^{-1})$.

Trajectories of a shell near the original surface and structure characteristics at the end of the calculations are shown in the density and temperature diagram of Figure 3. Simultaneously with material mixing, heat is transferred inwards (from shells of lower temperature to those of higher temperature) under a stable stratification, contrary to the usual direction of heat flow. This has a large effect in the outer shells at low densities, which are heated to higher temperatures as compared with the model without rotation-induced mixing. As the accreted layer deepens and approaches the shell of the largest κP at fixed temperatures, however, the inward heat flow encounters a bottleneck, and a steep temperature inversion develops around the bottleneck (Fujimoto *et al.* 1985); this strong stratification results in a large Richardson number. The transfer of angular momentum is, nevertheless, so effective that it brings the whole star into nearly uniform rotation; however, K_ρ becomes much smaller than ν, and heat transfer is not able to heat matter appreciably in the deep interior. Accordingly, the difference in the the thermal structure near the heat flow bottleneck becomes small between the models with and without the rotation-induced mixing. Since the maximum temperature occurs just above the bottleneck, the evolution leading to carbon ignition in the model with rotation-induced mixing will differ little from that in the model without it.

4. Conclusions and discussion

By numerical simulations of the rotation-induced mixing produced by nonaxisymmetric instabilities (Fujimoto 1993), we have demonstrated that:
(1) Rotation-induced mixing is efficient enough to redistribute the accreted angular momentum and bring the accreting star into nearly uniform rotation. It also produces composition mixing of the right order of magnitude to explain the observed enrichment of heavy elements in nova ejecta, and this may be taken as indirect support for the treatment of rotation-induced mixing which we have adopted.
(2) Composition mixing and the attendant heat transfer occur mainly in the outer layers and, hence, have little affect on the uniformly rotating interior.

The present results suggest that accreted angular momentum plays a role in the evolution of accreting white dwarfs, and hence, should be taken properly into account in future modeling of pre-nova evolution and of nova explosions.

In our computations, turbulent mixing has been treated as a diffusion process and we have therefore described only average properties, smoothing out fluctuations due to the irregular variability inherent in turbulence. Closer comparisons with the observations should take this inherent variability into account.

We have assumed that the energy for element mixing is extracted only from the kinetic energy of differential rotation. However, in the stellar interior, the timescale for heat diffusion is usually shorter than the molecular viscous timescale. Thus, the buoyancy of eddies due to temperature differences tends to be eliminated by heat diffusion while larger eddies decay into smaller ones. This effect may be important in the case of merging white dwarfs since there is essentially no difference in the chemical composition of accreted and white dwarf material.

REFERENCES

Endal, A., S. & Sofia, S. 1976, ApJ, 210, 184

Fujimoto, M. Y. 1987, A&A, 176, 53

———. 1988, A&A, 198, 163

———. 1993, ApJ, 419, 768

Fujimoto, M. Y., Hanawa, T., Iben, I., Jr., & Richardson, M. B. 1984, ApJ, 278, 813

Fujimoto, M. Y., & Iben, I. Jr. 1992, ApJ, 399, 646

Fujimoto, M. Y., Sztajno, M., Lewin, W. H. G., & van Paradijs, J. 1987, ApJ, 319, 902

Goldreich, P., & Schubert, G. 1967, ApJ, 150, 571

Iben, I. Jr., Fujimoto, M. Y., & MacDonald, J. 1992, ApJ, 388, 521

Iben, I. Jr., & Tutukov, A. V. 1984, ApJS, 54, 335

Kippenhahn, R., & Thomas, H.-C. 1970, in *Stellar Rotation*, ed. A. Slettebak, (Dordrecht: Reidel), p. 20

———. 1978. A&A, 63, 625

Kovetz & Prialnik 1985, ApJ, 291, 812

Kutter, G. S., & Sparks, W. M. 1989, ApJ, 340, 985

MacDonald, J. 1983, ApJ, 267, 732

Pinsonneault, M. H., Kawaler, S. D., Sofia, S, & Demarque, P. 1989, ApJ, 338, 424

Prialnik, D., & Kovetz, A. 1984, ApJ, 281, 367

———. 1995, ApJ, 445, 789

Townsent, A. A. 1958, J. Fluid Mech., 4, 361

Truran, J. W. 1990, in *Physics of Classical Novae, IAU Coll. No. 122*, eds. A. Castella & R. Viotti (Springer; Berlin), p. 373

Webbink, R. W. 1984, ApJ, 277, 355

Zahn, J. P. 1974, in *IAU Symp. No. 59, Stellar Instability and evolution*, eds. P. Ledoux, A. Noels, & A. W. Rogers (Dordrecht; Reidel), p. 185

Low Mass Binary White Dwarfs as Progenitors of AM CVn Variables and Type Ia Supernovae

By J. LIEBERT, D. ARNETT, AND W. BENZ

Steward Observatory, University of Arizona, Tucson, AZ 85721, USA

The possibilities that some of the newly-discovered, low mass white dwarf binaries could evolve into interacting AM CVn systems and/or supernovae of Type Ia (SNIa) are discussed. An increasing array of theoretical calculations suggests that the detonation of a degenerate helium layer atop a CO core of a white dwarf well below the Chandrasekhar mass may trigger an explosion of SNIa energy, while matching the observed diversity of light curves. Moreover, there is growing evidence that the diverse properties of SNIa explosions cannot be accounted for by a single mass (i.e., M_{ch}) being reached; instead, a range of mass in the progenitors is indicated.

We point out that, if the secondary's mass and binary mass ratio are sufficiently small, interaction of a low mass helium and a CO dwarf can result in stable mass transfer. The accreted helium does not necessarily ignite early in the interaction process. Driven hard by gravitational radiation, the initial mass transfer rate is super-Eddington, and the secondary's mass is drained to below $0.1\,M_\odot$ in $\sim 10^{6-7}$ years, after which time the accretion rate has dropped several orders of magnitude. The phase of slow mass transfer which follows may be identified with the observed properties of AM CVn binary systems. At these low mass transfer rates, the primary is allowed to cool – that is, the relatively massive helium envelope and CO core become more degenerate. We argue that these conditions may be favorable for helium ignition and the subsequent detonation of the carbon core. Theoretical work is currently underway to test these hypotheses.

Finally, we point out that, should the stable He + CO system fail to detonate, the accretion rate continues to decrease so that it is unlikely that AM CVn systems have time to evolve into DB white dwarfs, as is often proposed, unless some other angular momentum loss process (eg. magnetic braking) compensates for the decline in the gravitational radiation.

---

1. Introduction

Type Ia supernovae (SNIa) have unique properties including (1) occurrence in old populations, albeit perhaps at lower frequencies than in younger populations, and (2) a total absence of hydrogen observed in the spectrum. It has traditionally been believed that SNIa occur when a white dwarf reaches the Chandrasekhar mass limit ($M_{ch} \sim 1.45\,M_\odot$), with the approach to collapse somehow resulting in a catastrophic detonation. The increase in its mass is commonly attributed to either (1) accretion onto the white dwarf from a companion (the single degenerate SD scenario), or (2) dynamic merger of two white dwarfs (double degenerate DD progenitors). In any case, the type(s) of binary star serving as the progenitors has been unclear. In particular, it is well known that the average (single) white dwarf mass ($\sim 0.6\,M_\odot$) is far below M_{ch}, and it has been difficult to demonstrate that any proposed progenitor systems exist in sufficient numbers, if the mass of the exploding object must exceed the Chandrasekhar limit. This argument is applicable especially to ordinary cataclysmic variables, where the mass transfer rate from the low mass secondary is small, and the white dwarf loses at least most of this accreted material in nova outbursts (cf. Trimble 1982).

One of the more attractive of the proposed progenitor systems has been symbiotic accreting binaries, where the donor star – a red giant – could be losing mass at a high rate. If the mass transfer exceeds the order of 1–3 $\times 10^{-7}\ M_\odot\ \mathrm{yr}^{-1}$ (Paczynski & Zytkow 1978; Nomoto, Nariai, & Sugimoto 1978), the accreting white dwarf is expected to burn hydrogen in a stable shell, with the possibility that little hydrogen is ejected. It has also been argued that such systems are detected as the "supersoft" X–ray sources, a class of object found in *Einstein* and *ROSAT* observations (van den Heuvel *et al.* 1992; Rappaport *et al.* 1994). However, Kenyon *et al.* (1993) argued that accretion onto a symbiotic white dwarf could be very inefficient: In most symbiotics, the cool component is losing mass due to a wind, rather than from Roche-lobe overflow, and instabilities in the mass transfer are observed. Moreover, the appearance of Wolf-Rayet features indicates that the hot stellar component or its disk also loses mass at a substantial rate. Therefore, these authors concluded that the space density of symbiotic stars in the Galaxy is insufficient to account for the SNIa, if the latter must reach M_{ch} – see also Wheeler (1992).

Finally, we note that this type of object is also a candidate for the mechanism discussed in the next section (and the rest of this paper). However, the total absence of hydrogen features in the spectra of well-observed SNIa events such as SN1990M (Della Valle, Benetti, & Panagia 1996) may pose a problem for any model involving an explosion right next to secondary star with a large, massive, loosely-bound hydrogen envelope.

2. The Sub-Chandrasekhar Supernova Mechanism

In fact, the observed properties of SNIa's may be in conflict with any progenitor exploding upon reaching a single mass limit. A particular problem may be the apparent intrinsic inhomogeneity of SNIa properties (Phillips *et al.* 1992; Hamuy *et al.* 1994, 1996a). An intrinsic dispersion in the peak absolute magnitudes of ~ 0.8 mag in M_B and the finding that the peak absolute magnitude becomes brighter as the decline rate of the B light curve is slower (Phillips 1993; Hamuy *et al.* 1996b) suggest that the progenitors have a range of parameters. The most straightforward way to account for this correlation is to assume that the mass is the principal variable, and that the more luminous cases correspond to more massive progenitors (Hamuy *et al.* 1994; Pinto & Eastman 1996). Pinto and Eastman found that the total mass was the only parameter they could vary that could reproduce the observed relationship between M_B and the decline rate; moreover, they favor a wide mass range – approaching a factor of two. Such a requirement would appear to be incompatible with the assumption that the exploding objects must also reach or exceed the Chandrasekhar limit – although, for a different point of view, see Höflich, Khokhlov, & Wheeler (1995).

Recently, an alternative theoretical hypothesis for the SNIa explosion has been proposed in which the detonation of a object of sub-Chandrasekhar mass is proposed. For this mechanism to be possible, a substantial layer of helium must be assembled on top of a CO white dwarf core. If both layers are sufficiently degenerate, an ignition of helium near the base of the envelope may trigger the detonation of the carbon in the core, producing an incineration of SNIa energy. Preliminary theoretical calculations suggest that an SNIa explosion can occur for a total mass of this kind of white dwarf well below M_{ch}, for a substantial range of helium envelope and CO core masses (Livne 1990; Woosley & Weaver 1994; Livne & Arnett 1995). The viability of the mechanism remains controversial, however, and will be proven or disproven only after further calculations and observations (see, for example Branch *et al.* 1995 for a review). For the remainder of this discussion, we shall simply *assume* that, at least under some circumstances, a sub-Chandrasekhar white dwarf can explode, and outline a scenario under which this configuration might be assembled.

By making this assumption, we do not exclude the possibility that the diversity of SNIa properties is better satisfied by having some progenitors at or above the limit. In fact, the SNIa frequency may be substantially higher in populations of a younger mean age (Oemler & Tinsley 1979, van den Bergh 1990). Moreover, there is evidence that the most luminous SNIa events do not occur in elliptical and S0 galaxies (Hamuy *et al.* 1996b). Thus, the observations may be consistent with the assumption that the sub-Chandrasekhar mechanism is important in old, extragalactic populations.

3. Detached Helium DD Binaries and Their Formation Rate

While the first general surveys of field white dwarfs for close binary systems produced largely null results (cf. Robinson & Shafter 1987; Bragaglia *et al.* 1990; Foss, Wade, & Green 1991), a more systematic and successful way to find close DD pairs has recently been identified. In the spectroscopic analysis of a large sample of field DA white dwarfs by Bergeron, Saffer & Liebert (1992, BSL), a sequence of low mass white dwarfs was identified which arguably were safely below the $\sim 0.5\,M_\odot$ mass limit for core He ignition, and therefore must be composed of helium interiors. Other spectroscopic surveys also suggest that $\sim 10\%$ of DA white dwarfs are such objects, which cannot be explained by single star evolution.

Confirmation of the binary origin of the apparently low mass stars from BSL and the other surveys has come from radial velocity observations focussed on these candidates. Marsh, Dhillon & Duck (1995) and Marsh (1995, 1996) find that at least five of seven low mass stars from BSL are DDs, and a total of nine systems are now known from all samples. Additional low mass white dwarfs appear to have low mass main sequence companions. It is of great importance to note that 3 of 9 DD systems have orbital periods less than four hours; according to the standard gravitational radiation formula (cf. Landau & Lifschitz 1951), these may come into contact in a time less than the age of the Galactic disk ($\sim 10^{10}$ years). Observational selection may favor the discovery of short period systems. Theoretical modelling of the formation of close binaries (Yungelson *et al.* 1994, YLTS) suggests that the fraction of such short period systems should be somewhat less; a robust prediction of the expected period distribution is precarious, given that the prediction depends on α_{CE}, the efficiency parameter for the loss of orbital angular momentum in a common envelope.

These preliminary statistics on the frequency of DD systems among white dwarfs thus suggest that $\sim$ 10% are binaries with a visible helium degenerate component; for a preliminary exploration of the formation rate, we shall assume that half of these, $\sim$ 5%, have a degenerate companion (rather than an M dwarf). Guided by the YLTS predicted period distribution, we assume that only $\sim$ 1% of white dwarfs are DD with periods short enough to come into contact within 10^{10} years. Note that these assumptions, yet to be confirmed by further radial velocity followup observations, are nevertheless not in contradiction to the earlier spectroscopic surveys. Finally, we adopt the YLTS model fractions for He–He, He–CO and CO–CO white dwarf pairs of 50%, 30% and 20%, respectively. Since the selection is on the basis of a visible He white dwarf component, we assume that 3/8 of the selected DDs are composed of the relevant He–CO pairs in this calculation. The short period He–CO pairs would then account for 0.375% of field white dwarfs.

The local space density of white dwarfs is estimated to be 0.005 pc^{-3} (Liebert, Dahn & Monet 1988), corrected for white dwarfs missed in binaries with more luminous companions. So the local space density of He–CO DDs which will come into contact could be $\sim 2 \times 10^{-5}$. Now the expected rate that this kind of DD comes into contact is given by the space density divided by the lifetime of the DD before the interaction. Having restricted ourselves to DDs with periods less than 4 hours, let us simply assume that the lifetime is 10^{10} years, as before.

We estimate simply this potential contribution to an SNIa rate as

$$\dot{n}_{\rm SNIa} = \frac{2 \times 10^{-5}}{10^{10}} \quad {\rm pc}^{-3}\,{\rm yr}^{-1} \tag{3.1}$$

In a simple disk Galaxy with radius $= 10$ kpc, and height 0.6 kpc (total of both sides of the plane), the volume $\mathbf{V} \sim 2 \times 10^{11}\,{\rm pc}^3$ and (with no allowance for bulge or halo components)

$$\dot{N}_{\rm SNIa} = \dot{n}_{\rm SNIa} \times \mathbf{V} = 4 \times 10^{-4} \quad {\rm yr}^{-1} \tag{3.2}$$

which is within a factor of two of the low end of modern estimates of the SNIa rate, especially that applicable to a purely-old population, such as in elliptical galaxies. Given the severe uncertainties in the observed SNIa rate, this estimate is in an interesting range.

4. Interaction Phase

There are three key questions about the interaction of the He–CO DD binaries, once gravitational radiation has braked the orbit so that the lower mass component (M_2) fills its Roche lobe. First, is the mass transfer process unstable (dynamical) or stable (like a normal CV system)? Second, does the helium ignite early (hot and nondegenerate)? Third, if the answer to the preceding question is no, can the helium envelope and CO core achieve the conditions for the helium to ignite and trigger the carbon core?

Cameron & Iben (1986) studied the merging of two CO cores or one ONeMg primary and a CO secondary. They showed that, if the initial degenerate component masses have a ratio $q = M_2/M_1 \geq 0.6$, the interaction begins as a dynamical mass transfer event since in this regime, the radius of the secondary is growing faster than the size of the Roche lobe. Since the mass transfer rate becomes rapidly super-Eddington (Faulkner, Flannery & Warner 1972; Cameron & Iben 1988, their Fig. 3) an envelope will form which eventually engulfs both components, and an accretion disk may form around the primary star. In this case, the mass transfer probably remains dynamical until the secondary is (rapidly) used up. If the merger of a He–CO pair results in dynamical mass transfer, the helium is likely to ignite, the star swelling up to the helium "main sequence" (Iben & Tutukov 1985; Iben 1990).

For lower q, the mass transfer can be more sedate: an equilibrium situation arises in which orbital shrinkage due to gravitational wave radiation is exactly compensated by the receeding of the secondary following mass loss. The details of this process depend on the amount of angular momentum stored in the disk, the rate of change in secondary radius with mass loss, the friction in the common envelope, etc. (see Mochkovitch & Livio 1989, 1990, and especially Benz *et al.* 1990, their Figs. 14 & 15). No detailed multi-dimensional numerical calculations like those of Benz *et al.* are yet available for such lower mass, lower q cases. From Cameron and Iben (1985) we note, however, that for a secondary of $0.4\,M_\odot$ and sufficiently low q, the initial mass transfer rate is nearly $10^{-5}\,M_\odot\,{\rm yr}^{-1}$, proceeding as stable Roche lobe flow. Quickly the accretion rate drops rapidly as the orbital separation increases, and in 10^6–10^7 years, the secondary's mass is below $0.1\,M_\odot$, as the accretion rate dips below $10^{-8}\,M_\odot\,{\rm yr}^{-1}$.

This latter scenario thus results in the slow assembly of a fairly massive helium layer, atop a CO core. If the helium has not yet ignited by the time the slow accretion phase has been reached, both envelope and core will become more degenerate with time, as we discuss below. Even though the secondary might already have transferred the required amount of helium, no explosion has yet taken place as angular momentum may still provide partial support; but the "fuse" may have been lit.

5. The AM CVn Systems

At this point, the mass transfer rates and secondary masses of the stably-accreting He–CO pair match those of the known class of interacting helium DD binaries – the AM CVn (Canum Venaticorum) systems. These are interacting double degenerate (iDD) systems in which the donor star is composed of helium. They are a class of cataclysmic variable (CV) with the unique properties of (1) ultra-short orbital periods of $\sim$ 20–46 minutes, because both components are degenerate, and (2) helium-dominated spectra, showing absolutely no trace of hydrogen. These properties require (3) the secondary mass (M_2) to be $\leq 0.1\,M_\odot$ – all of this was worked out in the seminal paper of Faulkner, Flannery & Warner (1972). These properties apply to all six known members of the class (Warner 1995). Except for the longest-period member of the group, GP Com(= G 61–29) at 46 minutes, orbital periods from radial velocity measurements have not been obtainable, though less direct arguments place them in the range quoted above.

We argue that the existence of the interacting AM CVn binaries shows that (1) mass transfer from the helium donor star can occur in a stable fashion, as expected for low q, and that (2) the helium does not necessarily ignite during the initial phase of rapid mass transfer. Moreover, the properties of the observed class appear to be consistent with the expected secular decrease of the mass transfer rate with increasing orbital period.

The prototype AM CVn (also known as HZ 29), with a likely orbital period near 18 minutes, has a relatively high luminosity ($M_V \sim +7$) and estimated accretion rate of $3 \times 10^{-9}\,M_\odot\,\mathrm{yr}^{-1}$ (Warner 1995). The long period GP Com has a measured proper motion, and low luminosity appropriate to a nonaccreting DB or helium-atmosphere white dwarf ($M_V \sim +11$–12), and inferred accretion rate of the order $10^{-11}\,M_\odot\,\mathrm{yr}^{-1}$ or less. These two examples may thus represent fairly early and late stages of the interaction process. By the time the system resembles GP Com, several Gyr would be required for the last several hundredths of the secondary's mass to be used up. Yet the accretion rate should continue to decrease as long as the secondary remains degenerate – which should be the case at least to a mass below $0.001\,M_\odot$ (cf. Iben & Tutukov 1991); thus, the actual lifetimes of helium iDD systems could be greater than the age of the Galaxy.

However, we are suggesting here still a different fate for the evolving AM CVn systems. Since the primary star is accreting at well below the Eddington rate, its interior – meaning the presumed CO core and massive He envelope – is allowed to cool. Theoretical studies of the accretion–nova eruption cycle of the normal CV with a hydrogen-rich companion show that, at sub-Eddington accretion rates, the interior of the white dwarf primary cools at close to the normal rate, despite both the deposition of accretion energy and nova outbursts (Prialnik 1986). We may therefore conclude that the white dwarf primary in the docile iDD system should also cool in a normal fashion, decreasing slowly in entropy beneath the outermost $10^{-6}\,M_\odot$, if Prialnik's calculations are applicable. The cooling of the white dwarf should move it closer to the right conditions for detonation near the base of the helium layer. Of course, it is also possible that the AM CVn systems are examples of systems which have failed to produce a detonation, rather than being those which might still do so. The main point, however, is that it is likely that further cooling of the existing He envelope and CO core – not the few hundredths of a solar mass of additional material to be deposited – will make the detonation happen, if it is going to.

If there is the possibility that some fraction of AM CVn systems may yet be able to cool sufficiently (or add sufficient mass) to achieve a helium detonation, it is relevant to consider their formation rate in comparison to that estimated for SNIa systems, and for the proposed DD progenitors. Unfortunately, we are stymied by the fact that the lifetime of the iDD phase may be greatly shortened *if* the conditions for a great explosion do occur. These uncertainties also make it most difficult to estimate the lifetime and therefore the formation rate of AM CVn systems.

On the occasion of the "Ickofest" meeting in June 1996, we first became aware of the work of Tutukov & Yungelson (1996), which explores in detail the possible evolution of interacting binary systems with degenerate helium secondaries, in particular to produce AM CVn systems. The possible end state of a SNIa is given only cursory consideration, presumably in part because the authors might not have been willing to make the assumption we make in Section 2. Tutukov & Yungelson (1996) reach the same conclusion as stated above that the helium iDD phase should be very long, which forced these authors to the conclusion that the space density of AM CVn systems – dominated by the old, low luminosity cases like GP Com – should be comparable to the space density of the DB white dwarfs! These authors even suggest that some fraction of apparently-single DB white dwarfs are really iDD binary systems with very low accretion rates.

The analysis of field DB white dwarf by Beauchamp *et al.* (1996) probably lays to rest the possibility that a significant fraction of DB white dwarfs are interacting binaries with accretion rates comparable to GP Com (with strong emission lines of He I). In fact, we are aware of nearly 100 DB white dwarfs which now have excellent CCD spectrophometry, and show no evidence for lines in emission or with emission cores. Yet the comparable luminosities of GP Com and the DB white dwarfs mean that both can be found equally efficiently in magnitude-limited, UV-excess color surveys like the Palomar Green. We must therefore conclude that the space density of AM CVn is very conservatively *at least a factor of ten less* than that of DB white dwarfs (which in turn is about 1/5 of that of the DA white dwarfs). This leads us to conclude that the AM CVn systems may not live as long as suggested by the above reasoning of Tutukov & Yungelson (1996) and ourselves. One possible explanation, of course, is that a fuse can go off inside the primary star, changing its evolutionary state entirely.

6. Summary

In this paper we have discussed the hypothesis that the handful of known AM CVn variables are the products of the merger of the recently-discovered, detached low mass helium DD systems discussed in Section 3. In fact, the very existence of the AM CVn systems argues that contact can result in stable mass transfer without igniting the secondary's helium (Section 4). The low observed secondary masses and accretion rates are consistent with the hypothesis. The assembly of a massive helium envelope onto a sub-Chandrasekhar CO core under conditions allowing the accreting component's interior to cool would appear to (at least) approach the conditions desirable for the helium "trigger" to ignite the carbon in the core, as described in Section 2 and the papers referenced therein. The formation rates of helium DD systems also appear to be similar to the occurrence of SNIa events.

The hypothesis that a subset of helium DD and iDD encounters result in SNIa is consistent with several properties of these explosions discussed in the Introduction: First, they can certainly occur in older Galactic populations. Second, there is a total absence of hydrogen from both components by the time they reach the iDD phase. Third, the merging components should span a range of masses, which accommodates the diversity of SNIa properties, including the dispersion in peak luminosity and in decline time. Finally, the formation rates of these candidate progenitors appear to be large enough to be interesting. In fact, some elusive progenitors of the SNIa events may be observable in *not one but two* prior evolutionary phases!

Some caveats and cautionary statements have to be acknowledged as well. The sub-Chandrasekhar mechanism of Section 2 is controversial. The formation rate of DD pairs actually satisfying the necessary criteria is uncertain. The merger process of a DD system is not well studied. Finally, we reiterate that it is not necessary (or even desirable) for the old binary types discussed here to supply 100% of the SNIa events, given the growing evidence that the SNIa frequency may be higher in populations of a younger mean age, but that underluminous events may predominate in the older E-S0 galaxies. Thus, if our assumptions about the fraction of all white dwarfs which are suitable DD pairs, or the accretion rates of the iDDs are overly optimistic, they might still be of interest with respect to the lower SNIa rate in a very old stellar population.

In summary, our goal here is to put "into play" a new SNIa progenitor type and scenario. Probably much work remains to be done, both theoretically and observationally, before this idea can be proven or discarded. Given the problems plaguing other proposed progenitor binary systems, and the many questions yet to be resolved about the thermonuclear supernovae processes, we feel justified in advancing the ideas of this paper.

We thank Phil Pinto and Mario Hamuy for an advance copy of relevant papers. We acknowledge support from the the National Science Foundation grant AST 92–17961 (JL) and 94-17346 (DA), and NASA NAGW-2798 (DA, WB).

REFERENCES

Beauchamp, A., Wesemael, F., Bergeron, P., Lamontage, R., Liebert, J., & Saffer, R.A. 1996, in "Hydrogen-Deficient Stars and Related Objects," Bamberg, Germany, September 1995.

Benz, W., Bowers, R.L., Cameron, A.G.W., & Press, W.H., 1990, ApJ, 348, 647.

Bergeron, P., Saffer, R.A., & Liebert, J., 1992, ApJ, 394, 228. (BSL)

Bragaglia, A., Greggio, L., Renzini, A., & D'Odorico, S., 1990, ApJL, 365, L13.

Branch, D., Livio, M., Yungelson, L.R., Boffi, F.R., & Baron, E. 1995, PASP, 107, 1019.

Cameron, A.G.W., & Iben, I., Jr., 1986, ApJ, 305, 228.

Della Valle, M., Benetti, S., & Panagia, N., 1996, ApJL, 459, L23.

Faulkner, J., Flannery & Warner, B., 1972, ApJL, 175, L79.

Foss, D., Wade, R., & Green, R.F., 1991, ApJ, 374, 281.

Hamuy, M. *et al.* 1994, AJ, 108, 2226.

Hamuy, M., Phillips, M.M., Schommer, R.A., Suntzeff, N.B., Maza, J., & Aviles, R. 1996a, AJ, 112, 2391.

Hamuy, M., Phillips, M.M., Suntzeff, N.B., Schommer, R.A., Maza, J., Smith, R.C., Lira, P., & Aviles, R. 1996b, AJ, 112, 2438.

Höflich, P., Khokhlov, A.M., & Wheeler, J.C., 1995, ApJ, 444, 831.

Iben, I. Jr. 1990, ApJ, 353, 215.

Iben, I. Jr., & Tutukov, A.V., 1985, ApJS, 58, 661.

Iben, I. Jr., & Tutukov, A.V., 1991, ApJ, 370, 615.

Kenyon, S.J., Livio, M., Mikolajewska, J., & Tout, C.A., 1993, ApJ, 407, L81.

Landau, L. & Lifschitz, E. 1951, *The Classical Theory of Fields*, 2d ed. (Oxford: Pergamon), p. 367.

Livne, E. 1990, ApJL, 354, L53.

Livne, E. & Arnett, D., 1995, ApJ, 452, 62.

Marsh, T.R. 1995, MNRAS, 275, L1.

Marsh, T.R., Dhillon, V.S. & Duck, S.R. 1995, MNRAS, 275, 828.

Mochkovitch, R., & Livio, M., 1989, A&A, 209, 111.

Mochkovitch, R., & Livio, M., 1990, A&A, 236, 378.

Nomoto, K., Nariai, K., & Sugimoto, D., 1979, PASJ, 31, 287.

Oemler, A., & Tinsley, B.M., 1979, AJ, 84, 985.

Paczynski, B. & Zytkow, A.N., 1978, ApJ, 222, 604.

Phillips, M.M. 1993, ApJL, 413, L105.

Phillips, M.M., Wells, L.A., Suntzeff, N.B., Hamuy, M., Leibundgut, G., Kirshner, R.P., & Foltz, C.B. 1992, AJ, 103, 1632.

Pinto, P., & Eastman, R. 1996, submitted to Ap.J.

Prialnik, D. 1986, ApJ, 310, 222.

Rappaport, S. DiStefano, R., & Smith, M. 1994, ApJ, 426, 692.

Robinson, E.L., & Shafter, A.W., 1987, ApJ, 322, 296.

Trimble, V., 1982, *Observatory*, 102, 133.

Tutukov, A., & Yungelson, L. 1996, MNRAS, 280, 1035.

van den Bergh, S. 1990, PASP, 102, 1318.

van den Heuvel, E.P.J., Bhattacharya, D., Nomoto, K., & Rappaport, S.A., 1992, A&A, 262, 97.

Warner, B., 1995, Astr. & Spa. Sci., 225, 249.

Wheeler, J.C. 1992, in *Frontiers of Stellar Evolution*, 50th Anniversary of McDonald Observatory, ed. D. L. Lambert (ASP Conf. Series, vol. 20, Provo UT), p. 483.

Woosley, S.E. & Weaver, T.A., 1994, ApJ, 423, 371.

Yungelson, L.R., Livio, M., Tutukov, A.V. & Saffer, R.A., 1994, ApJ, 420, 336. (YLTS)

Presupernova Evolution in Massive Binaries

By PHILIPP PODSIADLOWSKI

Department of Astrophysics, University of Oxford, OX1 3RH, UK

We systematically discuss how binary interactions, in particular mass loss, mass accretion and merging, affect the presupernova structure of massive stars and the resulting supernova explosions. Since most massive stars are members of binary systems, binary interactions can explain a large variety of supernova (sub-)types (type II-P, II-L, II-b, Ib/Ic, SN 87A). The two nearby anomalous supernovae SN 1987A and SN 1993J are very probable products of binary evolution; the progenitor of SN 1987 most likely merged with a massive companion star in the not-too-distant past, while the progenitor of SN 1993J probably was a stripped supergiant, the natural end product of stable case C mass transfer. We estimate the frequencies of various supernova types predicted by binary evolution theory and show that the most common outcome of binary interactions is the production of helium stars, which are excellent candidates for type Ib/Ic supernovae. Finally, we compare the theoretical predictions with some observed helium-star binaries and discuss some of the outstanding issues of binary-evolution theory.

1. Introduction

Icko Iben and his collaborators have worked on almost any problem involving single and binary stars, indeed he often was the first to identify and define them. So it is not surprising that he has also looked at the question of how massive binary evolution affects supernova progenitors. In 1992, Tutukov, Yungelson and Iben applied binary population synthesis, a discipline they pioneered in the 80s, to massive binaries and predicted a large number of new supernova types (see their table 2), many of which still have to be discovered. In the same year, we published the results of a systematic study of how binary interactions affect the structure of supernova progenitors (Podsiadlowski, Joss & Hsu 1992; PJH). The main goal of our study was to explain the large variety of supernova (sub-)types that were being identified at the time (type II-P, II-L, II-b, SN87A, type Ib/Ic).

Hydrodynamical simulations of core-collapse supernova explosions show that the large variety of supernova light curves can be understood as a variation of envelope properties (e.g., Hsu et al. 1996), even if the conditions of core collapse were exactly the same for all supernovae. The most important factor is the mass in the hydrogen-rich envelope. Progenitors with large envelopes produce standard type II-P (plateau) supernovae. As the mass decreases, the light curves change to a linear type (II-L), then type II-b, and finally, when the hydrogen envelopes have been lost completely, to type Ib/Ic. A second important factor is the compactness of the progenitor. If the immediate supernova progenitor is a blue supergiant rather than a red supergiant (as the progenitor of supernova 1987A), more of the energy deposited in the envelope is required to expand the more tightly bound envelope and the resulting supernova is correspondingly dimmer.

Single-star evolution can naturally account only for some of this variability. For example, very massive stars (with initial masses larger than $25-35\,M_{\odot}$) lose their hydrogen-rich envelopes in a stellar wind and become Wolf-Rayet stars. While these are good candidates for the progenitors of type Ib/Ic supernovae, there are probably not enough Wolf-Rayet stars to explain the large number of observed type Ib/Ic supernovae (15–30%; see Podsiadlowski 1996). In addition, the supernova that produced the Cas A supernova remnant probably had a Wolf-Rayet star progenitor, but was substantially less luminous than a typical type Ib/Ic supernova, which may indicate that Wolf-Rayet stars produce another supernova type altogether.

On the other hand, we know that binary interactions can substantially alter the structure of stars. Since most massive stars are members of close binaries (see the references in PJH), it is immediately clear that binary evolution *must* play an important role in explaining the observed supernova variety. In this chapter, I will first summarize in section 2 how binary interactions affect the presupernova structure of massive stars, then apply these results in section 3 to two anomalous supernovae (SN 1987A and SN 1993), which probably had binary progenitors, and to helium-star binaries in section 4. Finally in section 5, I will discuss some of the open theoretical issues.

2. Binary interactions

2.1. *Mass-transfer types*

To understand binary interactions, one has to know that there are two qualitatively very different modes of mass transfer: more-or-less conservative mass transfer and dynamically unstable mass transfer.

(Quasi-)conservative mass transfer usually occurs when the mass donor has a radiative envelope and the mass ratio of the mass-accreting to the mass-losing component is not too small (e.g., de Loore & de Grève 1992). Then, most ($\gtrsim 0.5$) of the mass lost from the primary is accreted by the secondary. Thus in this case, both components of the binary are affected, one by losing mass, the other by accreting it. In addition, the orbital period tends to widen, if mass transfer is mainly conservative.

Dynamical mass transfer usually takes place when the secondary is a giant star with a deep convective envelope. In this case, mass transfer is dynamically unstable and leads to the formation of a common envelope surrounding the core of the giant and the secondary. Due to friction between this immersed binary with the common envelope, the orbit of the binary starts to shrink. Depending on how much energy is released in the orbital decay of the binary and is deposited in the envelope, two different outcomes are possible. If the deposited energy exceeds the binding energy of the envelope, the common envelope can be ejected, leaving a very close binary consisting of a helium star (or Wolf-Rayet star) and a normal companion star (which is hardly affected by the common-envelope phase). If the energy is not sufficient to unbind the envelope, the less dense component of the immersed binary will ultimately be tidally destroyed and the two components merge completely to form a single, but rapidly rotating giant.

In a typical binary scenario, a binary system may experience several different phases of mass transfer, which is the reason for the large variety of possible binary scenarios. In the following, we will concentrate on the main consequences of the various mass-transfer types for the structure of the immediate supernova progenitors (for a more detailed discussion see PJH and Hsu et al. 1996).

2.1.1. *Mass loss*

Some 20 – 40 % of all massive stars experience Roche-lobe overflow and mass loss at some point during their evolution. In most cases, they lose all of their hydrogen-rich envelopes and become helium stars or Wolf-Rayet stars, which are excellent candidates for type Ib/Ic supernovae. In some cases, however, it is possible for the primary to retain part of its hydrogen-rich envelope, in particular when mass transfer occurs very late during the evolution of the primary and when the initial mass ratio is very close to one. However, even in this case, at most a few solar masses can be retained in the envelope. The immediate supernova progenitor will be a *stripped supergiant* with a small envelope mass (Joss et al. 1988). It is interesting to note that the outer appearance of this star will be very similar to that of a star that has lost no mass at all, since the radius and the luminosity of the star are almost independent of the envelope mass (PJH). The resulting supernova will, however, be quite different. The light curve will show no extended plateau phase but resemble a type II-L or in the most extreme case a type IIb supernova (Hsu et al. 1996).

2.1.2. *Mass accretion*

Just as the structure and appearance of the mass-losing star can be strongly affected, the structure and further evolution of the accreting companion can also be dramatically altered by mass transfer. This depends, however, strongly on the evolutionary stage of the accreting secondary at the beginning of the mass-transfer phase.

Secondary on the main sequence
If the secondary is still on the main sequence at the beginning of the mass-transfer phase, the secondary is usually *rejuvenated* and will behave subsequently (after the mass-transfer phase) like a single, but now more massive star (Hellings 1983; PJH). However, as was recently shown by Braun & Langer (1995), this need not be the case if accretion occurs very late on the main sequence and if semi-convection is very slow (combined with the Ledoux criterion for convective instability), since the convective core will then not grow significantly as a result of accretion. The subsequent evolution will resemble the evolution of a star that accreted after the main-sequence phase (see below); in particular, the star may never become a red supergiant and spend the whole post-main-sequence phase in the blue-supergiant region of the Hertzsprung-Russell (H-R) diagram.

Secondary has completed hydrogen core burning
If the secondary has already left the main sequence before the beginning of the mass-transfer phase, the subsequent evolution will generally differ quite substantially from the evolution of a normal single star. Since the mass of the helium core will not grow as a result of mass accretion, the main effect of post-main-sequence accretion is to increase the envelope mass relative to the core mass. As was first shown by Podsiadlowski & Joss (1989) (see also De Loore & Vanbeveren 1992; PJH), this has the consequence that the star will now not become a red supergiant or, if it were a red supergiant at the time of the accretion phase, leave the red-supergiant region and spend its remaining lifetime as a blue supergiant (see also Barkat & Wheeler 1989 for a similar scenario). The final location of the star in the H-R diagram at the time of the supernova depends on how much mass has been accreted in the accretion phase. The star will be the bluer, the more mass it has accreted. The final supernova will of the SN 1987A variety (type II[blue]). Since a star has to accrete only a few percent of its own mass from a binary companion to be spun up to critical rotation, the progenitor is expected to pass through a phase where it is rapidly rotating. Rapid rotation may also induce large-scale mixing in the accreting star (even across chemical gradients). However, whether this happens depends critically on the angular-momentum transport inside and the structure of the accreting star.

2.1.3. *Binary mergers*

The most dramatic type of binary interaction is dynamical mass transfer leading to a common-envelope and spiral-in phase. If the orbital energy released during the spiral-in is sufficient to eject the common envelope, the end product is a short-period binary, consisting of a Wolf-Rayet primary and a normal stellar companion. The primary will eventually explode as a type Ib/Ic supernova. If the system remains bound, the secondary is also likely to experience mass loss in the future leading to a second mass-transfer phase, just as in the mass-accretion scenarios discussed in the previous subsection (the further evolution will also be similar).

If the common envelope is not ejected, the two components will merge completely to produce a more massive and rapidly rotating single star. This end product resembles in many ways the outcome of the post-main-sequence accretion models in section 2.1.2. In particular, it may end its evolution as a blue supergiant rather than as a red supergiant either because of the added mass in the envelope (Podsiadlowski, Joss & Rappaport 1990) or because of the dredge-up of helium (Hillebrandt & Meyer 1989) or both. The resulting supernova will again be of the SN 1987A variety.

The details of the merging process have not been studied in any detail in the past. Preliminary results from a study by Spruit & Podsiadlowski (1996) suggest that it is likely that the merging of two massive stars induces the mixing of He-rich material with the hydrogen-rich envelope and that it may also produce unusual nucleosynthesis (like s-processing, non-equilibrium CNO burning).

2.2. *Frequencies of supernovae in binaries*

In the previous subsection, we only discussed the main types of interactions. However, many binaries experience more than one phase of mass transfer. For example, low-mass helium stars, produced in a first mass-transfer phase, will expand again after core helium burning to become helium giants and may fill their Roche lobes for a second time (so-called case BB mass transfer; De Grève & De Loore 1977; Delgado & Thomas 1981). This still does not exhaust the whole variety of multiple mass-transfer scenarios. Nomoto et al. (1994) recently summarized various scenarios to produce bare CO cores as progenitors of type Ic supernovae, which involve up to *three* mass-transfer phases.

In order to assess the importance of binary interactions for the various supernova types, PJH have used Monte-Carlo simulations to estimate the frequencies of the various individual channels. Their results, however, must be considered very tentative because of the many observational and theoretical uncertainties involved in these estimates.

The main results of their study, which assumed a frequency of close binary systems between 1/3 and 2/3, were:

- A large fraction (20 % – 35 %) of all massive stars are affected by binary interaction.
- Some 15 – 30 % of stars lose all (most) of their hydrogen-rich envelopes and become helium stars, i.e., progenitors of type Ib/Ic supernovae; this estimate includes $\sim 5\,\%$ of stars with a very small hydrogen-rich envelope like the progenitor of SN 1993J (i.e., a type IIb supernova).
- Some 5 % of all stars explode as blue supergiants rather than as red supergiants with large hydrogen-rich envelopes (like SN 1987A because of accretion from or merging with a companion).
- Only in a few percent of all cases can a mass-losing star retain a significant hydrogen-rich envelope ($M_{\rm env} \gtrsim 0.5 M_\odot$); this fraction may, however, be substantially increased if enhanced wind mass loss before Roche-lobe overflow is taken into account.

3. Nearby anomalous supernovae

3.1. *SN 1987A*

SN 1987A in the Large Magellanic Cloud (LMC) was the first naked-eye supernova since Kepler's supernova in 1604. It has confirmed many long-held beliefs about the final stages in the evolution of massive stars, but has also provided many surprises. By now it has become clear that this supernova was a rather anomalous supernova (see Podsiadlowski 1992 for details).

The anomalies of SN 1987A

One of the major surprises of SN 1987A was the fact that its known progenitor, SK $-69°202$, was a blue supergiant rather than a red supergiant, the normally expected supernova progenitor.

A second surprise was the complex, but very axisymmetric nebula surrounding the supernova, which is formed out of material that was ejected from the progenitor in the not-too-distant past. The main geometry of the nebula consists of three rings, an inner ring centered on the supernova (Wampler et al. 1990; Jacobson et al. 1991) and two rings displaced to the South and the North, but in approximate alignment with the symmetry axis of the inner ring (Wampler et al. 1990; Burrows et al. 1995). This geometry implies an axisymmetric, but highly non-spherical structure of the envelope of the progenitor and/or its winds. The origin of this non-sphericity provides a severe constraint for models of the progenitor. A plausible mechanism to provide the required asymmetry is the flattening of the progenitor's envelope caused by rapid rotation (Chevalier & Soker 1989). However, straightforward angular-momentum considerations show that a single star which was rapidly rotating on the main sequence would be a slow rotator in any subsequent supergiant phase and could not possibly be significantly flattened at the time of the supernova explosion.

The third major surprise of the supernova involves a number of chemical anomalies in the hydrogen-rich envelope of the progenitor and in the presupernova ejecta. The most important of these are an enhancement of the helium abundance (by about a factor of two; Höflich 1988; Allen, Meikle & Spyromilio 1989; Wang 1991) and an enhancement of barium and other s-process elements (Williams 1987; Höflich 1988; Mazzali, Lucy & Butler 1992; Mazzali & Chugai 1995). These anomalies strongly suggest that most of the envelope of the progenitor above the CO core had been thoroughly mixed and that, in addition, s-processing has occurred in this unusual environment.

SN 1987A: a Binary Merger?

At present, single-star models fails to explain most of these anomalies, and the accumulated evidence now strongly suggests that the progenitor of this unusual supernova was a binary, at least in the not-too-distant past. Our preferred model at the moment is a merger model (Hillebrandt & Meyer 1989; Podsiadlowski et al. 1990; see also Chen & Colgate 1995 for a new variant of this scenario).

In our preferred picture, the system initially was a massive binary consisting of a primary of $15-20\,M_{\odot}$ and a significantly less massive secondary $(5-10\,M_{\odot})$ in a fairly wide orbit (with a period of several years), so that the primary started to fill its Roche lobe only on the asymptotic-giant branch (i.e., on its second ascent of the red-supergiant branch after helium core burning). The system then experienced dynamical mass transfer, leading to the complete merger of the two stars. The end product of this evolution is a single, but very rapidly rotating red supergiant, which has been thoroughly stirred up during the merging process (explaining the main chemical anomalies, provided that this environment allows for s-processing; see the discussion in Podsiadlowski 1992). As discussed in section 2.1.3, this star will now want to shrink to become a blue supergiant; but since it is initially rotating near break-up, it can do so only by losing a lot of angular momentum, most likely in an equatorial, disk-like outflow. Once the star has settled in the blue-supergiant region in the H-R diagram, it will turn on an energetic blue-supergiant wind which will start to sweep up all of the structures that formed previously; in particular, it will sweep up the excretion disk into the inner supernova ring. A possible explanation for the outer rings is that they could be the remnants of a swept-up biconical structure generated during an interacting-wind phase just preceding the merger phase (Podsiadlowski, Fabian & Stevens 1991), though another promising model has recently been suggested by Meyer (1996).

While some of the details of this model have not yet been calculated (e.g., the details of the formation of the triple-ring nebula and the nucleosynthesis in a rapidly rotating merged object), it is clear that such a model can quite naturally explain all of the major properties of SN 1987A (see also Podsiadlowski & Cumming 1995). This model must therefore be considered the best model for SN 1987A at the present time.

3.2. *SN 1993J*

SN 1993J in the spiral galaxy M81 is the brightest supernova since SN 1987A and appears to be another peculiar type II supernova which changed from a type II to a type Ib supernova (i.e., a type IIb). Its early light curve is characterized by a very sharp initial peak lasting for less than ten days followed by a less rapid secondary brightening. Photometric constraints on the progenitor (Aldering & Humphreys 1993; Podsiadlowski et al. 1993) suggest that the progenitor was a yellow/red supergiant with a mass of $\sim 15\,M_\odot$. The sharp peak in the light curve shows that the progenitor has lost all but a few tenths of a solar mass of its hydrogen-rich envelope prior to the supernova (e.g., Nomoto et al. 1993; Podsiadlowski et al. 1993; Woosley et al. 1994).

One of the most probable scenarios for such a progenitor is that the progenitor was a member of a close binary system and underwent stable case C mass transfer (i.e., as a red supergiant after helium core burning; for details see Podsiadlowski et al. 1994). As a result of Roche-lobe overflow, the progenitor will lose most of its hydrogen-rich envelope, of which a significant fraction will be accreted by the companion. However, when the mass in the hydrogen-rich envelope has decreased below a certain critical mass ($\sim 0.3\,M_\odot$ for a progenitor of $\sim 15\,M_\odot$), the envelope of the progenitor will shrink significantly so that the star will no longer fill its Roche lobe, and mass transfer will cease. Such a scenario produces very naturally and without significant fine-tuning the type of progenitor required for SN 1993J. Since the companion should still be present after the supernova (albeit most likely in an unbound orbit; Brandt & Podsiadlowski 1995), this model can be unambiguously verified once the supernova has sufficiently faded to reveal the companion (on a timescale of a few years). In addition, this would provide direct proof for the existence of stable case C mass transfer.

4. Helium-Star Binaries

As we showed in section 2, one of the most important outcomes of binary interactions is the production of a helium star in a binary. Unfortunately, very few such systems are known in nature. But those that exist provide some important constraints and also pose some new questions (for more details see Podsiadlowski 1996).

Close Wolf-Rayet binaries

The binaries HD97152 and HD152270 (St.-Louis et al. 1987) contain Wolf-Rayet primaries with O-star companions in very close orbits (with periods $\lesssim 10\,$d). These are excellent candidates for post-common-envelope systems, i.e., systems in which mass transfer was dynamically unstable and led to a common-envelope and spiral-in phase and the subsequent ejection of the envelope.

Binaries with helium-star secondaries

The binary Φ Persei with an orbital period of 127 d consists of a $21\,M_\odot$ primary and a $3.4\,M_\odot$ helium-star secondary (Poeckert 1981; Thaller et al. 1995). These parameters are most consistent with a conservative (case B) mass-transfer scenario.

Binaries with helium-star primaries

The binary υ Sgr with an orbital period of 138 d has a $\sim 2.5\,M_\odot$ helium-star primary with a $\sim 4\,M_\odot$ secondary (Dudley & Jeffery 1990). The primary has completed helium core burning and is now becoming a helium giant; it presently undergoes case BB mass transfer (Schönberner & Drilling 1983). Since the original mass of the primary must have been around $10\,M_\odot$, much larger than the present mass of the secondary, mass transfer must have been very non-conservative. On the other hand, the long orbital period suggests that there has not been significant spiral-in. This is somewhat puzzling in view of the discussion in section 2 and may suggest that the system experienced a contact phase (during early case B mass transfer), which lead to a large amount of systemic mass loss, but without significant loss of (specific) angular momentum.

5. Summary and some open issues

As we have shown in this chapter, binary interactions affect the presupernova structure of many stars (20 − 40%) and are at least in part responsible for the large variety of supernova (sub-)types.

However, there are still many gaps in our understanding of the relevant binary evolution physics. Some of the most important open issues involve:

- the occurrence of dynamical mass transfer, in particular the question of when it can be avoided and the criterion for common-envelope ejection,
- the details of the binary-merger process; does it produce mixing and unusual nucleosynthesis?
- the evolution of contact binaries (early case B evolution); can a spiral-in phase be avoided by an Eddington-driven wind from the accreting component? Would that explain systems like υ Sgr?

While Icko's work over the years has been a key in making stellar-evolution theory the mature science it is now, it is still encouraging for us younger people that there still remains quite substantial work to be done and that it will still take a few more year before the final book on stellar evolution can be written.

REFERENCES

Aldering, G., & Humphreys, R. M. 1994, AJ, 107, 662

Allen, D. A., Meikle, W. P. S., & Spyromilio, J. 1989, Nat, 342, 403

Barkat, Z., & Wheeler, J. C. 1989, ApJ, 342, 940

Brandt, W. N., & Podsiadlowski, Ph. 1995, MNRAS, 274, 461

Braun, H., & Langer, N. 1995, A&A, 297, 483

Burrows, C., et al. 1995, ApJ, 452, 680

Chen, K., & Colgate, S. 1995, preprint

Chevalier, R. A., & Soker, N. 1989, ApJ, 341, 867

De Grève, J.-P., & De Loore, C. 1977, ApSS, 50, 75

Delgado, A. J., & Thomas, H.-C. 1981, A&A, 96, 142

De Loore, C., & Vanbeveren, D. 1992, A&A, 260, 273

De Loore, C., & De Grève, J.-P. 1992, A&AS, 94, 453

Dudley, R. E., & Jeffery, C. S. 1990, MNRAS, 247, 400

Hellings, P. 1983, ApSS, 96, 37

Hillebrandt, W., & Meyer, F. 1989, A&A, 219, L3

Höflich, P. 1988, Proc. Astron. Soc. Australia, 7, 434

Hsu, J. J. L., Joss, P. C., Ross, R. R., & Podsiadlowski, Ph. 1996, submitted

Jacobson, P., et al. 1991, ApJ, 369, L63

Joss, P. C., Podsiadlowski, Ph., Hsu, J. J. L., & Rappaport, S. 1988, Nat, 331, 237

Mazzali, P. A., & Chugai, N. N. 1995, A&A, 303, 118

Mazzali, P. A., Lucy, L. B., & Butler, K. 1992, A&A, 258, 399

Meyer, F. 1996, preprint

Nomoto, K., et al. 1993, Nat, 364, 507

Nomoto, K., et al. 1994, Nat, 371, 227

Podsiadlowski, Ph. 1992, PASP, 104, 717

Podsiadlowski, Ph. 1996, in Hydrogen-Deficient Stars, ed. C. S. Jeffery & U. Heber (San Francisco, PASP), p. 419

Podsiadlowski, Ph., & Cumming, R. J. 1994, unpublished preprint

Podsiadlowski, Ph., Fabian, A. C., & Stevens, I. R. 1991, Nat, 354, 43

Podsiadlowski, Ph., Hsu, J. J. L., Joss, P. C., & Ross, R. R. 1993, Nat, 364, 509

Podsiadlowski, Ph., Hsu, J. J. L., Joss, P. C., & Ross, R. R. 1994, in Circumstellar Media in the Late Stages of Stellar Evolution, ed. R. E. S. Clegg, I. R. Stevens, and W. P. S. Meikle (Cambridge, Cambridge University Press), p. 179

Podsiadlowski, Ph., & Joss, P. C. 1989, Nat, 338, 401

Podsiadlowski, Ph., Joss, P. C., & Hsu, J. J. L. 1992, ApJ, 391, 246 (PJH)

Podsiadlowski, Ph., Joss, P. C., & Rappaport, S. 1990, A&A, 227, L9

Poeckert, R. 1981, PASP, 93, 297

Schönberner, D., & Drilling, J. S. 1983, ApJ, 268, 225

Spruit, H., & Podsiadlowski, Ph. 1996, in preparation

St.-Louis, N., Drissen, L., Moffat, A. F., Bastien, P., & Tapia, S. 1987, ApJ, 361, L23

Thaller, M. L., Bagnuolo, W. G., Jr., Gies, D. R., & Penny, L. R. 1995, ApJ, 448, 878

Tutukov, A. V., Yungelson, L. R., and Iben, I. , Jr. 1992, ApJ, 386, 197

Wampler, E. J., et al. 1990, ApJ, 362, L13

Wang, L. 1991, ApJ, 246, L69

Williams, R. E. 1987, ApJ, 320, L117

Woosley, S. E., Eastman, R. G., Weaver, T. A., & Pinto, P. A. 1994, ApJ, 429, 300

Stars at the Galactic Center

By K. Y. LO

Department of Astronomy, University of Illinois, 1002 W. Green St., IL 61801, USA

One of the problems Icko Iben worked on was the stability of supermassive (10^6–10^8 $M_\odot$) stars, at the time such objects were considered as energy sources for quasars and radio galaxies. Interestingly, recent observations of the stars at the center of the Galaxy have provided very strong evidence for the presence of a compact mass of 2.7×10^6 $M_\odot$, and an unprecedented view of the populations of stars within the central parsec.

1. Introduction

As chair of the Astronomy department of the University of Illinois, I am very pleased to represent my colleagues and my university to pay tribute to Icko Iben on the occasion of his 65th birthday. Icko of course was largely responsible for building up the Astronomy department at Illinois into a center for stellar astrophysics, by bringing to Illinois as faculty members distinguished astronomers and astrophysicists, including *D. Arnett*, R. Crutcher, *C. Folz, J. Gallagher*, T. Mouschovias, L.Smarr, L. Snyder, *J. Truran*, W. Watson and R. Webbink. (Those in italics have moved on to other institutions.) Icko's legacy at the University of Illinois is one among his crowning achievements.

While stellar astrophysics is not my specialty, I would like to describe some recent observations of stars at the center of our own Galaxy that has interesting bearings on some early work of Icko's on the stability analysis on supermassive stars.

At the first Texas Symposium in 1963 on "Quasi-Stellar Sources and Gravitational Collapse," Icko presented a paper entitled: "Massive Stars in Quasi-Static Equilibrium" (Iben 1963). The paper pointed out that understanding the behaviour of 10^6–10^8 $M_\odot$ "stars" requires the solution of the full dynamic equation of Genereal Relativity. His analysis showed that unlike the predicitions from classical hydrodynamics, "for stellar masses above 10^8 $M_\odot$, it does not appear likely that nuclear fuel can be burned rapidly enough to halt the collapse."

The interests in supermassive stars at the time were of course due to the proposal that the phenomena associated with then newly discovered quasars and radio galaxies might be explainable in terms of the gravitational energy released in the dynamical formation of "supermassive objects" with mass of 10^6–10^8 $M_\odot$ (Hoyle & Fowler 1963). An alternate mechanism of deriving luminosity from the release of gravitational energy from the accretion of external matter onto massive black holes was proposed independently ((Zeldovich 1964),(Salpeter, 1964)). Argument based on the Eddington luminosity limit suggests that the mass of the black holes needed to account for the luminosity of quasars ($\sim 10^{46}$ erg s^{-1}) is 10^8 $M_\odot$. While the supermassive black hole model has much success in accounting for a range of phenomena associated with quasars, radio galaxies and active galactic nuclei generally, direct evidence for the existence of a massive compact mass is not strong.

2. Sgr A*—Is it a Massive Black Hole?

Lynden-Bell and Rees (1971) suggested that the luminosity from the center of the Galaxy could be ultimately due to accretion of matter onto a massive black hole. They also anticipated observational consequences of their proposal, including the presence of a compact nonthermal radio source. Such a compact radio source was unambiguously first detected in 1974—Sgr A* (Balick & Brown 1974). Sgr A* has been observed intensively via VLBI since 1975 to determine its source structure, but as yet we can only set limits to its intrinsic source size which is extremely compact.

2.1. *Properties of Sgr A**

The properties of Sgr A* have been summarized (Lo 1994). The scale size of Sgr A* lies between the range of 10^{12} cm and 10^{13} cm. It has a rising spectrum between $\lambda = 1$ mm and $\lambda = 30$ cm. Its flux density varies with time but not by a large factor: $\Delta S/\langle S\rangle < 2$. Its radio luminosity of $\sim 3 \times 10^{34}\,\mathrm{erg\,s^{-1}}$ is $\sim 10^{-10}\ L_{\mathrm{Edd}}(10^6\,M_\odot)$. Its bolometric luminosity is less than $10^{36}\,\mathrm{erg\,s^{-1}}$. Taken all together its radio properties are unique in the Galaxy, but they are similar to compact radio sources in the nuclei of M81 and M104.

2.2. *Sgr A* = $10^6\,M_\odot$ Black Hole ?*

Given its extremely compact scale size of less than 1 AU (Rogers *et al.* 1994), a compact star cluster can be ruled out as the underlying energy source for Sgr A*. A highly dispersed pulsar would be extremely compact, but pulsar spectra are very steep, completely opposite to that of Sgr A*. While some binary stellar radio sources have many properties similar to Sgr A*, they typically have outbursts that increases 10^3-fold on very short time scales. A very young supernova would have a large expansion velocity which for Sgr A* cannot be larger than $30\,\mathrm{km\,s^{-1}}$(Lo *et al.* 1985).

The properties of Sgr A* can be accounted for by slow accretion onto a massive black hole (Rees 1980), although it has been argued that a $100\,M_\odot$ black hole would suffice (Ozernoy 1993). Other models include radial accretion of stellar winds by a $10^6\,M_\odot$ black hole at the center (Melia 1992) and advection-dominated accretion (Narayan *et al.* 1995).

However, given the very modest luminosity of Sgr A* ($\sim 10^{36}\,\mathrm{erg\,s^{-1}}$), and that the luminosity of the Crab nebula amounts to $10^{38}\,\mathrm{erg\,s^{-1}}$ which is powered by its central pulsar, a rotating neutron star cannot be ruled out as the underlying energy source. In fact, a model of Sgr A* as a neutron star powered wind which is confined by ram pressure has been proposed (Reynolds & McKee 1980).

2.3. *Mass of Sgr A* ?*

The missing piece of critical information is of course the mass of Sgr A*. An indirect constraint on the mass can be derived from the proper motion of Sgr A*. Measured against nearby background extragalactic compact radio sources, Sgr A* was found to have a transverse proper motion of $< 25\,\mathrm{km\,s^{-1}}$ (Backer & Sramek 1987). With the assumption of the equipartition of energy with the rest of the stars at the center, this implies a lower limit of $20\,M_\odot$ for the mass of Sgr A*.

Studies of the kinematics of ionized and neutral gas near Sgr A* have also led to constraints on the mass distribution at the center that may be consistent with the presence of a $10^6\,M_\odot$ compact mass (Lacy 1989). But, given the motion of the gas can be influenced by forces other than gravity, better test particles are necessary.

3. Stars at the Galactic Center

Because of the $A_V \sim 31^{\mathrm{m}}$ to the Galactic center, study of the stars at the center has to be done in the near-IR. With the advent of IR area detectors and the Hubble Space Telescope in the last few years, detailed observations of stars within the central parsec of the Galaxy have become possible.

3.1. *Population*

Besides the normally expected population of old stars at the Galactic center, there is a small cluster of blue stars with large luminosity and strong winds with He lines Allen (1987), (Krabbe *et al.* 1991) that are very massive stars, all found within the central 20″ (1.0 pc). These stars can account for the inferred bolometric luminosity of $10^7\,L_\odot$ and the observed amount of ionization. Given the relative lack of gas within the central pc of the center and the steep potential gradient, the formation of such a young cluster of stars in situ poses an interesting problem (McGregor *et al.* 1996).

Near-IR spectroscopy of the stars at the center has been used to measure the dispersion of radial velocities of the stars (Sellgren *et al.* 1990; McGinn *et al.* 1989; Krabbe *et al.* 1995) suggesting the presence of a concentration of mass within the central pc. The inference of a central point mass depends on how much of the mass distribution can be accounted for by the central star cluster (McGregor *et al.* 1996). The critical measurement is the stellar velocity dispersion well within the gravitational radius ($R \sim 2''$) of a $10^6\,M_\odot$ point mass at the Galactic center. Within 0.1″ of such a point mass, the Keplerian speed is $1000\,\mathrm{km\,s^{-1}}$.

The key requirement is therefore high resolution observations of the stellar motions at the sub-arc-sec level.

3.2. *HST and Speckle Observations of Proper Motions*

The Hubble Space Telescope (HST), with a resolution of 0.1″ at 1 μm, has been used to image the Galactic center. With the WFPC1, a 1 μm image shows the bright components of the 2 μm sources identifiable from ground-based observations and a "patch" of light centered roughly on Sgr A*. This image, despite its limitation by the spherical aberration of the HST at the time, indicates that the HST has the capability to study the proper motion of stars in the central star cluster within 0.1″ of the actual center and to provide the critical measurement of the large dispersion expected for a massive black hole at the Galactic center (Lo 1994).

Using the technique of speckle imaging, 0.15″ images at 2 μm of the Galactic center have also been obtained since about 4 years ago (Eckart *et al.* 1993). These images have been used to measure the proper motion of the stars at the center, and the latest results suggest that the stellar velocities display a Keplerian dependence on distance from the center, assumed to be Sgr A* (Eckart & Genzel 1996). The results are consistent with a compact dark mass concentration of $2.5 \times 10^6\,M_\odot$ within ~ 0.01 pc of Sgr A*.

This important result clearly should be confirmed independently. They will come from similar studies done with the Keck 10m telescope which has a diffraction limited resolution 3 times smaller than the NTT 3.5m telescope used and the near-IR NICMOS instrument to be placed on the HST in 1998.

4. Conclusions

The energy source for quasars, radio galaxies and other active galactic nuclei has been an outstanding issue for nearly three decades. Accretion of external matter onto a supermassive black hole has been largely accepted as the best explanation of the myriad of observed phenomena associated with active galactic nuclei, but direct observational evidence for a massive black hole has not been very stringent until recently. Observations of the proper motions of stars within 0.01 pc of Sgr A* constitute perhaps the best evidence to date for a $2.5 \times 10^6\,M_\odot$ black hole. The results will be definitively confirmed or disproved within the next three years.

We would like to acknowledge the support of Dean Jesse Delia of the College of LAS at the University of Illinois for supporting the travel to attend this meeting. We also thank Gene Ward, Icko's high school class-mate, who provided the Senior yearbook from which slides of Icko's high school activities were made. The slides, shown at the end of the talk, were very popular. Finally, we thank R. Genzel for communication of results before publication.

REFERENCES

D. Allen 1987, in AIP Conf. Proc. 155: The Galactic Center, ed. D. Backer, (New York: AIP), 1

R. Antonucci, J. Miller 1985, ApJ, 297, 621

D. Backer, R. Sramek 1987, in AIP Conf. Proc. 155: The Galactic Center, ed. D. Backer, (New York: AIP), 163

B. Balick, R. Brown 1974, ApJ, 194, 265

A. Eckart *et al.* 1993, ApJ, 407, L77

A. Eckart, R. Genzel 1996, Nature, 383, 415

F. Hoyle, W. Fowler 1963, in Quasi-Stellar Sources and Gravitational Collapse, eds. I. Robinson, A. Schild, E. Shucking, (The University of Chicago Press.: Chicago), 17

A. Krabbe *et al.* 1991, ApJ, 382, L19

A. Krabbe *et al.* 1995, ApJ, 447, L95

I. Iben 1963 in Quasi-Stellar Sources and Gravitational Collapse, eds. I. Robinson, A. Schild, E. Shucking, (The University of Chicago Press: Chicago), 67

J. H. Lacy 1989, in The Center of the Galaxy, eds. M. Morris, 493

D. Lynden-Bell 1969, Nature, 223, 690

D. Lynden-Bell, M. Rees 1971, MNRAS, 152, 461

K. Y. Lo 1994, in The Nuclei of Normal Galaxies: Lessons from the Galactic Center, eds. R. Genzel, A. I. Harris, 395

K. Y. Lo *et al.* 1985, Nature, 315, 124

E. Maoz 1995, ApJ, 447, L91

M. McGinn *et al.* 1989, ApJ, 338, 824

P. McGregor, J. Bignell, P. Saha 1996, in ASP Conf. Proc. 102: The Galactic Center, 4th ESO/CTIO Workshop, ed. R. Gredel, (San Francisco: ASP), 236

F. Melia 1992, ApJ, 387, L25

R. Narayan *et al.* 1995, Nature, 374, 623

L. Ozernoy 1993, Ann. NY Acad. Sci., 688, 596

M. Rees 1982, in The Galactic Center, eds. G. Riegler, R. Blandford, (New York: AIP), 166

S. Reynolds, C. McKee 1980, ApJ, 239, 893

A. E. E. Rogers *et al.* 1994, ApJ, 434, L59

K. Sellgren *et al.* 1990, ApJ, 359, 112

E. Salpeter 1964, ApJ, 140, 796

Ya. B. Zeldovich 1964, Akademii Nauk SSSR, 155, 67

The s-Process in AGB Stars: Its Operation and Theoretical Yields

By GRANT BAZÁN[1] AND JAMES W. TRURAN[2]

[1]Steward Observatory, University of Arizona, Tucson, AZ 85721, USA

[2]Department of Astronomy and Astrophysics, University of Chicago, 5640 S. Ellis Ave, Chicago, IL 60637, USA

We employ synthetic *interior* evolution models with synthetic *envelope* evolution models to examine the operation of the *s*-process in thermally pulsating, asymptotic giant branch (TP-AGB) stars of different initial mass and metallicity. Models are calculated for masses of 1.0, 1.2, 1.4, 1.6, 1.8, 2.0, 2.5, 3.0, 4.0, 5.0, and 6.0 $M_\odot$ and metallicities of $Z = 0.02$ and 0.007. The effects of 'hot-bottom burning' are included in the envelope via zoned nucleosynthesis with diffusive mixing. The *s*-process abundance trends of galactic disk MS, S, and N-type C stars can be explained if a ^{13}C-rich region extends over the upper 25–40% of the recycled portion of a given convective shell occurring in a thermal pulse. The dispersion in apparent total neutron exposure found in observations can be interpreted as due to a combination of the effects of dredge-up of *s*-process rich material which has not reached asymptotic abundances and either of initial metallicity, of ^{13}C abundance variations, or some combination of these latter two factors. Total yields for these stars are calculated and it is shown that the *s*-process production efficiency of TP-AGB stars decreases as a function of increasing metallicity for this prescription of the ^{13}C neutron source, dredge-up, and mass loss. The decrease in neutron exposure as a function of increasing metallicity demands that heavy *s*-only nuclei be predominantly produced at earlier epochs than lighter *s*-only nuclei, which might explain neutron capture element abundance trends in galactic chemical evolution.

---

1. Introduction

Burbidge et al. (1957) found that nuclei produced beyond the Fe-peak were generally the result of successive neutron captures. Captures occurring on timescales equal to or greater than the beta decay timescales of the lightest unstable isotopes were ascribed to the *s(low)*-process. Shortly thereafter, studies by Clayton et al. (1961) and Seeger et al. (1966) used existing neutron capture rates to show that repeated irradiations of material according to some decreasing distribution of integrated neutron flux best reproduced the relative solar abundances of '*s*-only' nuclei. Ulrich (1973) then explained how repeated irradiations of a 'recycled' fraction of a sample by a single integrated neutron flux could recreate this decreasing distribution of integrated neutron fluxes. Prior to some recent work on a possible site for the *s*-process in non-overlapping radiative regions (Straniero et al. 1995), this model provided the only viable explanation for the production of *s*-process nuclei with $A \geq 90$.

Since the first observations of Tc in their envelopes (Merrill 1952), AGB stars have generally been considered to be the site of the *s*-process. An observational sequence describing the conversion of generic solar composition M giants into highly *s*-process enriched C stars is now well established. This interpretation relies upon the 'dredge-up' together of *s*-process elements and ^{12}C into AGB envelopes from deep within stars, which has now been established from spectroscopic observations (Smith & Lambert 1990). The earliest signs of this transformation have been tentatively identified in some M giants without any apparent *s*-process enhancement other than Tc (Dominy & Wallerstein 1986, Vanture et al. 1991). MS stars show levels of *s*-process enhancement that can just be considered to be above systematic error ($\sim 0.3-0.5$ dex), while S stars exhibit significant enhancements of heavy elements up to 1.0 dex over solar proportions (Smith & Lambert 1985, 1986, 1990). At the end of this apparent sequence are N-type carbon stars, which on average have *s*-process enhancements of ≥ 1.0 dex (Utsumi 1970, Utsumi 1985), although there are large estimated errors in the abundance determinations (~ 0.4 dex). With the improvement in detector technology and abundance analysis, a serious effort should be made to measure abundances in a larger sample of bright, N-type C stars. R-type carbon stars are not considered part of this sequence, as they lie in a completely different region of the H-R diagram than TP-AGB stars. J-type carbon stars also do not fit in this evolutionary scheme, as they exhibit low $^{12}C/^{13}C$ ratios representative of CNO processing (Utsumi 1985).

The theoretical picture of the quantitative operation of the *s*-process remains quite uncertain. On the one hand, the 'classical model' of the *s*-process (Clayton et al. 1961, Seeger et al. 1966), on the basis of which the original picture of a decreasing distribution of neutron exposures for the *s*-process was developed, has provided useful constraints on the *s*-process site. The nuclear physics involved in the determination of variations in neutron capture rates across closed neutron shells is a sensitive diagnostic of the mean neutron exposure parameter τ_0 in the exposure distribution function. Competition between neutron captures and β-decays at so called 'branch points' leads to a characteristic value of the neutron density of the *s*-process. If these rates are temperature or electron density sensitive, characteristic values of these parameters may also be inferred. To date, such studies have yielded mean exposures in the range $0.2–0.3\,\mathrm{mb}^{-1}$, neutron densities from $10^7–10^9\,\mathrm{cm}^{-3}$, temperatures from $2–4 \times 10^8$ K, and electron densities around $10^{27}\,\mathrm{cm}^{-3}$ (Clayton et al. 1961, Ward et al. 1976, Howard et al. 1986, Bao & Kappeler 1987, Malaney 1988, Kappeler et al. 1990, Bauer et al. 1991, Beer et al. 1995, 1997).

The application of these schematic models to individual AGB star abundance anomalies has been somewhat successful in explaining *s*-process enrichments of galactic MS, S and N-type C stars. Relative fractions of *s*-process material in AGB envelopes have been determined by comparison of ad-hoc additions of different mean exposure compositions (Smith & Lambert 1990, Bazan 1991). Abundances across closed neutron shells indicate mean exposures from $0.1–0.4\,\mathrm{mb}^{-1}$ best match the observations, and these exposures seemingly increase as the level of enrichment increases along the MS-S-C star sequence. Note, though, that this interpretation requires that the *s*-process material mixed into the envelopes be at asymptotic values. Another useful guide to the nature of the *s*-process site comes from the required mixed fractions of 0.001–0.1 of *s*-process material in these envelopes, which imply that the *s*-process enrichment relative to solar *at the s-process site* should be in the range of 100–10000.

The theoretical connection of the *s*-process to AGB stars has been constantly changing. The first models mixed of protons from the convective envelope into the He shell, where, during a thermal pulse, a convective region would engulf the protons and mix them to temperatures where they would capture onto abundant ^{12}C nuclei and liberate neutrons via the $^{13}C(\alpha,n)^{16}O$ reaction (Sanders 1967). This mixing was eventually shown, by Icko Iben, not to occur, due to the effects of radiation pressure and the steep entropy jump across the hydrogen burning shell. Instead, an *s*-process was found to occur with a decreasing distribution of neutron exposures with $^{22}Ne(\alpha,n)^{25}Mg$ as the main neutron source (Iben 1975a, Iben 1975b, Iben 1976, Iben 1977, Iben & Truran 1977, 1978, Becker & Iben 1980). Unfortunately, this scenario has been shown both observationally and theoretically to be unlikely, due to the high neutron densities expected from this source in intermediate mass stars (Despain 1977, Cosner et al. 1980, Howard et al. 1986), the lack of any strong resonances in the ^{22}Ne reaction at temperatures characteristic of thermal pulses in low mass stars (Giesen et al. 1990, Wolke et al. 1989, Drotleff et al. 1993), and the lack of any observed enhancements of ^{25,26}Mg in Galactic MS and S stars (McWilliam & Lambert 1988).

More recently, ^{13}C has regained popularity as the main neutron source of the *s*-process. It has been shown that ^{13}C can be produced in a semi-convection zone beneath the convective envelope of a low metallicity, low mass AGB star (Iben & Renzini 1982, Hollowell & Iben 1988, Hollowell & Iben 1990). The inefficient mixing in such a semi-convective zone, created by enhanced opacity, allowed a limited number of protons to mix with ^{12}C-rich He shell material to make the ^{13}C. Follow up studies of *s*-process nucleosynthesis resulting from this neutron source can explain meteoritic *s*-process isotope anomalies (Kappeler et al. 1990, Gallino et al. 1990, Gallino et al. 1993, Wasserburg et al. 1994) and reconcile the problems of neutron density, exposure, and temperature that occurred when ^{22}Ne was considered as the sole neutron source (Hollowell & Iben 1990, Gallino et al. 1988, Kappeler et al. 1990). However, it should be noted that this particular formation of ^{13}C has only been shown to occur in one low metallicity, low mass model of low envelope mass, and not throughout all possible configurations of AGB models.

By assuming the production of ^{13}C to be ubiquitous in AGB stars in the interpulse phase, and varying the total amount of *s*-process elements produced, one can perform analyses of individual AGB stars. Given a certain total mass of ingested ^{13}C, the same values of neutron exposure seen via 'classical model' abundances is reproduced, although the lower mean exposures can be explained by mixing material which has not yet achieved asymptotic abundances (Busso et al. 1992, Busso et al. 1995). In addition, the *s*-process enrichment in Ba and CH star envelopes over a wide range of metallicities confirms that the amount of ^{13}C ingested in thermal pulses would have to be relatively constant as a function of metallicity to explain the apparent decline in effective mean exposure as a function of initial metallicity (Busso et al. 1995). It has also been found that the Rb overabundance in MS, S, and C star atmospheres is sensitive to neutron density, although the effects of neutron density time history and temperature stratification can influence *s*-process Rb production (Lambert et al. 1995).

While much progress has been achieved, we still have not come to an understanding of the total ramifications of stellar models for *s*-process nucleosynthesis. Meteoritic abundances reflect the sum total of all the *s*-process nucleosynthesis which occurred prior to the condensation of the meteorites. Thus, we must know the impact that stellar models have on *s*-process nucleosynthesis over both the range of initial masses which eventually ascend the AGB and the range of initial metallicities which comprise the sum of all stellar generations before the formation of solar system meteorites and the observable M-MS-S-C stars. To date, no *s*-process analysis of a theoretical AGB model undergoing dredge-up has yielded agreement with spectroscopic abundance determinations. This may be due to the restrictive nature of the studies, in which only one or two models have been explored.

To date, only the original study of Iben & Truran (1978) has examined the *s*-process over a sufficient range of initial stellar mass to have galactic chemical evolution implications. It utilized schematic evolution based on stellar models to arrive at final enrichments. Although only the ^{22}Ne$(\alpha,n)^{25}$Mg neutron source was employed, this study identified the mass range critical to the operation of the *s*-process and the creation of C stars (which led to the so-called 'C star problem' (Iben 1981). No such study has been performed with the added knowledge of AGB evolution that we now have available concerning the operation of the ^{13}C neutron source. Here, we adopt the notion from the ad-hoc mixing studies that the ^{13}C-rich region has a constant mass as a function of metallicity. The size of this region is constrained by the observations of galactic MS, S, and C stars. A set of yields is produced using schematic *interior* and *exterior* evolution, which can help to explain the relative neutron capture element trends seen in low metallicity stars (Gratton & Sneden 1994).

2. Models of the *s*-Process

2.1. *Nuclear Reaction Rates*

The nuclear data used in these *s*-process nucleosynthesis calculations includes β-decays and the rates of neutron production and capture, with associated α-capture reactions. Temperature dependent β^-, β^+ decay and e^- capture rates, for $T \geq 10^8$, are evaluated from fits (Howard et al. 1986) to the calculated rates in Takahashi & Yokoi (1987), at an electron density of $10^{27}\,\text{cm}^{-3}$. Tabulated rates in Takahashi & Yokoi (1987), which vary by more than 20% over the electron density range from 10^{26} to $3 \times 10^{27}\,\text{cm}^{-3}$, are interpolated by polynomial fits to determine rates at a given T and n_e. Below $T = 10^8$, the rates are calculated assuming thermal equilibrium between ground and first excited nuclear states, without an electron density dependence. Decay rates of long lived nuclear isomeric states are not treated as if they were in thermal equilibrium with the ground states; these nuclei are treated rather as separate nuclear species.

Maxwellian averaged neutron capture cross sections at $kT = 30\,\text{keV}$ are taken mostly from (Bao & Kappeler 1987), with revisions via the reanalysis of the standard ^{197}Au cross section (Ratynski & Kappeler 1988). Recent cross sections found in Beer et al. (1989) and Bauer et al. (1991) replace earlier values. The results of recent studies of ^{22}Ne ($\langle\sigma\rangle_{30} \sim 0.06\,\text{mb}$; Beer et al. 1989) and ^{12}C ($\langle\sigma\rangle_{30} \sim 0.0154\,\text{mb}$; Ohsaki et al. 1994) are also included. Temperature dependencies of neutron capture cross rates due to captures from and to excited states are estimated from the statistical Hauser-Feshbach method (Truran 1972, Holmes et al. 1976, Woosley et al. 1978). We have calculated neutron capture rates and temperature dependencies for the most neutron-rich nuclei in our network, using the nuclear model code SMOKER (Thielemann 1990). All (α, n), (α, γ), (p, γ), (p, α), (p, n) reaction rates are taken from Caughlan & Fowler 1988, with the exception of $^{22}\text{Ne}(\alpha, n)^{25}\text{Mg}$ and $^{13}\text{C}(\alpha, n)^{16}\text{O}$, which are taken from more recent work (Drotleff et al. 1993).

Convective mass motions are taken into account when evaluating effective reaction rates for *s*-process nucleosynthesis. A study of thermal pulses from some characteristic AGB models (Iben 1977, Hollowell 1988) confirms that most nuclear reaction timescales are significantly greater than the convective turnover time of the respective convective shells (2 days and 2 hours). Thus, we may average temperature and density dependent rates over the temperature and density gradients of the convective shell. These rates are then used in a nuclear network of 632 nuclei from H to ^{210}Po, which is solved by first-order backward differentiation (Arnett & Truran 1969).

2.2. *Synthetic Evolution*

Our prescription for synthetic AGB evolution is similar to those used in previous work (Iben & Truran 1978, Renzini & Voli 1981, Groenewegen & de Jong 1993, van den Hoek & Groenewegen 1996) with the addition of some relations for conditions in thermal pulse generated convective shells (Bazan 1991). These include the usual set of relations describing overlap, size, duration, peak temperature, and peak density of convective shells, as well as average interpulse luminosity and timescale. Special relations describing density and temperature histories at the base and top of the He-convective shell as a function of H-depleted core mass have also been developed, together with expressions for peak He luminosity during a thermal pulse. In an attempt to mimic the early thermal pulse behavior seen in stellar models, *non-asymptotic* relations are used prior to the point at which the asymptotic behaviors are achieved. An adiabatic structure is then assumed to define the interior mass coordinate dependencies of temperature and density, over which the rates are averaged.

For *s*-process studies, the key parameters are the dredge-up fraction, the total quantity of ^{13}C ingested, and the mass loss prescription. From the results of a sequence of AGB models which do show dredge-up, we find that the dredge-up fraction λ is a function of peak He luminosity the prior thermal pulse. In Figure 1, we show this dependence for a number of AGB models of different initial mass and metallicity. What is so striking, especially among the models from Lattanzio (1990), is that *there is no dependence on metallicity for a given stellar mass*, and *the only variables critical to dredge-up are the total mass and the peak He luminosity.* It remains to be seen whether more recent models also show this behavior. Actually, one might expect this behavior from a physical standpoint if dredge-up is a response of the envelope to the expansion, and later contraction, caused by the energy released in the thermal pulse. Since the deposition of energy occurs over a rather short timescale, the peak He luminosity serves as a good estimate of the total energy released into the envelope.

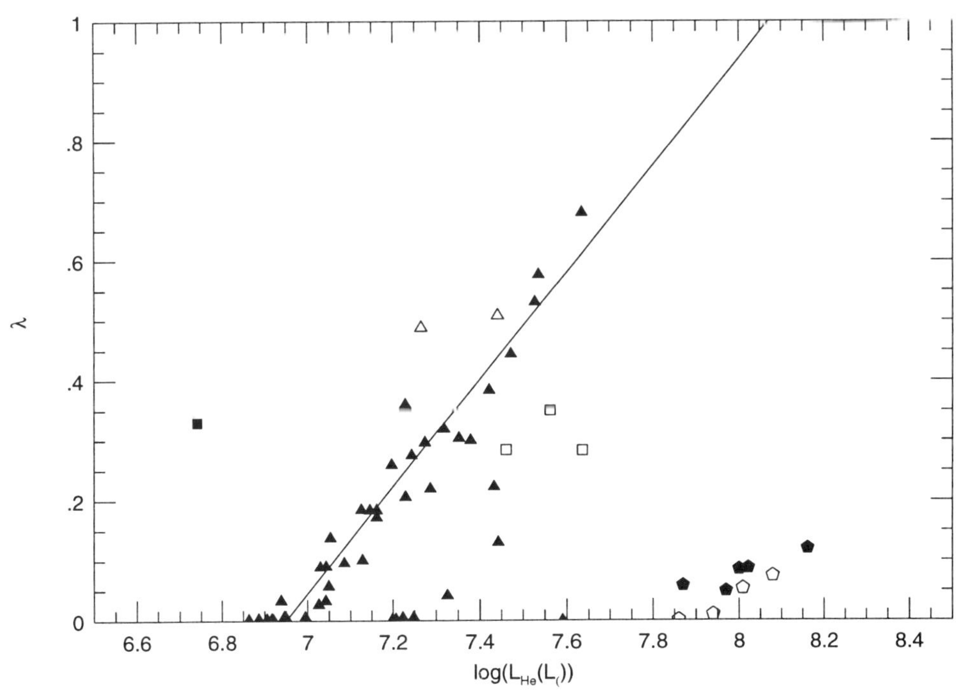

FIGURE 1. Dredge-up fraction as a function of peak He luminosity. The solid squares denote the 7 $M_\odot$ model of Iben 1975a, while the solid triangles mark various 1.5 $M_\odot$ models of different initial metallicities and helium fractions (Lattanzio 1986, Lattanzio 1990), the open triangles and squares mark the data from 5 $M_\odot$ and 3 $M_\odot$ models of Becker (1990), and the solid and open pentagons mark 1 $M_\odot$ and 0.7 $M_\odot$ models from Iben (1983).

The total ^{13}C ingested in a given thermal pulse is parameterized to mimic the distribution in the original works where it was found (Iben & Renzini 1982, Hollowell 1988), with one special consideration. Since the actual physical mechanism for the production of ^{13}C is still a subject of speculation, we allow the extent of the ^{13}C-rich region (and thus the total mass) to be a parameter determined by fitting to the spectroscopic data of MS, S, and C stars. We neglect any energy related changes to the convective shell which might result from the ^{13}C burning and neutron captures in the convective shell (Bazan & Lattanzio 1993, Mowlavi et al. 1995), and the possible consequences of any ^{13}C burning during the interpulse phase (Straniero et al. 1995). We instead adopt the philosophy that ^{13}C is produced continuously during the interpulse phase by protons which are mixed into the C-rich region via convective overshoot below the convective zone. This is sure to occur at some level, given the results hydrodynamic simulations of convection (Hurlburt et al. 1986, Hurlburt et al. 1994, Bazan & Arnett 1994, Bazan & Arnett 1997).

To describe envelope evolution, we employ similar relations to those of previous work (Iben & Truran 1978, Renzini & Voli 1981, Groenewegen & de Jong 1993 van den Hoek & Groenewegen 1996), with some notable exceptions. Our assumed initial elemental composition reflects the abundance evolution trends found in galactic disk and halo stars. Composition changes due to first and second dredge-up are taken from known relations based on stellar models (Lattanzio 1986, Becker & Iben 1979). The envelope temperature and density structure is determined by finding which combination of luminosity and effective temperature results in a completely convective envelope. 'Hot-bottom burning' adjustments to the luminosity, which follow initially from a core mass-luminosity relationship, are self-consistently determined.

Mass loss is the most crucial aspect of the envelope evolution. There is enough data on AGB mass loss to arrive at an empirical relation between mass loss and pulsational period (Bazan 1991, Vassiliadis & Wood 1993). We employ the Vassiliadis & Wood (1993) relation without the break at $2.5\,M_\odot$, since the presence of the Li-rich stars in the LMC somehow implies a superwind phase early in the existence of intermediate mass AGB stars. We also prescribe an upper limit for mass loss from the apparent upper limit of mass loss as a function of pulsation period. The fundamental mode period relation of Fox & Wood (1982) is used to calculate pulsational periods. The calculation terminates when a mass of $0.02\,M_\odot$ remains in the envelope.

3. *s*-Process in Individual Stars

3.1. *Comparison with Stellar Atmospheric Abundances*

We vary our two free parameters, scaling factors on the minimum L_{He} for dredge-up and the mass of ^{13}C-rich material, to match the observed trends in Galactic MS-S-C stars. By fitting the minimum luminosity at which carbon stars are seen to occur in the LMC and SMC (Lattanzio 1989), we can constrain the scaling factor on dredge-up; this takes values ~ 0.5. As in previous studies, the ^{13}C mass is set by the surface *s*-process element enhancements. Figure 2b shows the surface *s*-process abundance evolution for a $M = 1.6\,M_\odot$, $Z = 0.02$ model, when the fraction of ^{13}C rich mass in the recycled region is varied. The scatter in the observations can be explained either by a variation in the ^{13}C-rich pocket, or by the dredge-up of *s*-process rich material that hasn't reached its asymptotic abundance distribution (Busso et al. 1995; see, for example, the overabundance patterns as a function of time for this case, shown in Figure 3). A best fit model to the data occurs when the ^{13}C-rich region spans $\sim 25\%$ of the recycled region of the pulse. This means that extensive mixing of protons must take place below the convective envelope between thermal pulses, for this to occur.

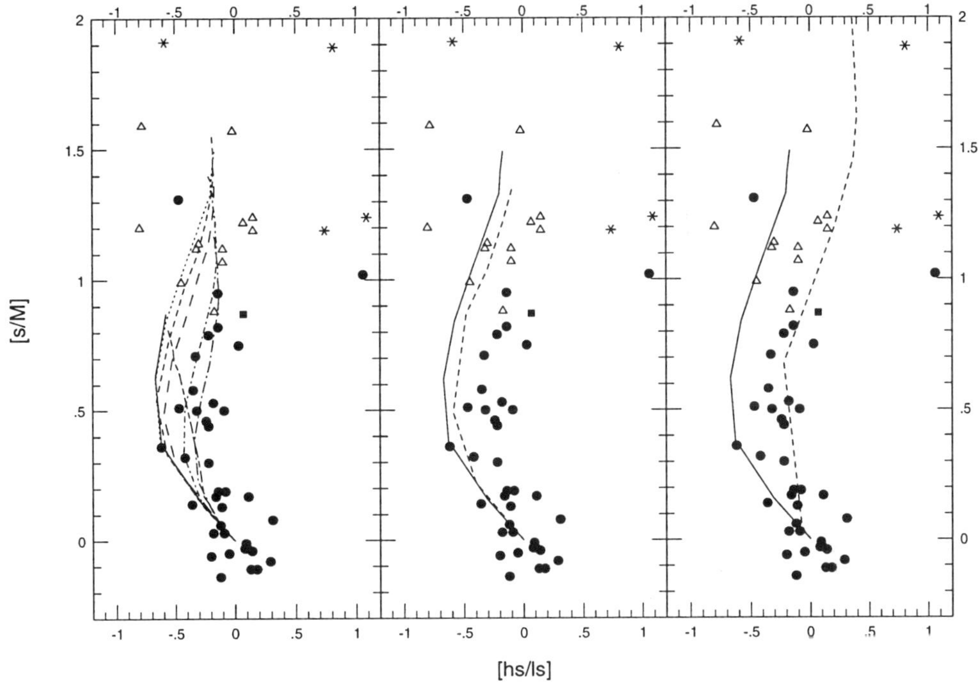

FIGURE 2. The *s*-Process envelope enrichment vs. neutron exposure abundance diagnostic of theoretical models is compared with observational data. Figure 2a shows models of 1.2, 1.4, 1.6, 1.8, 2.0, 2.5, 3.0, and 4.0 $M_{\odot}$, $Z = 0.02$, and $f_{13} = 0.25$. Figure 2b shows abundances in 1.6 $M_{\odot}$, $Z = 0.02$ models with $f_{13} = 0.25$ (solid line) and 0.4 (dashed line). Figure 2c shows abundances in 1.6 $M_{\odot}$, $f_{13} = 0.25$ models with $Z = 0.02$ (solid line) and $Z = 0.007$ (dashed line).

Figure 2a also shows the abundance evolution expected from models of different initial mass and $Z = 0.02$, with the 25% of the recycled material in a ^{13}C-rich pocket. Models having masses of 1.2, 1.4, 1.6, 1.8, 2.0, 2.5, 3.0, 4.0, 5.0, and 6.0 $M_{\odot}$ are shown. Little difference is seen for masses in the range $M = 1.4$–$3.0\,M_{\odot}$, due to the similar synthetic evolution parameters for $Z = 0.02$ stars in this mass range (Boothroyd & Sackmann 1988a, Boothroyd & Sackmann 1988b, Boothroyd & Sackmann 1988c, Lattanzio 1986, 1990). From 4–6 $M_{\odot}$, the characteristic exposure decreases, which explains the differences between models.

Figure 2c shows a comparison between $M = 1.6\,M_{\odot}$ stars of two different compositions, $Z = 0.02$ and $Z = 0.007$. We see that the $Z = 0.007$ model displays an abundance pattern representative of a higher neutron exposure. This is easily explained by a decrease in the number of heavy neutron absorbers relative to the primary neutron absorber ^{12}C; only at lower metallicities does this effect cease, since ^{12}C is the dominant neutron poison. This trend of decreasing exposure with increasing metallicity is consistent with observations (Busso et al. 1995): MS, S, Ba, and CH stars point to a ~ 0.5 dex increase in the heavy-*s* (Ba, La, Nd, Sm)/light-*s* ratio as [Fe/H] decreases from 0 to -0.5.

4. Stellar Yields

There is a definite dependence of stellar yield with both mass and initial metallicity. Figure 4 shows individual stellar yields as a function of initial mass for $Z = 0.02$. *s*-only nuclei are enriched by a factor of 1.6–1.8 dex from $M = 1.6$–$3.0\,M_{\odot}$ and decreases for

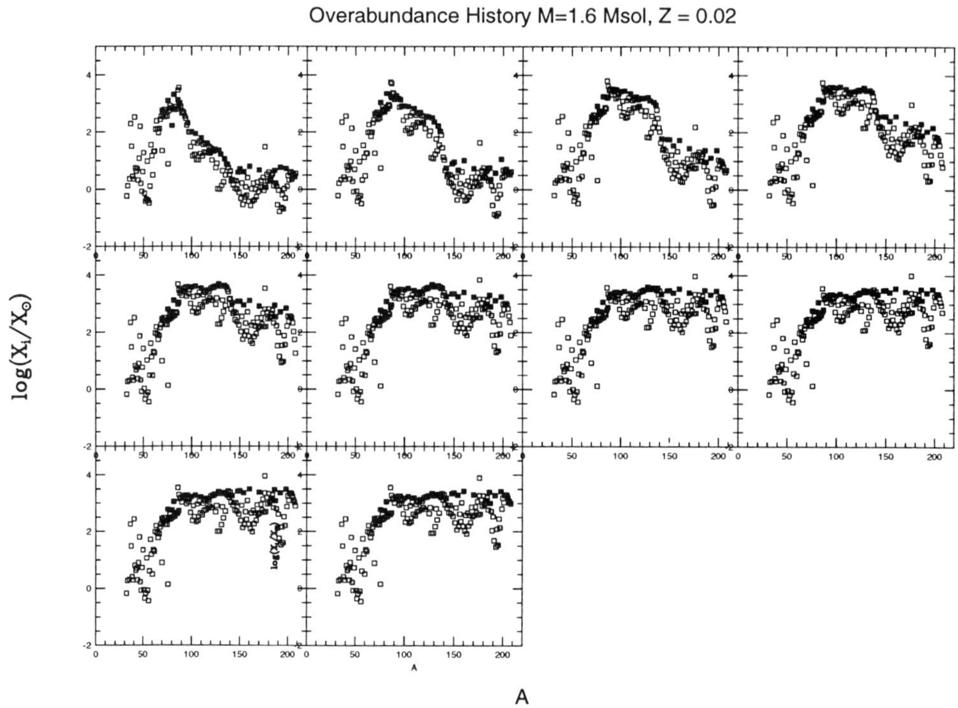

FIGURE 3. The overabundances of *s*-process nuclei in the dredged-up material of a $1.6\,M_{\odot}$, $Z = 0.02$ star are shown as a function of time, through the realization of the asymptotic abundance pattern.

masses both less than and greater than this range. The decrease at lower masses is due to the lower number of thermal pulses experienced before mass loss is complete, while the decrease at higher masses is due to a decrease in exposure and the lower over abundances in the dredge-up material.

Yields from LMC metallicity models (Figure 5) display the effects of a constant ^{13}C fraction. The most obvious properties are the level of enhancement and the reflection of the higher neutron exposure in the rising trend of overabundance with atomic number. The over-enhancement difference can be attributed to the level of higher overabundance in the dredge-up material, while the higher neutron exposure is the result of a decrease in available sink nuclei. It can be surmised that, since ^{12}C is produced *in situ* and is a major neutron poison, there is a metallicity at which ^{12}C becomes the dominant neutron sink and the exposure should near a constant value as a function of metallicity due to recycling into new ^{13}C. In addition, certain nuclei point to the rise in average neutron density. A comparison of Figures 4 and 5 shows that ^{96}Zr, while produced more than one would like in the solar models, increases drastically over this change in metallicity. A similar effect is also seen in the lower production of the *s*-only nucleus ^{142}Nd, which is due to the bypass of the branch at ^{141}Cs (Howard et al. 1986).

With the set of yields from individual stars, we can examine the overall impact of these stars via integration over an initial mass function. In Figure 6, we show the abundances obtained by integrating the stellar yields over an IMF with slope $= -1.5$ and dividing by the total stellar mass integrated over the same IMF. We see that the overall yield is heavily biased towards the yields coming from the low mass stars, which is not surprising since the yields are effectively constant over the range from 1.4–3 $M_{\odot}$ and the IMF is

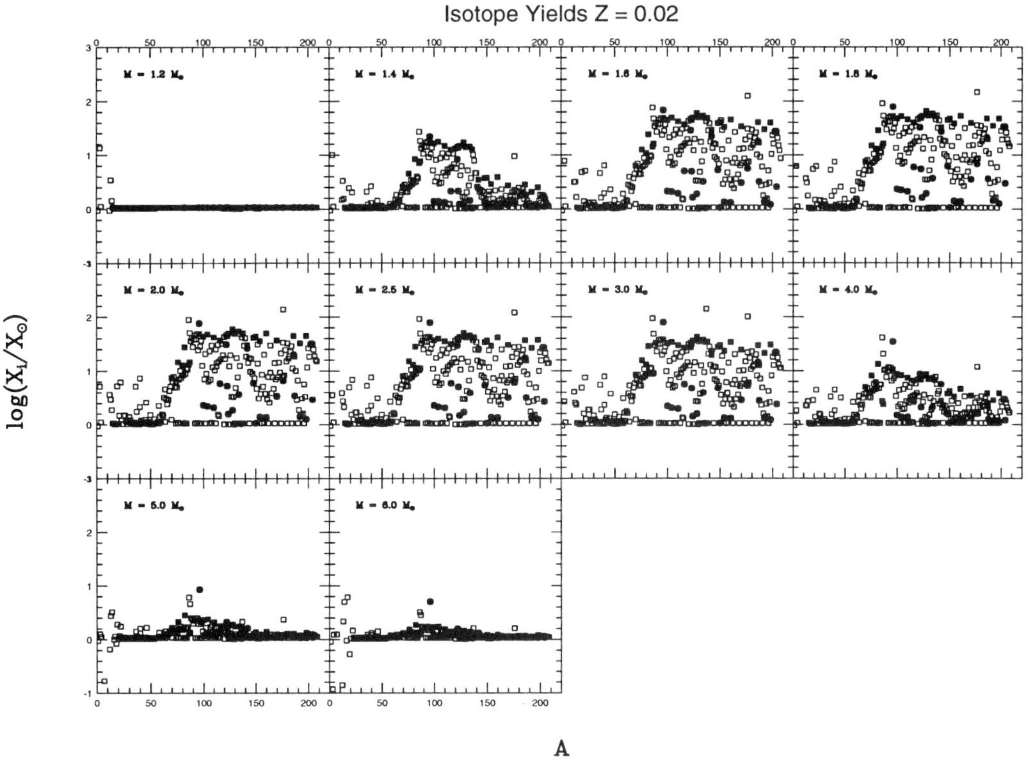

FIGURE 4. Star by star yields for $Z = 0.02$ stars. Filled squares denote *s*-only nuclei, empty squares mark *rs*-nuclei, and filled circles mark *r*-only nuclei.

biased towards lower mass stars. Figure 7 shows the same calculation for the LMC metallicity models, and reflects the same behavior as the solar metallicity models, in that the yields remain biased towards the yields coming from low mass stars.

5. Chemical Evolution Implications

The ultimate goal of works such as this is to explain the apparent chemical evolution of *s*-process elements. Neutron capture element abundances derived from high resolution spectroscopy of stars over a range of metallicity can serve as a historical record of the evolution of the *r*- and *s*-processes over the lifetime of the Galaxy (Truran 1981, Gilroy et al. 1988, Mathews et al. 1992, Gratton & Sneden 1994). This data indicates that, in the heavy element region from Ba to Eu, [x/Fe] values are ≥ 0 from $-2 \geq$ [Fe/H] ≥ -1, and that heavier elements are more overabundant than lighter ones. In fact, elements such as Dy and Pr become less overabundant as a function of increasing metallicity at the same time that light *s*-process elements such as Y and Zr become more abundant relative to solar (Gratton & Sneden 1994). This trend may be interpreted as the signature of the reduction of the neutron exposure in AGB stars, as a function of increasing metallicity, thus reflecting the behavior seen in our theoretical yields. More yields as a function of initial metallicity must be calculated and incorporated into detailed chemical evolution models, before this interpretation of the observational data can be taken for more than conjecture.

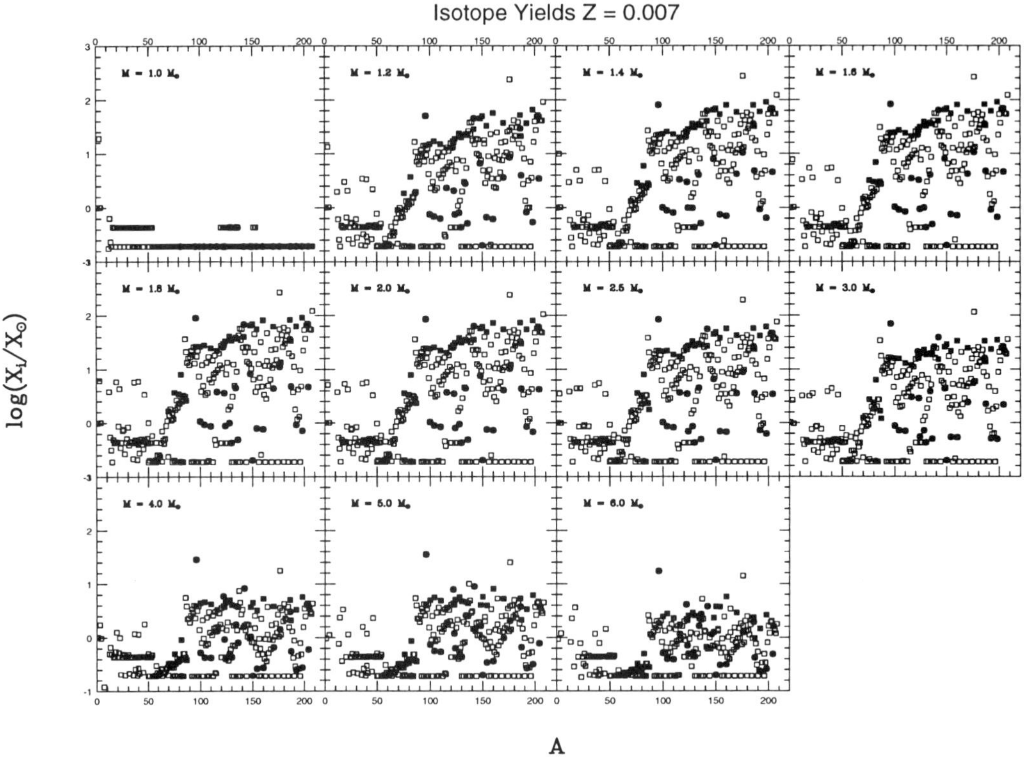

FIGURE 5. Star by star yields for $Z = 0.007$ stars. Filled squares denote *s*-only nuclei, empty squares mark *rs*-nuclei, and filled circles mark *r*-only nuclei.

REFERENCES

Arnett, W. D. & Truran, J. W. 1969, ApJ, 157, 339

Bao, Z. Y. & Kappeler, F. 1987, Atom. Nucl. Data Tables, 36, 411

Bauer, R. W., Bazan, G., Becker, J. A., Howe, R. E., & Mathews, G. J. 1991, Phys. Rev. C, 236, 2188

Bazan, G., Ph. D. Thesis, University of Illinois.

Bazan, G. & Arnett, W. D. 1994, ApJ, 433, L41

Bazan, G. & Arnett, W. D. 1997, ApJ, submitted

Bazan, G. & Lattanzio, J. C. 1993, ApJ, 409, 762.

Becker, S. A. & Iben, I. Jr. 1979, ApJ, 232, 831

Becker, S. A. & Iben, I. Jr. 1980, ApJ, 237, 111

Becker, S. A. 1990, *private communication*

Beer, H., Rupp, G., Vol, F., & Kappeler, F. 1989, in *5th Ringberg Workshop in Nuclear Astrophysics*, (Max Planck Institute Report MPA/P1), p. 6

Beer, H., Spettel, B., & Palme, H. 1995, in *AIP Conf. Proc. 327, Nuclei in the Cosmos III: Third International Symposium on Nuclear Astrophysics*, eds. M. Busso, R. Gallino, & C. Raiteri, (AIP Press, New York), 247.

Beer, H., Corvi, F., & Mutti, P. 1997, ApJ, 474, 843

Boothroyd, A. & Sackmann, I.-J. 1988a, ApJ, 328, 641

Boothroyd, A. & Sackmann, I.-J. 1988b, ApJ, 328, 653

Boothroyd, A. & Sackmann, I.-J. 1988c, ApJ, 328, 671

Burbidge, E. M., Burbidge, G. M., Fowler, W. A., & Hoyle, F. 1957, Rev. Mod. Phys., 29, 547

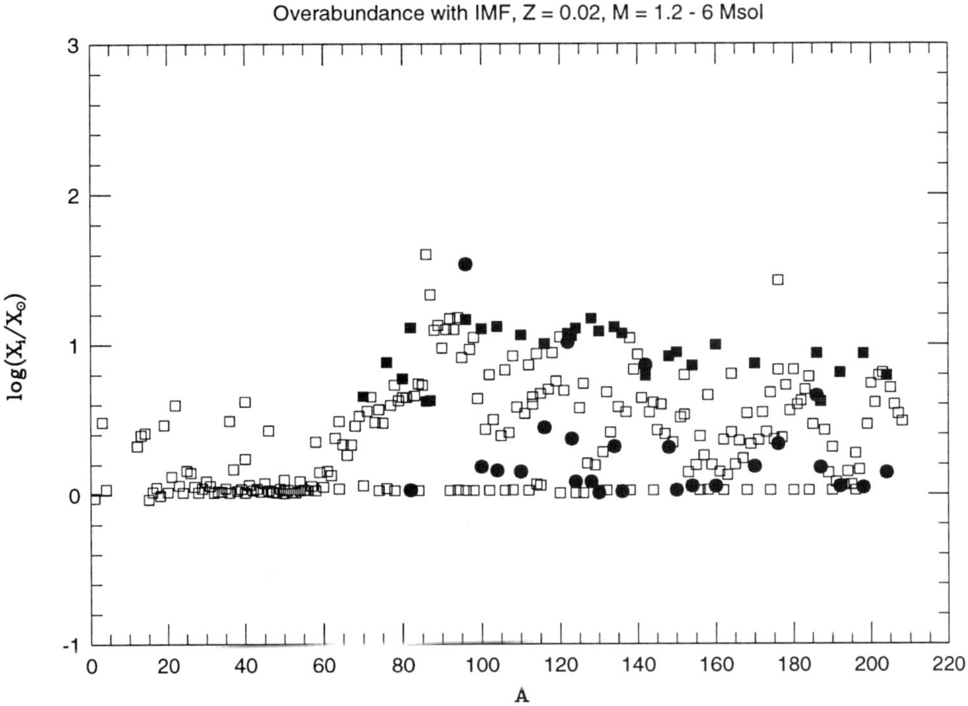

FIGURE 6. Integrated yields for *s*-process nuclei from 1.2–6 $M_\odot$, $Z = 0.02$ stars.

Busso, M., Gallino, R., Lambert, D. L., Raiteri, C. M., & Smith, V. V. 1992, ApJ, 399, 218.

Busso, M., Lambert, D. L., Beglio, L., Gallino, R., Raiteri, C. M., & Smith, V. V. 1995, ApJ, 446, 775.

Caughlan, G. R. & Fowler, W. A. 1988, Atom. Data and Nucl. Data Tab., 40, 283

Clayton, D. D., Fowler, W. A., Hull, T. C., & Zimmerman, B. A. 1961, Ann. Phys., 11, 121

Cosner, K., Iben, I. Jr., & Truran, J. W. 1980, ApJ, 238, L91

Despain, K. H. 1977, ApJ, 212, 774

Dominy, J. F. & Wallerstein, G. 1986, ApJ, 310, 371

Drotleff, H. W., Denker, A., Knee, H., Soine, M., Wolf, G., Hammer, J. W., Greife, U., Rolfs, C., & Trautvetter, H. P. 1993, ApJ, 414, 735.

Fox, M. W. & Wood, P. R. 1982, ApJ, 259, 198

Gallino, R., Busso, M., Picchio, G., Raitieri, G. M., & Renzini, A. 1988, ApJ, 334, L45

Gallino, R., Busso, M., Picchio, G., & Raitieri, G. M. 1990, Nature, 348, 298

Gallino, R., Raiteri, C. M., & Busso, M. 1993, ApJ, 410, 400

Giesen, U., Graff, S. M., Gorres, J., Wiescher, M., Azuma, R. E., King, J., Harms, V., & Kratz, K. L. 1990, *presentation at APS Nuclear Division Meeting*, Urbana, IL.

Gilroy, K. K., Sneden, C., Pilachowski, C. A., & Cowan, J. J. 1988, ApJ, 327, 298

Gratton, R. & Sneden, C. 1994, A&A, 287, 927

Groenewegen, M. A. T. & de Jong, T. 1993, A&A, 267, 410.

Harms, V., Wiescher, M. 1991, Zeis. für Physik C

Hollowell, D. E. 1988, Ph. D. Thesis, University of Illinois.

Hollowell, D. E. & Iben, I. Jr. 1988, ApJ, 333, L25

Hollowell, D. E. & Iben, I. Jr. 1990, ApJ, 349, 208

Holmes, J. A., Woosley, S. E., Fowler, W. A., & Zimmerman, B. 1976., Atom. Nucl. Data Tables,

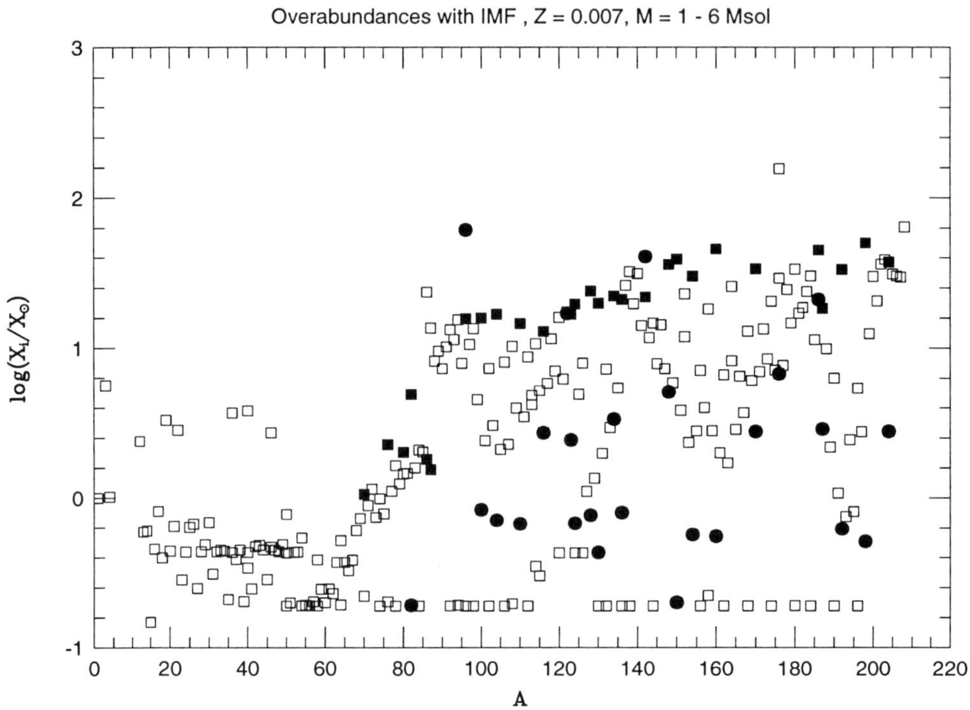

FIGURE 7. Integrated yields for *s*-process nuclei from 1.2–6 $M_\odot$, $Z = 0.007$ stars.

18, 306.

Howard, W. M., Mathews, G. J., Takahashi, K. & Ward, R. A. 1986, ApJ, 309, 633

Hurlburt, N. E., Toomre, J., & Massaguer, J. M. 1986, ApJ, 311, 563

Hurlburt, N. E., Toomre, J., Massaguer, J. M., & Zahn, J.-P. 1994, ApJ, 421, 245

Iben, I. Jr. 1975a, ApJ, 196, 525

Iben, I. Jr. 1975b, ApJ, 196, 549

Iben, I. Jr. 1976, ApJ, 208, 165

Iben, I. Jr. 1977, ApJ, 217, 788

Iben, I. Jr. 1981, ApJ, 246, 278

Iben, I. Jr. 1982, ApJ, 260, 821

Iben, I. Jr. 1983, ApJ, 275, L65.

Iben, I. Jr. & Truran, J. W. 1978, ApJ, 220, 980

Iben, I. Jr. & Renzini, A. 1982, ApJ, 259, L79

Iben, I. Jr. & Renzini, A. 1983, ARA&A, 21, 271

Kappeler, F., Gallino, R., Busso, M., Picchio, G., & Raiteri, C. M. 1990, ApJ, 354, 630

Lambert, D. L., Smith, V. V., Busso, M., Gallino, R., & Straniero, O. 1995, ApJ, 450, 302.

Lattanzio, J. C. 1986, ApJ, 311, 708

Lattanzio, J. C. 1989, ApJ, 344, L27

Lattanzio, J. C. 1990, *private communication*

Malaney, R. A. 1988, MNRAS, 231, 657

Mathews, G. J., Bazán, G. & Cowan, J. J. 1992, ApJ, 311, 300.

McWilliam, A. & Lambert, D. L. 1988, MNRAS, 230, 573

Merrill, P. W. 1952, Science, 115, 482.

Mowlavi, N., Jorissen, A., Forestini, M., & Arnould, M. 1995, in "AIP Conf. Proc. 327, Nuclei in the Cosmos," eds. M. Busso, R. Gallino, & C. M. Raiteri (New York:AIP), 403

Ohsaki, T., Nagai, Y., Igashira, M., Shima, T., Takeda, K., Seino, S., & Irie, T. 1994, ApJ, 422, 912.

Ratynski, W. & Kappeler, F. 1988, Phys. Rev. C, 37, 595

Renzini, A. & Voli, M. 1981, A&A, 94, 175

Sanders, R.H. 1967, ApJ, 150, 971

Seeger, P. A., Fowler, W. A., & Clayton, D. D. 1966, ApJS, 11, 121

Smith, V. V. & Lambert, D. L. 1985, ApJ, 294, 326

Smith, V. V. & Lambert, D. L. 1986, ApJ, 311, 843

Smith, V. V. & Lambert, D. L. 1989, ApJ, 345, L75

Smith, V. V. & Lambert, D. L. 1990, ApJS, 72, 387

Straniero, O., Gallino, R., Busso, M., Chieffi, A., Raiteri, C., Limongi, M., & Salaris, M. 1995, ApJ, 440, L85.

Takahashi, K. & Yokoi, K. 1987, Atom. Nucl. Data Tables, 36, 375

Thielemann, F.-K. 1990, *private communication*

Truran, J. W. 1972, Astro. Sp. Sci., 18, 308.

Truran, J. W. 1981, A&A, 97, 391

Truran, J. W. & Iben, I. Jr. 1977, ApJ, 216, 979

Ulrich, R. K. 1973, in *Explosive Nucleosynthesis*, ed. Schramm, D. N. & Arnett, W. D., (University of Texas:Austin), p.139

Utsumi, K. 1970, PASP, 22, 93

Utsumi, K. 1985, in *Cool Stars with Excesses of Heavy Elements*, eds. M. Jascheck & P.C. Keenan, (Dordrecht:D.Reidel), p.243

van den Hoek, L. B. & Groenewegen, M. A. T., preprint.

Vanture, A. D., Wallerstein, G., Brown, J. A., & Bazán,G. 1991, ApJ, 340, 340

Vassiliadis, E. & Wood, P. R. 1993, ApJ, 413, 641.

Ward, R. A., Newman, M. J., & Clayton, D. D. 1976, ApJS, 31, 33

Wasserburg, G. J., Busso, M., Gallino, R., & Raiteri, C. M. 1994, ApJ, 424, 412

Wolke, K., Harms, V., Becker, H. W., Hammer, J. W., Kratz, K. L., Rolfs, C., Schroder, U., Trautvetter, H. P., Wiescher, M., & Wohr, A. 1989, Zeis. f. Phys. A, 334, 491

Woosley, S. E., Fowler, W. A., Holmes, J. A., & Zimmerman, B. 1978, Atom. Nucl. Data Tables, 22, 371.

SN1987A: How a Massive Star Makes Rings

By DAVID ARNETT

Steward Observatory, University of Arizona, Tucson, AZ 85721, USA

Supernova 1987A is a Rosetta stone for study of mass loss in evolving massive stars and of its effect on the stellar environment. Models of the observed explosion constrain the nature of the star. The progenitor was observed, high resolution images of the ring stuctures are available, and the supernova ejecta will collide with the pre-explosion nebula with a few years, providing a wealth of new observations. Can we solve the puzzle first?

1. Introduction

Prof. Icko Iben, Jr., is a dominant figure in our attempt to understand how stars evolve. His work often shows a characteristic pattern: he defines a set of astronomical assumptions, of input physics, and then carefully solves in detail the problem at hand. One may argue with the assumptions (and, much more rarely, the treatment of the physics), but to the limits of the author's knowledge, the mathematics are always reproduceable. Unlike less careful simulations, this means that *both* positive and negative results are valuable. A second characteristic is his courage to attack big problems, problems that are big in both computational challenge and fundamental meaning.

This paper is an attempt to apply Icko's precepts to the problem of how the progenitor of SN1987A, the supernova in the LMC, built its spectacular three ring nebula (Burrows et al. (1995)); it is an outgrowth of work in collaboration with Dr. Crystal Martin (Martin & Arnett (1995), and in preparation). Because of the wealth of observational data and theoretical analysis, this event can provide unique clues about the evolution of massive stars. Because the nebula is nonspherical, analytic tools are limited in scope, and numerical simulations must be at least two dimensional. Because of the impending collision of the ejecta with the rings, there is a challenge for theorists to predict something useful *before* the event happens.

We have used a 1D stellar evolutionary calculation, with mass loss empirically estimated, to define an inner boundary condition for a 2D hydrodynamic calculation of the interacting red supergiant (RSG) and blue supergiant (BSG) winds. Hydrogen ionization and recombination are explicitly calculated, using the actual flux history of the stellar evolutionary model. The 1D model undergoes core collapse within the error box for the progenitor of SN1987A (Sk −69 202), giving a much more tightly constrained relationship between the various parameters determining the history of the interacting winds.

2. Preliminaries

While the simplest solution is not always the correct one, it is the best place to start. We examine the possibility that the progenitor was a single star (there is *no* observational evidence to the contrary as yet), and that the rings were formed by the interaction of the winds that such massive stars are observed to have. The key assumption is asymmetry in the RSG wind. Bjorkman & Cassinelli (1993) have shown that rotation naturally leads to enhanced flows in the equatorial plane, and least for B stars. The RSG phase coincides with the deepest extent, in mass coordinate, of the surface convection zone. This would allow the remaining angular momentum to be transported outward, spining up the surface. The rotational velocities would be low, but so is the wind velocity. It remains to be seen whether this leads to the asymmetrical wind needed. This scenario has the advantage of being testable with presently available resources.

2.1. *Stellar Evolution*

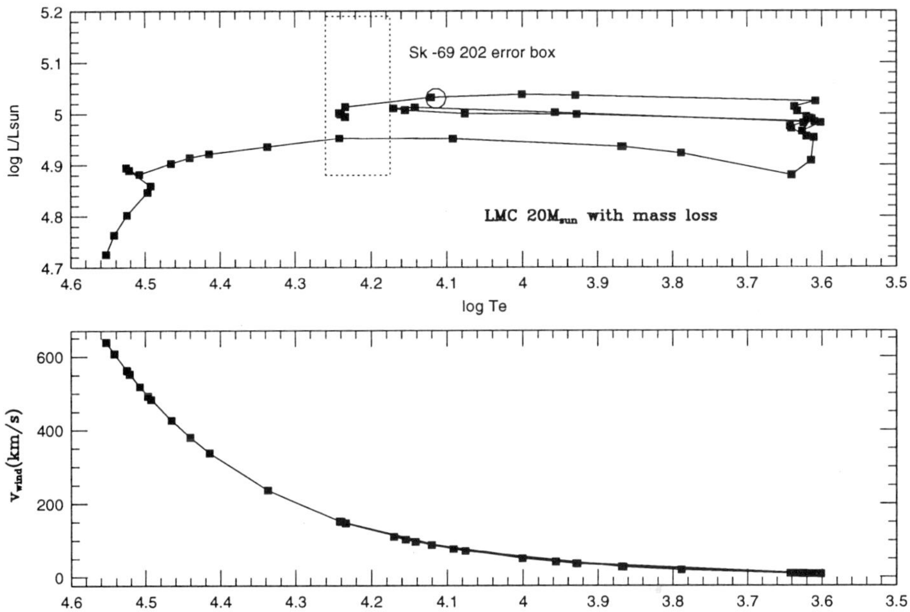

FIGURE 1. HR Diagram and Wind Velocities.

The top panel of Figure 1 shows the trajectory of a $20\,M_{\odot}$ star, with LMC abundances (taken as one third solar), with mass loss according to the de Jager algorithm quoted in Chiosi & Maeder (1986). Convective mixing was done using the Ledoux criterion. The solid boxes represent the points tabulated for use in interpolation in time for the inner boundary condition of the 2D calculation. The large dotted box is the error box in which the progenitor, Sk −69 202, was located. The star makes a first loop back to the "blue" during core helium burning, and a final loop to the "blue" for core carbon burning. It moves no more in the HR diagram until the explosion shock reaches the stellar surface. The large open circle denotes a timing feature of the 2D computation that will be discussed below.

It is convenient that the star actually dies in the Sk −69 202 error box, thus satisfying a necessary (not sufficient) condition. The occurrence of blue loops is well-known to be a sensitive issue (Kippenhahn & Weigert (1990)). Stars of this mass range die as a red supergiants if the Schwarzschild criterion for convection is used (e.g., Iben (1966), Lamb, Iben & Howard (1976)). This would then demand some additional hypothesis for the cause of moving to blue prior to explosion.

The bottom panel shows the estimated wind velocities along this path. These velocities are over 600 km/s for the O-star on the main sequence, and drop to 8 km/s on the Hayashi track (RSG limit). The intervening points are simply a logarithmic interpolation in the effective temperature T_e.

2.2. *Wind and Ionizing Flux*

The 1D model gives an estimate of mass loss $\dot{M}$ as a function of time. The interpolation just described gives the wind velocity in the radial direction, v_w. The inner boundary of the 2D grid is at a radius r_{in}. Thus the corresponding mass density is $\rho = \dot{M}/4\pi r_{in}^2 v_w$. The inner radius ($r_{in} \approx 10^{17}$ cm) is much larger than the stellar photospheric radius or that of the wind acceleration region.

The value of the temperature at this boundary is an interesting question. Free expansion would adiabatically cool it to tiny values; radiative heating from the star would oppose this. In these simulations, the thermal energy was taken to be 5×10^{-4} of the kinetic energy at the inner boundary (a "cold" wind).

The stellar model gives effective temperature and luminosity, so that the flux of ionizing radiation at the inner boundary can be estimated. A black body approximation was used (however, see below). This flux was propagated radially through the 2D grid, using a simple procedure based on Osterbrock (1974). The ionizing flux was attenuated by hydrogen ionization, using the "on the spot" approximation. The ionization balance was computed assuming only hydrogen was ionized (not a bad approximation for a B star or cooler). The thermal balance involved ionizational heating, and cooling at low temperature by bremsstrahlung and at high temperature as in Martin & Arnett (1995).

2.3. *Asymmetry*

In order to make a ring (Blondin & Lundqvist (1993), Martin & Arnett (1995)), an asymmetry should be introduced in the RSG wind. The disk-like component was defined by the same function of theta ($\sin^{64}\theta$) as in Martin & Arnett (1995), but the floor density was varied, with 0.01 and 0.0001 of maximum being used. This choice of shape and value was not constrained, and *did* make a difference.

Simulations in 1D (Owocki (1994), Feldmeier (1995)) show that radiation driven winds shed mass in shells because of instability; presumably in 3D these shells become clumps. Thus, our rotationally symmetric winds are probably still too smooth. This clumping not only affects our 2D hydrodynamics, but may even alter the wind parameters derived observationally (Hillier (1991)).

3. Two Dimensional Results

The 2D calculations were done with a variant of the PROMETHEUS code (for details, see Martin & Arnett (1995)). The principal improvement is the inclusion of transport of the ionizing flux, with ionization and thermal balance (see above).

3.1. *Large Scale Behavior*

A number of coarsely zoned (90x30) sequences were run, from the ZAMS to core collapse, to investigate the overall nebula structure. These extended out to $r \approx 10^{19}$ cm, and included the contact discontinuity with the original interstellar medium (taken to have a density $n \approx 10^{-3}$ cm^{-3} and temperature $T \approx 10^6$ K). For the region in which the observed rings exist, $10^{17} < r < 3 \times 10^{17}$ cm, this early evolution has little impact: the matter flows out to larger radii. This epoch is interesting in its own right, and has implications for larger structures, but for our purposes we simply note that we may begin our RSG wind in a steady BSG wind cavity. The rings are a product of the very last stages of the stellar lifetime.

We note in passing that moderate flows (10 km/s) along the symmetry axis give deviations from top/bottom symmetry of the same order as those observed in the rings—but in a different direction. Fully 3D simulations are needed to get the observed offset of the outer rings relative to the axis of the inner ring, but this result is encouraging.

3.2. *Ring Formation*

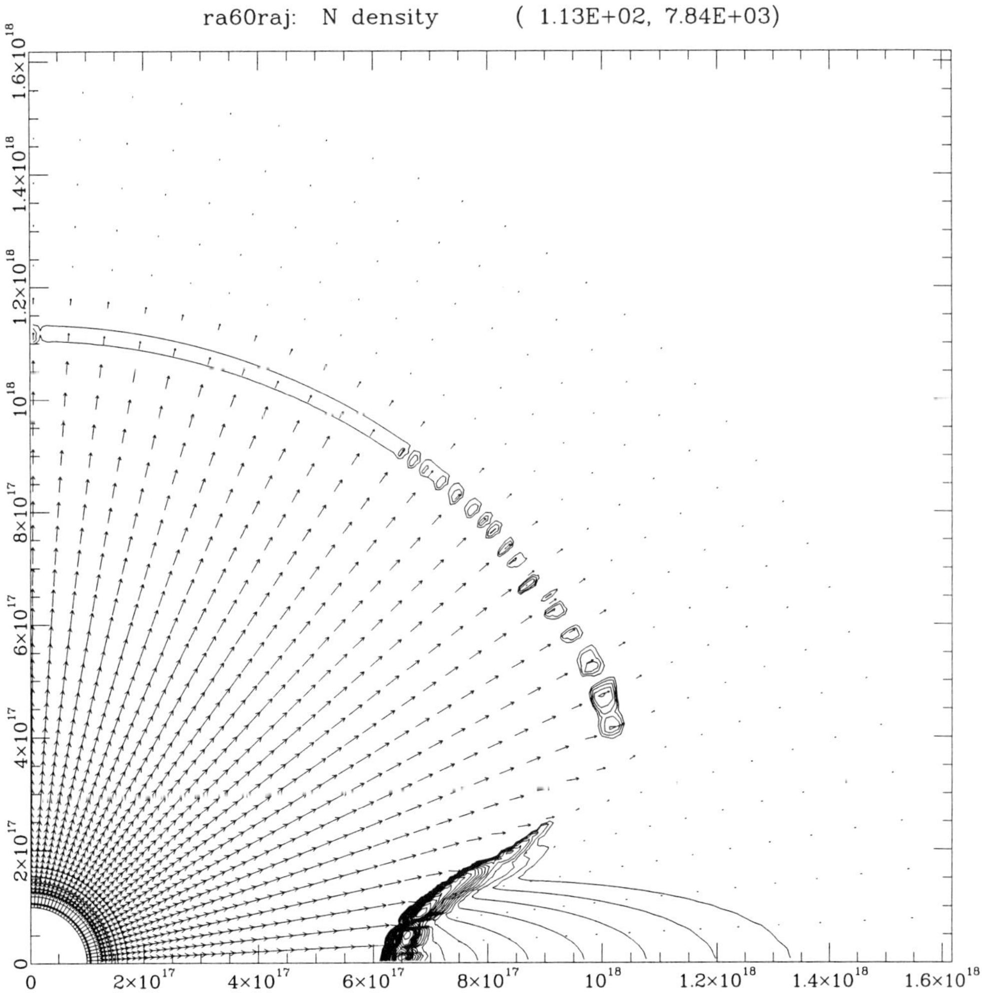

FIGURE 2. Density and Velocities for Sequence 0.

Two sequences will be discussed, both having good resolution (300x140), and covering the domain: $1.0 \times 10^{17} \leq r \leq 2.8 \times 10^{17}$ cm, and $0 \leq \theta \leq \pi/2$. Sequence 0 used a floor density of 0.01 of the maximum value during RSG stage. Sequence 1 used a lower floor density, of 0.0001 of the maximum value.

Figure 2 shows, for Sequence 0, the density contours (20 contours between densities of 7.83×10^3 and 1.13×10^2 nucleons cm^{-3}, and velocity vectors at a time when the inner ring radius is close to the observed value. A velocity vector is plotted for every $5 \times 5 = 25$ zones. This epoch was indicated by the large open circle in Figure 1. It lies just outside the error box for the progenitor because the ring expands too quickly, and arrives before the core has evolved to explosion (see below).

There is a prominent density enhancement in the equatorial plane, which can explain the inner ring. The shape may be too thick; presumably a better choice than $\sin^{64}\theta$ would fix this. There is a density enhancement at $r \approx 1.2\times10^{17}$ cm, the radius of the outer rings. It peaks at $n \approx 2.8\times10^2$ cm^{-3} at 67 degrees, compared to $n \approx 7.8\times10^3$ cm^{-3} in the inner ring. Thus the secondary peak is lower by a factor of almost 30. Sonneborn, et al. (1996) find an electron density of 2.6×10^3 cm^{-3} for the inner ring, which may be allowed by the flash ionization.

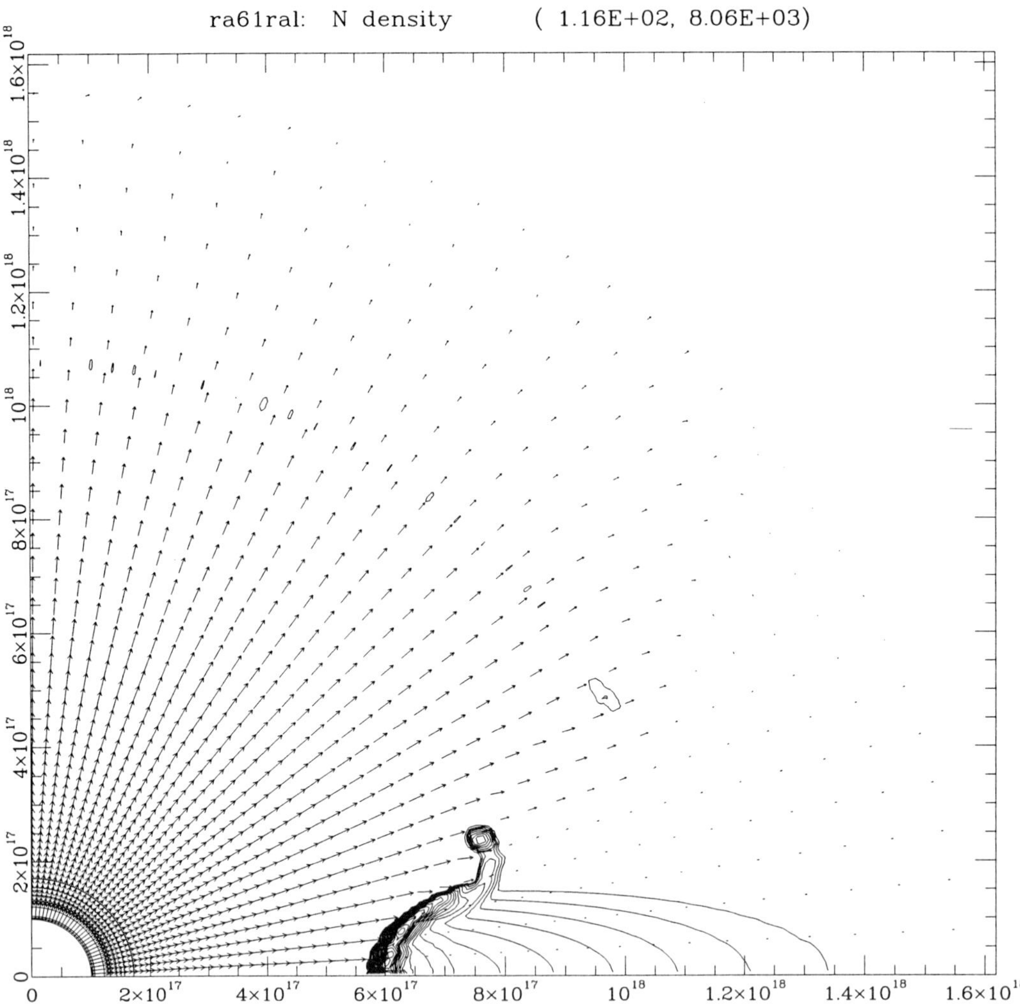

FIGURE 3. Density and Velocities for Sequence 1.

However, Panagia, et al. (1996) find that the [O II] electron density in the outer rings is ≈ 800 cm^{-3}, which is about one third the value in the inner ring. While this model will give outer rings at the observed position, they are at least a factor of four less dense than Panagia, et al. (1996) infer for SN1987A. The arc of decreasing density toward the symmetry (vertical) axis might not survive cleaning algorithms used on the observational images.

A qualtitative comparison of the images, both observational and computational, is considerably more complex (eg., Martin & Arnett (1995)), and in addition, involves issues of ionization and clumping.

Figure 3 shows, for Sequence 1, the density contours (20 contours between densities of 8.06×10^3 and 1.16×10^2 nucleons cm^{-3}), and velocity vectors at a time when the inner ring radius is close to the observed value. Again, this epoch was indicated by the large open circle in Figure 1; the two sequences both arrived early.

The outer arc has almost disappeared for this threshold of contouring. It is formed in both sequences by the shock from the BSG wind hitting the RSG wind at its floor density, and snowplowing. The lower floor density results in a lower density in the pile-up.

There is an new feature: the secondary maximum in density results from the break-up of the disk by instability. Its position (at least for these parameters) is roughly correct in angle, but its radius is about 30% less than the observed outer rings. While the peak density in the inner ring is almost unchanged for Sequence 0, the peak in the new secondary maximum is up to $n \approx 6.1 \times 10^2$ cm^{-3}.

4. Conclusions

There are several:

• We have not succeeded in doing an "Icko calculation" because we have developed more questions than answers!

• We do not yet have a properly posed problem in that there are many variables/parameters which are not adequately determined.

• Colliding wind models *do* make density enhancements (rings) of roughly the correct dimensions.

• Despite the fact that the winds and their histories are much more heavily contrained by the stellar evolutionary sequence, this seems to cause no difficulty for the simulations.

• Both sequences produce an inner ring of the correct dimensions slightly too early, *i.e.*, before they explode.

• A lot of work remains to be done, but the scenario already allows us to make quantitative comparison with a variety of observations, and seems to be testable.

What should we do?

An obvious improvement can be made in the estimate of the ionizing radiation, which was estimated using the black body approximation. Panagia (1973) has shown that this breaks down, even for supergiants, below log $T_e = 4.5$; for the crucial stages our flux should have been lower by a factor of 0.1 to 0.03. Calculations with this improvement are underway.

A related problem is the suggestion of Chevalier & Dwarkadas (1995) that the radio observations at $\approx$ 1500 days imply a lower mass loss rate for the BSG star just prior to explosion.

Both these changes will give a lower pressure than used here, and cause the inner ring to more slower, reducing the discrepancy between the times of explosion and of inner ring swelling to observed size. Martin & Arnett (1995) also found that a slight increase in stellar mass (20 to 20.5 $M_\odot$) would cause a significantly shorted time for the final BSG stage.

A more careful calculation is likely to improve agreement with observation. It appears that if we wish to succeed like Icko, we must work hard like Icko!

I wish to thank Dr. Crystal Martin and Dr. Douglas Miller for helpful conversations, and NSF 94-17346 and NASA NAGW-2798 for financial support.

REFERENCES

Burrows, C. J., et al., 1995, ApJ, 452, 680

Bjorkman, J. E., & Cassinelli, J. P., 1993, ApJ, 409, 492

Blondin, J. M. & Lundqvist, P., 1993, ApJ, 405, 315

Chevalier, R. A., & Dwarkadas, V. V., 1995, ApJ, 452, L45

Chiosi, C. & Maeder, A., 1986, Ann. Rev. A. & Ap., 24, 329

Feldmeier, A., 1995, A&A, 299, 523

Hillier, D. J., 1991, A&A, 247, 455

Iben, I., Jr., 1966, ApJ, 143, 516

Kippenhahn, R., & Weigert, A., 1990, *Stellar Structure and Evolution*, (Springer-Verlag: Berlin), p. 301

Lamb, S. A., Iben, Icko, Jr., & Howard, W. M., 1976, ApJ, 207, 209

Martin, C., & Arnett, D., 1995, ApJ, 447, 378

Osterbrock, D. E., 1974, *Astrophysics of Gaseous Nebulae*, (W. H. Freeman: San Francisco), p. 11–58.

Owocki, S. P., 1994, A&SS, 221, 3

Panagia, N., 1996, AJ, 78, 929

Panagia, N., Scuderi, S., Gilmozzi, R., Challis, P. M., Garnavich, P. M., & Kirshner, R. P., 1996, ApJ, 459, L17

Sonneborn, G., Fransson, C., Lundqvist, P., Casatella, A., Gilmozzi, R., Kirshner, R. P., Panagia, N., & Wamsteker, W., 1997, ApJ, 477, 848

Supernova Progenitor Constraints from Circumstellar Interaction: Type Ia

By PETER LUNDQVIST
AND ROBERT J. CUMMING

Stockholm Observatory, S-133 36 Saltsjöbaden, Sweden
(e-mail peter@astro.su.se, robert@astro.su.se)

Searching for the presence of a circumstellar medium (CSM) is a direct observational way to discriminate between different types of progenitor systems for Type Ia supernovae (SNe Ia). We have modeled whether such gas may give rise to detectable emission, especially in Hα, and compare the models with observations of SN 1994D. We obtain $\dot{M} \lesssim 2.5 \times 10^{-5}\, M_{\odot}\, \mathrm{yr}^{-1}$ for a wind speed of 10 $\mathrm{km\,s^{-1}}$. We find that X-ray observations in the range 5 – 10 keV, e.g., with *AXAF*, provide the most useful limits on the mass loss, while high-resolution optical spectroscopy offers the only direct way of identifying circumstellar hydrogen.

1. Introduction

SNe Ia are thought to be exploding white dwarfs in binary systems. The most likely type of progenitor system (Branch et al. 1995; see also Iben & Tutukov 1984) is a C-O white dwarf accreting H/He-rich gas from a companion, either from its wind or through Roche lobe overflow. Coalescing pairs of C-O white dwarfs are also possible, while accreting sub-Chandrasekhar-mass white dwarfs are less likely (Branch et al. 1995). In the non-coalescing scenarios, circumstellar gas will be present. The composition and geometry of this depend on the type of progenitor system. If the CSM emits detectable radiation, or absorbs radiation from the supernova, this can be used to distinguish between types of progenitor system. For example, any circumstellar lines of hydrogen would have to come from gas lost by the companion, and the luminosity of these lines are therefore particularly sensitive to the type of system.

The interaction of the ejecta with the putative CSM can generate radio (Boffi & Branch 1995) and X-ray emission (Schlegel & Petre 1993). If dusty and asymmetric, the CSM may also result in polarization of the supernova light (Wang et al. 1996). None of these studies have resulted in a detection. A different approach was taken by us in Cumming et al. (1996; henceforth CLSPK96). We used a high-resolution optical spectrum of SN 1994D around Hα taken only 6.5 days after the explosion to search for circumstellar hydrogen. The observations were compared with detailed photoionization calculations to establish a limit on the mass loss from the progenitor system. In addition, we discussed the effect of asymmetry of the CSM, and compared the sensitivity of optical studies to those at other wavelengths. Here we expand this discussion. In particular, we check the sensitivity of our results to the adopted maximum velocity of the ejecta, and demonstrate that absorption of soft ($\lesssim 1\!-\!2$ keV) X-rays by the CSM must be taken into account in interpreting X-ray limits. Following from this we reassess the X-ray limit from SN 1992A reported by Schlegel & Petre (1993).

2. Circumstellar excitation and expected circumstellar emission

There are four sources of radiation which could excite the CSM of SN Ia (CLSPK96): the radiation accompanying the supernova shock breakout, γ-rays from the decay of

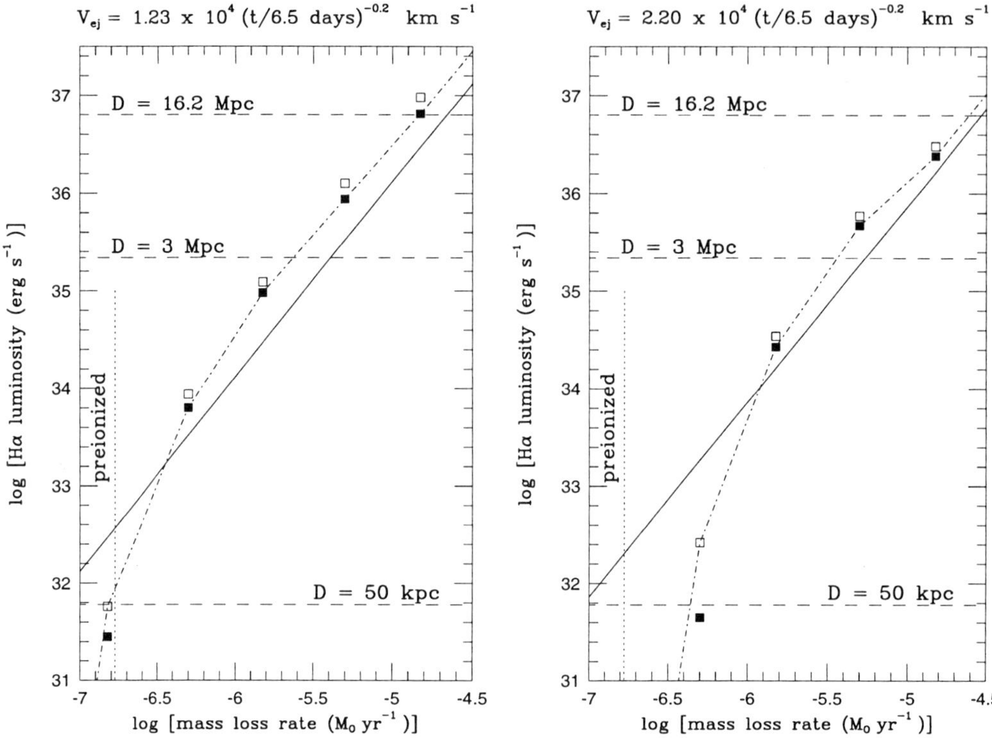

FIGURE 1. Luminosity of circumstellar Hα from a Type Ia supernova at 6.5 days after explosion, as a function of mass loss from the progenitor system, assuming a wind velocity of $10 \mathrm{km\,s^{-1}}$. Squares show model calculations where the wind is ionized by the radiation from the region of circumstellar interaction. The solid lines show the corresponding luminosities for a fully ionized wind at 2×10^4 K. The maximum velocity of the ejecta is given at the top of each panel. Preionization by the progenitor white dwarf is only important for $\dot{M}/u_{10} \lesssim 2 \times 10^{-7}\, M_\odot\, \mathrm{yr}^{-1}$. See text for details.

^{56}Ni, the radiation emitted by the progenitor prior to explosion (which can 'preionize' the CSM), and radiation from the interaction of the ejecta with the CSM itself. Of these, the first two are not important at all, while preionization is only observationally important for SN Ia in the Magellanic Clouds, or closer (CLSPK96). For more distant supernovae, the only important excitation mechanism is the radiation from the circumstellar interaction. In Figure 1, we present the calculated circumstellar Hα luminosity at 6.5 days as a function of mass loss from the progenitor system. We show results both for $V_{\mathrm{ej}} \approx 1.23 \times 10^9 \mathrm{~km\,s^{-1}}$ and $V_{\mathrm{ej}} \approx 2.20 \times 10^9 \mathrm{~km\,s^{-1}}$ at 6.5 days, where V_{ej} is the maximum velocity of the ejecta. For a wind with density falling off as $\rho \propto r^{-2}$, the interaction model of Chevalier (1982) predicts that the ejecta are slowed down at a rate $\propto t^{-1/(n-2)}$. Here n is the power law index for the density of the unshocked ejecta, $\rho_{\mathrm{ej}} \propto r^{-n}$. We use $n = 7$, which is a reasonable approximation to the commonly used W7 model of Nomoto, Yoki, & Thielemann (1984), resulting in $V_{\mathrm{ej}} \propto t^{-0.2}$. For the CSM we assume solar abundances.

The ionizing radiation in the interaction model comes from the C/O-rich ejecta shocked by the reverse shock propagating (in mass coordinate) into the supernova. The temperature of this gas, T_{rev}, at 6.5 days increases with the velocity of the ejecta, and is $\sim 1 \times 10^8$ K and $\sim 4 \times 10^8$ K for the two velocities shown in Figure 1. The spectrum of the ionizing radiation is therefore a free-free spectrum with $kT_{\mathrm{rev}} \sim 10$ keV and

$kT_{\rm rev} \sim 40$ keV, respectively. However, for low values of $\dot{M}$, electrons and ions in the shocked ejecta are not in energy equipartition, which affects the level and cutoff energy of the ionizing flux. Figure 1 shows the resulting Hα emission from the photoionized wind for two cases: full equipartition between electrons and ions, and an electron temperature which is a factor of 2 below the equipartition value. The dashed-dotted lines join the most likely models. A general feature of these models is that the circumstellar shock is preceded by an ionization precursor. The thickness of this photoionized region increases with $\dot{M}/u$ (CLSPK96), but decreases with $V_{\rm ej}$. The reason for the latter is mainly that the ratio of the dynamical time scale of the shock to the ionization time scale of the CSM decreases with increasing $V_{\rm ej}$. Using the Hα limit for SN 1994D (distance '16.2 Mpc' in Figure 1), and the fact that $V_{\rm ej}$ of the supernova at 6.5 days may have been as fast (Patat et al. 1996) as in the fastest model in Figure 1, we obtain $\dot{M}/u_{10} \lesssim 2.5 \times 10^{-5}\,M_\odot\,{\rm yr}^{-1}$. Here u_{10} is the wind speed in $10{\rm km\,s}^{-1}$. This limit is a factor ~ 1.7 higher than in CLSPK96, where we assumed somewhat slower ejecta, and close to the upper limit observed for mass loss rates in symbiotic systems. Observing SNe Ia in the Local Group (limit marked '3 Mpc' in Figure 1) for 10 000 s as early as 3 days after explosion, should take us down to a detection limit of $\dot{M}/u_{10} \sim 10^{-6}\,M_\odot\,{\rm yr}^{-1}$. Figure 2 demonstrates that the limit is rather sensitive to how soon after explosion the supernova is observed. Earlier than 3 days is probably observationally unrealistic.

3. Comparison with radio and X-ray limits

Radio limits have been presented for SN 1981B (Boffi & Branch 1995) and SN 1986G (Eck et al. 1995), using scaling arguments. In particular, for the close ($\sim$ 4 Mpc) SN 1986G, the range $10^{-7} - 3 \times 10^{-6}\,M_\odot\,{\rm yr}^{-1}$ was excluded. This limit is at the same sensitivity level as we can expect from early high-resolution optical observations (cf. §2). However, as we noted in CLSPK96, radio limits are subject to large systematic errors, because we do not know how efficiently the synchrotron radiation is generated. X-ray limits are potentially firmer and more sensitive than both radio and optical limits. Schlegel & Petre (1993) estimated $\dot{M} \lesssim (2-3) \times 10^{-6}\,M_\odot\,{\rm yr}^{-1}$ from *ROSAT* observations of SN 1992A (distance ~17 Mpc) at ~16 days after explosion. This is about an order of magnitude more sensitive than we can expect from Hα using the estimate in §2. However, Schlegel & Petre did not consider X-ray absorption in the CSM. In Figure 3 we show the photon energy below which the optical depth through the CSM is greater than unity (the cutoff energy, $\epsilon_{\rm cut}$) for the faster model in Figure 1. For the mass loss rates claimed to be excluded for SN 1992A, X-ray absorption below $\sim$ 1 keV is severe. Schlegel & Petre (1993) observed in the range 0.2–2.4 keV, with a sensitivity peak around $\sim$ 1 keV, so their limit on the mass loss rate was probably too optimistic. Furthermore, for $\dot{M}$ above $10^{-5}\,M_\odot\,{\rm yr}^{-1}$, the collapsed shocked ejecta probably block out all the flux in the *ROSAT* range (cf. SN 1993J at early times; Fransson, Lundqvist, & Chevalier 1996), unless the CSM is aspherical and the interaction region is viewed through the less dense part of the CSM. More reliable X-ray limits can be expected for photon energies in the range 5–10 keV.

Searches for circumstellar matter are beginning to provide interesting observational limits on the progenitor systems of SNe Ia. The most informative limits will come from X-ray telescopes like *AXAF*. High-resolution optical spectroscopy is less sensitive, but has the potential to provide information about velocities and abundances of the CSM. Radio observations are difficult to interpret. At all wavebands, observation as soon as possible after the explosion is highly desirable.

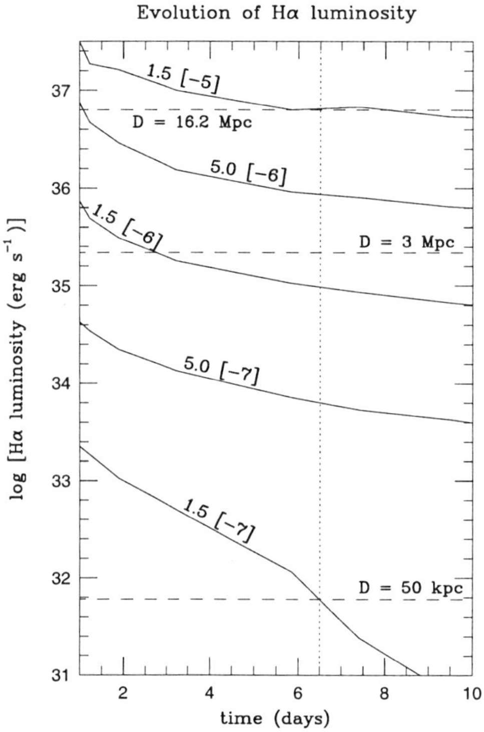

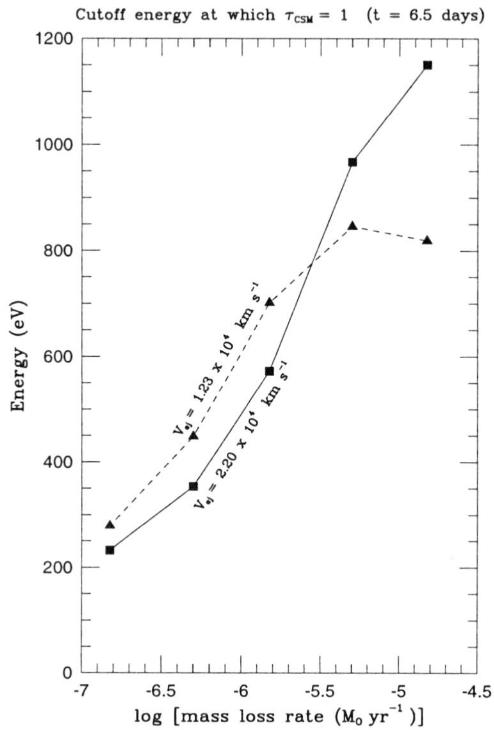

FIGURE 2. (left) Evolution of Hα luminosity for the models in Figure 1. Models are labeled with their $\dot{M}/u_{10}$. The vertical dotted line marks 6.5 days after explosion. The horizontal dashed lines mark the sensitivity of the SN 1994D observation of CLSPK96 at different distances. Figure 3. (right) Cutoff energy, $\epsilon_{\rm cut}$, at which the optical depth through the CSM is unity for the models in Figure 1. Below $\epsilon_{\rm cut}$ the CSM is opaque to X-ray emission from the supernova.

Acknowledgements: We thank David King, Linda Smith and Max Pettini for enlightening collaboration, and the Swedish Natural Science Research Council for support.

REFERENCES

Boffi F., Branch D. 1995, PASP, 107, 347

Branch D., Livio M., Yungelson L. R., Boffi F. R., Baron E. 1995, PASP, 107, 1019

Chevalier R. A. 1982, ApJ, 259, 302

Cumming R. J., Lundqvist, P., Smith, L. J., Pettini, M., King, D. L. 1996, MNRAS, 283, 1355 (CLSPK96)

Eck C., Cowan J., Branch D., Boffi F. 1995, ApJ, 451, L53

Fransson, C., Lundqvist P., Chevalier, R. A. 1996, ApJ, 461, 993

Iben, I., Jr., Tutukov, A. V. 1984, ApJS, 54, 335

Nomoto K., Thielemann F.-K., Yokoi K. 1984, ApJ, 286, 644

Patat, F., Benetti, S., Cappellaro, E., Danziger, I. J., Della Valle, M., Mazzali, P. A., Turatto, M. 1996, MNRAS, 278, 111

Schlegel E. M., Petre R. 1993, ApJ, 418, L53

Wang L., Wheeler J. C., Li Z., Clocchiatti A. 1996, ApJ, 467, 435

Supernova Progenitor Constraints from Circumstellar Interaction: Type II

By ROBERT J. CUMMING AND PETER LUNDQVIST

Stockholm Observatory, S-133 36 Saltsjöbaden, Sweden (e-mail robert, peter@astro.su.se)

All types of supernovae (SNe), except Type Ia (see Lundqvist & Cumming, this volume) have been observed to interact with their immediate circumstellar medium (CSM). This interaction can reveal their progenitor's histories, and in a broader sense constrain our ideas about the evolution of massive stars. Progress in this direction has concentrated itself in three areas. First, radio emission has been detected from a variety of core-collapse SNe, and modelling according to the simple model of Chevalier (1982) has given estimates of progenitor mass loss rates. For some of these SNe, especially SN 1993J, complementary information comes from X-ray observations. Second, multiwavelength observations of SN 1987A in the LMC have shown that the progenitor star passed through a phase as a red supergiant with a highly asymmetric and nitrogen-enhanced wind. Third, optical observations of some core-collapse SNe show that their emission to a large extent comes from circumstellar interaction, indicating that they are surrounded by a dense CSM.

As examples of the latest progress in these areas, we present some new results on two SNe: the famous 1987A, and the less well-known but rather intriguing 1994W. In the process we introduce Type IIn supernovae (SN IIn; Schlegel 1990) as an important new probe of massive star evolution.

1. SN 1987A: a 3D map of the triple-ring nebula

In Cumming *et al.* (1997a) we present and model high-resolution ($R \sim 50\,000$) spectra of SN 1987A's three circumstellar rings, taken with the échelle spectrograph UCLES at the Anglo-Australian Telescope between 1990 and 1993. We measure the expansion velocity for the inner ring to be 10.3 ± 0.4 km s^{-1}; here we present the most detailed measurements yet of the kinematics of the two outer rings (Figure 1).

The outer rings seem to be parallel to and equidistant from the inner ring. Each lies $0.4 \pm 0.1 (D_{\rm LMC}/50$ kpc) pc from the inner ring plane. Comparing with the continuum echo mapping of the dusty shell between the rings (Crotts, Kunkel & Heathcote 1995), we see that the separation of the rings is close to the distance to the polar edges of the shell, $0.6 (D_{\rm LMC}/50$ kpc) pc. Given the errors, the rings coincide with the outermost edges of the dusty shell. Our results strengthen the picture of the nebula as an open-ended, hourglass-shaped shell and provide an important constraint for hydrodynamic and stellar evolution models (e.g., Arnett, this volume).

2. SN 1994W: a type IIn supernova

We present observations from the first 201 days of the remarkable narrow-line Type II supernova 1994W in NGC 4041, taken at the *WHT*, *INT*, *JKT* and *NOT* on La Palma. A full analysis will appear in Cumming *et al.* (1997b). In Figure 2 we show the available photometry for SN 1994W. Figure 3 shows the spectral development.

The presence of the narrow P-Cygni lines seems quite consistent with the clumpy wind model presented by Chugai & Danziger (1994) and Chugai, Danziger & Della Valle (1995) for SNe 1988Z and 1978K. In this scenario, the wind consists of dense clumps embedded in a rarefied interclump wind. The broad component in H I is most easily

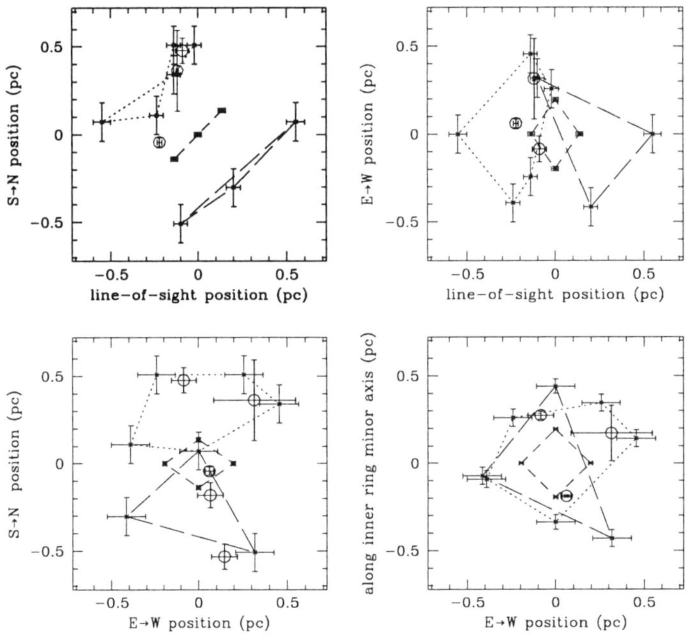

FIGURE 1. Four views of the three-dimensional structure of the triple-ring nebula, assuming homologous expansion. The morphology of the [N II] nebula hardly changed between 1036-1790d (Wang & Wampler 1992) and the ~2550-d *HST* imaging (Burrows *et al.* 1995). Using these images as a guide, we measured the velocities in different parts of the nebula from our position-velocity plots for days 1344 and 1397 together with the day 1402 observations of Crotts & Heathcote (1991). Then, assuming homologous expansion, we mapped each measured point to a position along the line of sight according to its velocity. Hydrodynamic models presented so far for the SN 1987A CSM have velocity fields not too different from this (e.g., Blondin & Lundqvist 1993). We followed the same procedure for the short-lived H I and He I blobs (Cumming & Meikle 1993 and references therein; open circles). Broken lines join points connected to each other in *HST* imaging.

explained using the shock model of Chevalier & Fransson (1985), in which the line wing emission comes from the cool, shocked ejecta close to the contact discontinuity between the shocked ejecta and the shocked circumstellar wind. In the clumps, the shock is slowed down, and high column density in the clouds can account for the observed narrow P-Cygni profiles. As long as the SN is covered by the clouds, we cannot see the broad Hα component normally seen in Type II SNe. The absence of normal ejecta lines in SN 1994W suggests both that the clumpy wind prevents us from seeing them, and that a substantial proportion of the SN's energy is reprocessed into circumstellar emission. We suggest that the drop of the light curve around day 120 marked the point at which the outgoing shock had passed most of the clumps. The simultaneous fading of the lines and the continuum indicates that *both* had a circumstellar origin.

SN 1994W's progenitor probably lost mass in a clumpy wind shortly before it exploded, a picture similar to the one suggested for SN 1984E (e.g., Gaskell & Keel 1988). However, SN 1984E showed normal ejecta lines both while its narrow lines were present and after they had disappeared. The resulting circumstellar shell round SN 1994W had, in contrast to 1984E, a clump filling factor high enough to conceal the SN ejecta entirely, and to reprocess most of its radiation.

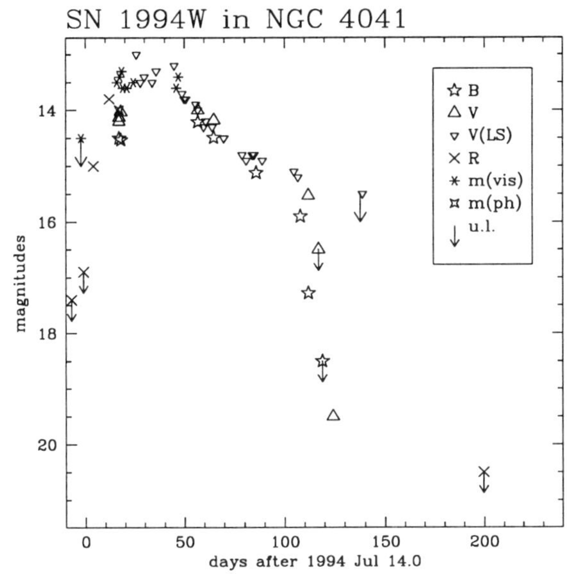

FIGURE 2. Light curve of SN 1994W. The inverted triangles are visual magnitude estimates by L. Szentasko which we have corrected to V by comparing a wide selection of his observations with contemporary photoelectric V measurements. The drop in the light curve after day 100 is the fastest ever seen in an SN, and is nicely bracketed by photometric points from Tsvetkov (1995) and our day 124 V point.

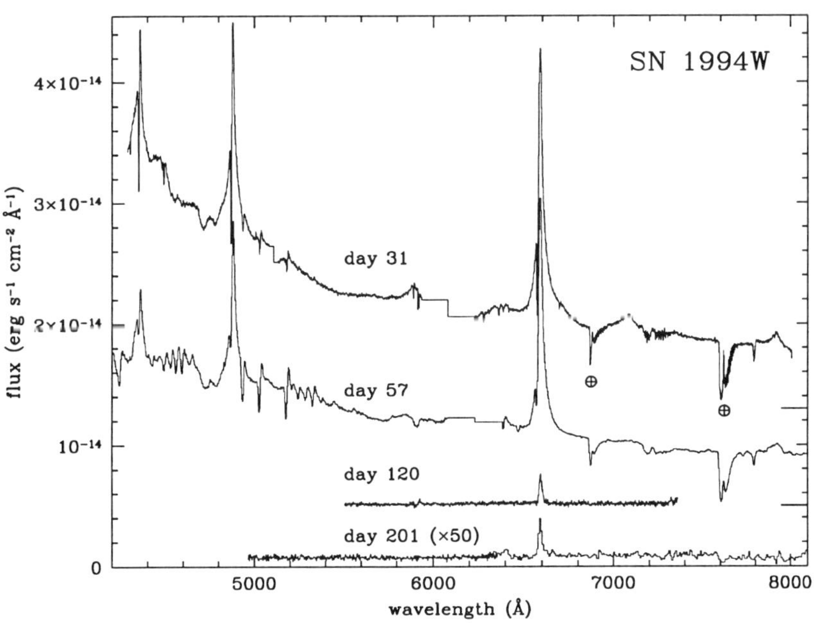

FIGURE 3. Spectral evolution of SN 1994W. On days 31 and 57 the spectrum was dominated by narrow ($\lesssim 10^3$ km sec^{-1}) H I lines with P-Cygni profiles and broad wings ($\sim \pm 10^4$ km sec^{-1}), which were accompanied by low-ionisation, mainly narrow P-Cygni lines — we identify He I (with triangular profiles), O I, Mg II Si II and Fe II, but no forbidden lines. The light curve drop coincided with a dramatic change in the spectrum. On day 120, just after the fading, Hα had narrowed and was present only in emission, accompanied by faint Na I and Ca II. By day 201, only Hα was left.

2.1. *The diversity of Type IIn supernovae: a picture emerges*

We suggest that the clumpy-wind model can be extended to account for the general features of *all* SN IIn, without abandoning spherical symmetry. It is the clumps that produce the narrow lines which distinguish these SNe as a class. The broad lines come from the SN ejecta's interaction with the interclump wind material. The prominence of the narrow lines relative to broad lines (either from the SN photosphere, as in SN 1984E, or from the interaction region, as in SNe 1988Z and 1994W) depends on the covering and filling factors of the clumpy wind, as well as the density ratio of the clump to interclump gas. Unlike SNe 1984E and 1994W (and probably SN 1987B; Schlegel *et al.* 1996), the progenitors of the slow SN IIn like SNe 1988Z, 1986J and 1978K (Leibundgut 1994, and references therein) had long-lived clumpy winds which still haven't been passed by the expanding shock.

In SN IIn, the flux evolution of the narrow lines measures the duration of clumpy mass loss ($\sim$500 yr for 1994W, for a 10- km sec^{-1} wind; much longer for 1988Z), while their widths measure the density contrast between the clumps and the interclump wind, if we know the original shock speed. That can be estimated from the width and evolution of broad lines at early times. From the early evolution of some SN IIn, like SN 1994W, we can probe the evolution of a massive star just prior to explosion. SN IIn are a whole class of supernovae that have the potential to provide us with just as detailed tests of stellar evolution as SN 1987A has.

Acknowledgements. We are happy to acknowledge stimulating collaboration with Peter Meikle and Jason Spyromilio (SN 1987A) and Jesper Sollerman and Miguel Pérez Torres (SN 1994W). We also thank Laszlo Szentasko for visual estimates, and Mike Breare, René Rutten and Phil Rudd for service spectra and photometry of SN 1994W.

REFERENCES

Blondin, J. M. & Lundqvist, P. 1993, ApJ, 405, 337

Burrows, C. J., et al. 1995, ApJ, 452, 680

Chevalier, R. A. 1982, ApJ, 259, 302

Chevalier, R. A. & Fransson C. 1985, in Supernovae as distance indicators, ed. N. Bartel (Berlin: Springer), p.123

Chugai, N. N. & Danziger, I. J. 1994, MNRAS, 268, 173

Chugai, N. N., Danziger, I. J., Della Valle M. 1995, MNRAS, 276, 530

Crotts, A. P. S. & Heathcote, S. R. 1991, Nat, 350, 683

Crotts, A. P. S., Kunkel, W. E., Heathcote S. R. 1995, ApJ, 438, 724

Cumming, R. J. & Meikle, W. P. S. 1993, MNRAS, 262, 689

Cumming R. J., Lundqvist P., Meikle, W. P. S., Spyromilio, J., Allen, D. A. 1997a, in preparation

Cumming, R. J., Lundqvist, P., Sollerman, J., Pérez-Torres, M. A. 1997b, in preparation

Gaskell, C. M. & Keel, W. C. 1988, in Proc. 4th George Mason Fall Workshop in Astrophysics, ed. M. Kafatos & A. G. Michalitsianos (Cambridge: Cambridge Univ. Press), p. 13

Leibundgut, B. 1994, in Proc. 34th Herstmonceux Conference, ed. R. E. S. Clegg, I. R. Stevens & W. P. S. Meikle, p. 100

Martin, C. M. & Arnett, D. 1995, ApJ, 447, 378

Schlegel, E. M. 1990, MNRAS, 244, 269

Schlegel, E. M., Kirshner, R. P., Huchra, J. P., Schild, R. E. 1996, AJ, 111, 2038

Tsvetkov, D. 1995, IAU IBVS, 4253

Wang, L. & Wampler, E. J. 1992, A&A, 262, L9

Chemical Abundance Constraints for the Evolution of 10 – 30 $M_\odot$ Stars

By Kim A. Venn†

Universitäts Sternwarte München, Scheinerstrasse 1, 81679 Munich, Germany, and Max-Planck-Institute-für-Astrophysik, Karl-Schwarzschild-strasse 1, 85740 Garching, Germany

Predictions of the CNO abundances in A-type supergiants differ between evolution scenarios since these elements are sensitive to mixing with H-burned (CNO-cycled) gas. Analysis of Galactic A-type supergiants show evidence of partial mixing only of the surface layers with CN-cycled gas from the interior. Thus, Galactic A-type supergiants do not appear to have visited the red giant branch where they would undergo the first dredge-up (a scenario predicted by many of the current stellar evolution models). New abundances for SMC A-type supergiants imply significantly different abundances of N from their Galactic counterparts, ranging from no mixing to extensive mixing with CN-cycled gas. These results, individually and together, support the new stellar evolution models that include rotation effects, i.e., Fliegner & Langer (1996) and Maeder & Meynet (1996). Additional evidence for partial mixing *on* the main-sequence in these stars is provided by new boron abundances in Galactic B-stars.

1. Introduction

Elemental abundances are a useful way to constrain evolution scenarios when reliable abundances of elements affected by evolutionary processes, i.e., He, C, N, O, can be determined reliably. The abundances of these elements are predicted to change after the first dredge-up on the red giant branch. However, do massive stars then undergo blue loops? Almost certainly low mass stars do, as evidenced by analyses of Cepheids (e.g., Luck & Lambert 1985). These blue loops are rather small, and apparently inhibited by the Cepheid instability strip. More massive stars, >10 $M_\odot$, are also expected to undergo blue loops in the course of their evolution, however these loops extend to very warm temperatures, AB-supergiants, and there is little observational evidence to support or refute them. Since these massive stars are the main sites for the nucleosynthesis and distribution of many of the chemical elements, then evolution scenarios for these stars are important for understanding the chemical evolution of galaxies.

A-type supergiants provide a unique test of the stellar evolution scenarios since reliable elemental abundances can be determined for these stars, and these stars are so bright that they can be seen in other galaxies to examine environmental effects, such as metallicity. Abundance analysis of A-supergiants in the Galaxy and the SMC are discussed here; the location of the target stars on an HR-diagram is shown in Fig. 1. It can be seen that the program stars straddle the ends of the blue-loops (they also straddle the Fitzpatrick & Garmany, 1990, ledge).

2. Galactic A-type Supergiant Abundances

An analysis of 22 A-type supergiants in the Galaxy by Venn (1995a,b) has shown that reliable abundances can be determined for light elements (C, N, O), alpha-elements (Na, Mg, Si, Ca), iron-group elements (Sc, Ti, Cr, Fe, Ni), and some s-process elements (Sr,

† Present address: Macalester College, Department of Physics and Astronomy, 1600 Grand Avenue, St. Paul, MN, 55105

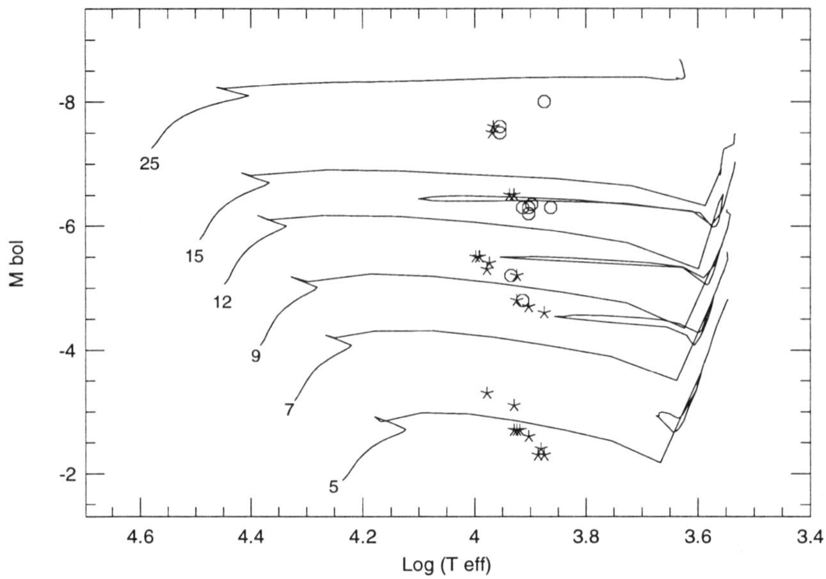

FIGURE 1. SMC (circles) and Galactic (asterisks) program stars. Evolution tracks for solar metallicity from Schaller *et al.* 1992.

Zr); see Fig. 2. Uncertainties in the model atmospheres analyses of these stars can be circumvented by tailoring the calculations to weak lines that form deep in the photosphere and by determining the atmospheric parameter from certain spectroscopic indicators.

The sample of Galactic target stars range in mass from 5 to 20 $M_\odot$, yet no significant differences in the abundances are found across this mass (thus luminosity) range. For most elements, an LTE analyses proved sufficient such that the same abundance is found over a range of lines, with different excitation potentials, line and oscillator strengths, and spectral wavelength location. NLTE line formation calculations were necessary for only a few elements; particularly detailed calculations were necessary for the NLTE corrections to the nitrogen N I abundances (see Lemke & Venn 1996, Venn 1995a).

Interpretations of the final abundances are complicated by uncertainties in the expected initial abundances. I found it necessary to adopt initial abundances based on spectral analyses of local B-stars and H II regions. This is because the Sun appears to be overabundant in each of CNO by 0.2 to 0.4 dex (see Venn 1995a for details). The initial abundances for (C, N, O) have been adopted as (8.2, 7.8, 8.7).

Thus, A-type supergiant abundances show,

1. NLTE nitrogen abundances are slightly enriched (~0.2 dex).

2. NLTE carbon and oxygen are slightly depleted (~0.1 dex). This small oxygen depletion includes a NLTE correction of −0.2 dex based on calculations by Baschek *et al.* (1977).

3. the sums of [C+N], relative to the B-stars, range from 0.0 to 0.2, within the uncertainties of the analysis to support that these abundances are from CN-mixing alone.

The C and N abundances strongly imply that these stars have *not* undergone the first dredge-up, which is expected to result in a much higher nitrogen enhancement, ~+0.5 dex. Instead, these abundances suggest that the 5 to 25 $M_\odot$ Galactic A-type supergiants have undergone *partial* mixing of CN-cycled gas from the H-burning layers of the star, and *are not evolving along blue-loops.*

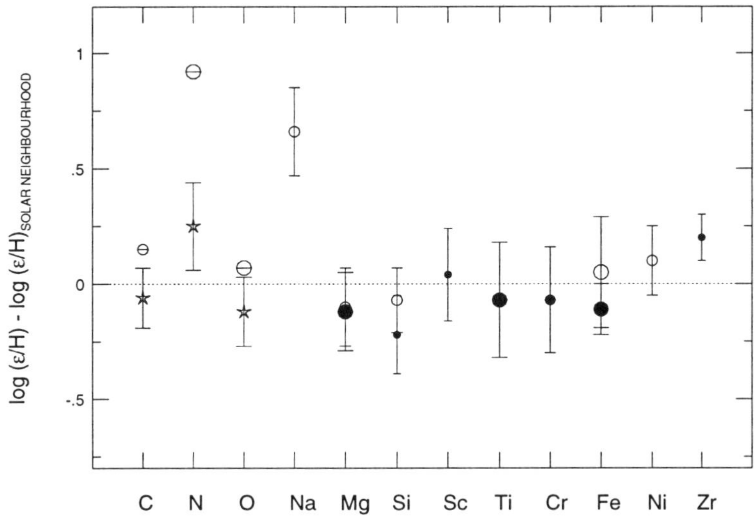

FIGURE 2. Galactic A-supergiant abundances relative to solar neighborhood abundances (=solar, except for CNO discussed in the text). Five-point stars represent the NLTE abundances for CNO, filled/empty circles represent LTE abundances from ions/neutral species. Symbol sizes are related to the number of target stars used in the average value.

3. Abundances in SMC A-Supergiants

Metallicity is an important parameter in stellar evolution models, affecting mass loss rates, nuclear burning rates, and interior opacities. Many calculations predict different evolutionary histories for similar stars with different metallicities (e.g., the evolution changes between the Geneva group's grid at solar metallicity versus that at the SMC metallicity).

Ten SMC A-supergiants have been analysed as the Galactic stars were analysed above. SMC data was taken with the ESO 3.6m telescope and CASPEC for high resolution and high signal-to-noise spectra. The only significant difference between the SMC and Galactic data sets is that the $\sim 5\ M_{\odot}$ were not observed in the SMC.

The first conclusion drawn from this project is that many of the published spectral types for these stars are too hot; as examples, AV 254 and AV 136 were classified as an anomalous A3 Ia and a normal A0 Ia (Humphreys *et al.* 1991), yet the atmospheric parameters imply temperatures and gravities typical of normal A7 Iab and A5 Iab stars.

The hotter spectral types are most likely a by-product of the difficulties encountered when applying the MK spectral classification system to metal-poor stars. This has been discussed previously by Fitzpatrick (1991) for LMC B-stars and Lennon (1996) for SMC B-stars.

Abundance calculations show that the metals in the SMC A-supergiants are 0.6 to 0.8 dex less than solar, which is consistent with the overall metal depletion of the SMC found from other analyses. Oxygen is depleted by $\sim$0.6 dex (including a -0.2 dex NLTE correction), also consistent with analyses of nebulae and other stars. Only upper limits could be determined for the carbon abundances, but these results are consistent with H II region and B-star carbon underabundances (which are substantial, ~ -1.3 dex relative to solar, see Table 1), but this means that no information is gleaned on any carbon depletions due to potential mixing events.

The nitrogen abundances are very interesting though, varying over a significant range

TABLE 1. SMC Nebular and B-Star Abundances

Elem	A-supers	Nebulae			B-stars			Adopted
		D84	RD90	G96	GAL	+R93	(=)	Initial
C	≤7.3	7.3*	...	7.4	8.2	−0.8	(7.4)	7.3
N	∼7.2	6.5	6.6	6.3	7.8	<-0.9	(<6.9)	6.5
O	8.1	8.0	8.1	8.1	8.7	−0.6	(8.1)	8.1

D84 = Dufour 1984 (* recalculated with new C atomic data; Dufour 1990); RD90 = Russell & Dopita 1990; G96 = Garnett *et al.* 1996; R93 = Rolleston *et al.* 1993.

from the (most likely) pristine values to (beyond) the first dredge-up values. One of the most difficult parts in the interpretation of these nitrogen abundances is in adopting an initial value. Nebular abundances have been determined and reviewed by several authors; see Table 1. The nitrogen abundance has been determined from N+/O+ (Dufour 1984) and N++/O++ (Garnett *et al.* 1996) with similar results, therefore it is unlikely to suffer significantly from neglected temperature fluctuation effects.

4. Rotational Mixing

Interpretation of the nitrogen abundances is easiest with a histogram. In Fig. 3, nitrogen abundances are plotted in bins covering 0.2 dex (to account for likely random uncertainties) for Galactic B-stars, Galactic A-type supergiants, and SMC A-type supergiants. The mean Galactic B-star abundance from ∼70 stars analysed by Gies & Lambert (1992), Kilian (1992), and Cunha & Lambert (1994) is adopted as the initial abundance for the A-supergiants. A slight offset of +0.2 dex in the mean of the Galactic A-supergiant abundance indicates partial mixing, but also that these stars have not undergone the first dredge-up which would predict a +0.5 dex enrichment.

The SMC nitrogen abundances vary over a much larger range. This range is not seen in a histogram plot for other elements, such as oxygen or iron. The nitrogen abundances imply that some stars have undergone no mixing while others have had extensive mixing of their surface layers with deeper CN-processed layers.

Evolution scenarios that include effects of rotation on the main-sequence can reproduce the CNO abundances; the most recent massive star evolution scenarios that include rotation are those by Fliegner & Langer (1996) and Maeder & Meynet (1996). In the Galaxy, the mean rotation rate of B-stars on the main-sequence is sufficient to mix the atmospheric layers through turbulent diffusive mixing; high rotation rates induce severe mixing, while no mixing occurs in slowly rotating stars. For the SMC stars, a variety of rotation rates are suggested by the nitrogen abundances, possibly favoring higher values than the average Galactic star rates.

5. Boron abundances

An additional test of the rotating stellar evolution models is the abundance of boron observed in the surface layers of main-sequence B-stars. In contrast to helium and CNO, the light elements (Li, Be, B) are destroyed upon exposure to temperatures more moderate than those in which the H-burning CN-cycle occurs, which makes them excellent

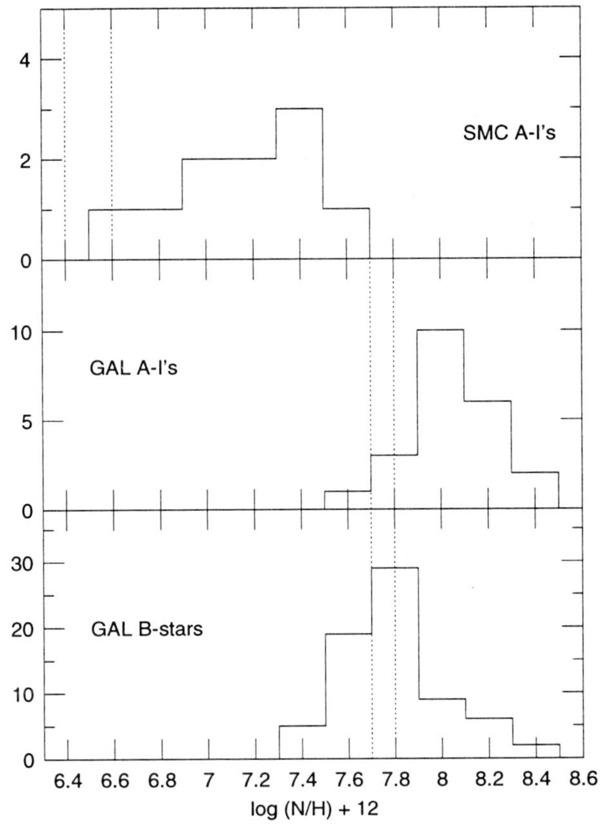

FIGURE 3. Histogram of nitrogen abundances for Galactic B-star, Galactic A-supergiants, and SMC A-supergiants (in bins of 0.2 dex). Dashed lines show mean N (range) for Galactic B-stars (lower and middle panels), and the SMC H II region N value range (top panel; see Table 1).

tracers of mixing processes between the cool stellar surface layers and hotter interior layers.

To examine mixing on the main-sequence, we need to study the light elements, yet only boron is observable through the B II and B III resonance lines in the UV. Additionally, boron is destroyed at temperatures of roughly 7 10^6 K, which occur sufficiently deep inside the stellar envelope of a 10 $M_\odot$ star that mass loss alone would not alter the surface boron abundance significantly. Thus, boron is a very sensitive test for the mixing processes in the upper stellar envelope, while CNO and helium trace the mixing with deeper layers.

Predictions from rotating stellar evolution models (Fliegner, Langer, & Venn 1996) show that boron could be depleted while the star is *on* the main-sequence. Nitrogen enrichments would lag behind the boron depletions, as mixing proceeded to deeper layers, yet by the end of the main-sequence lifetime, the nitrogen enrichments could be as much as the first dredge-up predictions (depending on the initial rotation rate); see Fig. 4.

Boron abundances have been estimated for a variety of field B-stars from IUE data (Venn, Lambert, & Lemke 1996), and more recently for four Orion B-stars from HST data (Cunha *et al.* 1996). Upper limits from the IUE data boron abundances, with nitrogen abundances determined by Gies & Lambert (1992) are consistent with this picture, see Fig. 4. NLTE boron abundances determined from the HST data are also in agreement since they find boron abundances that range from the interstellar abundance to 0.5 dex

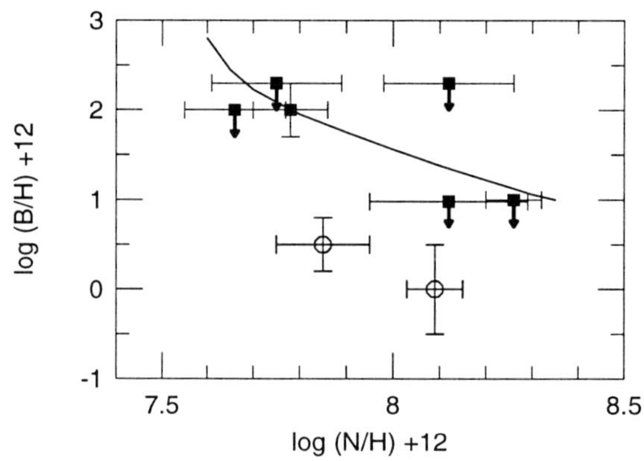

FIGURE 4. LTE boron (Venn *et al.* 1996) and NLTE nitrogen (Gies & Lambert 1992, Venn 1995a) abundances in Galactic B-stars (filled squares) and A-supergiants (circles). Rotating stellar evolution models by Fliegner & Langer (1996) predict that boron is depleted before significant enrichments in nitrogen occur during the main-sequence evolution (solid line).

less, yet where their nitrogen abundances are unchanged. Since the HST data is for very young Orion B-stars, we can expect little to no mixing with the interior layers.

REFERENCES

Baschek B., Scholz M., Sedlmayr E., 1977, A&A, 55, 375

Cunha K., Lambert D.L., Lemke M., *et al.*, 1996, ApJ, in preparation

Cunha K., Lambert D.L., 1994, ApJ, 426, 170

Dufour R.J., 1990, in *Evolution in Astrophysics*, ESA SP-310, 117

Dufour R.J., 1984, IAU Symp. 108, 353

Fliegner J., Langer N., 1996, A&A, in preparation.

Fliegner J., Langer N., Venn K.A. 1996, A&A, 308, L13

Fitzpatrick E. L., Garmany C.D., 1990, ApJ, 363, 119

Fitzpatrick E. L., 1991, PASP, 103, 1123

Garnett D.R., Skillman E.D., Dufour R.J., Peimbert M., Torres-Peimbert S., Terlevich R., Terlevich E., Shields G.A., 1996, ApJ, 443, 64

Gies D.R., Lambert D.L., 1992, ApJ, 387, 673

Humphreys R.M., Kudritzki R.P., Groth H., 1991, A&A, 245, 593

Kilian J., 1992, A&A, 262, 171

Lemke M., Venn K.A., 1996, A&A, 309, 558

Lennon D.J., 1996, A&A, in press.

Luck R.E., Lambert D.L., 1985, ApJ, 298, 782

Maeder A., Meynet G., 1996, A&A, 313, 140

Rolleston W.R.J., Dufton P.L., Fitzsimmons A., Howarth I.D., Irwin M.J., 1993, A&A, 277, 10

Russell S.C., Dopita M.A., 1990, ApJS, 74, 93

Schaller G., Schaerer D., Meynet G., Maeder A., 1992, A&AS, 96, 269

Venn K.A., 1995a, ApJ, 449, 839

Venn K.A., 1995b, ApJS, 99, 659

Venn K.A., Lambert D.L., Lemke M., 1996, A&A, 307, 849

Mass Function of the Stars in the Solar Neighbourhood

By NARAYAN CHANDRA RANA†
AND MRIDULA CHANDOLA

IUCAA, Post Bag 4, Ganeshkhind, Pune-411007, India

& Dept. of Physics, Pune University, Pune-411007, India

In the present work the PDMF of the stars in the solar neighbourhood has been derived considering the actual q values. Such a PDMF gives the stellar mass density of 40.1 $M_\odot\,\mathrm{pc}^{-2}$ as opposed to 39.5 $M_\odot\,\mathrm{pc}^{-2}$ for the photometrically corrected PDMF with assumed q distribution to fit some of the observations, and to only 31.3 $M_\odot\,\mathrm{pc}^{-2}$ as against the PDMF not corrected for the effects of stellar multiplicity. We have also calculated the IMF of the stars from the PDMF using the SFR with exponential time dependence, a form which gives a good fit to the data on chemical evolution. The IMF is found to be multimodal and a polynomial of order 9 fits it perfectly. The total surface mass density (68.8 $M_\odot\,\mathrm{pc}^{-2}$) predicted by the IMF is consistent with the presence of a very little or no non baryonic dark matter in the galactic disk in the solar neighbourhood. We have also used this IMF to calculate the iron abundances proposed by the model for the chemical evolution of the galaxies as given by Mihara and Takahara (1994) and found that the [Fe/H] so formed is larger than the value that one gets using the IMF proposed by Arimoto and Yoshii (1987).

1. Introduction

The single factor that decides the evolution of the star is its mass. The mass function of the stars i.e. their distribution according to the masses was first studied by Salpeter (1955) who studied the stars in the mass range of 1 to 10 $M_\odot$ in the solar neighbourhood and found that the function had a slope of -1.35 in the high mass end. The Present Day Mass Function or PDMF in short i.e. the number of stars per unit surface area of the disc per unit log range of mass is used to derive Initial Mass Function, or the IMF in short, which is the total number of stars ever produced by the disc per unit surface area per unit interval of log of mass. Knowledge of the IMF is essential to calculate the returned fraction, the chemical abundances etc.

The PDMF ($\xi(\log m)$) is derived from the Luminosity Function ϕ_{LF} (M_v) of the stars. Observationally the Luminosity function was first studied by Miller and Scalo (1979) who derived the PDMF from which they proposed a form of IMF ($\phi(\log m)$). They showed that slope of the IMF was much flatter for the low mass end than suggested by Salpeter while it was much steeper at the high mass end.

Garmany *et al.* (1982), Bisiacchi *et al.* (1983), vanBuren (1985) also derived the forms of IMF different from the Salpeter IMF, mainly for stars more massive than 20 $M_\odot$. Gusten and Mezger (1983), Larson (1986), recently Rana (1991) and Basu and Rana (1992) have given a form of IMF which is possibly bimodal in its mass distribution.

The luminosity function i.e. the number of stars in a given magnitude bin per unit surface area of the galactic disc is observationaly derived and is used for the calculation of the PDMF. Luminosity function is usually photometrically defined. So a multiple star system could appear as a single star of lower magnitude. This results in shifting

† N. C. Rana died on 22 August, 1996

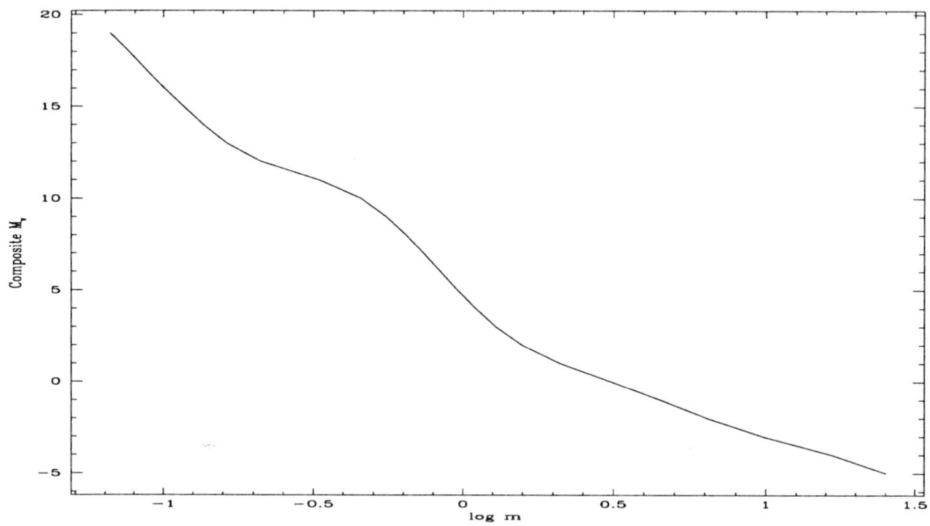

FIGURE 1. This is the composite M_v – log m relation used for this work. The mass is that of the primary alone while the M_v is total M_v of the system.

the mass function to the higher mass end. Rana (1991) had tried to take this effect into account while giving his form of IMF, by assuming that the stellar component of a multiple system have equal mass and that there is no need to correct the mass luminosity calibration for the effects of multiplicity while determining the mass corresponding to any luminosity while using the luminosity function. After that Basu and Rana (1992) have derived PDMF by assigning a function to the values of q i.e. the ratio of the mass of the secondary to the mass of the primary in a given binary system. In the present work we have used the actual observed q values to calculate the PDMF and hence the IMF.

The section 2 deals with the derivation of PDMF, the section 3 with that of IMF while the section 4 is devoted to the discussions and conclusions.

2. Derivation of PDMF

The derivation of PDMF involves a number of steps each of which rests on some observational analysis of data. The various parameters that go as input are the luminosity function ($\phi_{LF}(M_v)$), the scale height (H), and the magnitude of the slope of M_v–log m plot.

The PDMF is given as

$$\xi_{ms}(\log m) = \phi_{LF}(M_v)\,\frac{dM_v}{d\log m}\,2H$$

The luminosity function, as it is photometrically defined is a system luminosity function rather than a stellar luminosity function. Thus a multiple system could be mistaken to be a single more massive star. Thus the over all effect of overlooking the mass distribution in the multiple system results in shifting the PDMF towards the high mass end.

We have corrected the luminosity function for the presence of multiple stellar system. The total number of stars represented by luminosity function will increase due to the contribution from the secondary. As the mass of the secondary is not in general equal to

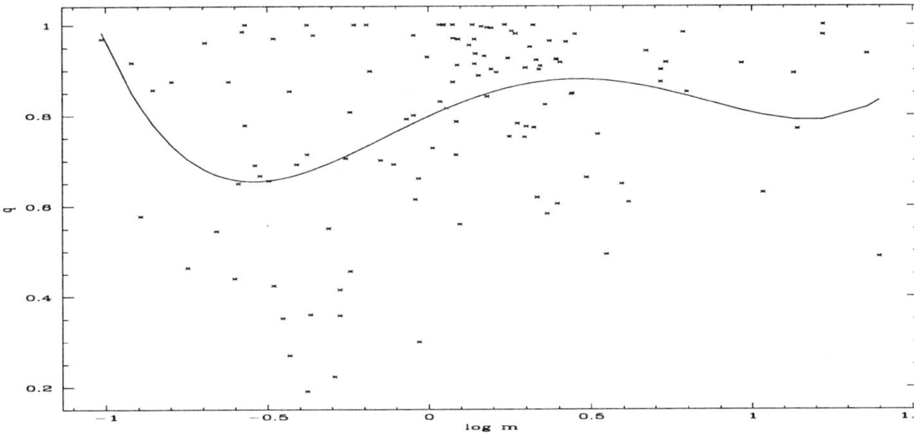

FIGURE 2. The crosses represent the actual q values while the solid line is the ploynomial fitted to it.

the mass of the primary, the contribution to the luminosity function from the secondary component of the system will not be to the same M_v bin to which primary belongs, but to a higher M_v bin.

A basic ingredient in the conversion of the luminosity function to the PDMF is the mass luminosity relation. Fig. 1 shows the result of binning the M_v–log m data in the bins of 0.1 in log m. The trend for the intermediate mass range is reasonably clear with a very little scatter. The low mass end of the M_v–log m relation is uncertain with a large scatter.

If the observed star is not a single star but a multiple system, the use of single star M_v–log m relation will result in assigning it a mass larger than the mass of its primary. This will in turn result in a spurious increase in the number of stars towards the high mass end. Therefore we have to construct a M_v–log m relation appropriate for the multiple system and use this composite M_v–log m relation to find the primary mass corresponding to each M_v bin. While calculating M_v–log m relation the mass we refer to is the mass of the primary while M_v is the total or the composite M_v of the system.

A lot of work has been done by a number of authors to determine the q distribution. Trimble (1974) found a bimodal distribution in q for a sample of spectroscopic binaries with peaks at 1 and 0.3. Lucy and Ricco (1979) found only a single peak at $q = 1$ for the double line spectroscopic binaries and Wolff (1978) found the same for short period binaries with a low rotational velocities. Abt and Levy (1978) favour a mass ratio around 1.0 while Halbawachs (1987) found only low q peak. Duquennoy and Mayor (1989) working with a sample of G-dwarf found a peak at 0.23. Complete subsamples of relatively bright, wide and close system show a distribution that is at best flat in q.

Thus the distribution of stars with respect to q is far from clear. Therefore in our present work we have taken the actual observed values of q. We have taken the data from the V/75 catalogue of astronomical parameters of binary systems as given by O. Yu. Malkov (1993).

We find that the q distribution is fairly random. As is obvious from the Fig. 3 there does not exist any preferential value for the mass of the companion star by the primary. We define the higher mass star as the primary and the lower mass star as the secondary.

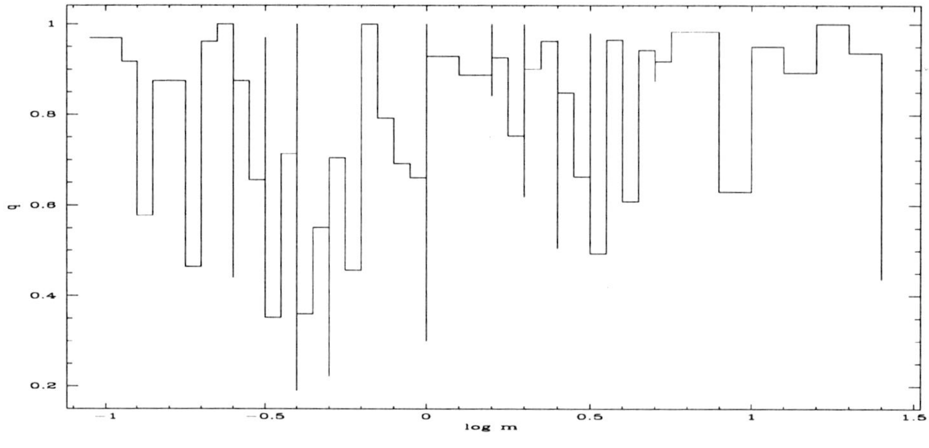

FIGURE 3. This is the histogram of the q values against the primary mass. There is no preference for the mass of the companion.

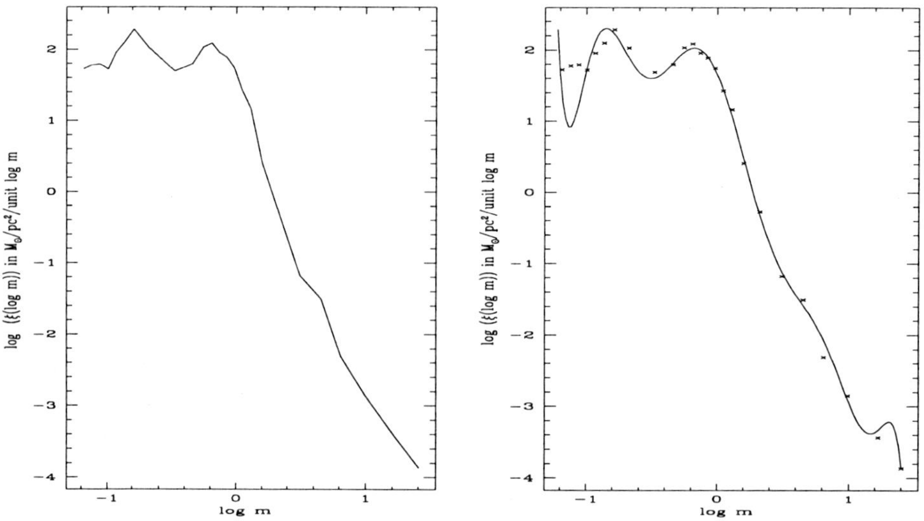

FIGURE 4. (a,b). Fig. 4a shows the PDMF. The shape of the PDMF shows that the IMF may possibly be multimodal. In Fig. 4b the bold line is the polynomial of order 9 fitted to the given PDMF.

We have tried to fit a polynomial of order 6 to the distribution (shown by the solid line in the Fig. 2) It is given as

$$q = -0.898(\log m)^6 + 0.843(\log m)^5 + 1.68(\log m)^4$$
$$-1.32(\log m)^3 - 0.712(\log m)^2 + 0.48(\log m)^1 + 0.81$$

PDMF is usually expressed in terms of number of stars per unit area rather than per unit volume, in a log interval of mass. We require the mean scale height of the stars to make the above conversion.

Stars of varying luminosity in the main sequence are found to have varying amount of random velocities which makes them oscillate about the mid galactic plane. The stars that have low mass and are hard to detect have a higher scale height while the young massive stars which are easily detected have a smaller value of the scale height. Thus the scale height depends on the magnitude i.e. the mass alone. Also they can be used as indication of the age (Twarog 1980, Twarog and Wheeler 1982, Carlberg et al. 1985).

Because the data for the value of the scale height is fairly random we have adopted the average scale height calculated from the $\psi(\mathrm{t})$ = constant as given by Basu and Rana (1992).

The derived PDMF is shown in the Fig. 4a. We have fit it with a polynomial of order 9 (Fig. 4b, curve shown by the solid line) given as

$$\log(\xi \log m) = -0.819(\log m)^9 + 9.82(\log m)^8 + 25.9(\log m)^7 - 29.2(\log m)^6 - 27.3(\log m)^5 + 28.1(\log m)^4 + 11.5(\log m)^3 - 11.0(\log m)^2 - 4.29(\log m)^1 + 1.65$$

3. Initial Mass Function

The IMF corresponds to the mass spectrum of the accumulated number of stars per unit area of the disc, that are born so far, where as the PDMF accounts for those stars which are still in their main sequence. Lifetime T_{ms} and the birth rate of the stars as a function of the time and the mass is essential for deriving the IMF from the PDMF. IMF is given as

$$\frac{\xi(\log m)}{\phi(\log m)} = \frac{\int_{T_d - T_{ms}}^{T_d} \psi(m,t)dt}{\int_0^{T_d} \psi(m,t)dt}$$

The main sequence lifetime of a star T_{ms} is defined as the life of the star spent in the hydrogen burning at its core and it is computed on the basis of the theoretical calculations of the stellar evolution. We have taken the stellar lifetime data from Maeder and Maynet (1989) and Maeder (1990). We take life of the galactic disc T_d as 13 Gyr.

The actual process of star formation is so far little understood. It possibly depends on the metallicity, pressure, and temperature of the gas, magnetic field and details of the trigger mechanism. We use the Star Formation Rate (SFR or $\psi(\mathrm{t})$) as given in Basu and Rana (1992). It simply suggests that Star Formation Rate (SFR) is proportional to the local surface density of the molecular hydrogen, raised to power of α and the molecular fraction of the gas by the mass is found to be corelated with the average metallicity of the region again fitted with a power law giving the index β for the Galaxy.

The values of α and β were chosen so that observed features of the solar neighbourhood and the rest of the galactic disc were reproduced when the abundance of iron was considered. Accordingly the value of α was taken as 1.15 while that of β as 1.25.

Thus

$$\psi = a\,\sigma_g^\alpha \left(\frac{Z}{Z_\odot}\right)^\beta$$

where σ_g is the surface density of the gas while Z is the metallicity and a is the constant of proportionality.

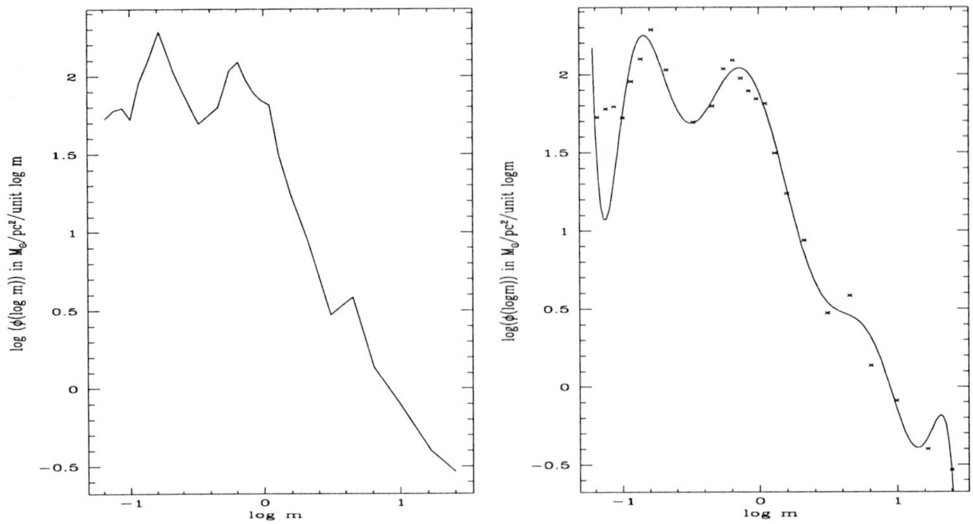

FIGURE 5. (a,b). Fig. 5a is the plot of IMF. As is clear the IMF is multimodal with three distinct peaks. In Fig. 5b we have fitted a polynomial of order 9 to our multimodal IMFF.

The derived form of the IMF is shown in the Fig. 5a. Again we have fitted it with a polynomial of order 9 (given as solid line in Fig. 5b) which is given as

$$\phi(\log m) = -6.14(\log m)^9 + 7.65(\log m)^8 + 19.1(\log m)^7 - 22.4(\log m)^6 - 19.6(\log m)^5 \\ +20.6(\log m)^4 + 7.88(\log m)^3 - 6.97(\log m)^2 - 2.17(\log m)^1 + 1.89$$

3.1. *Derived Quantities*

We have used the model as proposed by Mihara and Takahara (1994) for the chemical evolution of the galaxy. Their model not only accounts for the iron enrichment due to the type Ia Sn but also due to type II Sn and the loss during the entire main sequence due to the stellar wind. We have reproduced their results for the IMF as proposed by Arimoto and Yoshii (1987, hereafter AY). We have used this form of IMF because it is easy to handle and also it gives the expected results for most of the cases. We have also used our form of IMF to carry out the abundance calculations. The [Fe/H] turns out to be 1.25×10^{-02} using our IMF as against only 8.79×10^{-03} for AY. The difference in the two values is mainly due to the fact that our form of IMF includes more number of low mass stars rather the stars that may be type Ia Sn progenitors.

3.2. *Dark Matter*

We have estimated the total surface mass density in the solar neighbourhood to be $68.8\,M_\odot\,\mathrm{pc}^{-2}$. It must be noted that we have only considered stellar masses greater than $0.06\,M_\odot$ to calculate the total mass in the stars. We have not given any mass estimates for the stars of lower mass i.e. the brown dwarfs or the planets. The shape of the IMF at the lower mass end is debatable, but whatever the mass dependence, the very presence of the brown dwarfs will increase the estimate of the observed total surface mass density of the baryonic matter in the solar neighbourhoood. Thus at least in the solar neighbourhood we don't have to look for the nonbaryonic matter to explain the dark matter.

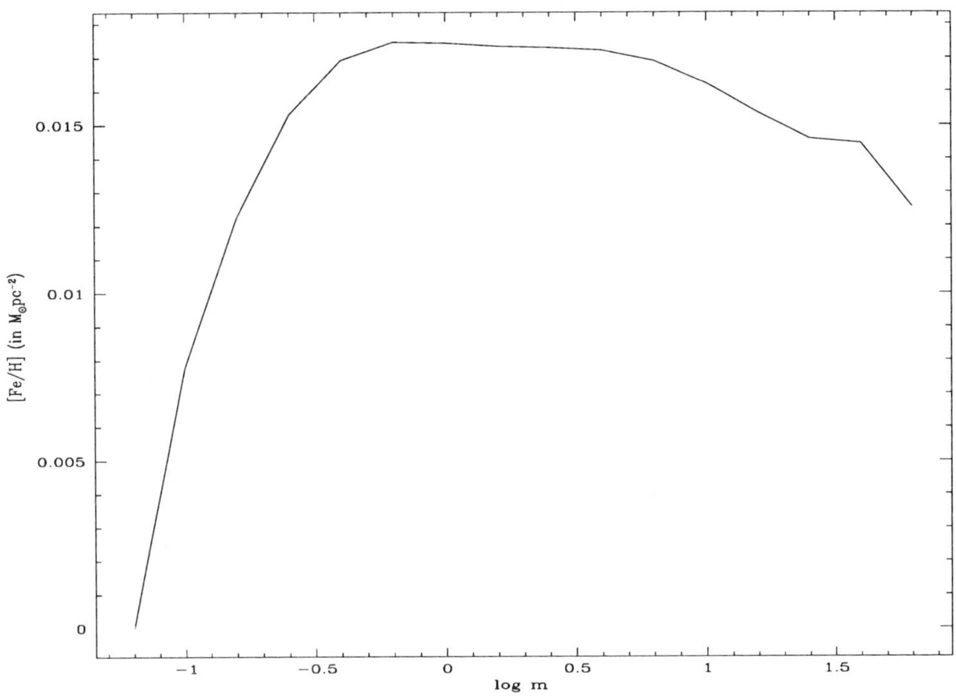

FIGURE 6. This shows the [Fe/H] against log m. The value so obtained is more than what we get from AY IMF

4. Discussions and Conclusions.

The present work was aimed to derive the mass function of the main sequence stars in the solar neighbourhood adopting a different set of new input data. Let us discuss the results and the conclusions that follow.

The IMF and PDMF have the same value for the mass below the cutoff mass. These are the stars that have a lifetime same as that of the galactic disc.

The PDMF is multimodal. It has been fitted with a polynomial of order 9. The dip in the low mass region (between 0.25 to 0.39 $M_\odot$) may not be the Wielen dip as the error bar in that region are large.

Whether the PDMF again rises towards the low mass side is unclear.

The IMF is also multimodal. The slope in the linear region changes at three distinct points, it being -2.5 at $\log m = 1.5$, -1.314 at $\log m = 0.9$, and -0.769 at $\log m = 0.65$.

The IMF may have an overall slope of -2 over the entire mass range 1.5 to 100 $M_\odot$.

The dip in the low mass region could be attributed to the error in observation, while the hump in the high mass end could indicate a new type of star formation or just a sudden burst of star formation activity resulting in stars of mass in a particular range. Thus the change in slope at the low mass end could be spurious, but at the high mass end IMF is defenitely multimodal.

In the high mass end the IMF is multimodal with a number of humps. The individual humps seem to occur at reasonably constant interval of $\log m \sim 0.8$. This explains why the individual determination of slopes of IMF over a narrow range appear to vary considerably widely ~ -0.7 to -2.5 depending upon the interval one chooses. These slopes often lead to misleading results when unduly extended to either side of the appropriate range.

The correction of PDMF of stars is important in the sense that it increases the total number of stars represented by the PDMF. The average correction for the different ratios of the secondary to primary star mass shifts the PDMF to the low mass side.

The present day surface mass density turns out to be 40.1 $M_{\odot}\,\mathrm{pc}^{-2}$ while the total surface mass density turns out to be 68.8 $M_{\odot}\,\mathrm{pc}^{-2}$. Thus at least in the solar neighbourhood most of the dark matter could be explained in the form of the baryons.

The present model seems to account for the total dynamical requirement for the total surface density of the matter in the solar neighbourhood in the form of stellar populations, dead or alive stars and interstellar gas only.

Because this IMF is derived by taking the actual observed values into account, we find that the type Ia supernova rate so predicted (following the model given by Mihara and Takahara, 1994) is much higher. This is mainly because more number of stars that could end into type Ia supernovae are predicted by this model. One of the reasons why the [Fe/H] ratio turns out to be higher, and much closer to the observed values.

REFERENCES

Abt H. A., Levy S. G. 1978, ApJS, 36, 241.
Arimoto N., Yoshii Y. 1987, A&A, 173, 23.
Basu S., Rana N. C. 1992, ApJ, 393, 373.
Bissiacchi G. F., Firmani C., Sermiento A. F. 1983, A&A, 119, 167.
Carlberg R. G., Dawson P. C., Hsu T., VandenBerg D. A. 1985, ApJ, 294, 674.
Duquennoy A., Mayor M. 1991, A&A, 485, 524.
Garmany C. D., Conti P. S., Choisi C. 1982, ApJ, 263, 772.
Gusten R., Mezger P. G. 1983, Vistas Astron., 26, 159.
Halbawachs J. L. 1987, A&A, 183, 234.
Larson R. B. 1986, MNRAS, 218, 409.
Lucy L. B., Ricco E. 1979, AJ, 84, 40.
Maeder A. 1990, A&AS, 84, 139.
Maeder A., Meynet G. 1989, A&A, 210, 155.
Mihara K., Takahara F. 1994, PASJ, 46, 447.
Miller G. E., Scalo J. M. 1979, ApJS, 41, 513.
O. Yu. Malkov 1993, Bull. Inf. CDS, 42, 27.
Rana N. C. 1991, ARAA, 121, 161.
Salpeter E. E. 1955, ApJ, 121, 161.
Trimble V. 1974, ApJ, 79, 967.
Twarog B.A. 1980, ApJ, 242, 242.
Twarog B. A., Wheeler J. C., 1982, ApJ, 261, 636.
vanBuren D. 1985, ApJ, 294, 567.
Wolff S. C. 1978, ApJ, 222, 553.

Synthetic Spectral Indices for Elliptical Galaxies

By LAURA GREGGIO

Universitaets Sternwarte, Scheiner Str.1, D-81679 Muenchen, FRG
& Department of Astronomy–Bologna, Italy

Synthetic Mg_2, $H\beta$, and Fe52 line strengths for stellar populations with a metallicity spread are presented. The metallicity distribution adopted is shaped like the predictions of the closed box model for the chemical evolution, but the minimum and maximum metallicity are treated as free parameters. It is found that the high values of Mg_2 and Fe52 indices measured in the nuclei of bright elliptical galaxies indicate that their inhabiting stellar populations are old, metal rich and with a small metallicity spread. It is suggested that in the process of galaxy formation the different dissipative properties of the newly formed stars and the collapsing gas result in a chemical separation within the galaxy: the gas enriched by SN explosions sinks to the center, where further star formation takes place. The metallic indices of bright ellipticals also show that their Mg to Fe ratio is likely above solar, which argues in favour of short time scales for the galaxy formation.

1. Introduction

The theory of stellar evolution finds a very exciting application in the problem of measuring the age and the metallicity of elliptical galaxies, a question which has a strong impact on cosmological models. The vast majority of ellipticals are too far to be resolved into stars, so that their inhabiting stellar populations have to be investigated by analysing their integrated light. During the last decades the problem of deriving age and metallicity for ellipticals from their spectrophotometric properties has been attacked by many authors with various approaches. For example, O'Connell (1980) and Pickles (1985) made use of an empirical population synthesis method, which consists in adding up stellar spectra of objects of different ages and metallicities to match the observed spectral energy distribution (SED) of a given galaxies. The best fit yields the average age and metallicity of the system under study. Bica & Alloin (1987), and Bica (1988) adopt entire clusters (as opposed to individual stars) as building blocks for the population synthesis, thereby including (empirically) the effects of stellar evolution. The evolutionary population synthesis (e.g. Buzzoni 1989, Bruzual & Charlot 1993, Worthey 1994, Weiss, Peletier & Matteucci 1995) consists in deriving the integrated spectra (or magnitudes and colors) summing up the contribution of stellar models distributed along theoretical isochrones. Each isochrone is characterized by an age and a metallicity, and the comparison of the theoretical SEDs with the observations allows to infer the values of these parameters which better represent the data. Finally, the theoretical modelling of formation and evolution of entire galaxies has been employed by Arimoto & Yoshii (1986a,1987), Guiderdoni & Rocca-Volmerange (1987), Barbaro & Olivi (1989), Bressan, Chiosi & Fagotto (1994), Vazdekis et al. (1996) and others, to compute the evolution of theoretical SEDs of composite systems. In these models, the chemical evolution is followed according to some prescriptions for the star formation and at each epoch the system is characterized by an age and a metallicity distribution. The related SED as a function of age is computed to be compared with the observational data.

In their early studies, different authors were reaching quite different results, some claiming intermediate ages (e.g. O' Connell 1986) and others old ages (e.g. Arimoto & Yoshii 1986b) for ellipticals. More recently, for the most massive ellipticals there seems to be a general consensus upon an old age and metallicity Z close to or larger than solar (hereinafter $Z_\odot$), while a substantial disagreement still persists on the parameters for the less massive ellipticals (Arimoto 1996). Small discrepancies among the results obtained by the various authors are attributable to the different prescriptions adopted in the theoretical modelling (especially the conversion from the theoretical log L, log $T_{\rm eff}$ plane to the observational parameters), as thoroughly investigated by Charlot, Worthey & Bressan (1996). However, there is an intrinsic difficulty in deriving age and metallicity from the spectrophotometry, related to the problem of the age-metallicity degeneracy (Renzini 1986). Indeed, the SEDs of two stellar populations, one with [Fe/H] $= -0.1$ and age of 5 Gyr and the other with [Fe/H] $= +0.2$ and age of 15 Gyr are almost the same (Worthey 1994), due to the fact that the SED essentially traces stellar temperatures, which are sensisitive to both age and metallicity. As a consequence, broad band colors, and low resolution SED as well, are of little use for deriving separately these two parameters.

Recently, the hope of disentangling age from metallicity effects has been entrusted to spectral indices: in essence, choosing two spectral features, one essentially controlled by age and the other by metallicity allows one to derive the two parameters independently. For example, considering the Hβ index (believed to trace the turn−off temperature) and a combination of Magnesium and Iron indices (tracing the total Z) Gonzales (1993) finds that the age of nearby ellipticals spans a large range. With an analogous approach, Buzzoni (1995b) reaches a quite different conclusion, with most ellipticals being old. These results have been obtained comparing the data to models computed for single age and metallicity stellar populations (SSP), but real galaxies host composite stellar populations, with a spread in both age and metallicity. In this paper I will present theoretical computations of the spectral indices Mg_2, Fe52 and Hβ (defined as in Burstein et al. 1984) for coeval stellar populations with a metallicity distribution, meant to describe systems in which the star formation occurs in a short initial episode. The comparison of these models to the nuclear line strengths observed in ellipticals will allow me to draw some hints on the major epoch and modalities of star formation in these galaxies. A more extensive report of these computations is given in Greggio (1996).

2. Spectral indices for composite stellar populations

The simplest way to describe the chemical evolution of galaxies is the *closed box model* (Tinsley 1980), whose final metallicity distribution follows the relation $f(Z) \propto e^{-Z/y}$. Here I consider a modification of this model, truncating the distribution so that Z varies between a minimum ($Z_{\rm m}$) and maximum ($Z_{\rm M}$) metallicity, and explore the consequences of letting vary these two parameters. In essence, when increasing $Z_{\rm M}$ one describes systems in which the chemical enrichment has proceeded to a higher level of completion. Models for the chemical evolution of galaxies including the effect of galactic winds (Arimoto & Yoshii 1987, Matteucci & Tornambé 1987, Bressan, Chiosi & Fagotto 1994) are characterized by closed box shaped metallicity distributions, with $Z_{\rm m} \simeq 0$ and $Z_{\rm M}$ increasing with the galactic mass. Increasing $Z_{\rm m}$, instead, is like assuming some degree of preenrichment in the gas at the onset of the chemical processing. Here, I consider a value for the yield of $y = 3\,Z_\odot$, which is consistent with the metallicity distributions in Arimoto & Yoshii (1987) models. The main results do not change considering $y = 2\,Z_\odot$, $4\,Z_\odot$ (see Greggio 1996 for more details). In the following, composite stellar populations with this kind of metallicity distributions will be referred to as CSPs.

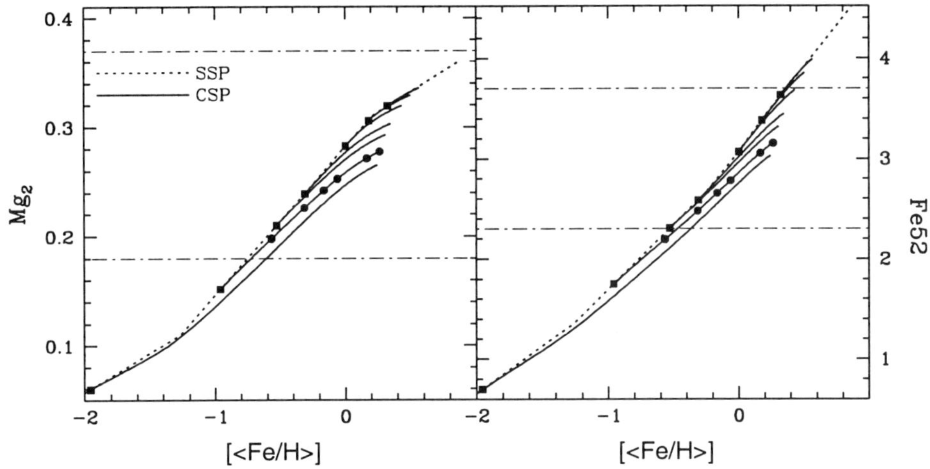

FIGURE 1. Model line strenghts using Buzzoni's 15 Gyr old SSPs, as functions of metallicity. The dotted lines are the loci of pure SSP models, while the solid lines display CSP models for various values of the $Z_{\rm m}$ parameter. The following cases are shown: $Z_{\rm m}/Z_{\odot}=$ (0.01,0.1,0.3,0.5,1,1.5,2), and the filled squares mark the values for SSPs with $Z = Z_{\rm m}$. Along each line, $Z_{\rm M}$ varies from $Z_{\rm m}$ up to 0.1. The filled circles mark CSP models with $Z_{\rm m} = 0.1 Z_{\odot}$, and $Z_{\rm M} = (0.5,1,1.5,2,4,6) \times Z_{\odot}$. The dot-dashed lines show the typical ranges spanned by the observational data for the nuclei of ellipticals.

Integrated line strenghts for CSPs can be obtained by summing up the contribution to the total flux in the appropriate wavelength range of the various SSPs at the different Z, whose spectral characteristics can be found in the literature (e.g. Worthey 1994, Buzzoni 1995a and references therein). Figure 1 shows the results of these computations for the Mg_2 and Fe52 indices, under different assumptions for the $Z_{\rm m}$ and $Z_{\rm M}$ parameters. Along a given solid line, the $Z_{\rm m}$parameter is kept constant, while $Z_{\rm M}$ increases, thereby increasing the average metallicity of the CSP. It appears that the Mg_2 and Fe52 indices of SSPs are systematically stronger than those of CSPs with the same (mass weighted) average metallicity. This is due to the fact that the low Z populations contribute more light in the optical bands, where these indices are measured. In much the same way, at any given Mg_2, the average metallicity of a CSP is larger than that of the SSP with the same value of the index. The difference between the two metallicities depends on the width of the Z distribution, and can be as large as ~ 0.3 dex for CSPs with Δ[Fe/H] = 2.5. This challenges the derivation of metallicity from integrated line strengths.

In Figure 1 the dot-dashed lines show the typical range covered by the observational data (e.g. Davies et al. 1987, Worthey, Faber & Gonzales 1992, Carollo, Danziger & Buson 1993), for the nuclear line strengths of ellipticals. The predictions of the wind models for the galactic chemical evolution ($Z_{\rm m} \simeq 0$) correspond to the lowest of the solid lines, and it can be seen that this class of models does not account for Mg_2 and Fe52 indices in excess of ~ 0.27, 3. respectively (see also Casuso et al. 1996, Worthey, Dorman & Jones 1996). Due to the Mg$-\sigma$ relations (Bender, Burstein & Faber 1992) these galaxies are also the most massive ellipticals. Instead, the metallic line strengths in the nuclei of the less luminous ellipticals are consistent with the expectations from the chemical evolution models with galactic winds.

2.1. *Metallicity distribution in the nuclei of the brightest ellipticals*

Looking at Figure 1, it can be seen that the high $Mg_2$and Fe52 values observed in the most massive ellipticals can be reproduced by assuming some degree of preenrichment in the gas which formed stars in their nuclei.

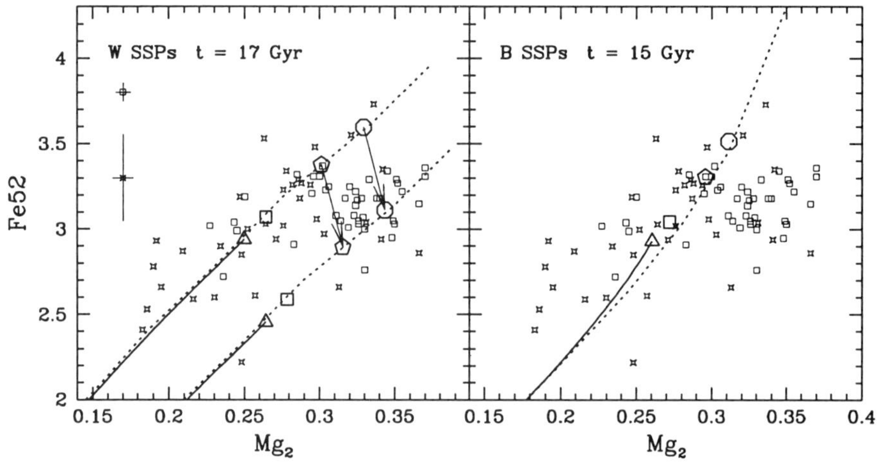

FIGURE 2. Dependence of Mg_2 on the parameter Z_m: the curves show the loci of CSP models, all characterized by a maximum value of the metallicity $Z_M = 3\,Z_\odot$. The curve labelled W have been computed with Worthey's (1994) 17 Gyr old models; those labelled B with Buzzoni's (1989, 1995) 15 Gyr old SSPs. Also plotted is the distribution of the Mg_2 values derived for early type galaxies from Davies et al. (1987).

In order to see which is the minimum metallicity allowed, within the CSP models considered here, I have computed integrated Mg_2 indices for a fixed (high) Z_M, and varying Z_m. The results are shown in Figure 2, together with the distribution of Mg_2 indices in early type galaxies from Davies et al. (1987) sample, which includes a total of 469 objects. The curves show how the Mg_2 index for CSPs increases with increasing Z_m, for two sets of SSP models (see captions). It appears that galaxies with Mg_2 higher than 0.3 (more than 40% of the total sample) require Z_m in excess of $\sim 0.5\,Z_\odot$, for both sets of SSP models. The allowed metallicity range gets narrower as Mg_2 increases, and only Worthey's SSP models account for the strongest Mg_2 ellipticals. The difference in the two curves is partly due to the different ages of the two sets of SSP models: using Worthey's 12 Gyr old SSPs, the Mg_2 index saturates at ~ 0.33. Buzzoni's models allows to extend the metallicity distributions up to $Z \simeq 6\,Z_\odot$. Taking this value for Z_M, 15 Gyr old CSPs with $Z_m = Z_\odot$ have $Mg_2 = 0.32$. Therefore, for both sets of models, the nuclei of the strongest Mg_2 ellipticals host both old and high metallicity SSPs, with a narrow metallicity dispersion.

2.2. [Mg/Fe] *overabundance in the brightest ellipticals*

The plot of the Fe52 versus Mg_2 indices allows one to check the internal consistency of the indications derived from the different line strengths on the composition of the stellar population. In Figure 3, CSP models (large open symbols) are compared to the data from Worthey, Faber & Gonzalez (1992), relative to the nuclear values of the indices in elliptical galaxies. The solid lines show the locus described in this diagram from the sequence of CSP models with $Z_m = 0.01\,Z_\odot$ and Z_M increasing up to $3\,Z_\odot$. It can be seen that in these diagrams SSPs add as vectors (see also Worthey, Faber & Gonzales 1992), in that the CSP indices lie on the SSP loci.

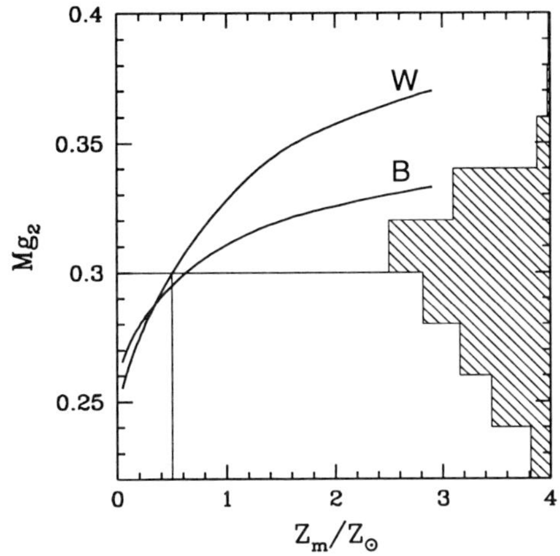

FIGURE 3. Comparison between model predictions and observations of the nuclear line strengths of ellipticals. The data points (small symbols) are from Worthey, Faber & Gonzalez (1992) (crosses) and Gonzalez (1993) (squares). The error bars quoted by the authors are shown in the left panel. SSP models by Worthey and Buzzoni are displayed as dotted lines. The large symbols show the line strengths for a selection of CSP models: $Z_{\rm m} = 0.01\,Z_\odot$ (triangle), $Z_{\rm m} = 0.1\,Z_\odot$ (square), $Z_{\rm m} = 0.5\,Z_\odot$ (penthagon) and $Z_{\rm m} = Z_\odot$ (octagon), all having $Z_{\rm M} = 3\,Z_\odot$. The sequence of CSP models obtained with $Z_{\rm m} = 0.01 Z_\odot$ and increasing $Z_{\rm M}$ is shown as a solid line. The effect of an α elements overabundance of [O/Fe] $\simeq 0.45$ is shown by the arrows.

The average location of galaxies with $\mathrm{Mg}_2 \leq 0.3$ is well reproduced by the models, but CSP models with $Z_{\rm m} = 0.01\,Z_\odot$ just encompass the range occupied by the weakest Mg_2 objects. Once again, galaxies with stronger metallic nuclear indices require some degree of preenrichment in their centers, within this class of CSP models. Notice that this is needed to account for both Mg_2 and Iron line strenghts.

Those objects characterized by Mg_2 larger than approximately 0.3 depart from the general (Fe52) − Mg_2 relation, exhibiting lower Iron indices than the model predictions, for a given Mg_2. Since the SSP models considered here have been computed for solar abundance ratios, this departure is interpreted as an evidence for a Mg/Fe ratio larger than solar in these (most luminous) ellipticals (Worthey, Faber & Gonzales 1992, Davies, Sadler & Peletier 1993, Matteucci 1994). According to the current understanding of how the chemical enrichment proceeds in galaxies, a [Mg/Fe] greater than 0 is likely to trace an overall α elements overabundance, due either to a flatter IMF, or to a short star formation timescale in the system. Although adequate SSP models, computed for non solar abundance ratios, would be necessary to derive the [Mg/Fe] ratio from the values of the indices, one can nevertheless get a rough estimate of the overabundance using models computed for solar ratios. The line streghts of a given SSP depend on the shape of its isochrone, and on the relations (fitting functions) between the line strength of the individual stars and the basic parameters controlling their spectrum, i.e. gravity, temperature and metallicity. When varying the [Mg/Fe] ratio, both the isochrone and the fitting functions change, in different ways. Taking consistently into account the various effects at work, Greggio (1996) finds that, indicating with $\zeta_{\rm Mg}$ and $\zeta_{\rm Fe}$ the contributions (by mass) to the total metallicity of Mg and Fe, two mixtures with the same total Z but different α element enhancements, have indices which differ by

$$\Delta(\mathrm{Mg}_2) \simeq a \times \log \zeta^{(1)}_{\rm Mg}/\zeta^{(2)}_{\rm Mg} \quad \text{and} \quad \Delta(\mathrm{Fe52}) \simeq b \times \log \zeta^{(1)}_{\rm Fe}/\zeta^{(2)}_{\rm Fe}$$

where a and b are the dependences on metallicity in the Mg_2 and Fe52 fitting functions. These relations hold if:
i) the shape of the isochrone (at given age and Z), and the distribution of stars along it, are the same for both (1) and (2) mixtures;
ii) the dependence of the fitting functions on the metallicity is linear in [M/H] and can be separated from the other dependences;
iii) a given index depends only on the abundance of the element which gives rise to the considered feature.

These three requirements are never strictly true; nevertheless within the current understanding they seem to be valid approximations. Using the results of chemical evolution models by Matteucci (1992), and the fitting functions adopted by Worthey (1994), which have $a \sim 0.19$ and $b \sim 1.5$, one derives $\Delta Mg_2 \sim 0.015$ and ΔFe52 ~ -0.37 for a mixture with [O/Fe] = 0.45. The corresponding SSP and CSP models with such an α element overabundance are shown in the left panel of Figure 3. It can be seen that, as long as [Mg/Fe] is fixed, the theoretical indices describe a locus parallel to the solar ratio sequence. Therefore, the average (flat) slope of the data points in the $Mg_2 > 0.3$ domain requires an [Mg/Fe] increasing for increasing Mg_2, up to an overabundance of ~ 0.4 dex. Similar conclusions have been reached by Weiss, Peletier and Matteucci (1995) with a different approach.

It is worth noticing that, since a Mg/Fe enhancement over the solar ratio has only a mild effect on the Magnesium index, the strongest Mg_2 ellipticals still require high Z and small metallicity dispersion in their nuclei. Actually, at a given total metallicity, α enhanced mixtures are characterized by a lower fraction of electron donors, leading to warmer RGBs, and, keeping fixed all the rest, to a lower Mg_2 index (Mg_2 decreases when the RGB temperature increases). Although this effect is counterbalanced by the increase of Magnesium absorption in the stellar spectra, a Mg/Fe enhancement (at constant total metallicity) will not lead to a large increase of the Mg_2 index.

2.3. *The Hβ index and the ages of ellipticals*

The Hβ line strength is mostly sensitive to the temperature of turn-off stars and can then be used to infer the age of the galaxies (see e.g. Gonzales 1993). In the left panel of Figure 4, I show the data for nearby ellipticals, together with Worthey (1994) models for different ages of the SSPs. It can be seen that the locus of SSP models of varying Z and constant (intermediate or old) age run too shallow with respect to the data: those galaxies with low Mg_2 index exhibit a strong Hβ line strength, suggesting a relatively young age for their central stellar populations. On the other hand, the strongest Mg_2 galaxies have low Hβ , compatible with old ages for these objects. Taking into account the Mg$-\sigma$ relation, the tight correlation between the Hβ and Mg_2 index would then imply that the nuclear populations of the most massive (and brightest) ellipticals are systematically older than those of the less massive galaxies (Faber et al. 1995), with all ellipticals hosting $\sim 3Z_\odot$ stars in their centers (with virtually no metallicity dispersion). Adopting SSP models by Buzzoni (right panel in Figure 4) a similar conclusion holds.

The need for invoking an age difference in these galaxies is related to the mild dependence of the Hβ line strength on the metallicity, which reflects the dependence on Z of the turn$-$off temperature. This is the case for SSP models with Red Horizontal Branches (RHB), irrespective of the metallicity, which I have considered until now. However, if low Z SSPs are characterized by Blue Horizontal Branches (BHB), Hβ becomes much more sensitive to Z. According to Buzzoni, Mantegazza & Gariboldi (1994), the presence of a BHB like the one observed in M3 leads to an enhancement of the Hβ index of ~ 0.7Å, relative to what obtained assuming a RHB. In the right panel of Figure 4, I

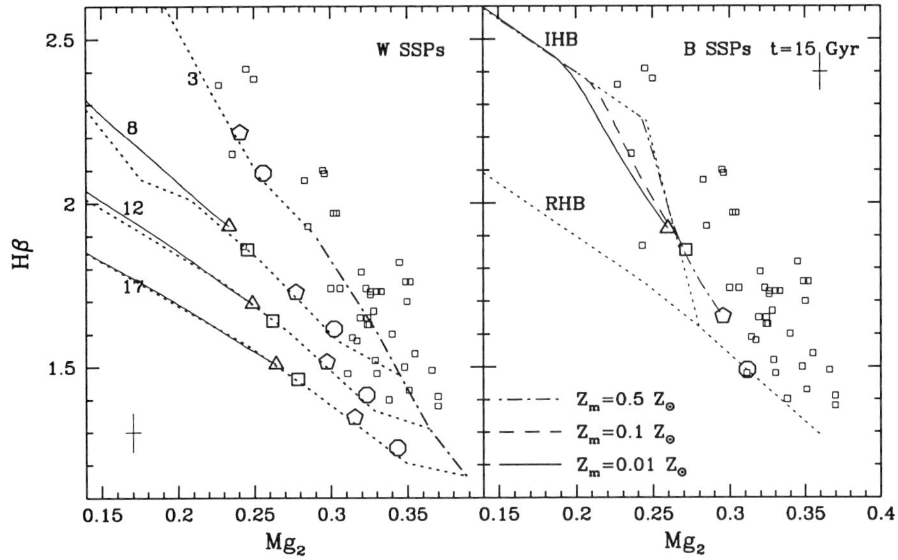

FIGURE 4. Comparison of data for the nuclei of nearby ellipticals from Gonzales (1993) (small squares) with models based on Worthey (left panel) and Buzzoni (right panel) SSPs. The quoted errors on the data are shown. Dotted lines indicate the loci of SSPs, with various ages, as labelled. CSP models are shown as large symbols, with the same encoding as in Figure 3. In the left panel, the solid line shows the sequence of CSP models with $Z_{\rm m} = 0.01\,Z_\odot$ and $Z_{\rm M}$ increasing up to $3Z_\odot$. In the right panel, the solid, dashed and dot-dashed lines show sequences of CSP models computed assuming an enhancement of 0.5Åin the Hβ index of SSPs with $Z \leq 0.6\,Z_\odot$ (so as to mimic the presence of IHB in the low Z tail of the distribution). A constant shift of 0.015 (for Worthey's models) and 0.004 (for Buzzoni's) has been applied to the model Mg_2 indices to account for the [Mg/Fe] overabundance.

show the impact of this effect (see caption): if the low metallicity tail of the CSPs had BHB, the larger the population in this Z range, the stronger the Hβ index. Therefore, the Hβ index may trace either the age, or the fraction of the low metallicity component in an old CSP.

3. Conclusions

Models for composite stellar populations with a spread in metallicity show that the nuclei of brightest ellipticals should host stars with i) large total metallicity; ii) a non solar elemental ratio ([Mg/Fe] > 0); iii) small metallicity dispersion. Condition i) is met by processing the gas through multiple stellar generations, while condition ii) requires that the star formation process was completed within a short time scale (e.g. Matteucci 1994). This implies that, during the galaxy formation, the SNII rate was very high, and, in order to proceed with the chemical processing, the gas should have been located within a deep potential well. The high Mg_2 values of the brightest ellipticals further suggest that the nuclei of these objects are old, and their formation should have taken place at high redshift. Condition iii) requires that the gas out of which the galaxy centers formed was substantially preenriched. This is indeed the result of galaxy formation in which some dissipation played an important role: once formed (in the outer regions of the galaxy), the stars stop participating of the general collapse, but pollute the gas, which continues to flow to the center. If this was the case, the low metallicity component missing in the nuclei of the brightest ellipticals should be found in their outer regions: the closed box models Z distribution would then apply to the galaxy as a whole.

REFERENCES

Arimoto N., 1996, in Leitherer C., Fritze-von Alvensleben U., Huchra J., eds, ASP Conf. Ser. 98, From Stars to Galaxies: the impact of Stellar Physics on Galaxy Evolution, p. 287

Arimoto N., Yoshii Y., 1986a, A&A, 164, 260

Arimoto N., Yoshii Y., 1986b, in Chiosi C., Renzini A., eds, Spectral Evolution of Galaxies. Kluwer, Dordrecht, p. 309

Arimoto N., Yoshii Y., 1987, A&A, 173, 23

Barbaro G., Olivi F.M., 1986, ApJ, 337, 125

Bender R., Burstein D., Faber S.M., 1992, ApJ, 399, 462

Bica E., 1988, A&A, 195, 76

Bica E, Alloin D., 1987, A&A, 186, 49

Bressan A., Chiosi C., Fagotto F., 1994, ApJS, 94, 63

Bruzual G., Charlot S., 1993, ApJ, 405, 538

Burstein D., Faber S.M., Gaskell C.M., Krumm N., 1984, ApJ, 287, 586

Buzzoni A., 1989, ApJS, 71, 817

Buzzoni A., 1995a, ApJS, 98, 69

Buzzoni A., 1995b, in Buzzoni A., Renzini A., Serrano A., eds, ASP Conf. Ser. 86, Fresh Views of Elliptical Galaxies. p. 189

Buzzoni A., Mantegazza L., Gariboldi G., 1994, AJ 107, 513

Carollo C.M., Danziger I.J., Buson L., 1993, MNRAS, 265, 553

Casuso E., Vazdekis A., Peletier R.F., Beckman J.E., 1996, ApJ, 458,533

Charlot S., Worthey G., Bressan A., 1996, ApJ, 457, 625

Davies R.L., Burstein D., Dressler A., Faber S.M., Lynden-Bell D., Terlevich R.J., Wegner G., 1987, ApJS, 64, 581

Davies R.L., Sadler E.M., Peletier R.F., 1993, MNRAS, 262, 650

Faber S.M., Trager S.C., Gonzalez J.J., Worthey G., 1995, in Van Der Kruit P.C., Gilmore G., eds, Proc. IAU Symp. 164, Stellar Populations. Kluwer, Dordrecht, p. 249

Gonzalez J.J., 1993, PhD Thesis

Greggio L., 1996, MNRAS, in press

Guiderdoni B., Rocca-Volmerange B., 1987, A&A, 186, 1

Matteucci F., 1994, A&A, 288, 57

Matteucci F., Tornambé A., 1987, A&A, 185, 51

O'Connell R., 1980, ApJ, 236, 430

O'Connell R., 1986, in Chiosi C., Renzini A., eds, Spectral Evolution of Galaxies. Kluwer, Dordrecht, p. 321

Pickles A.J., 1985, ApJ, 296, 340

Renzini A., 1986, in Chiosi C., Renzini A., eds, Spectral Evolution of Galaxies. Kluwer, Dordrecht, p. 194

Tinsley B.M., 1980, Fund. Cosmic Phys., 5, 287

Vazdekis A., Casuso E., Peletier R.F., Beckman J.E., 1996, ApJS, in press

Weiss A., Peletier R.F., Matteucci F., 1995, A&A, 296, 73

Worthey G., 1994, ApJS, 95, 107

Worthey G., Faber S.M., Gonzales J.J., 1992, ApJ, 398, 69

Worthey G., Dorman B., Jones L.A., 1996, AJ, 112, 948

The Far-Ultraviolet Radiation from Elliptical Galaxies

By BEN DORMAN

Laboratory for Astronomy & Solar Physics,
NASA/GSFC, Greenbelt, MD 20771
and
Department of Astronomy, University of Virginia,
Charlottesville, VA 22903-0818, USA

Since the discovery of the Ultraviolet Upturn Phenomenon ("UVX") in early-type galaxies it has been clear that the stellar populations of such systems contain an unexpected hot component. Recent work has provided strong circumstantial evidence that the stars radiating at short wavelengths ($\lambda < 2000$ Å) is in fact due to hot horizontal branch, post-HB stars and post-AGB stars. We summarize the arguments in favour of this hypothesis. We then derive an estimate for the fraction of all HB stars that must be contributing to the UV upturn phenomenon in the strongest UVX galaxy, NGC 1399, and derive a hot star fraction $f_H \sim 0.16$. The implication is that UVX arises from a minority fraction of the dominant stellar population. We conclude that the mechanism that produces the UVX is not one that can be explained naturally by the presence of an extremely metal-rich or metal-poor population.

1. Introduction

The Ultraviolet Upturn Phenomenon was first found by the OAO-2 spacecraft in the late sixties (Code 1969). It consists of a 'UV rising branch' at wavelengths shorter than $\lambda < 2500$ Å, which varies in amplitude amongst the galaxies observed. The amplitude of the UVX phenomenon varies by a factor of about 10 from NGC 1399 [$(15 - V) = 2.05$] to M32 [$(15 - V) = 4.50$] with very similar slopes in the IUE spectral range ($\lambda < 1200$ Å).The question was investigated in further detail by Burstein *et al.* (1988, hereafter B3FL) who presented a sample of galaxies observed by the International Ultraviolet Explorer (IUE). They determined the UV/Optical color $m_\lambda(1550) - V$ [hereafter $(15 - V)$] from ground-based photometry. They plotted these colors against the absorption line index Mg_2, a reliable metallicity indicator for the Galactic globular cluster system (Burstein *et al.* 1984; Brodie & Huchra 1990), and the central velocity dispersion σ_0. For the galaxies that are quiescent, i.e. which contain no evidence of recent star formation and nuclear activity there is a non-linear correlation between Mg_2 and the UVX amplitude (see also Faber 1983). The UVX was also found to correlate, albeit not as tightly, with σ_0, which is an indicator of luminosity through the Faber-Jackson relation. Thus either the amplitude increase is directly related to the metallicity (or at least magnesium abundance) of the galaxies, or there is a tendency for bright galaxies to have strong UV upturns. Enhancements in α-capture elements such as Mg have been found to be a feature of the spectra of bright ellipticals (Worthey, Faber, & Gonzalez 1992) and imply something about the early enrichment history of massive systems.

This contribution is organized as follows: in the next section we discuss the recent observational evidence pertaining to the UVX. Section 3 summarizes the problems associated with modelling the Horizontal Branch. In §4 we give an overview of evolutionary population synthesis, and use it to derive estimates of the fraction of hot stars present as a function of the UV upturn strength. §5 discusses these estimates in the light of population models for the UVX.

2. The 'Old Population' Model

2.1. *UV sources in Old Populations*

The potential sources of UV radiation in old stellar populations have been reviewed in detail by Greggio & Renzini (1990), to which the reader is referred for a full discussion. Briefly, the sources that are likely to produce the largest contribution to the UV output from a galaxy are evolved stars powered by stable nuclear reactions. The production of a hot star on the HB requires that much or all of the hydrogen rich envelope has been stripped from the star prior to arrival on the HB. The sources are (see Dorman, Rood, & O'Connell 1993, hereafter DRO93 and references therein):

- Post-Asymptotic Giant Branch (P-AGB) stars, which are bright ($L/L_\odot \sim 1,000 - 10,000$) but short-lived ($t \lesssim 25,000\,\mathrm{yr}$) . These are likely the most common 'exit channel' for metal-rich stars dying as white dwarfs after going through thermal pulses on the AGB.
- Post-Early Asymptotic Giant Branch stars, which miss the thermally- pulsing stage of AGB evolution, and cross the HR diagram at $L/L_\odot \sim 1,000$ in $\sim 10^6$ yr.
- Hot Post-HB stars, termed AGB-Manqué stars, which are intermediate in brightness ($L/L_\odot \sim 100 - 500$) and lifetime ($t \sim 2 - 4 \times 10^7$ yr).
- Hot Horizontal Branch (HB) stars, which are fainter ($L/L_\odot \sim 10-50$) but long-lived ($t \sim 10^8$ yr). For convenience, these are grouped as either "Intermediate Blue HB stars" (IBHB) if their envelopes are large enough to allow ascent up the AGB, and "Extreme HB stars" (EHB) if not.

All of these sources have their observational counterparts in the Galactic field or in globular clusters (see DOR95 and Dorman 1997a,b for summaries). P-AGB stars are present as UV-bright stars or the nuclei of planetary nebulæ. Their total UV lifetime radiation (and contribution to the integrated spectrum) in this phase decreases strongly with increasing luminosity, since both the total fuel available to them decreases and its rate of consumption increases. It is well-known that hot HB stars are produced in large numbers in metal-poor globular clusters. The Galactic field hot subdwarfs (designated sdB, sdO and variants; see Greenstein & Sargent 1974; Saffer *et al.* 1994) dominate the UV excess sources in surveys of our Galaxy (Green, Schmidt, & Liebert 1986 and others). They are kinematically related to the old disk (Saffer & Liebert 1995) which is predominantly metal-rich compared to the halo clusters. The sdB stars occupy the same HR diagram location as the EHB stars. The AGB-Manqué stars, progeny of the EHB population, are similarly associated with the sdO population.

The elliptical galaxies are dominated by metal-rich populations (probably of solar metallicity or above). Until recently, the only evidence for hot HB stars in a metal-rich environment was in the Galactic field. There were no unambiguously metal-rich populations that contained them i.e. in the coeval, homogeneous Galactic clusters. However, Liebert, Saffer, & Green (1994) found a number of sdB stars in the super-metal-rich ([Fe/H] ~ 0.2; Tripicco *et al.* 1995) Galactic old open cluster NGC 6791. Very recently, Rich *et al.* (in preparation; see also Piotto *et al.*, these proceedings) found well-populated blue tails in the metal-rich clusters NGC 6388 and NGC 6441. While the process that gives rise to these hot stars is unknown (and may not be relevant to galaxies), taken with the field sdB/sdO objects, these clusters provide an "existence theorem" for hot helium burning stars in metal-rich populations.

2.2. *The Nature of the UVX Phenomenon*

The most obvious originating stellar population for the UVX is a small population of young, massive stars, and several studies based on this hypothesis appeared. However, later work with with the resolving power of HST (King *et al.* 1993) appears to rule out this hypothesis for M31. King *et al.* found an underlying smooth flux distribution increasing toward the nucleus with none of the discreteness found for OB associations, which in any case had no concentration toward the centre. Also, some of the flux ($\sim$ 20%) emanated from resolved P-AGB stars.

Hills (1971) first suggested that post-asymptotic giant branch stars might be the source of the upturn found in the bulge of M31 [$(15 - V) = 3.5$]. Brocato *et al.* (1990) and Greggio & Renzini (1990) both showed that the maximum possible UV flux given by the numbers of P-AGB stars expected from standard stellar population models was insufficient to account for the amplitude of the strongest UV upturns. Ferguson & Davidsen (1993) contrasted Hopkins Ultraviolet Telescope (HUT) spectra of the strongest UVX system, NGC1399 [$(15-V) = 2.05$] and that of M31 and concluded that (a) in the IUE spectral range ($\lambda > 1250$ Å) much of the flux emanated from sources with characteristic temperature of $\sim 25,000$ K and (b) the spectral energy distributions (SEDs) did not match at the short end of the spectrum ($\lambda < 1200$ Å), with the flux from M31 being hotter. This is consistent with the notion that a larger proportion of the UV flux from M31 comes from the P-AGB stars, which have a much higher time-averaged effective temperature. These conclusions are corroborated by the later studies of Brown, Ferguson, & Davidsen (1995) and Brown *et al.* (1997), who found similar characteristic temperatures in six other elliptical/S0 galaxies. Their spectral fits to the Astro-2 HUT data suggest, under a large range in assumptions about the P-AGB stars in the galaxies, that at least some EHB stars are apparently present in all systems. Thus the model that appears to be consistent with the observations is of an old population where the UV flux emanates partly from the P-AGB stars and partly from EHB stars.

3. The Horizontal Branch Mass Dispersion

The HB stars are the hottest potential source of UV radiation in an old ($t \gtrsim 2$ Gyr) stellar population and have moderate intrinsic luminosities with appreciable stellar lifetimes. However, it is well-known that modelling HB populations in Galactic globular clusters is subject to a major difficulty, *viz.* that of the mass dispersion. Iben & Rood (1970) first realized that the observations implied a scatter in HB properties, and Rood (1973) explicitly showed using 'synthetic' HB sequences that a dispersion in HB envelope masses could reproduce the observations. However this quantified rather than explained the dispersion in mass, likely due to mass loss on the red-giant branch, by mechanisms still poorly understood to this day. The hottest HB stars are produced by extreme mass loss (Ciardullo & Demarque 1977; Caloi 1989; Castellani & Tornambè 1991; DRO93). Also, as clusters age, the models imply that the RGB mass decreases so that less mass needs to be stripped to produce hot HB stars in older populations. Synthetic HB models thus show a passive drift of the HB sequences blueward as a cluster ages, which mimics the effect of higher mass loss.

Iben & Rood (1970) also noted the variations of HB stellar properties with envelope mass, metallicity, helium abundance Y (see also Sweigart & Gross 1976), the CNO element fraction, as well as the core mass, although the variations in the latter are constrained by energy arguments. As a result, cluster observations are subject to many possible interpretations (aside from observational problems such as determinations of reddening, calibration of blue stars, etc etc).

In Galactic clusters it is possible to identify the resolved hot stars as members of the HB. In studying the stellar populations of galaxies, however, we must start by asking a more basic question, i.e. 'which stars are actually responsible for the UV radiation?'

The following step is to explain the observed UVX correlation in terms of the galaxy properties, but the interpretation is still not straightforward (see §5).

In a detailed study of the possible explanations for the UVX, Greggio & Renzini (1990) suggested that an increase in the mass loss with metallicity would produce blue HB stars. The assumptions involved both an increase in mass loss with metallicity (almost universally acknowledged to happen, although without quantitative theoretical backing), and an extrapolation of the helium enhancement vs. metallicity relations derived at low metallicity to the super-solar metallicity regime. By using simple analytical fits to the behavior of the HB stars and the HB morphology with increasing metallicity, they showed how the UVX-metallicity correlation might be produced naturally, by properties of stellar evolution rather than of the galaxy environment. Environmental effects such as galaxy size, density or previous history are necessarily more difficult to quantify, but may be relevant to the problem: we regard these as 'stellar populations' inputs to the problem as distinct from those of stellar evolution (*cf.* Dorman 1996).

4. Modelling of Ultraviolet Bright Old Populations

The unresolved stellar populations of galaxies are studied using the methods of evolutionary population synthesis (*cf.* Tinsley 1980; Renzini & Buzzoni 1986; see also Bruzual & Charlot 1993). Evolutionary population synthesis attempts to model the integrated spectrum of a stellar population by summing the contributions at each individual wavelength arising from the population components assumed to be present. We sketch here the derivation of values for UV/Optical colours such as $(15 - V)$.

4.1. *Population Synthesis in a Nutshell*

Since the evolutionary connection between the early and late stages of evolution is broken by the indeterminate nature of the RGB mass loss process it is convenient to break the problem into pre-He flash $\mathcal{L}_\lambda^{\rm E}$ and post-He flash $\mathcal{L}_\lambda^{\rm L}$ stages. We consider the integrated light from each component with fixed composition $\mathbf{X} = (Y, Z_\alpha, Z_{\rm CNO}, Z_{\rm Fe-peak})$ etc. For populations old enough to produce RGB sequences, the equation for the luminosity in any given bandpass designated by λ is (see DOR95 for more details)

$$\begin{aligned}\mathcal{L}_\lambda(\mathbf{X}) &= \mathcal{L}_\lambda^{\rm E} + \mathcal{L}_\lambda^{\rm L}\\ &= \int d\tau \int dm' L_\lambda(m';\mathbf{X})\Psi(m',\mathbf{X},\tau)dm'\\ &\quad + \dot{n}_0(t)\mathcal{L}_{\rm V}^{\rm E}\Delta V_0 \int_0^{M_{\rm RGB}-M_c} P(m';\mathbf{X})E_\lambda(m';\mathbf{X})\, dm'. \end{aligned} \tag{4.1}$$

where

$$E_\lambda(M,\mathbf{X}) = \int_{ZAHB}^{WD} L_\lambda(M,\mathbf{X})dt. \tag{4.2}$$

The first term here is the contribution of the early (pre-RGB) evolutionary phases, and is shown here schematically as an integral over time and mass. The time integral is over the interval where the star formation rate is significant, and the mass integral goes from the lower end of the main sequence to the RGB tip. The luminosity L_λ of a star at λ is modelled using the relation

$$L_\lambda(M,t,\mathbf{X}) = \alpha_\lambda(g,T_{\rm eff},\mathbf{X})L_{\rm bol}(M,t,\mathbf{X}) \tag{4.3}$$

where α_λ is the 'bolometric correction,' i.e. the proportion of the total flux radiated in the band designated by λ. This is quite general: λ can be a narrow pass (a few angstroms) for the construction of synthetic spectral energy distributions, or a broadband color such as a Johnson filter. The values of α_λ are supplied using either an empirical flux library (more realistic, but never complete and hard to calibrate over long wavelength baselines), or synthetic stellar fluxes (e.g. Kurucz 1991). The latter have the great advantage of completeness, but their accuracy particularly at short wavelengths is difficult to test at different metallicities. Their use also presupposes the accuracy of temperature/SED relations derived from the library of model stellar fluxes.

As for the late stages, the terms in the equation are the integrated lifetime energy output from the ZAHB to the end of the white dwarf stage. This term is simplified by the fact that the core mass changes almost negligibly with cluster age (for $t \gtrsim 5\,\mathrm{Gyr}$), so that the properties of the ZAHB are invariant with time. The unknown mass distribution of the ZAHB stars due to variable mass loss on the RGB is denoted by $P(m,t,\mathbf{X})$. The lifetime energy flux [equation (4.2)] is multiplied by the 'specific evolutionary flux,' denoted as $\dot{n}_0(t,\mathbf{X})$, which is the input rate *per unit magnitude of pre-HB population* arriving on the HB. $\dot{n}_0$ was approximated by Renzini & Buzzoni (1986) and DOR95 as the number of stars leaving the main sequence, but it is also possible to derive this number from the RGB evolutionary tracks or empirically (O'Connell 1980). $\mathcal{L}_{\mathrm{V}}^{\mathrm{E}}$ is the V band luminosity from the pre-He flash stages, and ΔV_0 is a constant, the width of the V filter (= 872Å).

Potentially, both the IMF Ψ and the HB mass distribution function P are affected by factors beyond stellar evolution in single stars: this is the 'stellar populations' component to the problem. The effects of the initial conditions (such as rotation, composition inhomogeneities, dynamics and environment) are not understood. In the case of the IMF, we can show that for low mass stars, the power law functions fit to cluster and field populations do not much affect the output model spectra. The same is not true for the HB mass distribution where small changes can produce large effects in the ultraviolet output. It is thus somewhat misleading to write $P(M)$ as an implicitly single-valued function of age and composition, as would be correct if only stellar evolution had an impact on the problem. P can however be constrained empirically and the model tested for consistency against its other observational predictions.

4.2. *The Energy Function $E_\lambda(M)$*

Figure 1 shows the lifetime energy radiated by HB stars as a function of envelope mass at 1500 Å and in the V band. Here we have assumed that red HB stars evolve through the P-AGB channel that is the most favourable (i.e., lowest mass) for UV production. The models used are from DRO93 with the P-AGB flux from the Schönberner (1983) $M_{bol} = -3.2$ ($M = 0.546\,M_\odot$) model. The thin lines show the energy radiated up to the core He exhaustion point. The regimes of M_{env} labelled show (a) red HB stars (with far-UV radiation only in the P-AGB stage); (b) the IBHB stars, which reach the AGB after core helium exhaustion: these are the stars that populate the part of the HB usually thought of as the 'blue HB' in globular clusters such as M3; and (c) the EHB stars whose total post-HB energy radiated is similar to that emitted before core He exhaustion. The pre–core-He exhaustion energy radiated by the EHB stars is seen to be approximately constant with envelope mass and similar to the maximum post-HB output. Models for other metallicities (DRO93, DOR95) show that the total far-UV radiation from all EHB stars from (Y, [Fe/H]) = (0.25, −2.26) to (0.46, 0.58) varies by less than 50%. Also, metal-rich compositions produce the IBHB stars for only a small range in mass (DRO93, D'Cruz *et al.* 1996), which is the primary cause for the 'first parameter'

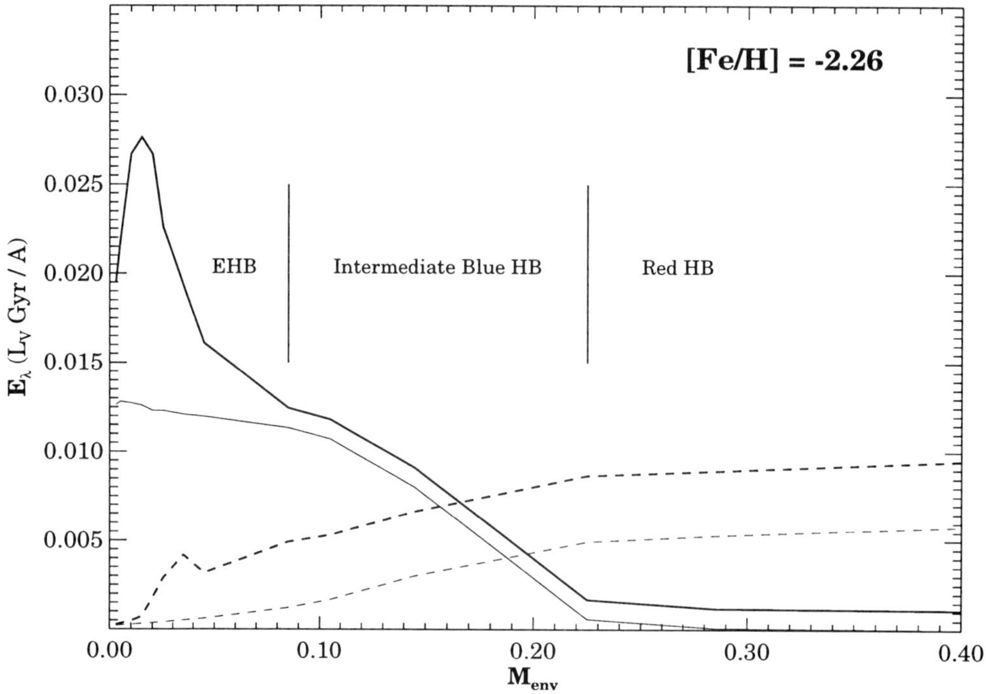

FIGURE 1. The total energy radiated (E_λ) by HB stars at 1500 Å, and in the V band as a function of envelope mass for [Fe/H] $= -2.26$. The vertical axis is in units of the solar V band luminosity/Gyr/Å. The data has been derived by integrating the DRO93 HB evolutionary tracks and using the Kurucz 1991 model stellar fluxes. For the more massive stars, the 1500 Å flux is radiated entirely in the P-AGB phase of evolution, and the flux from the lowest mass Schönberner (1983) evolutionary track has been adopted (this gives an upper bound to the P-AGB flux; see DRO93).

effect in globular clusters. This also means that far-UV radiation from purely metal-rich populations (above what can be produced by P-AGB stars) requires mass loss in RGB stars to be sufficiently vigorous to strip (almost) the entire envelope.

The other factor that influences the UV production from a stellar population is the rate of evolution $\dot{n}_0(t)$ which is primarily a function of Y, being roughly double the rate at $Y_{\mathrm{ZAMS}} = 0.45$ as it is at $Y_{\mathrm{ZAMS}} = 0.27$. This reflects the faster evolution of helium rich models.

4.3. *Simple Models of UV Populations*

Equation 4.1 can be simplified to obtain some important bounds on the UV-radiating population. Here, we focus on the number of hot stars that must be present for consistency with the observed far-UV radiation. We assume a simple mass distribution, initial mass function and metallicity distribution as follows. The model consists of:

- an instantaneous starburst at a single metallicity $\Psi = \psi(M)$
- a bimodal HB mass distribution with some fraction f_H of the stars becoming hot.

With the first assumption, our conclusions will of course not give precise metallicity and age information but will instead give luminosity-weighted mean quantities. This problem is more relevant to the study of radiation in the mid-UV (Worthey, Dorman & Jones 1996).

The HB mass distribution used is justified as follows: the red HB stars of metal-rich populations enter the integrated light mainly in the V-band where their total energy output varies little with envelope mass (this result can also be seen in the metal poor models of Fig. 1, and explicitly in DOR95). The fraction f_H making up the EHB UV emitters realistically is spread over a finite mass distribution (i.e. $f_H = \int_{\rm EHB} P(m')dm'$). This mass distribution cannot be easily distentangled from integrated light, although Brown *et al.* (1997) find that a small mass range of EHB stars produces the best models for the HUT far-UV spectra. In order to avoid adding parameters to the problem that cannot be constrained DOR95 used either single evolutionary sequences and uniform distributions of EHB stars to model the hot contribution. The 'single evolutionary sequence' models can provide an lower bound to the size of the hot population present, since a realistic distribution always produces a redder colour than the hottest stars present. Thus we may choose the peak of the $E_\lambda(M)$ function to derive a bound.

Magnitudes and UV/Optical colours may now be derived from Equation 4.1. The result is

$$m_\lambda - V = -2.5\log_{10}\left(\frac{(\mathcal{L}_\lambda/\mathcal{L}_V)^E + \dot{n}_0(t)\Delta V_0[f_H\mathcal{E}_\lambda^H + (1-f_H)\mathcal{E}_\lambda^C]}{1+\dot{n}_0(t)\Delta V_0[f_H\mathcal{E}_V^H + (1-f_H)\mathcal{E}_V^C]}\right). \tag{4.4}$$

where

$$\mathcal{E}_\lambda^H = \int_{\rm EHB\ masses} E_\lambda P(m')dm', \tag{4.5}$$

and similarly $\mathcal{E}_\lambda^C$ is the energy output integrated over the red HB stars. Thus for any given observed $F_{UV} = 10^{[-0.4(15-V)]}$, we obtain (since $\mathcal{L}_{1500}^E \approx 0$)

$$f_H = \frac{F_{UV}(1+\dot{n}_0(t)\Delta V_0\mathcal{E}_V^C) - \dot{n}_0(t)\Delta V_0\mathcal{E}_{1500}^C}{\dot{n}_0(t)\Delta V_0\left[\mathcal{E}_{1500}^H - \mathcal{E}_{1500}^C + F_{UV}(\mathcal{E}_V^C - \mathcal{E}_V^H)\right]} \tag{4.6}$$

For the entire galaxy population, which is composite in abundance, we need to use in this formula the average specific evolutionary flux $\overline{\dot{n}_0(t)}$. A simple but accurate approximation is

$$f_H = \frac{F_{UV}}{\overline{\dot{n}_0(t)}\Delta V_0\left[\mathcal{E}_{1500}^H - \mathcal{E}_{1500}^C + F_{UV}(\mathcal{E}_V^C - \mathcal{E}_V^H)\right]} \tag{4.7}$$

For the specific case where the mass function is taken to be the Salpeter law, $\psi(M) = M^{-2.35}$, DOR95 have tabulated the quantities $(\mathcal{L}_\lambda/\mathcal{L}_V)^E$ and $\dot{n}_0(t)$, and also the values of E_{1500} and E_V for tracks computed by DRO93. Apart from the constant ΔV_0, the terms in the denominator of equation (4.7) are the mean input rate to the HB and the (mean) lifetime energy radiated by the hot stars.

5. Discussion

We have focussed here on deriving limits on the population size responsible for the UV upturn. This is an important prediction of the theory, since is is the most easily tested for consistency against other observational consequences. In particular, for the case of NGC 1399 [$(15 - V) = 2.05$] we obtain $f_H \sim 0.16$. This conclusion implies that the UVX sources do not arise from a trace population more prevalent in UVX-strong galaxies, and instead favours the notion that a sizeable minority of the dominant population is contributing to the UVX.

The work of Park & Lee (1997) and Bressan, Chiosi, & Fagotto (1994) have assumed that a fraction of stars at the low and high end of the metallicity distribution contributes the UV radiation from the galaxies. In the case of the metal-poor hypothesis, a 16% fraction of the population in stars with [Fe/H] < -1 would give a large enough contribution to the mid-ultraviolet radiation to produce a discrepancy with observation (Worthey, Dorman, & Jones 1996; Bressan *et al.* 1994). Quantitative estimates employing the 2500 Å fluxes in DOR95 in equation (4.7) show that bimodal metallicity distributions, in which the majority population contributes relatively little in the mid-ultraviolet, are necessary to reconcile that longer wavelength flux with observations.

The hypothesis that the UVX stars are entirely metal-rich clearly does not suffer from this problem. However, the large fraction of stars necessary to account for the observed flux seems to imply that a spread in metallicity must give rise to the UVX rather than the most extreme composition. This is contrary to the work of Bressan *et al.* who suggest that the UVX arises from the effects of stellar evolution at large ages. They invoke high metallicity, strongly helium enhanced models with RGB mass loss similar to what is inferred from the globular cluster system.

Assuming that it is a fraction of the dominant population that gives rise to the UVX accords better with the observational record we have for Galactic populations. We do not need to hypothesise that *all* RGB stars of a given metallicity have the same degree of mass loss, which we do not observe in any other context (little information about mass distributions can be deduced from red HB clumps). Unfortunately one cannot test the initial metallicity of the field sdB stars easily since their spectra are affected by diffusion in the surface layers. However, the question of how the UVX arises remains open, at least for the time being (but see Dorman 1997a for empirical uses of the far-UV radiation).

It should be stressed that both 'metal-poor' and 'metal-rich' models as presented by these authors rely on the hypothesis that age is established as a 'global' second parameter for HB morphology in old stellar systems. They use the results of simple, unimodal synthetic HB models to derive ages of galaxies. We argue that the Galactic record in resolved populations does not support the use of these models and cannot therefore constrain unresolved stellar populations.

We close with a caveat that applies to the interpretation of the UVX as a metallicity-driven phenomenon, on which the 'metal-rich' hypothesis relies heavily. We caution that this assumption should not be regarded as incontrovertible. The argument for a metallicity driven mechanism is weakened by the discovery that the Mg_2 indicator does not trace the heavy-element abundance in galaxies and by the observation that Mg_2 line strengths are correlated with σ_0 but not with iron indicators (Worthey, Faber, & Gonzalez 1992). The UVX may thus be uncorrelated with iron abundance (Dorman 1997a and references therein). Another definite prediction of the metallicity relation is that the UVX gradient with galactocentric radius should be correlated with optical line-index gradients. This question is currently under study using Ultraviolet Imaging Telescope data (Ohl *et al.*, in preparation). Thus, the Ultraviolet Upturn Phenomenon remains, like many others in astronomy, an unsolved problem which requires spatially resolved spaceborne observations for further study.

Acknowledgements: B.D. acknowledges support from NASA grants NAG5-700 and NAGW-4106, and healthy discussions with Robert T. Rood and Robert W. O'Connell. He is also glad to report that Icko Iben Jr., who we honour with this volume, did not find anything wrong with his presentation.

REFERENCES

Bressan, A., Chiosi, C., & Fagotto, F. 1994, ApJS, 94, 63

Brocato, E., Matteuci, F., Mazzitelli, I., and Tornambè 1990, ApJ, 349, 458

Brodie, J. P. & Huchra, J. 1990, ApJ, 362, 503

Brown, T. M., Ferguson, A. F., & Davidsen, A. F. 1995, ApJ, 454, L15

Brown, T. M., Ferguson, A. F., Davidsen, A. F., & Dorman, B. 1997, ApJ, in press

Bruzual, G. & Charlot, S. 1993, ApJ, 405, 538

Burstein, D., Bertola, F., Buson, L. M., Faber, S. M., & Lauer, T. R. 1988, 328, 440 (B3FL)

Burstein, D., Faber, S. M., Gaskell, F., & Krumm, N. 1984, ApJ, 287, 586

Caloi, V. 1989, A&A, 221, 27

Castellani, M. & Tornambè A. 1991, ApJ, 381, 393

Ciardullo, R & Demarque P. 1997, Trans. Yale U. Obs., 33, 1

Code, A. D. 1969, PASP, 81, 475

D'Cruz, N. L., Dorman, B., Rood, R. T., & O'Connell, R. W. 1996, ApJ, 466, 359

Dorman, B. 1996 in *Stellar Evolution: What Should Be Done?* eds. A. Noels, D. Fraipont-Caro, M. Gabriel, N. Grevesse, & P. Demarque, (Liège: Institut d'Astrophysique), 291

Dorman, B. 1997a in *The Nature of Elliptical Galaxies* eds. M. Arnobaldi, G. S. Da Costa, & P. Saha, (San Francisco:ASP), in press

Dorman, B. 1997b in *The Third Conference on Faint Blue Stars* ed A. G. D. Philip, (Schenectady:L.Davis), in press

Dorman, B., O'Connell, R. W., & Rood, R. T. 1995, ApJ, 442, 105 (DOR95)

Dorman, B., Rood, R. T., & O'Connell, R. W. 1993, ApJ, 419, 596

Faber, S. M. 1983, Highlights Astron., 6, 165

Ferguson, H. C. & Davidsen, A. F. 1993, 408, 92

Green, R. F., Schmidt, M., & Liebert, J. W. 1986 ApJS, 61, 305

Greenstein, J. L., and Sargent, A. I. 1974, ApJS, 28, 157

Greggio, L. and Renzini, A. 1990, 364, 35

Hills, J. G. 1971, A&A, 12, 1

Iben, I., Jr., & Rood, R. T. 1970, ApJ, 161, 587

King, I. R. et al. 1993, ApJ, 413, L117

Kurucz, R. L. 1991 im *Stellar Atmospheres: Beyond Classical Models* eds. L. Crivellari, I Hubeny, and D. G. Hummer, (Dordrecht:Kluwer), p.441

Liebert, J. W., Saffer, R. A., & Green, E. M. 1994, AJ, 107, 1408

O'Connell, R. W. 1980, 236, 430

Park, J.-H. & Lee, Y.-W. 1997, ApJ, 476, 28

Renzini, A. and Buzzoni, A. 1986 in *Spectral Evolution of Galaxies* eds C. Chiosi and A. Renzini (Dordrecht:Reidel), 195

Rood, R. T. 1973, ApJ, 184, 815

Saffer, R. A., Bergeron, P., Koester, D., & Liebert, J. W. 1994, ApJ, 432, 351

Saffer, R. A. & Liebert, J. W. 1995 in *Proceedings of the 9th European Workshop on White Dwarfs* eds. D. Koester & K. Werner (Berlin:Springer) p.221

Schönberner, D. 1983, 272, 708

Sweigart, A. V. & Gross, P. G. 1976, ApJS, 32, 367

Tinsley, B. 1980, Fund. Cos. Phys., 5, 287

Tripicco, M. J., Bell, R. A., Dorman, B., & Hufnagel, B. 1995 AJ, 109, 1697

Worthey, G. S, Dorman, B, & Jones, L. A. 1996, AJ, 112, 948

Worthey, G. S, Faber, S. M., & Gonzalez, J. J. 1992, ApJ, 398, 69

Evolutionary Synthesis of Stellar Population: A Modular Tool

By CLAUDIA MARASTON

Department of Astronomy, University of Bologna, Via Zamboni 33, 40126 Bologna, ITALY

An innovative tool for the construction of Evolutionary Synthesis models of Stellar Populations is presented. It is based on three independent matrices giving respectively 1) the fuel consumption during each evolutionary phase as a function of stellar mass, 2) the typical temperatures and gravities during such phases, and 3) colors and bolometric corrections as a function of gravity and temperature. The modular structure of the code allows us to easily assess the impact, on the synthetic spectral energy distribution, of the various assumptions and model ingredients, such as, for example, uncertainties in the stellar evolutionary models, mixing length, temperature distribution of the horizontal branch (HB) stars, AGB mass loss, color-temperature transformations,etc. As an illustrative example, a solar model ($Y = 0.27$, $Z = 0.02$), with an age ranging between 30 Myr and 15 Gyr and various choices for the slope of the IMF is presented. Broad band colours, as a function of age, are compared with Magellanic Clouds Clusters data and the relative contributions of the various stellar stages to the total L_V to the data relative to the Bulge clusters NGC 6528 and NGC 6553 (see also Lanteri *et al.* in this volume). Then, the evolution of the stellar mass-to-light ratios for the various luminosities has been computed: in this, the global mass of a stellar population at a certain age, i.e., mass of "living" stars plus mass in remnants (Renzini & Ciotti, 1993), has been taken into account. Finally, the various model outputs has been also compared with those of other authors.

---

1. Introduction

During the last years, many authors have spent their research time in computing Evolutionary Synthesis models for Stellar Populations, simple, like globular clusters, as well as complex ones, such as galaxies. In doing this, two main approaches have been followed. The first, known as "Isochrone Synthesis" technique, simply consists in summing up the contributions to the luminosities in the various passbands of all the mass-point along an isochrone, having assumed an Initial Mass Function. The integration starts from a fixed lower limit mass and terminates to the latest point on the isochrone itself, that usually coincides with the end of the so-called Early Asymptotic Giant Branch phase. Later stellar phases are then added, in an analitycal or empirical way, with some sort of individual receipts. This method is followed, among the others, by Bruzual & Charlot and by the Padua group. An alternative approach is that introduced by Renzini & Buzzoni (in *Spectral Evolution of Galaxies*, 1986) and based on the so-called **Fuel Consumption Theorem** (Renzini & Buzzoni, 1986). In this case, the main ingredient for the synthesis is the nuclear fuel (in hydrogen and/or helium masses) available for a star in a peculiar evolutionary stage. This quantity, obtainable from the stellar tracks, is then converted to bolometric luminosity and, finally, in the luminosities in the various passbands, thanks to a temperatures/gravities-colours set of tranformations. At present, this method has been used by Buzzoni (1989) and Worthey (1994), but they both offered models only in a narrow age range, precisely from 4 and 1.5 Gyr, so missing young and intermediate age populations. In this work we, following Renzini's approach, have computed models suitable for simple populations from relatively young (30 Myr) ones to old (15 Gyr) ones, to allow a comparison with the globular cluster family of the Magellanic Clouds.

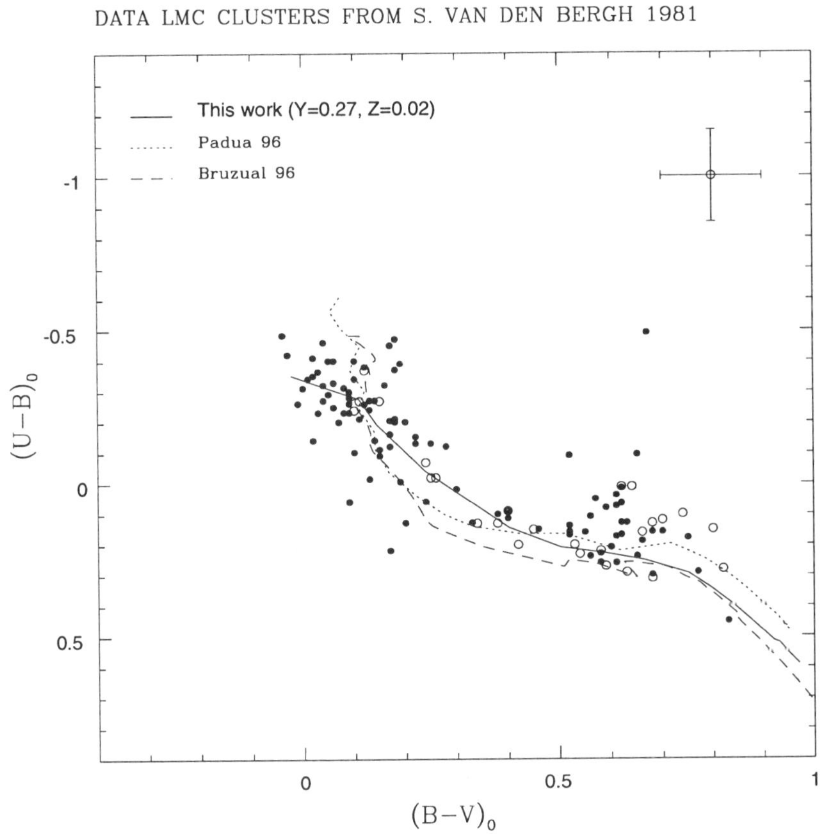

FIGURE 1. $(U-B)_0$ vs $(B-V)_0$: models aren't significative for $(B-V)_0 > 0.5$ owing to the increasing difference in metallicity of the old clusters

2. A brief description of the model

2.1. *Ingredients*

The code is based on three independent matrices giving respectively 1) the fuel consumption during each evolutionary phase as a function of stellar mass, 2) the typical temperatures and gravities during such phases, and 3) colors and bolometric corrections as a function of gravity and temperature. The first matrix allows us to calculate the relative contribution of each evolutionary phase to the bolometric light of the population. The other two matrices allow us to distribute such contributions in the various passbands, and thus to compute the synthetic colors of the population as a function of age. All the relevant stellar stages have been included, until the end of the Thermally Pulsing Asymptotic Giant Branch (**TP-AGB**). For the treatment of this last phase, in particular, no-one analytical approximation has been used, but a semi-empirical recipe, that makes use of the theoretical fuel comsumption together with the empirical distribution of the **AGB** M and C *type* stars, as observed in the Magellanic Clouds clusters by Frogel, Mould & Blanco (1990; for the details, we refer to the forthcoming paper). For the Main-sequence, we have used a set of isochrones computed by Castellani, Chieffi & Straniero (1992); to taking into account the contribution of the low-main sequence stars too, we have properly linked these with the unevolving low-mass tracks of VandenBerg (1983): in this way, our synthetic populations contains stars until 0.1 in mass. The at-

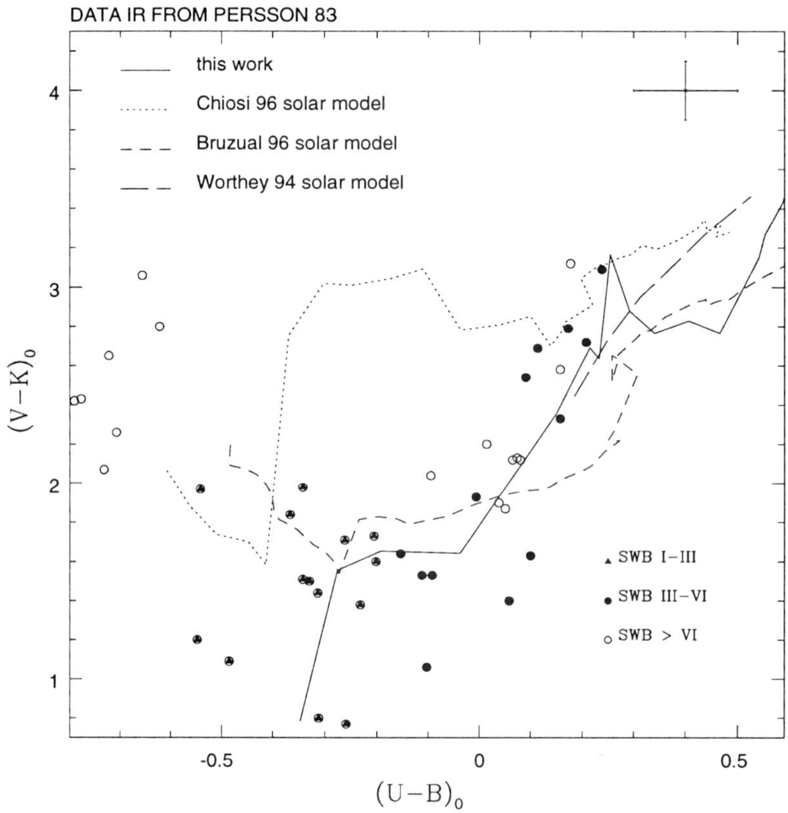

FIGURE 2. $(V-K)_0$ vs $(U-B)_0$; also plotted are models from Tantalo, Chiosi *et al.* (1996), Bruzual & Charlot (1996, in preparation), Worthey (1994)

mospheres models are those of Kurucz (1992), for the UBV bands; for the infrared range, we have used the classical empirical tables of Johnson (1966), together with the models of Bessel (1989) for M *type* stars .

2.2. *Model outputs*

The main outputs of the code are broad band colors $(U-V)_0$, $(U-B)_0$, $(B-V)_0$, $(V-R)_0$ and $(V-K)_0$ (but this choice is not a restrictive one) and the relative contributions of the various stellar phases to the total luminosities in the various passbands, all computed as a function of age under different assumptions concerning the Initial Mass Function (IMF) slope (in particular, single-slope, double slope and Scalo multislope IMF). Finally, we have computed the evolution of the mass-to-light ratios as referred to the U,B,V,R,K passbands plus the bolometric. In these, the real global mass of a stellar population at a certain age has been taken into account, i.e., mass of "living" stars plus mass in remnants (white dwarfs, neutron stars or black-holes, depending on the initial mass of the star) using the recipe of Renzini & Ciotti (1993). This is, in our opinion, an innovative approach, as all the other authors simply assume an initial constant total mass or include only the mass of the "living stars" at a certain age (that's, at about, the turn-off mass at that time).

Evolutionary phase	NGC6553	NGC6528	10 Gyr SSP	15 Gyr SSP
MS + SGB	0.36	0.38	0.41	0.43
RGB	0.36	0.32	0.346	0.35
HB	0.20	0.22	0.153	0.136
AGB	0.09	0.08	0.09	0.084

TABLE 1.

3. Main conclusions

We have performed a comparison between our results and those of some other authors, precisely Bruzual & Charlot and Tantalo, Chiosi *et al.* (1996) last models, Worthey (1994) and Buzzoni (1989). In the case of UBV indexes, the three sets of models evolve in much the same way in time and show quite good agreement with the observations. There are minor discrepancies due to the different choices of the reference stellar tracks (see Fig. 1). The infrared colours behave very differently. In particular, it's found that, except for our model, all fail in reproduce the observed locus for the Magellanic Cloud clusters in a two-colour diagram (Fig. 2). We believe that this is due to our correct inclusion of the **TP-AGB** phase, because this stellar stage has a strong impact on the integrated $(V-K)_0$ colour at intermediate age. This is confirmed by the relative bolometric contribution of the AGB in a synthetic stellar population (SSP), as a function of age, when compared with observational data (see Fig. 3): the points are the percentage of AGB bolometric luminosity measured for a sample of Magellanic Clouds clusters by Frogel, Mould & Blanco (1990). The ages for the observational data have been computed on the basis of the SWB parameter (Searle, Wilkinson & Bagnuolo, 1980), adopting the relation Log(t/yr) = 0.5 SWB *type*(t) + 6.8 (see also Fall & Frenk (1983)). The disagreement between model and observational data, clearly visible at later SWB types ($\log t > 9.4$), is mainly due to the increasing difference in metallicity between these clusters and the model. Moreover, the data don't include stars with $M_{BOL} \geq -3.6$, so that some, less bright, EARLY-AGB stars could be lost in the observational sample, while they are included in the theoretical model. But, without any doubt, it is a fairly good agreement. Another possible test concerns the relative contributions of the evolutionary phases to the various total luminosities: we have performed this comparison with the empirical total L_V as obtained for the two Bulge clusters NGC 6528 and NGC 6553 (see also the Lanteri *et al.* contribution). Table 1 shows the results: theoretical values are for a 10 Gyr, 15 Gyr - Scalo IMF - SSP; to allow a better comparison, the completeness limit of samples at $V \simeq 22$ (Renzini, *private communication*) has been taken into account. The agreement is fairly good, expecially for the RGB and AGB; observational MS + SGB values are affected by incompleteness at the lower end while the theoretical HB is probably too red. Anyway, this comparison suggests an old age for these clusters and consequently for the Bulge. We can conclude that this new code, founded on the Fuel Consumption Theorem, is able to correctly reproduce the observational integrated properties of simple stellar populations. A detailed description of the various other results will be given in a forthcoming paper.

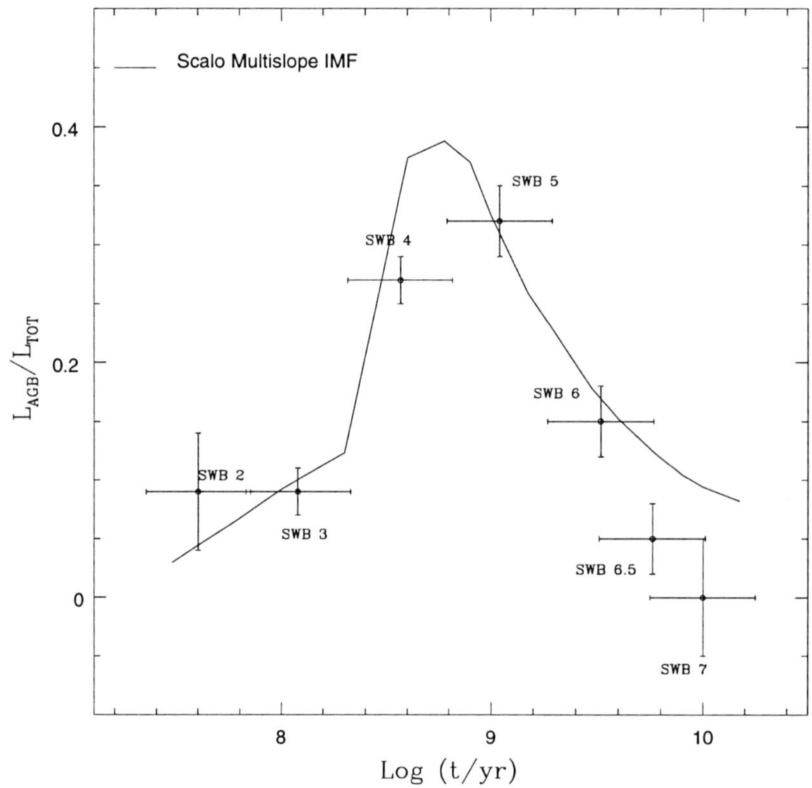

FIGURE 3. The relative bolometric contribution of the AGB in a SSP, compared with observational data

Capabilities of HUBE, the Hopkins Ultraviolet Explorer

By RICHARD C. HENRY

Department of Physics & Astronomy, The Johns Hopkins University, Baltimore, MD 21218-2695, USA

In April 1996, the Hopkins Ultraviolet Background Explorer (HUBE) was selected by NASA as a Medium-class Explorer *Alternate* Mission. (An Alternate Mission undergoes Pre-Phase-A study, and also undergoes Definition Study, should the primary mission not succeed in its own Definition Study phase.) HUBE, as a Medium-class Explorer, will open an important new field of astrophysical research: doing for diffuse UV radiation what IRAS and COBE have already done for the diffuse infrared. Although optimized for the study of the diffuse UV background, HUBE, as the first sensitive survey instrument in the UV, will discover vast numbers of point sources. For example, HUBE will discover all of the quasars that are exploitable by HST and FUSE. Please visit the HUBE homepage, http://msx4.pha.jhu.edu

---

1. Introduction

HUBE will perform an all-sky survey of the cosmic diffuse ultraviolet background radiation and ultraviolet point sources in the wavelength ranges 850 − 1200 Å and 1230 − 2000 Å. Both imaging and spectroscopy will be employed to produce a wealth of significant new information on the following important astrophysical problems:

- the nature of the hot component of the interstellar medium (ISM) and the galactic halo,
- the distribution of molecular hydrogen in the ISM,
- the distribution of interstellar dust and the scattering properties of the interstellar grains,
- the integrated light of galaxies, for study of star formation and galaxy evolution,
- the possibility of radiation from an ionized intergalactic medium, or (more speculatively) from the decay of neutrinos or more exotic particles of non-zero rest mass,
- the location and ultraviolet brightness, of faint stars, galaxies, and quasars, and
- spectra of faint point sources, 912 1800 Å.

2. The HUBE Team

HUBE Co-Investigators are: A.F. Davidsen, P.D. Feldman, H. Ford, J. Kruk, J. Murthy, D.A. Neufeld (JHU), L.J. Paxton, K. Peacock (JHU/APL), J. Atkins, E. Hammond (Morgan State U), G. Carranza (Córdoba, Argentina), P.A. Charles (Oxford), M. Clampin (STScI), E. Conway (Sykesville Middle School), P. Jakobsen (ESTEC), R.A. Kimble (GSFC), R.W. O'Connell (UVa), A. Sandage (Carnegie), and C. Vaz (U. Algarve, Portugal).

3. The HUBE Concept

HUBE has been described in a preliminary way by Kimble et al. (1990). An overview of diffuse ultraviolet background radiation is given by Henry (1991), while recent research papers on this subject include Henry (1995), Henry & Murthy (1993, 1994), Murthy & Henry (1995), and Murthy et al. (1993, 1994).

Survey of the State of our Study of the Universe at All Astronomical Wavelengths

Wave-length	Radio	μ-wave	Sub-mm	Infra-red	Visible	**Ultra-violet**	Soft X-ray & EUV	X-Ray	γ-ray
Detailed Study of Point Sources	VLBI	(ground)	SWAS	ISO & SIRTF	HST	HST, FUSE & IUE	Einstein	Einstein & AXAF	GRO
Point Source Sky Survey	(ground)	(ground)	DIRBE	IRAS & WIRE	(ground)		Uhuru & EUVE	Uhuru & HEAO	GRO
Diffuse Background Sky Survey	RAE-A	COBE	DIRBE	IRAS & WIRE	UVISI		Rosat & EUVE	HEAO	GRO

Hube will fill *both* the remaining blanks in our inventory of the Universe

FIGURE 1. HUBE, the Hopkins Ultraviolet Background Explorer, when implemented by NASA will fill the last remaining gaps in humankind's initial survey of our Universe. It will carry out a mapping mission for the diffuse ultraviolet background radiation, as well as a point source survey that will find millions of ultraviolet-emitting objects.

HUBE will operate in two extremely dark regions of the electromagnetic spectrum: shortward of Lyman α, and between the intense geocoronal Lyman α line at 1216 Å and the zodiacal light, which brightens greatly longward of 2000 Å. The sky background in the HUBE band is $\sim$ 100 times fainter than that in the visible, even from space. As a result, the potential for HUBE to detect entirely new classes of objects is very great. HUBE will, for example, be ideal for detecting large, extremely low surface brightness objects, such as Lyman α emitting clouds out to redshifts ~ 0.5.

Three small (two 20 cm aperture, one 8 cm aperture) complementary, co-aligned instruments are employed by HUBE to elucidate the relative contributions of the possible sources listed above. All instruments incorporate photon-counting microchannel plate detectors virtually identical to detectors now flying successfully on EUVE. The three instruments are:

- an $f/3.2$ imager (1350 – 2000 Å and 1350 – 1600 Å), 2.5° field of view, 35″ resolution,
- FUVS, an $f/2$ spectrograph (1230 – 1800 Å) with 5 Å spectral resolution
- EUVS, an $f/5$ spectrograph (850 – 1200 Å) with 3 Å and 17 Å resolution.

The HUBE instruments will be mounted on a near-duplicate of the SWAS spacecraft to be built by Goddard Space Flight Center. The all-sky survey will require, nominally, 217 days of operation, allowing, in a two-year mission, selection of a large number of additional "deep-pointing" targets for intensive examination. HUBE is inexpensive because it is small and simple. High sensitivity for diffuse radiation requires short focal lengths and fast f/ratios. For this reason instruments such as IUE and HST are com-

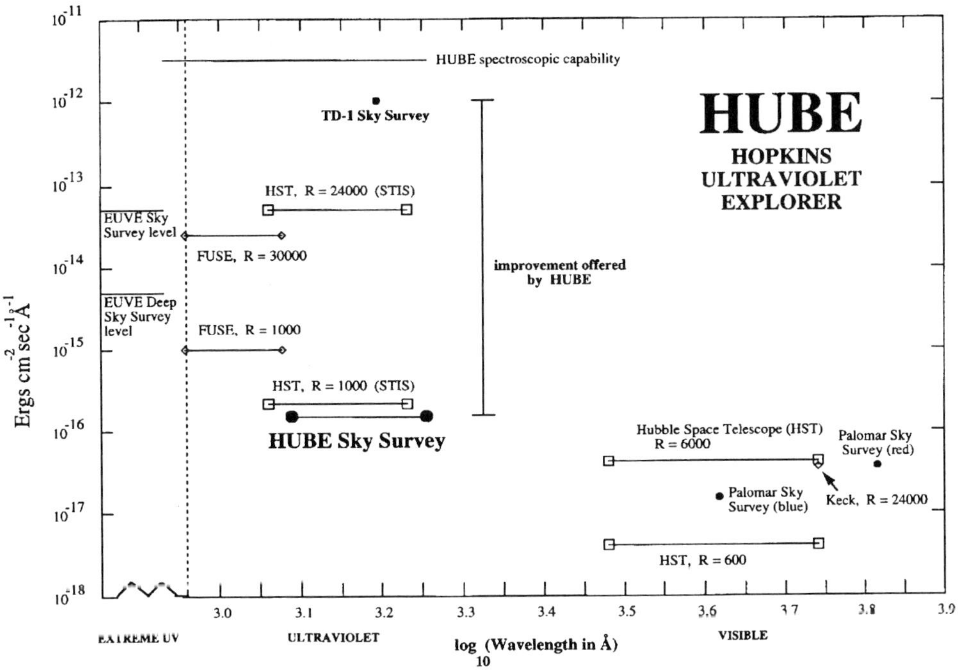

FIGURE 2. In any wavelength band, one wishes to have available a sky survey that is well-matched to the attainable magnitude limits of the most powerful large telescopes in that same wavelength band. For example, in the visible, the Palomar Sky Survey is well-matched to the capabilities of Keck and HST. In marked contrast, the only all-sky survey in the ultraviolet, TD-1, is ludicrously ill-matched to the power of HST-STIS and FUSE. Our figure shows how HUBE will remedy this situation.

pletely insensitive to the UV background. The optics, detectors, pointing, mechanical, and thermal requirements for HUBE pose no significant new technical challenges. The required spacecraft can be provided economically by reproducing the Goddard SWAS design. As a result, the proposed cost for HUBE is exceptionally well-founded.

We particularily emphasize that while HUBE will break new ground scientifically, in all other respects it is extremely conservative. There is no detector development involved; the HUBE detectors are simple adaptations of currently-flying (EUVE) detectors, and these detectors are of a kind with which the Johns Hopkins Applied Physics Laboratory has substantial direct experience. The spacecraft maximizes investment return on the Goddard Space Flight Center development of the SWAS spacecraft. The instruments are simple and straightforward. The weight and power and cost estimates are clear and firm.

REFERENCES

Henry, R. C. 1991, ARA&A, 29, 89

Henry, R. C. 1995, in The Physics of the Interstellar Medium and Intergalactic Medium, ASP Confererence Proceedings, 80, ed. A. Ferrara, C. Heiles, C. McKee and P. Shapiro, (San Francisco: ASP), p. 561

Henry, R. C., & Murthy, J. 1993, ApJ Letters 418, L17

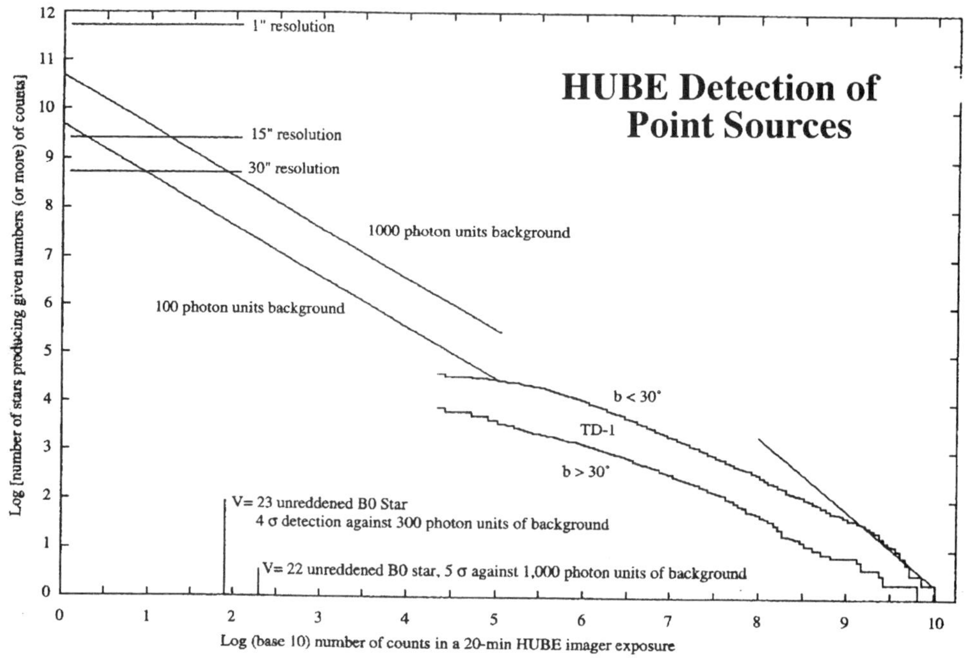

FIGURE 3. Point Sources as Detected by the HUBE Imager: along the bottom scale, the logarithm of the number of counts the HUBE imager will obtain for one star in a 20 minute exposure. The limits for detection of these stars against the cosmic diffuse background are shown as short vertical lines. The vertical axis gives the logarithm of the number of stars that will produce that number of counts (or more); that is, this is a standard log N - log S plot. The actual stars that are plotted are two samples, the TD-1 stars at latitudes higher than $|b| = 30$ degrees (lower curve), and the TD-1 stars at lower latitudes (the upper curve). The short straight line tangent to the bright end of the upper curve is what would be expected for a uniform distribution of equal-brightness stars throughout space. It matches the upper curve for the brightest stars, which are essentially unreddened, but note the immediate, strong, and increasing effects of interstellar reddening for fainter stars. The data plotted go about a factor 10 in brightness fainter than the level to which TD-1 is believed to be complete; this incompleteness is apparent at the faint ends of the two lines. It is interesting to speculate from these data as to how many fainter stars HUBE will detect. Limits on what HUBE might find for fainter stars are provided by existing measurements of the cosmic diffuse background. If one assumes that the measurement or upper limit is due to an unresolved population of faint point objects (eg, white dwarfs or halo stars), one gets the distributions shown by the straight lines labelled "100 [1000] photon units background." It is clear that the $35''$ spatial resolution of HUBE is adequate to avoid any likely source-confusion problems. UIT and FAUST experience strongly confirms this idea.

Henry, R. C., & Murthy, J. 1994, in Extragalactic Background Radiation, ed. D. Calzetti, M. Fall, M. Livio, and P. Madau (Cambridge: Cambridge University Press), p. 51

Kimble, R. A., Henry, R. C., & Paresce, F. 1990, in IAU Symposium No. 139, The Galactic and Extragalactic Background Radiation, ed. S. Bowyer & C. Leinert (Dordrecht: Kluwer Academic Publishers), p. 441

Murthy, J., Im, M., Henry, R. C., & Holberg, J. B. 1993, ApJ 419, 739

Murthy, J., Henry, R. C., & Holberg, J. B. 1994, ApJ 428, 233

Murthy, J., and Henry, R. C. 1995, ApJ, 448, 848

Other Papers

The following papers were presented at the workshop but have/will be published elsewhere.

E. Brocato , V. Castellani, F. Ferraro, A.M. Piersimoni, V. Testa 1996 *The Age of the old magellanic cloud clusters: NGC1786 - NGC1841 - NGC2210*, MNRAS, 283, 614

Allen Sweigart 1997, *Effects of Helium Mixing on the Evolution of Globular Cluster Stars*, in Faint Blue STars III, ed. A. G. D. Philip (San FRancisco: ASP); also ApJ, 474, L23

Enzo Brocato 1997, *New CCD photometry of M 80 and the ranking of the C-M Diagrams of globular clusters*

Ted Stecher 1997, *Ultraviolet Observations of Evolved Stars in Clusters and Galaxies*

Stefano Sandrelli 1997, *Hot Bottom Burning*

M. A. Dopita, E. Vassiliadis, P. R. Wood, S. J. Meatheringham, J.P. Harrington 1997, *HST Observations of Magellanic Cloud Planetary Nebulae*, ApJ, 460, 320

Frank Alberts, Onno Pols 1996, *The Age of PSR J1012+5307*, Nature, 380, 676

Onno Pols 1997, *Formation and evolution of helium stars in binaries*

Takashi Iijima 1997, *Spectral Evolution of Nova Cas 1995*

Z. Kraicheva, V. Stanishev, L. Iliev, A. Antov, V. Genkov 1997, *TT Ari: photometric variability from 1985 to 1994*, A&AS

Saul Rappaport 1995, *The Evolution of Wide Binary Millisecond Pulsars*, cf. MNRAS, 273, 731

Izumi Hachisu, Mariko Kato, & Ken'ichi Nomoto 199, *A New Model for Progenitors of Type Ia Supernovae*, ApJ, 470, L97

T.P. Idiart, F. Thévenin, J. A. de Freitas Pacheco 1996, *Calibration of Metallicity Indices: The Chemical History Effects*, cf. AJ, 111, 1169; 112, 2541

R. Michael Rich 1997, *HST Results for the M31 Halo & Metal Rich Galactic Globular Clusters*

Taft Armandroff 1996, *HST Observations of Andromeda I: a dSph Galaxy in the Halo of M31* , AJ, 112, 257